Manufacturing Facilities Design and Material Handling

FIFTH EDITION

Matthew P. Stephens
Fred E. Meyers

Purdue University Press
West Lafayette, Indiana

Credits and acknowledgments borrowed from other sources and reproduced, with permission, in this textbook appear on appropriate page within text. Unless otherwise stated, all figures and tables belong to the authors.

This book was previously published by: Pearson Education, Inc.

Cataloging-in-Publication data on file at the Library of Congress.

ISBN-13: 978-1-55753-650-1
ISBN-10: 1-55753-650-3

To my son Ethan

Preface

The fifth edition of *Manufacturing Facilities Design and Material Handling* embraces the same practical approach to facilities planning as the previous editions. Building on the same systematic approach, it expands upon an important and relevant topic of lean manufacturing. In addition to a rich collection of discussion questions and problems that follow each chapter, a comprehensive case study has been added. This case study is presented as an Appendix and clearly illustrates the step-by-step approach in facilities planning as explained in the textbook, leading to the development of a complete example of a facility design and layout.

Layout-iQ, a state-of-the-art facilities planning and simulation software package is introduced in this edition, and access to the software is included for purchasers of the book.

The goals of this project-oriented facilities design and material handling textbook are to provide students and practitioners with a practical resource that describes the techniques and procedures for developing an efficient facility layout, and to introduce some of the state-of-the-art tools such as computer simulation.

This how-to book leads the reader through the collection, analysis, and development of vital and relevant data to produce a functional plant layout. Our systematic and methodical approach allows the novice to follow along step-by-step. However, the textbook has been structured so that it may also be used easily and productively by more experienced planners and serve as a useful guide and reference.

The mathematical background and requirements have been intentionally kept at the level of high school algebra. Although quantitative analyses and the manipulation of numbers are extremely important for planning an efficient facility, these skills can be developed without confusing the process with obscure mathematical procedures.

Some experience with computers and computer-aided design (CAD) software packages will prove beneficial for the facilities planner and for other professionals in manufacturing and technology. Those techniques are discussed and emphasized.

On the average, a manufacturing facility will undergo some layout modification and change once every 18 months. Furthermore, the efficiency, productivity, and profitability of any given enterprise are directly correlated with the efficiency of the layout and the material handling systems. Thus, individuals with skills in this area are in demand and are well compensated.

The design of the facility and material handling systems starts with collecting data from various departments. Chapter 2 describes the sources and the significance of this information. The marketing department provides data on various customer requirements that determine production volume and various manufacturing capabilities. The product engineering department supplies engineering drawings and bills of materials, and assists with equipment requirement determination. Inventory and investment policies are determined according to management policies which in turn dictate space requirements, make or buy decisions, production start dates, and so on.

Among the most basic and fundamental data are principles of time and motion economy and time standards. On the basis of this information, machine and personnel requirements are calculated, assembly lines are balanced, and workload in manufacturing cells are leveled. Chapter 3 introduces the reader to the concepts of motion and time study.

Chapter 4 describes the development of route sheets, the sequence of operations, assembly charts, assembly line balancing, and fraction equipment calculation. Use of computer simulation has also been added. Chapter 5 analyzes material flow to ensure proper placement of machines and departments to minimize costs. Seven techniques are discussed in the chapter, as well as the use of computer-aided flow design and analysis.

Chapter 6 describes the activity relationship diagram. The importance of relationships among departments, people, offices, and services, and their effect on the layout is explored. The activity relationship leads to the creation of the dimensionless block diagram.

Space calculation and ergonomic considerations are major and significant aspects of facilities planning. Chapter 7 discusses workstation design, Chapter 8 covers auxiliary services' space requirements, Chapter 9 discusses employee services' space requirements, and Chapter 12 covers office layout techniques and space requirements.

The dimensionless block diagram, which was developed in Chapter 6, is used as a guide to area allocation and is discussed in Chapter 13. The area allocation procedure results in an area allocation diagram. At this point, a plot plan and a detailed layout are created. Chapter 14 discusses various layout construction techniques.

Many other functions require space. Some of these areas need as much space as the production department. The stores and warehouse departments are good examples. Good analysis and knowledge of design criteria can save much space and promote efficiency of both personnel and equipment. Other functions and spaces such as receiving, shipping, lunchroom, restrooms, first-aid rooms, and offices need careful consideration by the facilities planner. The location and size of each activity can have an effect on the overall operational efficiency. Chapters 8, 9, and 12 are dedicated to these topics.

Material handling systems are discussed in Chapters 10 and 11. The reader is introduced to new and exciting material handling concepts and equipment. Application of automatic identification and data capture (AIDC) and ergonomic considerations are emphasized. The reader is encouraged to integrate material handling with other functions to increase productivity and efficiency.

Chapter 15 discusses the concept of simulation and introduces the reader to various applications and the power of computer simulation in the facilities planning arena. Some state-of-the-art simulation software packages are introduced to the reader, and case studies are discussed. As stated earlier, access to Layout-iQ is provided for hands-on application and use of layout design software.

Chapter 16 covers selling the layout through a project report and oral presentation, an important part of any project.

The resultant facility design is only as good as the data and the data analyses upon which the plan has been based. Probably no single factor affects the operational efficiency and safety of an enterprise more than its layout and material handling system.

Matthew P. Stephens
Fred E. Meyers

Acknowledgments

I would like to express my gratitude to the reviewers and the wonderful staff at Purdue University Press whose generous help, efforts, and guidance has made the fifth edition of the *Manufacturing Facilities Design and Material Handling* a reality. A very special note of gratitude goes to Mr. Nelson E. Lee and Rapid Modeling Corporation for generously and kindly providing the users of this edition with links to Layout-iQ, a state-of-the-art simulation and planning software. I would like to acknowledge and thank Manny Cuevas, Michael Thoma, Bryan Orozco, Jarrett Hullinger, and Ben Unger for their hard work and efforts in developing the S. S. Turbo Manufacturing case study. I would like to express a heartfelt "thank-you" to Mr. Shaharyar Masood for his tireless and invaluable assistance with the necessary research and development of this edition.

Matthew P. Stephens

About the Authors

Matthew P. Stephens, Ph.D., CQE, is a Professor and Faculty Scholar in the Department of Technology Leadership and Innovation at Purdue University, where he teaches graduate and undergraduate courses in facilities planning, statistical quality control, and total productive maintenance (TPM). Dr. Stephens holds undergraduate and graduate degrees from Southern Illinois University and the University of Arkansas, with specialization in operations management and statistics.

Prior to joining academe, Dr. Stephens spent 9 years with several manufacturing and business enterprises, including flatbed trailer, and washer and dryer manufacturers. He has been extensively involved as a consultant with a number of major manufacturing companies.

Dr. Stephens has numerous publications in the areas of simulation, quality and productivity, and lean production systems. He has served various professional organizations, including the Association of Technology, Management, and Applied Engineering (ATMAE) and the American Society for Quality (ASQ), where he obtained his Certified Quality Engineering and Six Sigma Black Belt training. Dr. Stephens is also the author of the *Productivity and Reliability-Based Maintenance Management* textbook (Purdue University Press, 2010).

Fred E. Meyers, PE, is president of Fred Meyers and Associates, an industrial engineering management consulting company. Mr. Meyers is a registered professional industrial engineer and a senior member of the Institute of Industrial Engineers. He is a Professor Emeritus in the College of Engineering at Southern Illinois University–Carbondale.

Contents

CHAPTER 11 *MATERIAL HANDLING EQUIPMENT* *251*

CHAPTER 

Introduction to Manufacturing Facilities Design and Material Handling

OBJECTIVES:

Upon the completion of this chapter, the reader should:

- Understand the importance of a systematic approach to facilities planning
- Be able to define facilities planning and material handling
- Understand the relationship between facilities planning and lean thinking
- Be able to identify various types of waste, "muda"
- Understand the goals of facilities planning and material handling
- Understand the systematic layout procedure

THE IMPORTANCE OF MANUFACTURING FACILITIES DESIGN AND MATERIAL HANDLING

Facilities planning is a multi-faceted process, influenced by numerous factors and variables which are not always necessarily in concert and at times may even have contradictory impact on the decision-making process. One of the fundamental aspects of facilities planning is site selection or the location strategy. This decision is usually made at the highest corporate level and may be more influenced by such factors as economics, i.e. tax incentives, or geopolitical considerations that may have very little or no relationship with engineering principles such as proximity to raw material or transportation systems that an industrial engineer may consider to be guiding factors in site selection.

The factors that may influence the location strategy can vary from the availability of resources such as raw material, energy, and so on, to abundance of human resources and lower labor costs. A manufacturing site may be selected based on proximity to sources of raw material, markets, and transportation systems such as highways, railroads, or waterways. It may seem desirable to locate a research facility near a think-tank environment such as a research university. However, factors influencing location selection are not always quite as altruistic. Incentives to attract production facilities vary from corporate tax abatement and low-cost land, to relaxed environmental regulations. These attractions not only impact plant locations within the United States, but also result in migration of these facilities outside of the country.

Global economy probably has had its biggest impact on the location of manufacturing facilities. Due to various incentives offered by local and state governments, not only have we seen a steady migration of manufacturing facilities in a southwardly direction in the United States, but also, as trade barriers are eased or completely removed, this migration has continued beyond the borders to such far places as India or China. It can perhaps be argued that in the past the product market location might have been a secondary factor, whereas labor costs and other incentives may have been an overriding factor in the plant location decision-making process. With the current soaring cost of energy and the resulting expenditures to transport the products to the market, it might be interesting to observe at what point the location strategy equation may be rewritten. Further discussion of this topic is more appropriate for a political science or economics class and is beyond the scope of this text.

Manufacturing facilities design is the organization of the company's physical assets to promote the efficient use of resources such as people, material, equipment, and energy. **Facilities design** includes plant location, building design, plant layout, and material handling systems. As stated above, plant location strategy decisions are made at the top corporate level, often for reasons that have little to do with operation efficiency or effectiveness, and may not always be an engineering decision.

Manufacturing facilities design and material handling affect the productivity and profitability of a company more than almost any other major corporate decision. The quality and cost of the product and, therefore, the supply/demand ratio are directly affected by the facility design. A plant layout project (facility design) is one of the most challenging and enjoyable projects that an industrial or manufacturing engineer will ever have. The project engineer or, at a higher level, the project manager, after receiving corporate approval, will be responsible for spending a great deal of money. The project manager will also be held responsible for the timely, cost-effective achievement of the goals stated in the project proposal and cost budget. The responsibilities of a project manager approach those of a company's president, and only project managers who achieve or beat the stated goals will be given bigger projects.

Building design is an architectural job, thus the architectural firm's expertise in building design and construction techniques is extremely important to the facilities design project. The architectural firm will report to the facility design project manager.

Layout is the physical arrangement of production machines and equipment, workstations, people, location of materials of all kinds and stages, and material handling equipment. The plant layout is the end result of a manufacturing facility design project and is the main focus of this book. In addition to the need for developing new manufacturing facilities, existing plants undergo some changes continually. Major relayouts of plants occur on the average of every 18 months as a result of changes in product design, methods, materials, and process.

Material handling is defined simply as moving material. Improvements in **material handling** have positively affected workers more than any other area of work design and ergonomics. Today, physical drudgery has been eliminated from work by material handling equipment. Every expense in business must be cost-justified, and material handling equipment is no exception. The money to pay for material handling equipment must come from reduced labor, material, or overhead costs, and these expenses must be recovered in 2 years or less (50 percent return on investment [ROI] or higher). Chapters 10 and 11 will discuss material handling systems, procedures, and equipment. Material handling is so entwined with the physical layout of equipment that the two subjects, facilities planning and material handling, are usually treated as one subject in practice. As a result, material handling is part of nearly every step of a facility design process and material handling equipment choice will affect the layout.

New manufacturing plant construction is one of the largest expenses that a company will ever undertake and the layout will affect the employees for years to come. The cost of the plant's products will be affected as well. Continuing improvements will be needed to keep the company current and competitive. The need for continuous improvement and implementation of lean manufacturing concepts is discussed throughout the text.

It is said that if you improve the flow of material, you will automatically reduce production costs. The shorter the flow is through the plant, the better the reduction costs are. Material handling accounts for about 50 percent of all industrial injuries and from 40 to 80 percent of all operating costs. The cost of equipment is also high, but a proper ROI can be obtained. Keep in mind that many industrial problems can be eliminated with material handling equipment. In no area of industrial history has more improvement been made than by the use of material handling equipment. Today, material handling systems can easily be incorporated with cutting edge technologies in automatic data capture equipment and automatic inspection systems for a variety of quality and productivity purposes. Item tracking and inventory control systems can be implemented as part of the material handling procedures.

The **cost reduction formula** is valuable when working with manufacturing facilities design and material handling. Some examples of a cost reduction formula follow:

Ask	*For Every*	*So We Can*
Why	Operation	Eliminate
Who	Transportation	Combine
What	Inspection	Change sequence
Where	Storage	Simplify
When	Delay	
How		

Facilities planners ask the six questions (column one) about everything that can happen to a part flowing through the manufacturing facility (column two) to eliminate steps, combine steps, change sequence of steps or simplify (column three). This requires studying the company's products in depth to identify every step in the process. The best advice is not to take shortcuts or to skip steps in the proposed manufacturing facility design procedure. There are many tools and techniques to help identify the steps in the process. These are described in detail in the following sections.

Implementing the five (5) S's and five why's will also help reduce costs. The **5 S's principles** are

1. *Sifting (organization).* Keeping the minimum of what is required will save space (affects the facility layout), inventory, and money.
2. *Sorting (arrangement).* Everything has a specific place, and everything in its place is a visual management philosophy that affects the facility layout.
3. *Sweeping (cleaning).* A clean plant is a result of a facility layout that has been thought to provide room for everything.
4. *Spick and span (hygiene).* A safe plant is a result of good layout planning.
5. *Strict (discipline).* Following the procedures and standardized methods and making them a habit will keep the plant operating efficiently and safely.

The **five why's** will ensure that the solution to a problem is not a symptom of the problem, but rather, the base cause. For example: A machine broke down.

1. Why?
2. The machine jammed up. Why?
3. The machine was not cleaned. Why?
4. The operator didn't clean it out at regular intervals. Why?
5. Was it because of lack of training? Why?
6. The supervisors forgot. They make a written instruction to be mounted on the machine. It will not happen again.

The planners could have asked six or seven why's. The important thing is to arrive at a final solution that will eliminate the problem from occurring again.

LEAN THINKING AND LEAN MANUFACTURING

A new vocabulary has developed in the past few years that stems from the **Toyota production system** and a book titled *Lean Thinking* by James Womack and Daniel Jones. **Lean manufacturing** is a concept whereby all production people work together to eliminate waste. Industrial engineers, industrial technologists, and other groups within management have been attempting this since the beginning of the industrial revolution, but with a well-educated, motivated production workforce, modern manufacturing management has discovered the advantage of seeking the workforce's help in eliminating waste. The Japanese word for waste is *muda,* which is the focus of much attention all over the world. Who knows better than the production employee—who spends 8 hours a day on a job—how to reduce waste? The goal is to tap this resource by giving production employees the best tools available.

Muda (waste) is defined as any expense that does not help produce value. There are **eight kinds of muda:** overproduction, waiting, transportation, processing, inventory, motion, rework, and poor people utilization. The goal is to try to eliminate or reduce these costs. One of the techniques for doing this is asking "why" five times (five why's). Asking "why" about any problem or cost at least five times attempts to get to the **root cause** of the problem.

Toyota's employees are encouraged to stop the production line or process if a problem exists. A lighted visual indicator board (called an **andon**) is located above the production line. When operations are normal, a green light is on. A yellow light indicates an operator needs help, and if the operator needs to turn off the line, a red light flashes. The term **autonomation (jidoka)** has been coined to indicate the transmission of the human element into automation. An example is the employee turning off the production line if a problem is detected.

In the culture of continuous improvement, **kaizen** is another effective tool that can be easily applied to different aspects of facilities planning and material handling. *Kaizen* is the Japanese word for constant, or continuous, improvement. The main element of kaizen is the people involved in the improvement process. Kaizen touches upon all levels of the organization and requires the participation of all employees—from the top management throughout various levels of the organizational chart and production teams. Every person in the company is encouraged to search for new ideas and opportunities to further improve the organization and its processes including reducing waste.

One of the requirements of kaizen that has been found particularly effective is the need to begin improvements immediately other than waiting until there is a sound plan in place. Kaizen differs from reengineering by the level of change that happens at one time; there are no major breakthroughs with kaizen. Some criticize kaizen because the process makes only small improvements at a time which may, in some cases, lead to further problems.

Kanban is another technique that affects manufacturing facilities design. **Kanban** is a signal board that communicates the need for material and visually tells the operator to produce another unit or quantity. The kanban system, also referred to as a "pull" system, differs from the traditional inventory "push" systems such as just-in-time (JIT) or material requirements planning (MRP). With push systems, parts are produced only when the need arises and they have been requested or there is "pull" from production operations.

Value-stream mapping (VSM) is a major waste reduction and productivity improvement tool that an organization can employ to evaluate its processes. *Value-stream mapping* can be defined as the process of assessment of each component or the step of production to determine the extent to which it contributes to operational efficiency or product quality. Value-stream mapping is clearly linked with and is an important component of lean manufacturing. Using the tools and resources of VSM, a company can document and develop the flow of information and material through the system as an aid in eliminating non-value-added operations or components, reducing costs, and making the necessary improvements. This continuous improvement process goes through three repeating stages: assessment, analysis, and adjustment. Through these three stages, changes and modifications can be made to further improve the process and eliminate waste.

The advantages in using value-stream mapping are numerous. They include improved profitability, efficiency, and productivity for the company or institution. Particular to facilities design and material handling, VSM can clearly reduce or eliminate excessive material handling, eliminate wasted space, create a better control of all forms of inventories (e.g., raw materials, in-process, and finished goods), and streamline various production steps.

THE GOALS OF MANUFACTURING FACILITIES DESIGN AND MATERIAL HANDLING

A good set of goals ensures a successful facility design. Without goals, facilities planners are without direction and a primary mission statement is the first step. A well-thought-out mission statement ensures that the project engineer or manager and the company's management share the same visions and objectives for the project. It also opens communication lines between management and designer: Feedback and suggested changes at this early stage save much work and even headaches later on.

A **mission statement** communicates the primary goals and the culture of the organization to the facilities planner. The mission statement defines the purpose for the existence of the enterprise. The statement should be short enough so that its essence is not lost and can be easily remembered, and it must be timeless so that it is easily adaptable to the organizational changes. For the most part, the mission statement is a philosophical statement that sets the cultural tone of the organization. The mission of a corporation must go beyond expectation of profits and profitability for its shareholders; as a member of the society, it should strive to extend these benefits to its customers and employees. A company may state its mission as follows: "ACME is dedicated to the pursuit of manufacturing the safest, highest quality, and the most reliable bicycles while maintaining the lowest possible price and a strongest commitment to customer satisfaction. ACME recognizes that it is only through strong commitment to our employees that we can achieve our mission."

Although the mission statement is developed by the corporate management, it provides a clear signal and a guiding light for developing strategies at all levels of the company activities including the design of the physical facilities. For example, a mission statement that signals a strong commitment to employee development and training, communicates the need for such facilities in the overall design of the plant layout.

Production goals and objectives that are consistent with the mission of the corporation can then be derived from the mission statement.

Subgoals are added to help achieve specific goals. Potential goals may include

1. Minimize unit and project costs.
2. Optimize quality.
3. Promote the effective use of (a) people, (b) equipment, (c) space, and (d) energy.
4. Provide for (a) employee convenience, (b) employee safety, and (c) employee comfort.
5. Control project costs.
6. Achieve the production start date.
7. Build flexibility into the plan.
8. Reduce or eliminate excessive inventory.
9. Achieve miscellaneous goals.

A mission statement should be simple and should be used to keep the facilities planner on track and to help in all project decisions. As planner, your goal is to provide a specific number of quality units per period of time at the lowest possible cost—not to show off your advanced manufacturing knowledge or to have a showplace for your computers and robots. The mission statement intends to remind you to stay on track, and will assist your decision making along the process.

Let us take a closer look at the subgoals:

1. *Minimize unit and project costs.* This means that every dollar expended in excess of the most economical method of getting into production must be cost-justified. It *does not* mean buying the cheapest machine because the most expensive machine may produce the lowest unit cost. When products are new, production volume may be low. Not much can be spent on advanced manufacturing technology, but you still need equipment. This is when you buy the cheapest ones available.

2. *Optimize quality.* Quality is critical and difficult to measure. Everyone knows that a near-perfect car is available—a Rolls-Royce—but how many can you sell? You can make a better product if you buy better materials, machine closer tolerances, add additional options, and the like. But is there enough of a market for this high-quality, high-cost item?

Mass production is made possible by providing products that the masses can afford. This calls for lowering the designed strength of material, cost of production, and, therefore, the actual quality of the finished product. Top management of the auto industry might state this as a quality standard:

> Let's design a utility automobile that will last 100,000 miles. If we wanted higher quality, why not design it for 200,000 miles? Cost is "why." How many people can afford this more costly automobile?

Once the design criteria have been established, the product designers will design every part with these goals in mind. They may state more clearly that 95 percent of the autos will last 100,000 miles or more. The average, therefore, would be higher, but any cost spent to create any one part of better quality will be money misspent. Manufacturing facilities designers strive to achieve the design criteria by selecting equipment, designing workstations, and establishing work methods that produce quality parts and assemblies. Quality and cost are the two primary competitive fronts. Controlling one without the other will lead to failure. You must constantly balance cost and quality. In manufacturing facilities design and material handling, the planner must consider quality in every phase and do nothing to harm quality. Space must be provided for quality control facilities.

3. *Promote the effective use of people, equipment, space, and energy.* This is another way of saying "reduce costs," or "eliminate muda." People, equipment, space, and energy are a company's resources. They are expensive and you want to use them effectively. Productivity is a measure of use and is the ratio of output over (divided by) input. To increase productivity, you need to increase output, reduce input, or a combination of the two. The location of services like restrooms, locker rooms, cafeterias, tool cribs, and any other service will affect employee productivity and, therefore, the employees' utilization or effectiveness. It is said that you can run pipe and wire, but you cannot run people. Providing convenient locations for services will increase productivity.

Equipment can be very expensive and the operating costs must be recovered by charging each part produced on that machine a portion of the cost. The more parts run on one machine, the lower is the unit cost that each part must carry. So to achieve the second supporting objective, namely, to reduce cost, you must strive to get as much out of each machine as possible. Calculate how many machines are required in the beginning for maximum machine use. Remember, machine location, material flow, material handling, and workstation design all affect equipment usage.

Space is also costly, thus designers need to promote effective use of the space. Good workstation layout procedures will include everything required to operate that workstation, but no extra space. Normally, planners can do a good job of using floor space, but what about the other spaces?

a. Under the floor (basements) is a good place for utility tunnels, walkways between buildings, under-the-floor conveyors for material delivery or trash removal, and tanks under the floor for storage. Use your imagination and save expensive floor space.
b. Overhead (from 7 feet to rafters) is usable space. This space can be used for overhead conveyors, pallet racks, mezzanines, shelves or bins for storage, balcony offices, pneumatic delivery systems, dryers, ovens, and so on. Again, use your imagination and save floor space.
c. Ceiling space under the roof (in the rafters or trusses) is space that can be used for utilities, heating and cooling, sprinkler systems, catwalks, and some storage.
d. On the roof, space can be used for parking, weather testing of a product, utility units, ovens, golf driving range, tennis court, and so on.

As stated, designers want to promote the use of *all* the space in the plant. This concept is known as "utilizing the building cube." The idea consists of utilizing the vertical dimensions as well as the horizontal dimensions of the facility. Keep in mind that whereas land is purchased based on square units, space is obtained in cubic units. Oftentimes management asks industrial engineering to help justify more building space, and after the initial study, finds there is plenty of space if they just go vertical. Floor space gets the most attention, but there is so much more space available. Planners must use their imaginations and create space, focusing first on using the present space more efficiently.

Energy costs can be excessive: Million-dollar utility budgets are common. You can promote energy efficiency by good facilities design techniques. Opening dock doors allows heating and cooling energy to escape. Placing hot equipment where energy can be contained could reduce energy requirements. A "silly" example would be running the air conditioning while having a fire in the fireplace; however, this is what manufacturing facilities are doing all the time with hot operations. Isolating these operations and controlling heat can save big dollars. Another example is that heat rises, thus dryers can be placed near the ceiling to reduce the heat needed. Electricity, gas, water, steam, oil, and telephone must all be used efficiently. The plant layout will greatly affect these costs.

4. *Provide for the convenience, safety, and comfort of our employees.* Although convenience has already been discussed, in addition to being a productivity factor, it is also an employee relations concern. If you design plants with inconvenient employee services, you are telling the employees every day that the company does not care about them. Drinking fountains, parking lot design and location, employee entrances, as well as restrooms and cafeterias must be convenient to all employees.

Employee safety is a moral and legal responsibility of the manufacturing facilities designer. The weight of tools and products, the size of aisles, the design of workstations, and housekeeping all affect the safety of the employees. Every decision made in manufacturing facilities design and material handling design must include safety considerations and consequences. Material handling equipment has reduced the physical demands of work and, therefore, has improved industrial safety.

But material handling equipment itself can be dangerous. The industrial safety statistics indicate that 50 percent of all industrial accidents occur on shipping and receiving docks while handling material. Designers must continue the fight to reduce injury by every means possible.

Good housekeeping means having a place for everything and having everything in its place. The term "everything" is all-inclusive—tools, materials, supplies, empty containers, scrap, waste, and so on. If the manufacturing facilities design hasn't considered any one of these items, a housekeeping problem will result and this clutter is dangerous and costly.

"Comfort" is a term that could communicate plush, costly surroundings, but in workstation design and ergonomics, it refers to working at the correct work height, having sufficient lighting, and standing or sitting alternately, and so on. You don't want to add fatigue to the operator. When operators are on a break, you want to provide comfortable surroundings so that they can recuperate and return to work refreshed and, therefore, more productive.

5. *Control project costs.* The cost of the facility design and material handling project must be determined before presenting the plan to management for approval. What top management approves is the "spending" of money. Once the project is approved, the project manager is authorized to spend the budgeted monies. Going back for more money could be harmful for your career. Budgeting and then living within the budget are two things that successful managers and engineers learn to do early in their careers.

6. *Achieve the production start date.* Early in the life cycle of a new product, the production start date is set. The success of the project depends on whether the product gets to the market on time. Thus, the planner must meet these goals. If there is a late start, the employees may not be able to make up for the lost production. This is especially true for seasonable products. In fact, if you miss the season, you miss the whole year. Fast-moving consumer products companies, like toy companies, will set the production start date and schedule backward to establish a product schedule. Figure 1–1 shows such a schedule. Each important project milestone is identified and listed in the first column. Subsequent columns are used to track each product. The product number, product name, and the responsible project engineer in the column heading identify each product. For example, the third column is used to track product 1810, known by the product name Gizmo. The project engineer for this product is identified as Stephens. For each product, the scheduled completion date is listed across from each project step. For example, for product 1810, all time standards are to be set by April 5, shown as 4-5. Upon the completion of each step, an *X* is placed next to the scheduled completion date.

In this example, steps 10 and 11 are behind schedule for product number 1670, known as Wizbang. Notice that the date of this report is March 11. Both steps 10 and 11 for product 1670 were to have been completed on March 10 according to the scheduled completion date. The absence of the *X* by the scheduled completion date indicates that these steps are behind schedule for this product. On the other hand, steps 5 and 6 are ahead of schedule for product 1810 as indicated by the presence of the *X* next to the scheduled completion dates. Note that for this product, the scheduled completion date for steps 5 and 6 is listed as April 1, hence ahead of schedule as compared with the current (report date) of March 11.

Work schedules such as the one shown in Figure 1–1 are used to keep upper management informed. If anything is behind schedule, management will want to know what you are doing to catch up. If you need help, ask for it, but do not miss the production start date. It cannot be overstated that schedules *must* be met.

7. *Build flexibility into the plan.* It is certain that things will change, and designers need to anticipate where they are going to expand, select equipment that is versatile and movable, and design buildings that will be able to support a wide variety of uses.

Engineer ______		Date 3/11/XX		
1. Obtain product number	1670	1810	1900	1700
2. Create product name	Wizbang	Gizmo		
3. Select project engineer	Meyers	Stephens		
4. Determine production rate per shift	1,500	1,750		
5. Complete manufacturing plan	3-1 X	4-1 X		
6. Complete material handling plan	3-1 X	4-1 X		
7. Set time standards	3-5 X	4-5		
8. Determine number of	3-6 X	4-6		
a. fabrication machines needed	3-6 X	4-6		
b. assembly stations needed	3-6 X	4-6		
9. Issue flow chart	3-10 X	4-10		
10. Design workstations	3-10	4-10		
11. Select material handling equipment	3-10	4-10		
12. Prepare budget plan	3-12	4-15		
13. Prepare layout plan	3-14	4-15		
14. Present to management	3-15	4-15		
15. Write work orders to build stations	3-25	4-15		
16. Issue purchase orders	3-15	4-15		
17. Develop quality control requirements	4-1	5-1		
18. Try out first workstations	4-1	5-1		
19. Install equipment	4-14	5-14		
20. Write work methods sheets	4-14	5-14		
21. Run production pilot	4-15	5-15		
22. Start production	5-1	5-30		
23. Recheck everything				

Note: *X* means this step is complete.

Figure 1–1 New product work progress report to be completed by one engineer.

8. *Reduce or eliminate excessive inventory.* Inventory costs a company about 35 percent a year to hold. These costs include

a. The cost of space and its supporting cost
b. The cost of money tied up in inventory
c. The cost for employees required to move and manage the inventory
d. The loss due to damage, obsolescence, and shrinkage
e. The cost for material handling equipment

All these costs add up to big money, so minimize all forms (raw material, work in process, finished goods) of inventory.

9. *Achieve miscellaneous goals.* These include additional goals and objectives of the facilities and material handling plan. These should be added as you and the management decide something is important. For example, you may want to

1. Restrict operator lifting to one part. This will require the designer to select material handling equipment that will eliminate the operator lifting boxes to a work area and off the station. This will also be a payback with lower back injury problems.
2. Use work cells. This will reduce inventory and material handling.
3. Use plug-in-plug-out equipment to allow operators to move equipment easily for flexibility.
4. Minimize work in process because inventory is expensive.
5. Build kanban (signal board or instruction card) or just-in-time inventory philosophy into the manufacturing facility design.
6. Build visual management systems into the design to improve factory management.
7. Design for first-in-first-out inventory for inventory control.

Whatever you think is important and want to accomplish by your new facility design should be stated as a goal. Goals are to strive for but not always to achieve perfectly. However, without goals designers have much less chance to achieve all they want to do. Two last thoughts about goals: They should be measurable and achievable.

THE MANUFACTURING FACILITIES DESIGN PROCEDURE

The quality of a manufacturing facility design (the plant layout blueprint) depends on how well the planner collects and analyzes the basic data. The blueprint is the final step of the design process and the step during which novice planners want to start. This is like reading the last page of a book first. Resist jumping into the layout phase before collecting and analyzing the basic data. If you have faith and follow the procedure, a great design will, like magic, automatically appear. The following is a systematic way of thinking about a project:

1. Determine what will be produced; for example, a toolbox or swing set or lawn mower.
2. Determine how many will be made per unit of time; for example, 1,500 per 8-hour shift.
3. Determine what parts will be made or purchased complete—some companies buy out all parts, and they are called assembly plants. Those parts the company makes itself require fabrication equipment and considerably more design work.
4. Determine how each part will be fabricated. This is called *process planning* and is usually done by a manufacturing engineer, but in many projects, the

manufacturing facilities designer is also responsible for tool, equipment, and workstation design.

5. Determine the sequence of assembly. This is called *assembly line balancing*. The topic is covered in depth throughout this book.
6. Set time standards for each operation. It is impossible to design a plant layout without time standards.
7. Determine the plant rate **(takt time).** This is how fast the facility needs to produce. For example, it needs to make 1,500 units in 8 hours (480 minutes), so 480 minutes divided by 1,500 units equals .32 minute. The speed of the plant and every operation in the plant must make a part every .32 minute (about three parts per minute).
8. Determine the number of machines needed. Once you know the plant rate and the time standard for each operation, divide the time standard by the line rate and the number of machines results. For example, you have an operation with a time standard of .75 minute and a line rate of .32 minute. How many machines are needed (.75 divided by .32 equals 2.34 machines)? You will need to purchase three machines. If you buy only two, you will never produce 1,500 units per shift without working overtime. This will cause a bottleneck.
9. Balance assembly lines or work cells. This is dividing work among assemblers or cell operators according to the line rate. Try to give everyone as close to the same amount of work as possible.
10. Study the material flow patterns to establish the best (shortest distance through the facility) flow possible.

    a. string diagram
    b. multiproduct process chart
    c. from-to chart
    d. process chart
    e. flow process chart
    f. flow diagram

11. Determine activity relationships—How close do departments need to be to each other to minimize people and material movement?
12. Lay out each workstation. These layouts will lead to department layouts, and then to a facilitywide layout.
13. Identify needs for personal and plant services, and provide the space needed.
14. Identify office needs and layout as necessary.
15. Develop total space requirements from the above information.
16. Select material handling equipment.
17. Allocate the area according to the space needed and the activity relationships established in item 11 above.
18. Develop a plot plan and the building shape. How will the facility fit on the property?
19. Construct a master plan. This is the manufacturing facility design—the last page of the project and the result of all the data collected and the decisions made over the past months.

20. Seek input and adjust. Ask your peer-level engineers and managers to review your plan to see if they can punch holes in your design before you present it to management for approval.
21. Seek approvals, take advice, and change as needed.
22. Install the layout. At this stage, the plan comes together and is one of the most rewarding times, as well as one of the most stressful.
23. Start production. Anticipate that many things will go wrong. No one has started up any production line without some problems; don't expect you will be the first. You will get better each time, but it will never be perfect.
24. Adjust as needed and finalize project report and budget performance.

Many engineering professors and industrial consulting firms are trying to develop a computer formula for manufacturing facilities design. So far, they have developed computer algorithms and simulations for parts of the analysis. Facilities planners will use these tools like any other tool, but the quality of the design will depend on how well they analyze the data, not the computer's ability to solve problems. Therefore, it is best to take a systematic approach, one step at a time, and to add information at each step. When completed this way, the results appear like magic (a great plant layout result). A mature layout technician knows that a good layout is inevitable if the procedure is followed.

The manufacturing facilities design procedure is a general plan of the project. Each step will include some techniques that will not be used in every situation. Skipping steps is permissible if considered and determined not to be necessary. The 24-step procedure just presented is the basic outline for the remainder of the book. If you are doing a layout project, you might use this list as an outline.

TYPES AND SOURCES OF MANUFACTURING FACILITIES DESIGN PROJECTS

1. *New facility.* This is by far the most fun and where the most influence can be made with a new manufacturing facility project. There are fewer restrictions and constraints placed on a new project because you do not have to be concerned about old facilities.
2. *New product.* The company sets aside a corner of the plant for a new product. The new product must be incorporated into the flow of the rest of the plant, and some common equipment may be shared with old products.
3. *Design changes.* Product design changes are always being made to improve the cost and quality of the product. The layout may be affected by these changes, and the facility designer should review every design change.
4. *Cost reduction.* The plant facility designer may find a better layout that will produce more products with less worker effort. Others within the company may make suggestions for improvements and cost reductions that will affect the layout. All must be considered.

5. *Retrofit.* Because many old plants are poorly laid out, older manufacturing facilities designers may spend most of their time working on making these facilities more productive. The procedure for **retrofit** is the same as for a new plant—except there are more constraints. These constraints include: existing walls, floor pits, low ceilings, and any other permanent fixtures that may pose an obstacle to an efficient material flow.

Every area of human activity has material or people flows. Disney World's flow is people; hospitals have flows of patients, medical supplies, and food service; stores have flows of customers and merchandise; kitchens have flows of cooks and foods. If designers study the flow, they can improve it by changing the facilities layout. Opportunities are everywhere.

Although it is said that only death and taxes are certain, there exists a third certainty—a plant layout will change. Some industries are more changeable than others. For example, a toy company may have new products added to its product line every month. Plant layout work would be continuous in such a company. In a paper mill, the layout would change very little from year to year, so plant layout work would be minimal.

COMPUTERS AND SIMULATION IN MANUFACTURING FACILITIES DESIGN

Computer simulation and modeling are rapidly becoming important in the manufacturing and service segment of American industry. As a result of market dynamics and fierce global competition, manufacturing and service enterprises are forced to provide a better quality product or service on a more cost-effective basis while trying to reduce the production or service lead time. The quest for the competitive edge requires continuous improvement and changes in the processes and implementation of new technologies. Unfortunately, even the most carefully planned, highly automated, sophisticated manufacturing systems are not always immune from costly design blunders or unanticipated failures. Among the common examples of these costly mistakes are insufficient space to hold in-process inventory, mismatches in machine capacities, inefficient material flow, and congested paths for automatic guided vehicles (AGVs).

Although computer simulation and modeling are not new to solving complicated mathematical problems or to providing insights into sophisticated statistical distributions, the power of the new generation software has dramatically increased the application of computer modeling as a problem-solving tool in the facilities design arena. Simulation packages currently available no longer require a strong background in mathematics or computer programming languages in order to perform real-world simulations. There are a number of user-friendly advanced simulation packages available that allow the user to simulate either the working of a factory, a just-in-time inventory environment, a warehousing and logistics problem, or the behavior of a group technology system. These simulation packages have been demonstrated to be a valuable aid in the decision-making processes. They also require a relatively small investment of time on the part of the novice in order to develop a working knowledge of the simulation process.

Simulation can be used to predict the behavior of a manufacturing or service system by actually tracking the movements and interaction of the system components and aiding in optimizing such systems. The simulation software generates reports and detailed statistics describing the behavior of the system under study. Based on these reports, the physical layouts, equipment selection, operating procedures, resource allocation and utilization, inventory policies, and other important system characteristics can be evaluated.

Simulation modeling is dynamic, in that the behavior of the model is tracked over time. Second, simulation is a stochastic process, meaning random occurrences can be studied.

In the scope of facilities planning and design, computer simulation can be utilized to study and optimize the layout and capacity, JIT inventory policies, material handling systems, and warehousing and logistics planning. Computer simulation allows the comparison of different alternatives and studies various scenarios in order to select the most suitable setup.

Currently, there are a number of user-friendly advanced simulation packages available to assist the facility planners in achieving the best possible results. Computer simulation and its application are discussed in greater detail in Chapter 15.

ISO 9000 and Facilities Planning

ISO 9000 and other quality standards have become a major contributing factor in the operations of many manufacturing and service enterprises. The ISO series of international standards were first published in 1987 by the International Organization for Standardization (ISO). All or part of these standards can be adopted by an organization depending on the size and the scope of the operation of the enterprise. A large number of corporations demand that their vendors be registered under this or other similar quality standards. Indeed, such registration is now a primary prerequisite for many vendors. ISO 9000 standards and requirements can have a direct influence on facilities design. Considerations must be given during the initial planning of the facilities in order to incorporate and facilitate the implementation of these standards. The latest revision of the ISO 9000 standard emphasizes the "process approach" to the organization of the enterprise. When analyzing the facilities planning from a macroscale approach, each and every aspect of the enterprise—from receiving to shipping, with all the intermediate and support functions of the facility—should function as an integrated and a cohesive system supporting the process. Some of the specifics are as follows.

A facility layout is only as effective as the management team and the plan that the team follows to run the company. An effective quality management system reinforces and complements the physical aspects of the facilities and aims to maximize the return on the investments in the organization's physical assets such as production equipment. The company must develop, document, implement, and maintain an effective quality management system. This system needs to outline critical processes and records that are to be maintained. The documented quality system needs to be controlled to ensure that the company is operating on current standards and correct procedures. The company must have upper management's commitment to producing a quality product. Personnel's responsibilities must be outlined and understood at all levels. Top management must ensure that customer requirements are determined and met to promote customer satisfaction.

Management must regularly review the company's quality management system to ensure that the current practices indeed adhere to the stated policies and that the current standards are adequate for the company's capabilities. This includes capability analyses of equipment, personnel, and space resources of the organization. Management must continually monitor operations for improvement opportunities.

In addition, the company must ensure that it has adequate resources. These resources are, but are not limited to, qualified personnel, proper equipment, and sufficient levels of inventory. The company must determine and provide adequate resources needed to implement and maintain the quality management system and enhance customer satisfaction. The work environment needs to be suitable to achieve conformity to product and customer requirements. The responsibility and role of the facility planner in determining the required level of these resources is of paramount importance.

A company must have a well-defined and a structured system for managing its inventory in order to ensure that parts are being completed on schedule and within customer specifications. The organization must have a written and well-documented plan on how products and components will be traced from receipt, through all stages of processing, and finally, delivery. When batch or product traceability is required, data collection capabilities can be built into material handling equipment and also incorporated as part of the workstation design. Handheld or stationary scanners, for the purposes of data collection and item tracking, should be designed as part of the workstation design and facilities planning.

The company must plan and develop the processes needed for product realization. Customer requirements need to be considered and specific processes must be determined to achieve customer satisfaction. These customer requirements must be reviewed and approved prior to acceptance to ensure that the equipment and process capabilities that are required to meet these requirements do exist.

The design and development process must be considered as well. Beginning with customers' specifications to the outputs from the facility, all procedures and processes must tie back to achieving customer satisfaction. The company is required to ensure that production of the product is maintained under controlled conditions. This requirement can be tied directly to JIT, MRP, kanban, or other production control systems. In addition, consideration must be given by the planners in the initial stages of facilities design as to how to incorporate procedures for quality assurance or verification at receiving, work in process (WIP), and finally, during the final stage of production.

There are specific processes that need to be measured and analyzed to adhere to customer requirements. An example would be testing the hardness of steel to ensure it conforms to stated customer requirements. Those processes need to be identified and the analysis documented. Internally, the company must monitor its processes and procedures to ensure they coincide. This is handled through the internal auditing process. This process also enables top management to identify opportunities for improvement, whether it be in updating equipment or changing processes to increase efficiency. The ISO standards stress continual improvement, which implies that the quality management system will continually change, as the company changes and opportunities to improve arise.

Furthermore, at any given stages, procedures must be in place to handle all nonconforming process or products. Systems should be developed to identify, document, evaluate, and segregate nonconforming occurrences. The means of handling products that are nonconforming and adequate staging facilities must be provided until proper disposition has been determined. Such disposition may include rework or acceptance with or without further rework, or the item may be rejected and scrapped.

Adequate mechanisms should be in place to ensure proper handling, storage, packaging, preservation, and delivery of the product.

Facilities planners have many opportunities to incorporate these procedures into the initial planning stages of the plant design.

GLOSSARY OF SOME MAJOR TERMS AND CONCEPTS IN FACILITIES PLANNING

andon This is the line stop method indicator board located above the production line that serves as a visual control. When operations are normal, the green light is on. A yellow light is turned on when an operator wants to adjust something or to call for help. If the line must stop to solve the problem, a red light is turned on. See also *line stop concept.*

autonomation (jidoka) Autonomation, or automation with a human touch, means transferring human intelligence to a machine. Devices capable of making judgments are built into a machine. In the lean system, this concept applies not only to the machinery but also to the production line and the operators. If a problem occurs, an operator is required to stop the line. Defects are prevented on the production line, allowing the situation to be investigated.

cost reduction formula This is a way of thinking about removing waste (muda) from the process by asking why, what, where, when, and how of every operation, transportation, inspection, storage, and delay to eliminate, combine, change routing, or simplify.

eight kinds of muda (waste) Types of muda include (1) overproduction, (2) waiting, (3) transportation, (4) processing, (5) inventory, (6) motion, (7) rework, and (8) people utilization. The idea of improvement is to work easier, faster, cheaper, smarter, and safer. In trying to eliminate waste, ask if you can eliminate it, then combine it with another cost, change the routing, or simplify it.

facilities design This includes the site selection, building design, plant layout, and material handling. Often facilities design is used synonymously with plant layout.

This is the organization of the company's physical facilities to promote the efficient use of the company's resources such as people, equipment, material, and energy.

five why's Every time a problem occurs, "why" is asked five times or more. When "why" is repeated five times, the root cause of the problem, as well as its solution, becomes clear instead of just a symptom of the problem.

five (5) S's principles These concepts are used to describe in more detail what proper housekeeping means: (1) organization sifting, (2) arrangement sorting, (3) cleaning or sweeping, (4) hygiene—making things spick and span, and (5) strict discipline.

ISO 9000 This is a series (ISO 9000, ISO 9001, ISO 9002, ISO 9003, and ISO 9004) of international standards that were first published in 1987 by the International Organization for Standardization (ISO). They were intended to be used in two-party contractual settings; however, after their adoption by the European community, the standards have become universally accepted. All or part of the standards can be adopted by an organization depending on the size and the scope of the operation of the enterprise.

kaizen Kaizen means continuous improvement. Kaizen is done by a team of employees or by a single employee. It is the constant search for ways to improve the present situation.

lean manufacturing Lean manufacturing is a continuance of *lean thinking* where less of everything is better. Value added is the guiding philosophy, by which all elements of cost that do not add to the value of the end product are eliminated.

kanban A kanban (signal board) is a simple and direct form of communication that is always located at the point where it is needed. The kanban is usually a small card inserted into a plastic envelope. On the card is written information such as the part number, the quantity per container, the point of delivery, and so on. The kanban card tells the operator to produce the quantity withdrawn from the earlier process. The card is a tool used to manage and ensure JIT production. Containers or a kanban square can be used instead of kanban cards to accomplish the required results.

line stop concept (andon) The concept allows an operator to stop the production line if necessary. Whenever a problem occurs, the operator stops the line, identifies the problem, resolves the problem, and regains the flow as soon as possible. This approach calls for discipline in responding and solving problems quickly.

manufacturing facilities design See *facilities design.*

material handling This means handling material, and includes both the principles and the equipment.

mission statement This is a statement of the primary goal of the project and will include subgoals.

pokayoke (fool-proofing) To ensure 100 percent quality products, defects must be prevented from occurring. Pokayoke are innovations that are made to tools and equipment in order to install devices for the prevention of defects. Some examples are

1. When an operation is forgotten, the next operation will not start.
2. Problems in the earlier operations are checked in the later operations to stop the defective product.

3. When there are problems with the material, the machine will not start.
4. Tooling and fixturing are designed to accept a part in the correct direction only.

production leveling Also referred to as line balancing, fluctuations in the product flow increase waste. To prevent this waste, fluctuations at the final assembly line must be kept to zero. Production is leveled by making one model, then another model, and so on.

retrofit This means reworking the facility plan and is a part of the ongoing improvement plan (kaizen) or the single major effort made when the situation gets out of control.

root cause In all problems, symptoms will result that keep the root cause of the problem from being discovered. Asking "why" five times can assist in finding the root cause. Otherwise, actions cannot be taken and problems will not be truly resolved.

simulation This is a means of experimenting with a detailed model, featuring the characteristics of real systems, in order to determine how the system will respond to various changes to its components, environment, and structure. For our purposes, a system may be defined as a work cell, an assembly line, a group of machines, or an entire manufacturing facility. Simulation provides the opportunity to play a series of what-if games and to observe the effects of various changes or manipulations to the model in order to optimize or improve the real system.

standardization This is recording the method and procedures to arrive at the same result consistently. Standardization is very important to an improvement program. Without standardization, things will revert to the old processes. Once standard methods are established, they must be revised to reflect improvement activities.

takt time Takt time, or *R* value, is determined on the basis of the periodic production requirements and the amount of operating time during the period. Setting takt time for each process is the key to bringing all the different parts together at all stages of assembly at exactly the right time. Every workstation in the plant needs to keep up with the takt time. If each process makes things according to its takt time, production will amount to exactly what is needed when it is needed. Producing by takt time ensures that all production will be matched to the final assembly process.

$$\text{Takt time} = \frac{\text{Total daily operating time}}{\text{Total daily production requirement}}$$

Toyota production system This is the beginning of the lean thinking concept and lean manufacturing.

value-added work This is work that actually transforms materials, changing either form or quality, through such activities as assembly, milling, welding, heat treating, or painting. In a typical factory, it is often found that 95 percent of an operator's time is not used in adding value to the product. These questions can be asked when considering value-added work: (1) Are the activities absolutely necessary for the production activities? (2) Are the activities adding value to the product instead of cost? (3) Are the activities related to items that the customer may see or even care about?

value-stream mapping (VSM) A pictorial representation of a process that allows a systematic assessment of each component or step in the process.

QUESTIONS

1. What is a plant layout?
2. What is a facility design?
3. What is material handling?
4. Explain how the cost reduction formula is used in the manufacturing design process.
5. Material handling accounts for what percent of injuries and what percent of operating costs?
6. List the goals of manufacturing facilities design and material handling.
7. What is a mission statement?
8. What two items in Figure 1–1 are behind schedule?
9. What is the value of a manufacturing facilities design procedure?
10. List the manufacturing facilities design procedure.
11. What are the five types of manufacturing facilities design projects?
12. What is the difference in procedure between a new facility design and a retrofit?
13. What are lean manufacturing and lean thinking?
14. Define *muda, kaizen, kanban,* and *andon.*
15. Define *simulation* and explain why you think a simulation can be an important tool in facilities design.
16. Explain how you would incorporate various ISO 9000 requirements into the facilities planning process.
17. What do you believe is meant by a "random" process? Give an example of a random occurrence on the factory floor and how simulation can be helpful in understanding such a phenomenon.
18. Are you familiar with any automatic data capture technology? Where and how do you see that such technology may be applied in the facilities planning process?
19. On the average, a facility undergoes some "layout design changes" once every 18 months. What would necessitate such changes?

C H A P T E R

Sources of Information for Manufacturing Facilities Design

OBJECTIVES:

Upon the completion of this chapter, the reader should:

- Be able to identify various sources of information for facilities planning
- Understand the concept of and be able to calculate takt time
- Be able to define and construct a flat and an indented bill of material
- Be able to understand the role of management policy in facilities planning

The facility design depends on basic information that the facility designer must obtain from other sources. Much of this basic information comes from other departments within the company. Sometimes getting this information is like "pulling teeth," but the designer needs reliable information, and others are the best source. The larger the company is, the less data are actually produced by the designer. Some companies have several subdepartments within manufacturing and industrial engineering. Examples include

1. The processes section would establish the routing and select the machine to be used.
2. The tool design section would design the fixtures and specify the specific tools.
3. The time standards application section would set the time standards for each operation.
4. The quality department would specify inspection procedures and require space for tools and people.
5. The safety department will want to review and input its requirements.
6. Inventory and production control department policies will affect the space needs as well as the procedures.

All this information will affect the facilities design.

This chapter will discuss those sources of information from outside the manufacturing department, Chapter 3 will discuss time standards, and Chapter 4 will discuss the other information required from within the manufacturing department. The manufacturing facilities designer will always need to get the external information from someone else, but in smaller companies and in consulting situations, he may need to produce the manufacturing department's information himself.

There are three basic sources of information outside the manufacturing department:

1. Marketing
2. Product design
3. Management policy

THE MARKETING DEPARTMENT

The marketing department provides a research function that analyzes what the world's consumers want and need. It searches out ways to fill these potential customers' demands. Some of the information that marketing provides is (1) the selling price, (2) volume (How many can we sell?), (3) seasonality (Is it a summer or winter product?), and (4) replacement parts that an older product may require.

Determining the selling price is not an exclusive function of the marketing department. The industrial engineering organization may supply the cost data for pricing, but the selling price has a direct influence on the number of units the company sells. Every customer makes a value analysis on each purchase. The lower the price is, the more customers will choose the product. Pricing is very complicated and the marketing, production, and finance departments are all part of this decision, but marketing needs that information before it can ask the customers, "How many do you want to buy?"

Volume comes down to how many units the company wants to build per day. The marketing department may take some model shop samples to a few important customers and ask their opinions. If these customers like the new product, they would be asked how many they would buy. Typically, 20 percent of customers buy 80 percent of total production (an interesting statistic based on *Pareto analysis*[1]).

---

[1] Pareto analysis states that 80 percent of activity (problems and opportunities) comes from 20 percent of the sources. This statistic in facilities design is important when you consider the product (80 percent of sales are from 20 percent of the product line) and customers (80 percent of sales are to 20 percent of the customers). Another useful example is the cost of production parts: 80 percent of the material cost of the product will make up only 20 percent of the total part numbers. It also means that 80 percent of the people or machine problems come from 20 percent of the people or machines. The result of knowing this is that a company can identify these most important components and can manage them more closely.

Therefore, if a small group of customers say that they will buy 125,000 units that represent 50 percent of annual sales, 250,000 units will be needed. If the plant works 250 days a year (50 weeks times 5 days per week), then 1,000 units would be needed every day.

The number of units required per day is a very important number for the facility designer because it determines the number of machines and people for which he needs to provide space. To achieve this goal, he must determine the plant rate (how fast every machine and workstation need to work to meet the goal).

Determining Takt Time or Plant Rate

To meet the production goal or a present production volume, every machine and operation must keep a certain pace. For example, if the schedule calls for the production of 1,500 grills per day or per shift, then at the end of that production period, the company must have 1,500 completed grills, packed and ready for shipment. To meet this schedule, essentially the plant must produce enough parts and components per period to keep up with the production demand. If each grill requires one cooking grid, then the production capacity must meet the demand of producing 1,500 grids per period.

However, if each grill has *two* side tables, then the plant must have the capacity of producing 3,000 of these side tables during the same period. Stated differently, the rate of production for these side tables must be twice the rate of the grids. In other words, each side table must be produced in one-half the time of that of the grid. Keep in mind that we are not saying that the time requirement for the production of a side table is only one-half that of a grid. What we are saying is that because for every grid two side tables are required, the company must produce the side table at a faster rate. This rate of production is called ***takt time*** or production rate, as discussed in Chapter 1, or simply, the R value.

The plant rate or takt time (a German word in common use today having the same meaning) is the rate at which operations, processes, parts, components, and so on must run in order to meet the production goal.

To calculate takt time, you must know the production goal, the amount of time allotted for the production of these units (e.g., one 10-hour shift, or two 8-hour shifts, etc.), and any nonproductive time that is taken away from production such as breaks, team meetings, lunch, and so forth. Furthermore, a general knowledge of the overall plant efficiency as determined by unplanned downtimes, inventory stock-outs, absenteeism, and so on is necessary in order to calculate the takt time.

The following example illustrates the calculation of the takt time.

> *Example:* Let us assume that you need to produce and ship 1,000 units of product from the plant in an 8-hour shift. Thirty minutes for lunch, 10 minutes for break, and 8 minutes for team meetings are allotted during each shift. Furthermore, let us assume that the plant is operating at 90 percent efficiency. (Calculating plant efficiency is beyond the scope of this discussion; however, suffice it to say that expectations of 100 percent efficiency are unrealistic and in this example, 90 percent efficiency is quite respectable.)

Given the data above, how much time do you have to produce one unit of product?

8-hour shift × 60 minutes = 480 minutes production time

480 minutes production time − (30 minutes for lunch) − (10 minutes for breaks) − (8 minutes for meetings) = 432 minutes actual work time

432 minutes actual work time × 90 percent efficiency = 389 minutes effective (productive) time

Therefore, you have *only* 389 minutes to produce 1,000 units of product.

$$\text{Takt time OR } R = \frac{389 \text{ minutes}}{1{,}000 \text{ units}} = .389 \text{ minute per unit}$$

$$\text{Takt time OR } R \text{ (plant rate)} = .389 \text{ minute per unit}$$

A finished, packed-out unit must come off the assembly and packout line every .389 minute, or just about 2.5 units per minute. This means every workstation and every machine in the plant will need to produce about 2.5 parts or sets of parts per minute.

$$\text{Units per minute} = \frac{1 \text{ Unit}}{.389 \text{ minute per unit}} = 2.57 \text{ units per minute}$$

If you need two parts (like axles on a toy wagon or two side tables per grill) per finished unit, then you would need 5.14 parts per minute.

Of course, this rate assumes that no scrap parts are produced and no rework time is necessary. Although both scrap and rework are undesirable occurrences, they do occur and take up production time and resources. Plant rate, or takt time, must be adjusted to reflect this, as illustrated in the following discussion.

Calculating Scrap and Rework Rates

Although undesirable, manufacturing operations do produce *scrap* or unusable parts. Furthermore, often there is a need to redo an operation simply because the part was not produced within the desired specifications the first time. This is called *rework*. Scrap and rework result in an inefficient and wasteful use of the facilities' resources. Every effort should be made to eliminate such waste. However, as long as the plant has to deal with scrap and rework, it cannot ignore their demand on production time.

Quality and production departments have historical data that indicate the level of rework and scrap for each operation. In determining the plant rate, or takt time, you must include scrap and rework rates into your calculations. Indeed, it is also prudent to add into these calculations the need for spare or replacement parts.

To illustrate, let us assume in the preceding example the press operation produces 3 percent scrap. Therefore, to end up with 1,000 finished wagon bodies, you must start with a larger number so that after scrapping 3 percent, you will have

1,000 good parts. If you designate the finished parts with the letter *O* for output, you can calculate the input, *I*, as follows:

$$I = \frac{1{,}000}{(1 - .03)} = 1{,}031 \text{ units}$$

This is the number of blanks with which you need to start your process. Keep in mind that if additional operations are performed, and each operation produces additional scrap, further adjustment to the input volume is necessary. For example, assume in addition to the stamping operation, two other steps are performed. One step has a 2.5 percent scrap rate and the other produces .5 percent scrap. You can calculate input as follows:

$$I = \frac{1{,}000}{(1 - .03)\ (1 - .025)\ (1 - .005)} = 1{,}063 \text{ units}$$

The general formula is stated as

$$I = \frac{\text{output}}{(1 - \%\text{scrap1})\ (1 - \%\text{scrap2})\ (1 - \%\text{scrap3}) \ldots (1 - \%\text{scrap } n)}$$

You no longer are concerned with the required pieces of equipment to produce 1,000 wagon bodies. The takt time is now calculated on the basis of 1,063 units.

$$\frac{389 \text{ minutes}}{1{,}063 \text{ units}} = .366 \text{ minute/unit}$$

Therefore, the adjusted take time, or plant rate,

$$= .366 \text{ minute/unit}$$

The plant rate is one of the most important numbers in manufacturing facilities design. It is used to calculate the number of machines and workstations, the conveyor speed, and the number of employees required by the facility design.

Seasonality is important to the facilities design because the plant may be required to produce the total year's product needs in a few months, so a larger facility will be needed. Customers want space heaters and sleds in the winter, gas grills and swimming pools in the spring and summer, and toys in the stores for Christmas. If you waited until just before the season to start producing your product, you would either need a great deal of extra machines or miss your market time. If you produce all year long just for the Christmas season, you would need 10 to 12 months of warehouse space. Determining how soon to start and how much to make per day is a compromise between inventory carrying cost and production capacity cost. The objective is to minimize total cost. Production and inventory control is a manufacturing extension of the marketing department, and will probably be your source of volume information.

The subject of production and inventory control goes hand in hand with manufacturing facilities design, and production and inventory control policies will have a big effect on your design.

Recognize the need for replacement parts. If you have been in business for any period of time, your product will begin to wear out. Customers may call you for replacement parts that have worn out or have broken. This business requires not only that you build extra inventory but also that you have storage and shipping areas for such customer service. Again, production and inventory control will tell you how much to increase your volume to account for replacement parts on a part-by-part basis.

THE PRODUCT DESIGN DEPARTMENT

Blueprints, a bill of materials, assembly drawings, and model shop samples inform the facilities designer of the prime mission—a detailed description of what needs to be accomplished. The product design department is the source of this valuable information. The first question anyone would ask when assigning a new facility design project is, "What are we going to make?" The output of the product design department tells you exactly what you are going to manufacture.

Blueprints, sketches, pictures, CAD (computer-aided design) drawings, and model shop samples all communicate the idea of what the company wants to build. (See Figure 2–1, Figure 2–2, and Figure 2–3.) There will be drawings of each individual part of the product as shown in Figure 2–1. Figure 2–1a shows the individual parts in a toolbox handle. For illustration purposes, Figure 2–1b is presented here to show a more detailed drawing of a clap and plate subassembly. These drawings tell you the size, shape, material, tolerances, and finish. Assembly drawings (see Figure 2–2) show many parts (if not all parts) and how they fit together. An exploded drawing (Figure 2–3) is especially useful for the facilities designer because it helps visualize how the parts fit together. In Figure 2–3, two different exploded views are presented. Figure 2–3a is an exploded view of the toolbox. Figure 2–3b displays the exploded view of a common automobile battery cable clamp. Centerlines are used to separate parts, and the parts are aligned to show the assembly relationship. These give the facilities designer clues to the sequence of assembly.

When the designer is working on the assembly line layout, the exploded drawing will be the guide. The designer cannot get started without blueprints or sketches.

Either a parts list or a bill of materials will be provided to the facilities designer by the product engineering department with each new product (see Figure 2–4 on p. 35). The parts list or bill of materials list all the parts that make up a finished product. This list includes part numbers, part names, the quantity of each part, what parts make up subassemblies, and may include material specifications, parts and raw material unit costs, and make or buy decisions. The decision to either make or buy a part is up to top management, not just the product engineering department, but the parts list is a good place to indicate that decision.

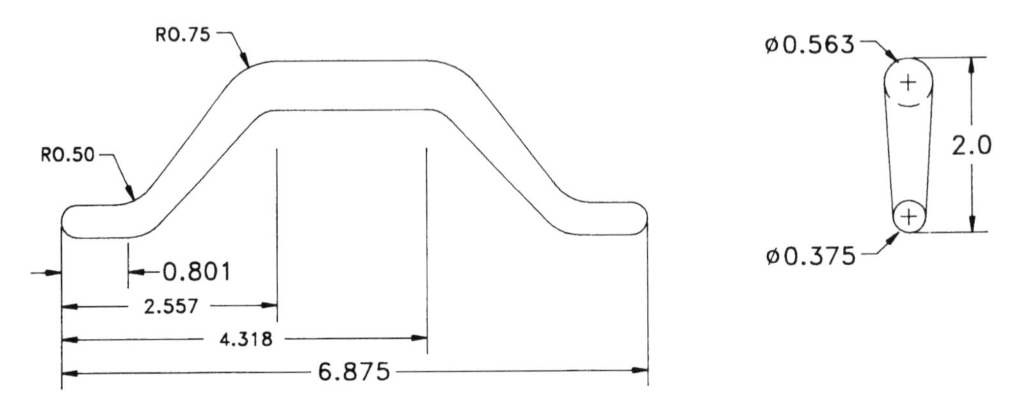

Figure 2–1a Sample blueprint for toolbox handle.

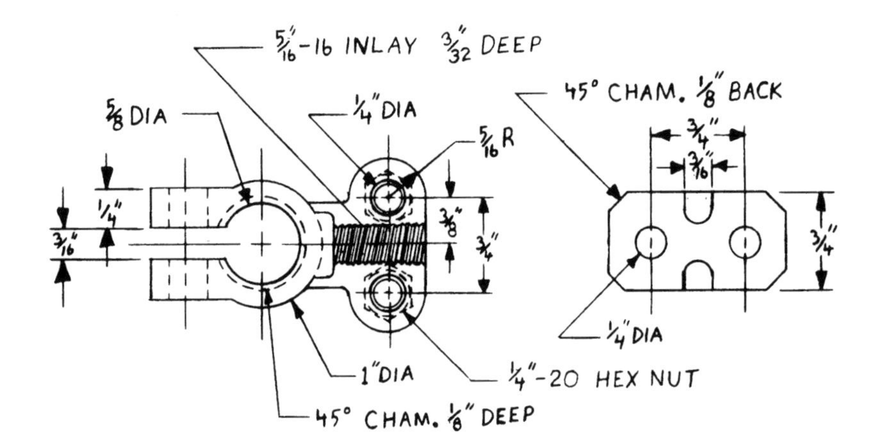

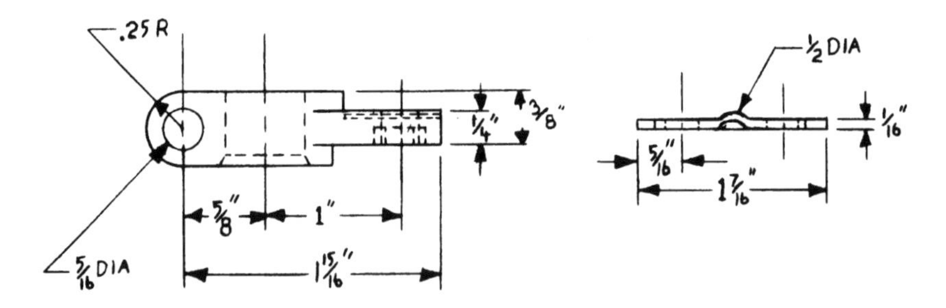

Figure 2–1b Part drawing for clamp and plate.

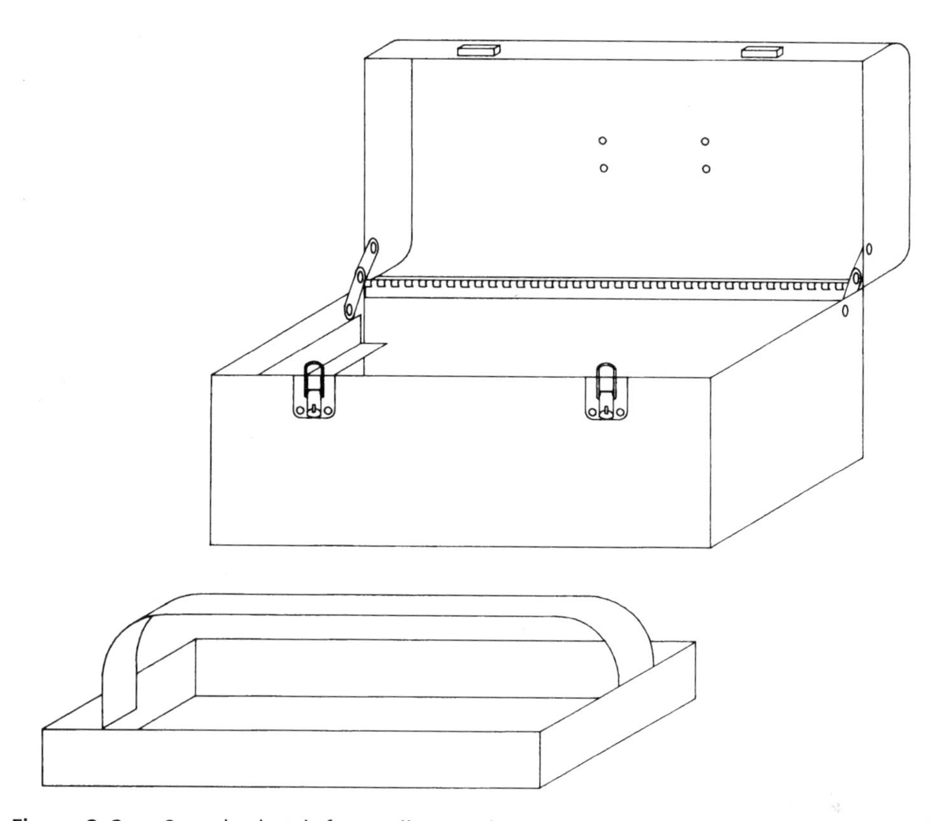

Figure 2–2a Sample sketch for toolbox and tray.

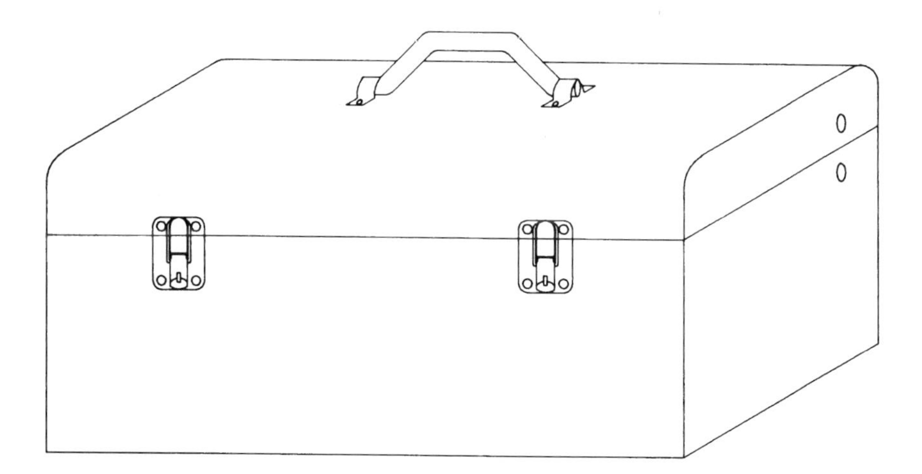

Figure 2–2b Sample sketch for toolbox—Three-dimensional (3D) view.

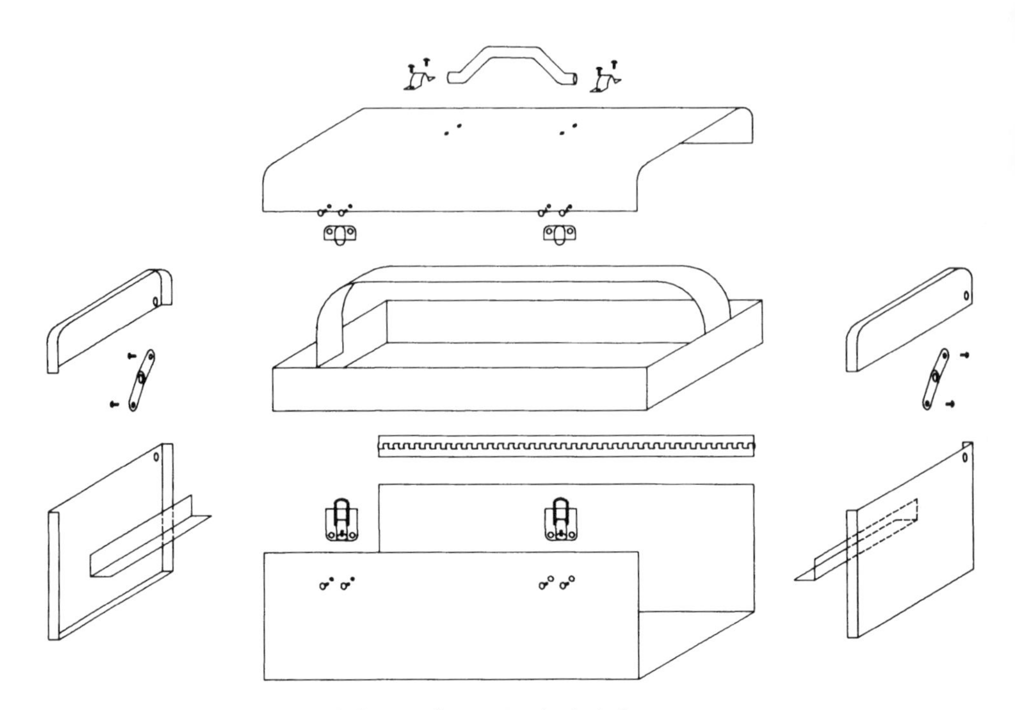

Figure 2–3a Sample sketch for toolbox—Exploded view.

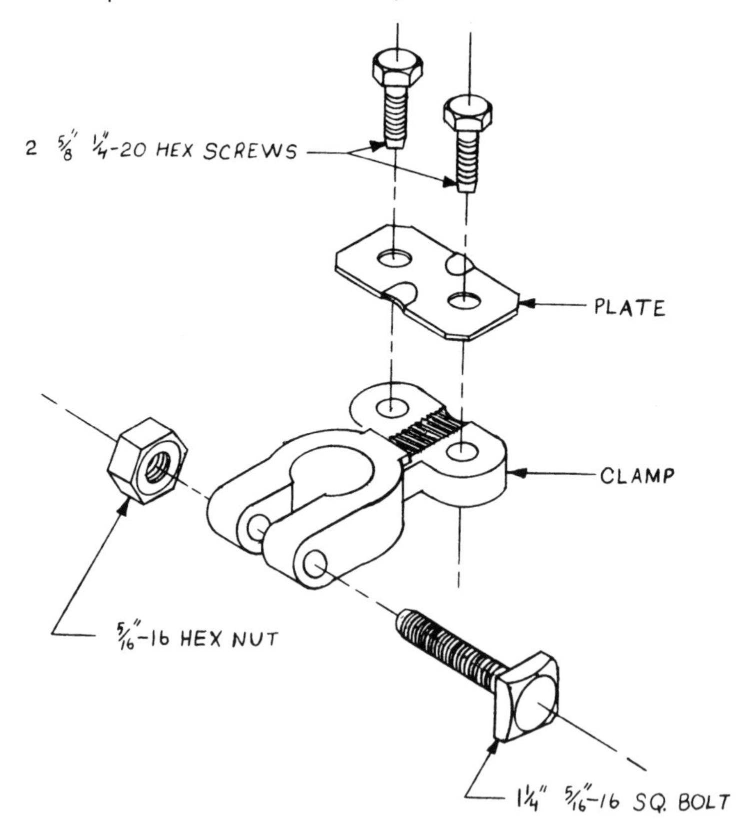

Figure 2–3b Battery cable clamp—Exploded drawing.

Part No.	*Part Name*	*Quantity Needed per Unit*
1	Body end	2
2	Tray bracket	2
SSA1	Body end assembly	2
3	Body	1
SA1	Body assembly	1
SSA2	Cover end	2
5	Cover	1
SSA3	Cover handle assembly	1
SA2	Cover assembly	1
17	Hinge	2
7	Tray end	2
8	Tray body	1
9	Tray handle	1
A2	Tray assembly	1
10	Paint	AR*
11	Handle	1
12	Clip	2
13	Rivet	16
A1	External body assembly	1
14	Catch	2
15	Strike	2
A2	Tray assembly	1
A3	Parts bag	1
19	Packing list	1
20	Registration card	1
21	Name tag	1
22	Divider	4
23	Plastic bag	1
24	Carton	1
25	Tape	24"

* As required.

Figure 2–4 Toolbox parts list.

The Indented Bill of Material

The indented bill of material is also an important aid in the design of the facility and configuration of the work cells and assembly lines. An indented bill of material provides the same basic information as the parts list. However, the indented bill of material presents the hierarchical structure of the product by identifying each assembly, subassembly, and the required or subordinate parts of each assembly or subassembly. An indented bill of material is shown in Figure 2–5 (below). The highest level of the product, or the finished assembly, appears on the top of the list and is given level number zero. Under this are listed the major assemblies and each is assigned as level one (.1). The period before the numeral *1* indents the major subassemblies from the main assembly. Under each subassembly, the required components that comprise that subassembly are listed and numbered level two (..2). In turn, under each component, subordinate parts are listed and each is numbered as level three (...3). If a given level three itself is comprised of multiple parts, those parts would be listed following the given level three part and would be numbered level four (....4), ad infinitum. The purpose of the leaders or periods before each level number is to offset or indent (hence indented bill of material) each level in order to enhance readability.

Figures 2–6 and 2–7 further illustrate the parent–child or hierarchical relationship in the indented bill of material. Figure 2–6 shows the indented bill of material whereas Figure 2–7 shows the assembly chart for the toolbox. The final product as identified by the last step of the assembly chart (see Figure 2–7) is the finished toolbox. In Figure 2–6, the toolbox, being the finished product or the highest level, is represented as the level zero. Further examination of the assembly chart (Figure 2–7) reveals that three assemblies, A1, A2, and A3, are the last major assemblies prior to final product assembly. These three assemblies are numbered as level (.1) in the

Company: ACME, Inc. Prepared by: M.P.S.

Product: Super Gismo Date:

Level	*Part No.*	*Part Name*	*Drwg. No.*	*Qty. per Unit*	*Make or Buy*
0	0012	Super gismo	0012	1	Make
.1	0034	Main frame	0034	1	Make
.1	0421	4′ bracket	0421	2	Make
..2	0344	Strap ties	0344	4′	Buy
.1	0113	1/4″ insert	0113	2	Make
..2	0123	Tube	0123	1	Make
...3	0014	Clear paint		1 gal/100	Buy
.1	0019	Brace	0019	3	Make
..2	0177	1/4-20 nut	0177	4	Buy
..2	0192	3/16″ collar	0192	2	Make
.1	0330	Cylinder	0330	1	Buy

Figure 2–5 Indented bill of material.

Level	*Part Number*	*Part Name*	*Qty. per Unit*	*Make / Buy*
0		**Toolbox**	1	Make
.1	**A1**	**Exterior Body Assembly**	1	Make
..2	SA1	Body Assembly	1	Make
...3	SSA	Body End Assembly	2	Make
....4	1	Body End	2	Make
....4	2	Tray Bracket	2	Make
...3	3	Body	1	Make
..2	SA2	Cover Assembly	1	Make
...3	5	Cover	1	Make
...3	SSA2	Cover End	2	Make
....4	13	Rivets	4	Buy
....4	17	Hinge	2	Buy
...3	SSA3	Cover Handle Assembly	1	Make
....4	11	Handle	1	Make
....4	12	Clips	2	Buy
....4	13	Rivets	4	Buy
..2	14	Catch	2	Buy
..2	15	Strike	2	Buy
..2	13	Rivets	8	Buy
..2	10	Paint		Buy
.1	**A2**	**Tray Assembly**	1	Make
..2	7	Tray End	2	Make
..2	8	Tray Body	1	Make
..2	9	Tray Handle	1	Make
.1	**A3**	**Parts Bag**	1	Make
..2	19	Packing List	1	Buy
..2	20	Registration Card	1	Buy
..2	21	Name Tag	1	Buy
..2	22	Dividers	4	Buy
..2	23	Plastic Bag	1	Buy

Figure 2–6 Indented bill of material for the toolbox.

indented bill of material. Each assembly contains other minor subassemblies or components. These subordinate items are listed under each appropriate assembly. For example, by looking at Figure 2–7, you can see that packing list, registration card, name tag, dividers, and plastic bag are "parts" of the A3 assembly. Hence, in Figure 2–6, these items are under the A3 level and are assigned level (..2). Now try to follow the structures under the assembly A1. You will see that A1 has two subassemblies labeled SA1 and SA2. In Figure 2–6 these subassemblies are assigned level (..2) and each is further broken down into smaller sub-subassemblies and components that are numbered accordingly.

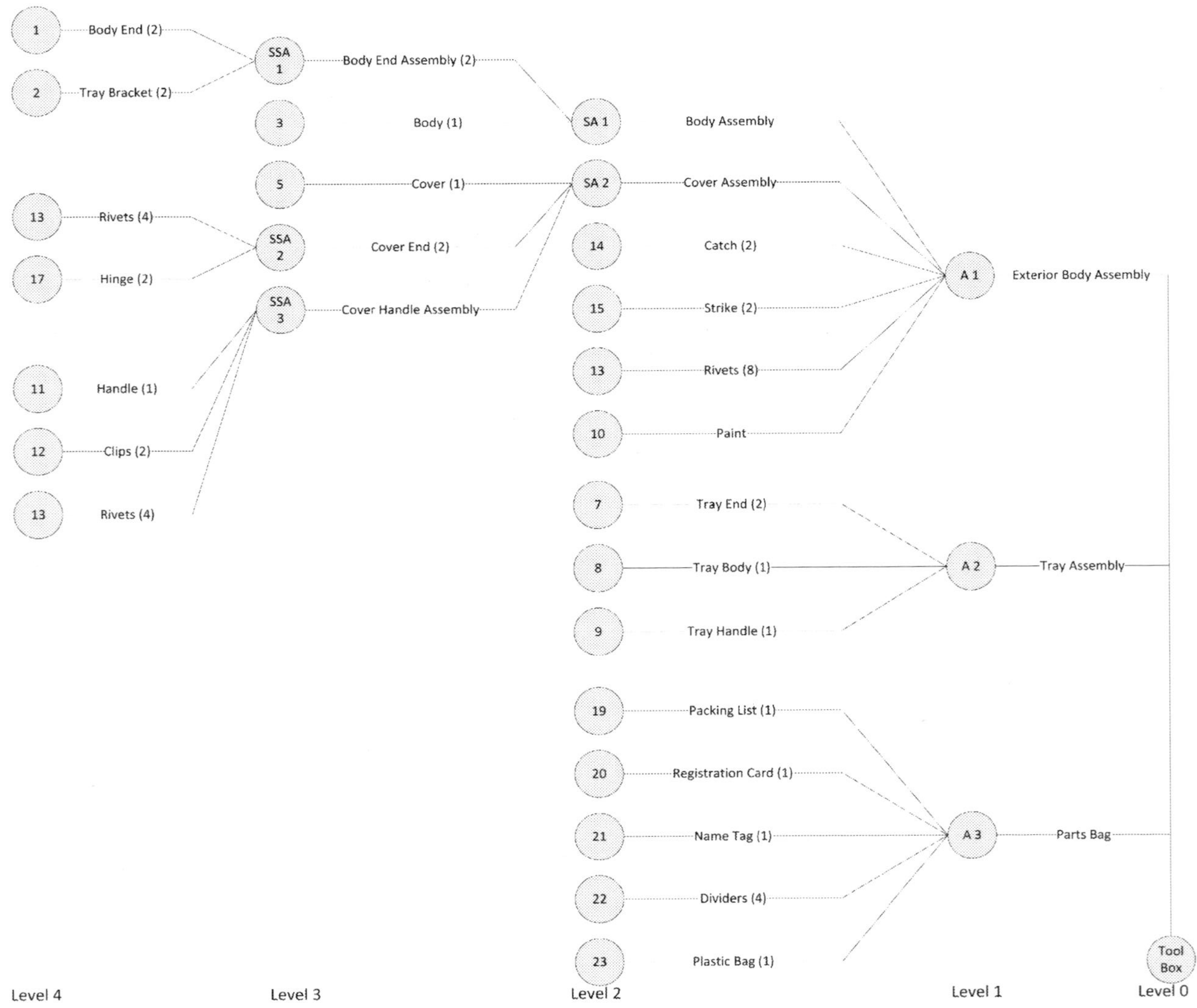

Figure 2–7 The assembly chart for the toolbox.

The indented bill of material provides not only data regarding the composition of the final assembly but also valuable insight into the flow of parts and components in the final assembly.

Companies themselves do not fabricate every part of their products. The parts that are purchased complete are called *buyouts* and can be fabricated cheaper by someone else. Some companies purchase every part complete from outside. These companies are called *assembly plants*. The parts that a company "makes" are basic requirements for the fabrication end of the facility.

Model shop samples, or *prototypes*, are handmade, very costly, and exact models of what the product engineering department wants to produce. These prototypes are not always available, but if they were, they would be very helpful. The ability to "feel" the parts, to take them apart and reassemble them, to look at each part and consider how to make that part furthers your understanding of the product. A systematic disassembly of a prototype, or even that of a finished product produced by a competitor, is a helpful process in determining the logical and proper steps for assembling the product. The disassembly process, often referred to as "reverse engineering," is most helpful in visualizing the order and steps of the process and should plant the initial seeds for the arrangement of the facility. The reverse engineering process can also assist with the product design and development activity. It can point to steps or parts that seem to be superfluous or awkward or difficult to fit or join with other components. The basic steps are outlined below.

A model shop sample would be used as follows:

1. Unpack the unit, noting the sequence of unpacking. This will be the basic information for the packout line. Be sure to keep good notes; photographs are also useful.
2. Play with the finished product to see how it functions. A good understanding of the finished unit's purpose is very helpful.
3. Disassemble the product carefully. Keep good notes again. Then, reassemble the product. This will be your basic information about the assembly line.
4. Disassemble and look at each part. Decide which parts you are going to make in the plant and which parts are going to be purchased complete (buyouts).
5. The "make" parts—those parts that are going to be made by the plant—will need further study to determine how they will be fabricated from raw material. This is the subject of the next chapter.

Without a model shop sample, the facilities planner would still need to follow the above steps—except the information will come from drawings, prints, or sketches. The model shop sample makes the process easier and ensures better results.

The product engineering department can be very helpful to the plant designer. It can point out special manufacturing problems and critical relationships, dimensions, and functions. The product designer and the facilities designer need to work closely together. This open communication and cooperation between the two designers is a fairly new concept in engineering. The traditional approach to product design has often been referred to as the "over-the-wall" approach, in which ideological and territorial barriers often hindered communication between different segments of an organization. When dealing with product development, a breakdown in communication may well result in designing a product that the customer does not want; in specifying standards and specifications that the manufacturing department is not capable of meeting; or in requesting material and components that the purchasing department is not able to procure in a timely fashion. Concurrent engineering attempts to bring all aspects of the product design, development, and manufacturing together so that problems can be detected early and solutions can be sought during the planning stages.

MANAGEMENT POLICY INFORMATION

Management refers to upper-level employees who are responsible for the financial performance of a company. Such information as (1) inventory policy and lean thinking, (2) investment policy, (3) startup schedules, (4) make or buy decisions, (5) organizational relationships, and (6) feasibility studies, will all have an effect on the plant facility design. Facility designers must understand these policies up front; otherwise, they may waste a lot of time.

Inventory Policy

The company's inventory policy could be as simple as "provide space for a one-month supply of raw materials, work in process, and finished goods." Those inventories would require space and facilities, but once the quantity to be stored has been determined, calculating the space requirement is easy. Stores and warehouse layouts will be discussed in Chapter 8. Just-in-time (JIT) and kanban philosophies reduce inventory and, therefore, space, facilities, and cost. Work in process (WIP) requires space, and less inventory means less of everything—which is the definition of *lean manufacturing* and *lean thinking*.

Lean Thinking and Muda as Part of Management Policy

Taiichi Ohno (1912–1990), the Toyota executive who was an adamant foe of waste, developed the Toyota production system upon which lean manufacturing philosophy is based. *Muda,* which means "waste," specifically refers to any human activity that absorbs resources but creates no *value.* Ohno identified the first six types of muda:

1. Mistakes that require rectification; any rework is a good indication of muda.
2. Production of inventory that no one wants at this time wastes space and promotes product damage and obsolescence.

3. Useless process steps that can be eliminated without harming the value of the end product is muda.
4. Any movement of people or inventory that does not create value is muda. Move material as short a distance as possible, or do not move people at all.
5. Idle people waiting for inventory is an indication of a plant that is not in balance. Workers must expend approximately the same amount of effort or bottlenecks are created.
6. Goods produced for which there is no customer demand is muda. If you manufacture too far ahead, you take a chance that there will be no demand for your product because something better came along.

There are more causes and examples of muda all around you; you just need to be aware of them. Fortunately, there is a powerful cure for muda: *lean thinking* and *lean manufacturing*. Lean thinking and lean manufacturing encourage designers to think about value and to create actions in the best sequence, conduct these activities without interruption whenever someone requests them, and perform them more and more effectively. In short, lean thinking provides a way to do more with less: less human effort, less equipment, less time, and less space.

As stated, lean thinking is an important part of the facilities design process, especially in the reduced inventory levels, less movement of material and people, and better balance of workload among workers.

Investment Policy

The corporate investment policy is communicated in terms of *return on investment (ROI)*. *Return* is another way of saying "the savings," and *investment* is the cost of implementing the idea to get those savings. If a project saves a large enough percentage of the cost, then it is a good idea. For example, a facilities design project might be approved with a 33 percent ROI. Thirty-three percent is also a 3-year payback period. Facilities design projects are one of the few investments that management will allow with such a long payback period. Most cost reduction work requires an ROI of greater than 100 percent or a payback period of less than one year.

When presenting a manufacturing facility design proposal to management for approval, what you are really seeking is approval to spend the budgeted money. The project engineer must combine estimated costs from suppliers, vendors, maintenance personnel, and the like, and then prepare a budget. As discussed earlier, it is critical that the facilities planner stays under budget.

Startup Schedule

Assume you are asked to design a facility to produce a new product. Typically, you would be told something like this:

> Provide a production facility to manufacture 1,200 gas grills per day, beginning on November 15 of this year.

All the work needed to accomplish the task will be backdated from November 15. Here is an example of the schedule:

Process Step	*Completed By*	*Date Actually Complete*
Start production	November 15	
Install equipment	November 1	
Obtain approval for and order equipment	October 1	
Complete master plan	September 15	
Develop plot plan and allocate area	September 1	
Select M.H. equipment	August 25	
Develop total space requirement	August 20	
Lay out workstation	August 15	Used for control
Determine activity relationships	August 10	
Identify office needs	August 5	
Identify personnel and plant service needs	August 1	
Develop flow requirements	July 25	
Balance assembly lines	July 15	
Determine number of machines	July 15	
Set standards and plant rate	July 10	
Determine assembly sequence	July 6	
Develop route sheets	July 5	
Determine make or buy decisions	July 2	
Determine what and how many will be made	July 1	

Make or Buy Decisions

Does the company make this part (fabricate from raw materials), or does it buy this part complete from a supplier who specializes in this kind of part? (See Figure 2–8.) The decision is normally quite straightforward and easy. If you are an existing company with a product line, you know what you can and what you cannot make. If you are a new company, you may buy out all the parts and become an assembly operation only. As you get going, you might start making a few parts yourself. No plant would make its own nuts, bolts, screws, tires, gauges, bearings, tapes, and the like; but someone might have special equipment that makes that part faster and better than you can and at a lower cost than you could ever achieve. The fabrication section of your manufacturing department is always in competition with your purchasing department because the cheapest way to provide the part to the assembly department is the best source.

The fabricated parts are the subject of fabrication layout. If you make no parts, no fabrication department layout is needed. If you make lots of parts, a large layout project results.

Part No.	*Part Name*	*Quantity Needed per Unit*	*Make or Buy*
1	Body ends	2	M
2	Tray bracket	2	M
SSSA1	Body end	2	
3	Body	1	M
SSA1	Body assembly	1	
4	Cover end	2	M
5	Cover	1	M
SSA2	Cover assembly	1	
6	Hinge	18″	M
SA1	Toolbox	1	
7	Tray ends	2	M
8	Tray body	1	M
9	Tray handle	1	M
SA2	tray	1	
10	Paint	AR	B
11	Handle	1	B
12	Clips	2	B
13	Rivet	4	B
A1	Toolbox		
14	Catch	2	B
15	Strike	2	B
16	Rivets	8	B
A2	Toolbox		
17	Hinges	2	B
18	Rivets	4	B
A3	Toolbox		
19	Packing list	1	B
20	Registration card	1	B
21	Name tag	1	B
22	Dividers	4	B
23	Plastic bag	1	B
SA3	Parts bag	1	
24	Carton	1	B
25	Tape	24″	B
Final packout		1	1

Figure 2–8 Toolbox make or buy decision.

Organizational Relationships

An organizational chart communicates much to the facilities designer (see Figure 2–9). The number of employees determines the size of many areas such as cafeterias, restrooms, office space, and medical facilities. The relationships among the various functions determine the closeness requirement for each department to every other department.

Feasibility Studies

Many new product ideas are recommended to management. These ideas need to be evaluated before they are accepted as new manufacturing facilities design

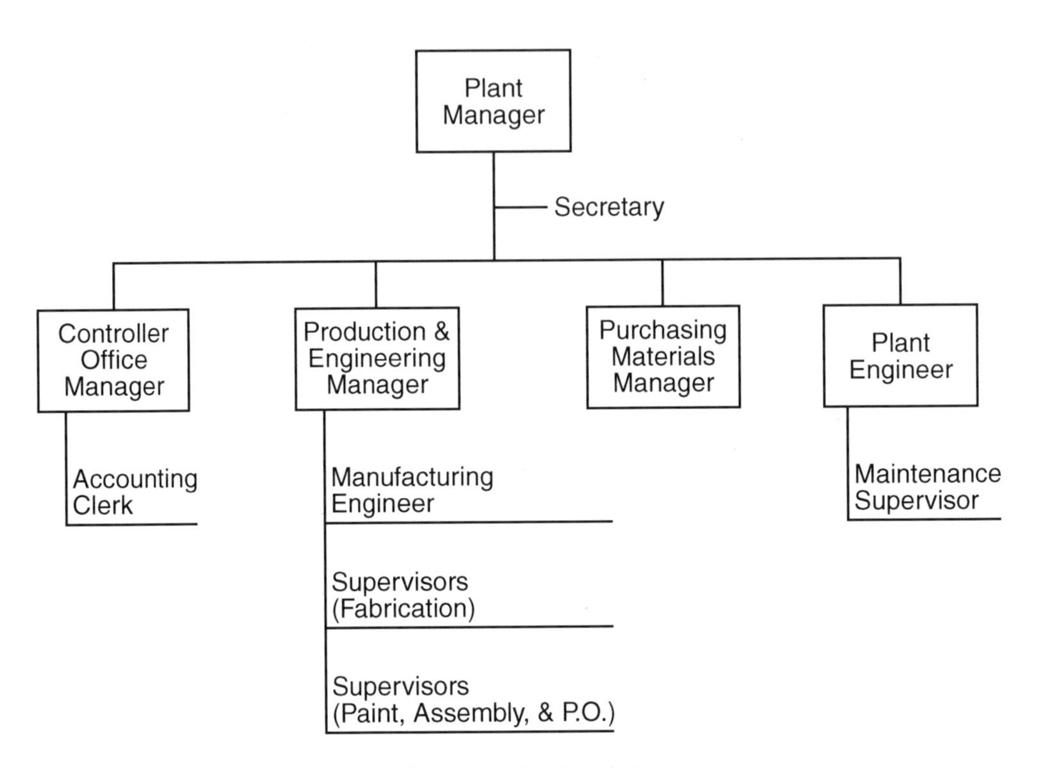

Figure 2–9 Toolbox manufacturing plant organizational chart.

projects. One method used to determine whether a project idea is workable is a *feasibility study*. Feasibility studies are generally performed by the highest level project managers and engineers. Out of the many feasibility studies conducted by a company, only a small number of projects typically result. At a specific toy company, for example, out of four proposals only one project was approved.

CONCLUSION

The product engineering design department provides blueprints and a bill of materials to help facilities designers understand which parts will be made inside the plant and which will be purchased from outside sources. Those parts that are made (manufactured) inside the plant will require manufacturing plans, as will be discussed in Chapters 3 and 4.

The marketing department researches the potential market demand for new or redesigned products and determines a quantity to produce per period of time. The designer breaks down the quantity into units per day to determine the number of machines and the number of people needed.

Management policy communicates the company's attitudes and decisions. The return on investment, the inventory policy, the startup dates required, and other factors all have significant effects on the manufacturing facilities design and material handling project. The designer cannot proceed without this information.

A plant layout or material handling project cannot begin until the necessary information is provided by other departments. For example, designers need input on product design, marketing, and management policy before initiating a project.

QUESTIONS

1. As you embark upon the design of your manufacturing facilities, what data would you seek from the following sources, and how would this information affect your planning? Briefly discuss
   a. Marketing
   b. Product design
   c. Management policy
2. What information does marketing provide?
3. What is takt time (R value)?
4. What is included in a takt time (R value) calculation?
5. Why is the R value so important?
6. What information do designers get from the product design department?
7. What information does management policy provide?
8. What is a make or buy decision?
9. Who is in competition with the fabrication department? Why?
10. What are six causes of muda?
11. Explain how to use the indented bill of materials.
12. What is the difference between an indented and flat bill of materials?
13. Why is the additional information provided by the indented bill of materials important?
14. Explain the concept of concurrent engineering. How does it relate to facilities planning?

CHAPTER 3

Time Study

OBJECTIVES:

Upon the completion of this chapter, the reader should:

- Understand the concepts of time study and time standards
- Understand the importance of time standards in facilities planning
- Be able to identify various techniques for establishing time standards
- Be able to identify the steps for performing a stopwatch time study
- Understand the difference between observed, normal, and standard times and be able to calculate each
- Understand the concepts of normalizing and allowances

Time standards are among the most basic yet important pieces of information required by the facilities planner. Time or labor standards are used for a variety of purposes in an organization. The uses include cost and budget allocation and control, production and planning and inventory management, performance evaluation and incentive pay where applicable, and evaluation of alternative methods of operation. For a facilities planner, the standard time is the primary input for determining the required number of people and workstations needed to meet the production schedule, and for calculating the number of machines, work cells, assembly line balancing, and staffing. Ultimately, this information is used to calculate the space requirements for all work centers and the overall production facilities space requirements.

This chapter is composed of four parts:

1. The definition of time study and time standards
2. The importance and uses of time standards
3. The techniques of time study
4. Time standards for manufacturing facilities design

WHAT IS A TIME STANDARD?

Before you can understand the importance and uses of time study, you must understand what the term *time standard* means. A time standard is defined as "the time required to produce a product at a workstation with the following three conditions: (1) a qualified, well-trained operator; (2) working at a normal pace; and (3) doing a specific task." These three conditions are essential to your understanding of time study, and so further discussion follows. The process of setting time standards is time study.

A *qualified, well-trained operator* is required. Experience is usually what makes a qualified, well-trained operator, and time on the job is the best indication of experience. The time needed to become qualified varies with the job and the person. For example, sewing machine operators, welders, upholsterers, machinists, and many other high-technology jobs require long learning periods. The greatest mistake made by new time study personnel is time studying someone too soon. A good rule of thumb is to start with a qualified, fully trained person and to give that person 2 weeks on the job prior to the time study. On new jobs or tasks, predetermined time standards systems (PTSSs) are used. These standards seem hard to achieve at first because the times are set for qualified, well-trained operators.

Normal pace is the pace at which a trained operator, under normal conditions, performs a task with a normal level of effort. A normal level of effort is one in which the operator can maintain a comfortable pace—not too fast and not too slow. Only one time standard can be used for each job, even though individual differences in operators cause different results. A normal pace is comfortable for most people. In developing the time standards for a task, 100 percent of the time of the normal pace is used as the normal time. When the pace is judged to be either slower or faster than the normal pace, adjustments are made accordingly. Some common examples of the normal pace are:

1. Walking 264 feet in 1.000 minute (3 miles per hour)
2. Dealing 52 cards into four equal stacks in .500 minute (at a bridge table)
3. Filling a 30-pin board in .435 minute (using two hands)

Training films for rating are also used to develop this concept.

A *specific task* is a detailed description of what must be accomplished. The description of the task must include

1. The prescribed work method
2. The material specification
3. The tools and equipment being used
4. The positions of incoming and outgoing material
5. Additional requirements such as safety, quality, housekeeping, and maintenance tasks

The time standard is good for only this one set of specific conditions. If anything changes, the time standard must change. The written description of a time standard is important, but the mathematics is even more important. If a job takes 1.000

Time Standard Minutes	*Pieces per Hour*[a]	*Hours per Piece*[b]	*Hours per 1,000 Pieces*[c]
1.000	60	.01667	16.67
.500	120	.00833	8.33
.167	359	.00279	2.79
2.500	24	.04167	41.67
.650	—	—	—
.050	—	—	—

[a]Pieces per hour is calculated by dividing the time standard minutes into 60 minutes per hour.

[b]Hour per piece is calculated by dividing the pieces per hour into one hour ($1/x$).

[c]Hours per 1,000 pieces is calculated by multiplying the hour per piece by 1,000 pieces.

Figure 3–1 Mathematical calculations practice for developing time standards.

standard minute to produce (Figure 3–1), you can produce 60 pieces per hour, and it will take .01667 hour to make one unit or 16.67 hours to make 1,000 units. The time in decimal minutes is always used in time study because the mathematics is easier. The following three numbers are required to communicate a time standard:

1. The decimal minute (always in three decimal places, e.g., .001)
2. Pieces per hour (rounded off to whole numbers, unless less than 10 per hour)
3. Hours per piece (always in five decimal places, e.g., .00001)

Many companies use hours per 1,000 pieces because the numbers are more understandable or meaningful.

Figure 3–2 is a time standard conversion table that may be useful for a quick reference when needed. It can be used when either the minute per unit, hours per unit, units per hour, or units per 8 hours are known, and you need to find the other three numbers pertaining to that standard. It can also be used to set goals for assembly lines or work cells. An interesting additional use is when you add jobs together and require a new standard for the combined jobs. Play with this table to understand the relationship between the different numbers that make up the term "standard time." For example, if two jobs that need to be combined used to have a standard of .72 minute per piece, or 83 pieces per hour, and .28 minute per piece, or 214 pieces per hour, what is the new standard? Add .72 and .28 to get 1.00 minute, or 60 pieces per hour combined.

Now that you understand what a time standard is, let's look at why time standards are considered to be one of the most important pieces of information produced in the manufacturing department.

THE IMPORTANCE AND USES OF TIME STUDY

The importance of time standards can be shown by three statistics: 60, 85, and 120 percent performance. An operation that is not working toward time standards typically works 60 percent of the time. Those operations working with time standards

Standard Minutes	*Standard Hours*	*Units per Hour*	*Units per 8 Hours*	*Standard Minutes*	*Standard Hours*	*Units per Hour*	*Units per 8 Hours*
480	8.000	.125	1.0	.98	.01633	61.22	489.80
240	4.000	.250	2.0	.96	.01600	62.50	500.00
160	2.667	.4	3.0	.94	.01567	63.83	510.64
120	2.000	.5	4.0	.92	.01533	65.22	521.74
96	1.600	.6	5.0	.9	.01500	66.67	533.33
80	1.333	.8	6.0	.88	.01467	68.18	545.45
70	1.167	.9	6.9	.86	.01433	69.77	558.14
60	1.000	1.0	8.0	.84	.01400	71.43	571.43
50	.833	1.2	9.6	.82	.01367	73.17	585.37
48	.800	1.2	10.0	.8	.01333	75.00	600.00
45	.750	1.3	10.7	.78	.01300	76.92	615.38
40	.667	1.5	12.0	.76	.01267	78.95	631.58
38	.633	1.6	12.6	.74	.01233	81.08	648.65
35	.583	1.7	13.7	.72	.01200	83.33	666.67
32	.533	1.9	15.0	.7	.01167	85.71	685.71
30	.500	2.0	16.0	.68	.01133	88.24	705.88
28	.467	2.1	17.1	.66	.01100	90.91	727.27
26	.433	2.3	18.5	.64	.01067	93.75	750.00
25	.417	2.4	19.2	.62	.01033	96.77	774.19
24	.400	2.5	20.0	.6	.01000	100.00	800.00
23	.383	2.6	20.9	.58	.00967	103.45	827.59
22	.367	2.7	21.8	.56	.00933	107.14	857.14
21	.350	2.9	22.9	.54	.00900	111.11	888.89
20	.333	3.0	24.0	.52	.00867	115.38	923.08
19	.317	3.2	25.3	.5	.00833	120.00	960.00
18	.300	3.3	26.7	.48	.00800	125.00	1,000.00
17	.283	3.5	28.2	.46	.00767	130.43	1,043.48
16	.267	3.7	30.0	.44	.00733	136.36	1,090.91
15	.250	4.0	32.0	.42	.00700	142.86	1,142.86
14	.233	4.3	34.3	.4	.00667	150.00	1,200.00
13	.217	4.6	36.9	.38	.00633	157.89	1,263.16
12	.200	5.0	40.0	.36	.00600	166.67	1,333.33
11	.183	5.5	43.6	.34	.00567	176.47	1,411.76
10	.167	6.0	48.0	.32	.00533	187.50	1,500.00
9	.150	6.7	53.3	.3	.00500	200.00	1,600.00
8	.133	7.5	60.0	.28	.00467	214.29	1,714.29
7	.117	8.6	68.6	.26	.00433	230.77	1,846.15
6	.100	10.0	80.0	.24	.00400	250.00	2,000.00
5	.083	12.0	96.0	.22	.00367	272.73	2,181.82
4	.067	15.0	120.0	.2	.00333	300.00	2,400.00
3	.050	20.0	160.0	.18	.00300	333.33	2,666.67
2	.033	30.0	240.0	.16	.00267	375.00	3,000.00
1	.017	60.0	480.0	.14	.00233	428.57	3,428.57
				.12	.00200	500.00	4,000.00
				.1	.00167	600.00	4,800.00
				.08	.00133	750.00	6,000.00
				.06	.00100	1,000.00	8,000.00
				.04	.00067	1,500.00	12,000.00
				.02	.00033	3,000.00	24,000.00

Figure 3–2 Time standard conversion table: Minutes, hours, pieces per hour, pieces per eight hours.

work at 85 percent of normal performance. This increase in productivity is equal to about 42 percent. In a small plant of 100 people, this improvement is equal to 42 extra people, or about a million dollars per year in savings. Not only is time standard very important but it is also very cost-effective. One hundred and twenty percent is the average performance of industrial plants on incentive pay plans.

The time standard is used to

1. Determine the number of machine tools to buy. In facilities design, how would you otherwise calculate this important piece of information for a manufacturing facility design?
2. Determine the number of production people to employ. Again, this is a very important piece of information when you are determining facility space requirements.
3. Determine manufacturing costs and selling prices.
4. Schedule the machines, operations, and people to do the job and to deliver on time with smaller inventories. This is what lean thinking and lean manufacturing are all about.
5. Determine the assembly line balance and the conveyor belt speed, load the work cells with the correct amount of work, and balance the work cells. This information determines the work cell and assembly line layouts.
6. Determine individual worker performance and identify and correct problematic operations. This is the basic philosophy behind kaizen.
7. Pay incentive wages for outstanding team or individual performance.
8. Evaluate cost reduction ideas and pick the most economical method based on cost analysis, not on opinion.
9. Evaluate new equipment purchases to justify their expense.
10. Develop operation personnel budgets to measure management performance.

A discussion of each of these uses of time study follows. As the company's facilities planner, how would you answer the following questions without time standards?

1. How Many Machines Do We Need?

One of the first questions raised when setting up a new operation or starting production on a new product is, "How many machines do we need?" The answer depends on two pieces of information:

a. How many pieces do we need to manufacture per shift?
b. How much time does it take to make one part? (This is the time standard.)

1. The marketing department wants you to make 2,000 wagons per 8-hour shift.
2. It takes .400 minute to form the wagon body on a press.
3. There are 480 minutes per shift (8-hour shift times 60 minutes per hour).
4. Subtract 50 minutes downtime per shift (breaks, cleanup, etc.).
5. There are 430 minutes per shift available at 100 percent.
6. Based on history or expectation, we can assume 75 percent performance ($.75 \times 430 = 322.5$).

7. There are 322.5 effective minutes left to produce 2,000 units.

8. $\frac{322.5}{2{,}000 \text{ units}} = .161$ minute per unit or 6.21 parts per minute.

The .161 minute per unit is called the takt time, or plant rate. (As you recall, takt time is the available minutes divided by the desired output.) Every operation in the plant must produce a part every .161 minute; therefore, how many machines do you need for this operation?

$$\frac{\text{Time standard} = .400 \text{ minute per unit}}{\text{Plant rate: } .161 \text{ minute per unit}} = 2.48 \text{ machines}$$

This operation requires 2.48 machines. If other operations need to use a machine of this type, you will add all the machine requirements together and round up to the next whole number. In the preceding example, you would buy three machines. (Never round down on your own. You will be building a bottleneck in your plant.) This information is critical to the facility design.

2. How Many People Should We Hire?

Look at the operations chart shown in Figure 3–3. This chart lists the time standard for every operation required to fabricate each part of the product and each assembly operation required to assemble and pack the finished product.

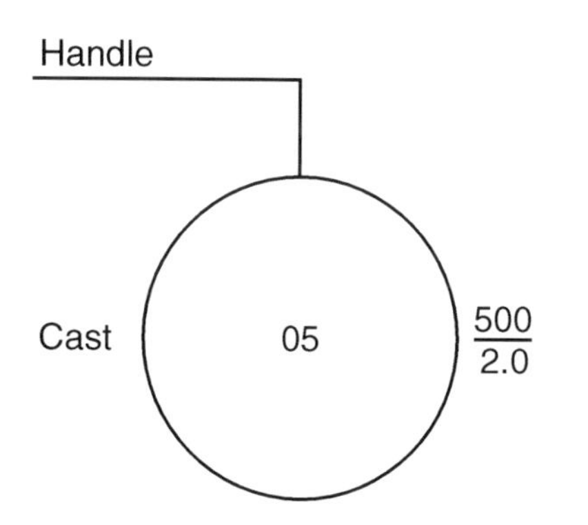

In this operation (casting the handle), the *05* indicates the operation number. Usually, 05 is the first operation of each part. The *500* is the pieces per hour standard. This operator should produce 500 pieces per hour. The *2.0* is the hours required to produce 1,000 pieces. At 500 pieces per hour, it would take 2 hours to make 1,000 pieces. How many people would be required to cast 2,000 handles per shift?

$$\begin{array}{r} 2{,}000 \text{ units} \\ \times\ 2.0 \text{ hours per } 1{,}000 \\ \hline 4.0 \text{ hours at standard} \end{array}$$

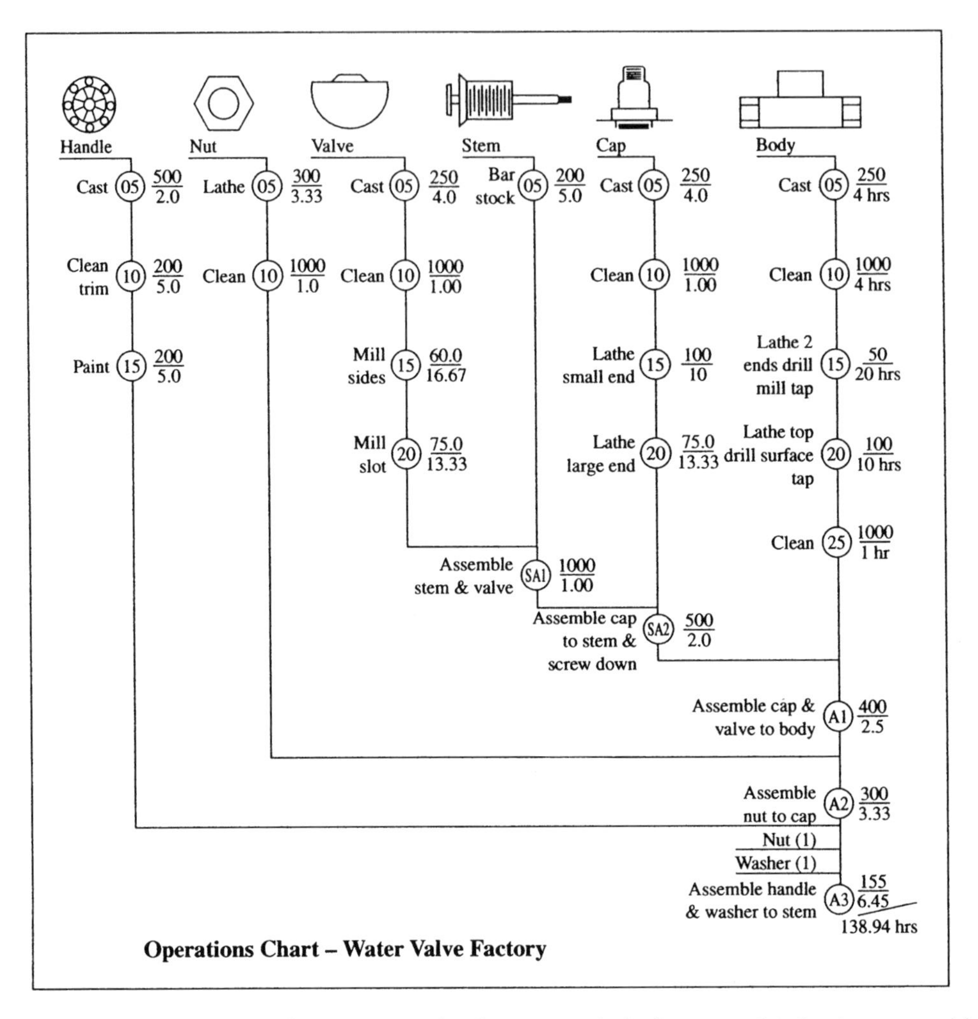

Figure 3–3 Operations chart for a water valve factory: A circle for every fabrication, assembly, and packout operation.

Not many people, departments, or plants work at 100 percent performance. How many hours would be required if they work at the rate of 60, 85, or 120 percent?

$$\frac{4 \text{ hours}}{60 \text{ percent}} = 6.66 \text{ hours} \quad \frac{4 \text{ hours}}{85 \text{ percent}} = 4.7 \text{ hours} \quad \frac{4 \text{ hours}}{120 \text{ percent}} = 3.33 \text{ hours}$$

Therefore, depending on anticipated performance, you will budget for a specific number of hours. Either performance history or national averages will be used to factor the 100 percent hours to make them practical and realistic.

Look again at the operations chart shown in Figure 3–3. Note the total number of hours (138.94) at the bottom right-hand side. The operations chart includes every operation required to fabricate, paint, inspect, assemble, and pack out a product. The total hours is the total time required to make 1,000 finished products. In the water valve factory, employees must work 138.94 hours at 100 percent to produce 1,000 water valves. If this is a new product, you could expect 75 percent performance during the first year of production. Therefore,

$$\frac{\text{138.94 hours per 1,000}}{\text{74 percent performance}} = \text{185 hours per 1,000}$$

where 75 percent = .75.

The marketing department has forecasted sales of 2,500 water valves per day. How many people are needed to make water valves?

$$\text{185 hours per 1,000} \times \text{2.5 (1,000)} = \text{463 hours per day needed}$$

Divide this by 8 hours per employee per day, which equals 58 people.

Management will be judged by how well it performs this goal. If fewer than 2,500 units are produced per day with the 58 people, management will be over budget, and that is unforgivable. If it produces more than 2,500 units per day, management is judged as being good at managing, and the managers are promotable.

Most companies produce more than one product. The problem of how many people to hire to produce each product is the same. For example, how many direct labor employees would you need for a multiproduct plant?

Product	*Hours per 1,000*	*No. of Units Needed per Day*	*Hours at 100%*	*Actual %*	*Actual Hours Needed*
A	150	1,000	150.0	70	214
B	95	1,500	142.5	85	168
C	450	2,000	900.0	120	750
					Total 1,132 hours

Per day, 1,132 hours of direct labor are needed. Each employee will work 8 hours; therefore,

$$\frac{\text{1,132 hours}}{\text{8 hours per employee}} = \text{141.5 employees}$$

Thus, you will budget for 142 employees. Without time standards, any other method of calculating labor needs would be a guess. Management doesn't want to be judged and compared to unattainable time standards or production goals.

3. How Much Will Our Product Cost?

At the earliest point in a new product development project, the anticipated cost must be determined. A feasibility study will show top management the profitability of a new venture. Without proper, accurate costs, the profitability calculations would be nothing but a guess.

Product costs may include the following:

		Typical %
Manufacturing costs 50%	Direct labor	8
	Direct materials	25
	Overhead costs	17
	Plus	
Front-end costs	Sales and distribution costs	15
	Advertising	5
	Administrative overhead	20
	50% Engineering	3
	Profit	7
		100%

Direct labor cost is the most difficult component of product cost to estimate. Time standards must be set prior to any equipment purchase or material availability. Time standards are set using predetermined time standards or standard data from blueprints and workstation sketches, and are compiled in a chart such as the one shown in Figure 3–3. The bottom right-hand side of the water valve operation chart indicates that it takes 138.94 hours to produce 1,000 units:

$$\frac{\text{138.94 hours per 1,000 units}}{\text{85 percent anticipated performance}} = \text{163.45 hours per 1,000}$$

\$163.46	hours per 1,000 water valves
× \$7.50	per hour labor rate
\$1,225.94	per 1,000 or 1.23 each

Direct material is the material that makes up the finished product and is estimated by calling vendors for a bid price. Normally, 50 percent of the manufacturing cost (direct labor + direct materials + factory overhead) is direct material cost. For this example, we will use 50 percent. On the operating chart, raw materials are introduced at the top of each line. Buyout parts are introduced at the assembly and packout station.

Factory overhead costs are all expenses of running a factory, except the previously discussed direct labor and direct material. Factory overhead is calculated as a percentage of direct labor. This percentage is calculated using last year's actual costs. All manufacturing costs for last year are divided into three classifications:

Direct labor	$1,000,000
Direct material	3,000,000
Overhead	2,000,000
Total factory costs	$6,000,000

The factory overhead rate for last year is

$$\frac{\$2{,}000{,}000 \text{ overhead cost}}{\$1{,}000{,}000 \text{ direct labor costs}} = 200 \text{ percent overhead rate per labor dollar}$$

Thus, each dollar in direct labor cost has a factory overhead cost of $2.00.

Example:

Labor	$1.23	from time standards
Overhead	$2.46	200 percent overhead rate
Material	$3.69	from suppliers
Total factory cost	$7.28	
All other costs	7.38	from ratio
Selling price	$14.76	

4. When Should We Start a Job, and How Much Work Can We Handle with the Equipment and People We Have? Or, How Do We Schedule and Load Machines, Work Centers, Departments, and Plants?

Even the simplest manufacturing plant must know when to start an operation for the parts to be available on the assembly line. The more operations there are, the more complicated the scheduling is.

Example: One machine plant operates at 90 percent.

Backlog Job	*Hours per 1,000*	*Units Required*	*Hours Required*	*Backlog (Cumulative Hours)*	*Backlog (Days)*
A	5	5,000	27.8	27.8	1.74
B	2	10,000	22.2	50.0	3.12
C	4	25,000	111.1	161.1	10.07
D	3	40,000	133.3	294.4	18.40

The chart in Figure 3–4 shows the same information as the preceding data. This plant operates a single machine 16 hours per day, 5 days per week. There are 294.4 hours of backlog at 16 hours per day, which equals 18.4 days of work in the backlog.

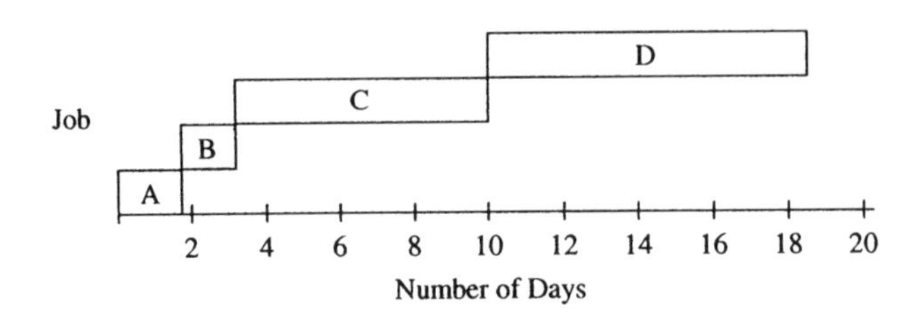

Figure 3–4 A picture of a machine's or department's time schedule of work.

What if a customer comes in with a job and wants it in 10 days? The job is estimated to take only 48 hours of machine time. Can you deliver? What about the other four jobs? When have you promised them?

One scheduling philosophy is that operating departments are compared to buckets of time. The size of the bucket is the number of hours that each department can produce in a 24-hour day. The following chart illustrates this concept:

Department	*No. of Machines*	*Hours per Day (Two Shifts Available)*	*Historical Department Performance %*	*Hours Capacity*
Shears	2	32	85	27.2
Presses	6	96	90	86.4
Press breaks	4	64	80	51.2
Welding	4	64	75	48.0
Paint	3	48	95	45.6
Assembly line	1	80	90	72.0

The scheduler can keep adding work to any department for a specific day until the hour capacity is reached; then it spills over to the next day.

Without good time standards, manufacturing management would have to carry large quantities of inventory to avoid running out of parts. Inventory is a huge cost in manufacturing; therefore, knowledge of time standards will reduce inventory requirements, which will reduce cost. Production inventory control is an area of major importance in industrial and manufacturing management, and time standards are a prerequisite.

5. How Do We Determine the Assembly Line Balance and the Conveyor Belt Speed, Load the Work Cells with the Correct Amount of Work, and Balance the Work Cells?

The objective of assembly line balancing is to give each operator as close to the same amount of work as possible. Balancing work cells has the same objective. It is of no value that one person or one cell has the ability to outrun the rest of the plant by 25 percent because one person cannot produce more than the amount that comes to her or more than the subsequent operations can use. If the person has extra time, she could be given some of the work from a busier workstation.

Assembly line balancing or work center loading can be accomplished only by breaking down the job into tasks that need to be performed and by reassembling them into jobs or cells of close to the same length of time. There will always be a workstation or cell that has more work than the others. This station is defined as the 100 percent loaded station, or bottleneck station, and will limit the output of the whole plant. To improve the assembly line (reduce the unit cost), you will concentrate on improving the 100 percent station. If you reduce the 100 percent station, as in the following example, by 1 percent, you save an additional 1 percent for each person on the line because each person now can go 1 percent faster. You can keep reducing the 100 percent station until another workstation becomes the 100 percent station (busiest station). Then your attention turns to this new 100 percent station for cost reduction. If you have 200 people on an assembly line and only one 100 percent station, you can save the equivalent of two people by reducing the 100 percent station by just 1 percent. You can use this multiplier to help justify spending large sums of money to make small changes. (Assembly line balancing is discussed in detail in Chapter 4.)

6. How Do We Measure Productivity?

Productivity is a measure of output divided by input. If you are talking about labor productivity, then you are developing a number of units of production per hour worked.

Example:

$$\text{Present} = \frac{\text{output} = 1{,}000 \text{ units per day}}{\text{input} = 50 \text{ people @ 8 hours per day}} = \frac{1{,}000}{400} = 2.5 \text{ units per work hour}$$

$$\text{Improved} = \frac{\text{output} = 2{,}000 \text{ units per day}}{\text{input} = 50 \text{ people @ 8 hours per day}} = \frac{2{,}000}{400} = 5.0 \text{ units per work hour}$$

or a 100 percent increase in productivity (a doubling of production).

You could also increase productivity by maintaining the output constant or reducing the number of people.

$$\text{Improved output} = \frac{1{,}000 \text{ units per day}}{40 \text{ people @ 8 hours per day}} = \frac{1{,}000}{320} = 3.125 \text{ units per work hour}$$

These examples are good for plants or whole industries, but for individuals, use

$$\frac{\text{Earned hours}}{\text{Actual hours}} = \text{percent performance}$$

Earned hours are the hours of work earned by the operator based on the work standard and the number of pieces produced by the operator. For example, if an operator worked

8 hours and produced 1,000 units on a job with a time standard of 100 pieces per hour, you would have the following:

A. Earned hours $= \frac{\text{1,000 pieces produced}}{\text{100 pieces per hour}} = \text{10 hours}$

B. Actual hours = 8 hours
Actual hours are the real time the operator works on the job (also called the time clock hours).

C. Percent performance $= \frac{\text{earned hours}}{\text{actual hours}} = \frac{10}{8} = \text{125 percent}$

Industrial engineers will improve productivity by reporting performances of every operation, operator, supervisor, and production manager every day, week, month, and year. Performance reports are based on daily time cards filled out by operators and extended within the computer performance control system. All of the following five functions must be in place to have a functioning performance control system:

1. Setting goals (setting time standards)
2. Comparing actual performances with the goals
3. Tracking results (graphing)
4. Reporting variances larger than acceptable limits
5. Taking corrective action to eliminate causes of poor performances

A performance control system will improve performance by an average of 42 percent over performance with no control system. Companies without performance control systems typically operate at 60 percent of standard. Those companies with performance control systems will average 85 percent performance. This is accomplished by (1) identifying nonproductive time and eliminating it, (2) identifying poorly maintained equipment and fixing it, (3) identifying causes for downtime and eliminating them, and (4) planning ahead for the next job.

Performance control systems hold problems up to the "light of day," and facilities planners fix the problems. In plants that do not have standards, the employees know that no one cares how much they produce. Management's reactions to problems speak louder than its words. How can supervisors know who is producing and who is not if they do not have standards? How would management know the magnitude of problems such as downtime for lack of maintenance, material, instruction, tooling, services, and so on if downtime is not reported?

7. How Can We Pay Our People for Outstanding Performance?

Every manufacturing manager would like to be able to reward outstanding employees. Every supervisor knows whom to count on to get the job done. Yet, only about 25 percent of production employees have an opportunity to earn increased pay for increased output.

A 400-plant study by an industrial engineering consultant, Mitchell Fein, found that when employees are paid via incentive systems, their performance improves by 41 percent over measured work plans, and 65 percent over no standards or no performance control system.

Stage I. Plants with no standards operate at 60 percent performance.
Stage II. Plants with standards and performance control systems operate at 85 percent performance.
Stage III. Plants with incentive systems operate at 120 percent performance.

A small company with 100 employees could save about $820,000 per year ($20,000 per year salary times 41 employees) on labor costs going from no standards to a performance control system.

A National Science Foundation study found that when workers' pay was tied to their efforts, productivity improved, cost was reduced, workers' pay increased, and workers' morale improved.

8. How Can We Select the Best Method or Evaluate Cost Reduction Ideas?

A basic rule of production management is, "All expenditures must be cost-justified." A basic rule of life is, "Everything changes." Planners must keep improving or become obsolete. To justify all expenditures, the savings must be calculated. As discussed previously, this is called the return. The cost of making the change is also calculated, which is called the investment. When the return is divided by the investment, the resulting ratio indicates the desirability of the project. This ratio is called the ROI, or return on investment. To provide a uniform method of evaluating the ROI, annual savings are used; therefore, all percentages are per year.

Example:

You have been producing product A for several years and look forward to several more years of sales at 500,000 units per year or 2,000 units per day. The present method requires a standard time of 2.0 minutes per unit or 30 pieces per hour. At this rate, it takes 33.33 hours to make 1,000 units. All production will run on the day shift.

A. *Present method and costs.* With a labor rate of $10.00 per hour, the labor cost will be $333.30 to produce 1,000 units. The cost of 500,000 units per year would be $166,665.00 in direct labor.

$$\frac{1{,}000 \text{ pieces}}{30 \text{ pieces per hour}} = 33.33 \text{ hours per } 1{,}000 \text{ units}$$

B. *New method and costs.* You have a cost reduction idea. If you buy this new machine attachment for $1,000, the new time standard would be lowered to 1.5 minutes per unit. Will this be a good investment?

First, how many attachments will you have to buy to produce 500,000 units per year?

$$\frac{500{,}000 \text{ units per year}}{250 \text{ days per year}} = 2{,}000 \text{ units per day}$$

480	minutes per shift
−50	minutes per shift downtime
430	minutes per shift 100 percent
@80 %	expected efficiency
344	effective minutes available to produce 2,000 units per shift

$$\frac{344 \text{ minutes}}{2{,}000 \text{ units}} = .172 \text{ minute per unit}$$

To produce 2,000 units per shift, you need a part every .172 minute.

$$\text{Number of machines} = \frac{1.50 \text{ minutes per cycle}}{.172 \text{ minute per unit}} = 8.7 \text{ machines}$$

You will purchase nine attachments at $1,000 each. Your investment will be $9,000 (9 times 1,000).

Second, what is your labor cost?

$$\text{Pieces per hour} = \frac{60 \text{ minutes per hour}}{1.5 \text{ minutes per part}} = 40 \text{ parts per hour or } 25 \text{ hours per } 1{,}000$$

$$25 \text{ hours per } 1{,}000 \times \$10.00 \text{ per hour wage rate} = \$250 \text{ per } 1{,}000$$

$$500{,}000 \text{ units will cost } 500 \times \$250 = \$125{,}000$$

New labor costs will be $125,000 per year.

C. *Savings.* Direct labor dollars.
Old method $166,665 per year
New method $125,000 per year
Savings $41,665 per year

$$\frac{\text{Return (savings) } \$41,665 \text{ per year}}{\text{Investment (cost) } \$9{,}000} = 463 \text{ percent}$$

$$\text{ROI} = 463 \text{ percent}$$

$$463 \text{ percent} = .216 \text{ year or } 2.59 \text{ months to pay off}$$

D. *Return on investment.* This investment will pay for itself in less than 3 months. If you were the manager, would you approve this investment? Of course you would, as would anyone.

Cost reduction programs are important to the well-being of a company and the peace of mind of the industrial engineering department. A department that shows a savings of $100,000 per employee per year doesn't have to worry about layoffs or elimination. A properly documented cost reduction program will update all time standards as soon as methods are changed. Every standard affected must be changed immediately.

Cost reduction calculations can be a little more complicated than our example, which did not account for

1. Taxes
2. Depreciation
3. Time value of money
4. Surplus machinery—trade-in
5. Scrap value

9. How Do We Evaluate New Equipment Purchases to Justify Their Expense?

The answer to this question is the same as the answer to question 8. Every new machine is a cost reduction. No other reason is acceptable.

10. How Do We Develop a Personnel Budget?

This question was answered in question 2 when you determined the number of people to hire. Budgeting is one of the most important management tools, and the manager must understand it fully to manage effectively. It is said that you become a manager when you are responsible for a budget, and that you are a promotable manager when you come in underbudget at the end of the year. Budgeting is a part of the cost-estimating process. Labor is only one part of the budget, but it is one of the most difficult to estimate and control. Without time standards, it would be a very expensive guess.

How can managers make such important decisions as those discussed in this chapter? Much of manufacturing management has received no formal training in making these decisions. It will be your job to show management the scientific way to manage its operations.

TECHNIQUES OF TIME STUDY

This section presents an overview of time study techniques. Further study will be required if you want to set time standards or to apply any of these techniques. Time study (setting time standards) covers a wide variety of situations. At one time, a job must be designed, workstations and machines built, and a time standard set before the plant is built. In this situation, a PTSS (predetermined time standards system), methods time measurement (MTM), or standard data would be the techniques used to set the time standard. Once a machine or workstation has been operated for a while, the stopwatch technique is used. Some jobs occur once or twice a week, whereas others repeat thousands of times per day. Some jobs are very fast, whereas others take hours. Which technique do you use? The job of an industrial engineer and technologist is to choose the correct technique for each situation and correctly apply that technique.

New manufacturing facilities design requires that you set the work method and time standard before work begins. This requires using the PTSS, or standard data. Once production starts, you can check yourself with the stopwatch time study technique. Retrofit projects can use stopwatch time study to time the existing methods, but new methods or new equipment will require estimating the time standard using the PTSS or standard data. We will discuss all five techniques of time standard development in this text:

1. Predetermined time standard systems
2. Stopwatch time study
3. Work sampling
4. Standard data
5. Expert opinion standard and historical data

A brief description of these five techniques is presented in this chapter. Each technique will be developed fully in its own chapter.

Predetermined Time Standards Systems

When a time standard is needed during the planning phase of a new product development program, the PTSS technique is used (see Figure 3–5). At this stage of new product development, only sketchy information is available, and the technologist must visualize what is needed in the way of tools, equipment, and work methods. The technologist would design a workstation for each step of the new product manufacturing plan, develop a motion pattern, measure each motion and assign a time value. The total of these time values would be the time standard. This time standard would be used to determine the equipment, space, and people needs of the new product and its selling price.

Frank and Lillian Gilbreth developed the basic philosophy of predetermined time motion systems. They divided work into 17 elements:

1. Transport empty
2. Search
3. Select
4. Grasp
5. Transport loaded
6. Preposition
7. Position
8. Assemble
9. Disassemble
10. Release load
11. Use
12. Hold
13. Inspect
14. Avoidable delay
15. Unavoidable delay
16. Plan
17. Rest to overcome fatigue

These 17 work elements are known as *therbligs*. Each therblig was reduced to a time table, and when totaled, a time standard for that set of motions was determined.

Stopwatch Time Study

Stopwatch time study (see Figure 3–6) is the method that most manufacturing employees think of when talking about time standards. Fredrich W. Taylor started using the stopwatch around 1880 for studying work. Because of its long history, this technique is a part of many union contracts with manufacturing companies.

Fred Meyers & Associates		**Predetermined Time Standards Analysis**
Operation No. 25	Part No. 2220	Operation Description:
Date: 1-21-xx	Time:	Assemble brackets to body using 4 bolts
By I.E.: Meyers		

Description–Left Hand	Freq.	LH	Time	RH	Freq.	Description–Right Hand	Element Time
To next body		R30	18	M30		Aside completed	
Grasp		G2	6	RL		Release	
Move body to fixture		M30	18	R30		To L.H.	
			2	G2		Grasp body in L.H.	
Place in fixture		AP1	5	AP1		Place in fixture	
			49				.049
Get and place 2 brackets on body							
		R12	9	R12		To bracket	
		G2	6				
			4	R2		To bracket	
Same as R.H.			6	G2		Grasp bracket	
		M12	9	M12		To body	
		AP2	10				
			5	AP2(1/2)		On body	
			49				.049
Get and assemble 4 bolts – hand tighten							
		R8	7	R8		To bolt	
		G3	9				
			9	G3		Grasp bolt	
Same as R.H.		M8	7	M8		To body	
		AP1	5	AP1		On body	
	10	G4	40	G4	10	Turn 10 times	
		SF	5	SF			
			82		2	3rd and 4th bolt	.164

Time	Study	Cycle
25	1.20	2.11
50	1.42	2.35
74	1.65	
97	1.89	
Total		2.35
Occ.		10
Avg. occ.		.235
Lev. fact		110
Norm. time		.258

Cost:

Hours per unit	.00480
Dollars per hour	$15.00
Dollars per unit	$.072

Total Normal Time in Minutes per Unit	.262
+ 10% Allowance	.026
Standard time	.288
Hours per unit	.00480
Pieces per hour	208

Figure 3–5a PTSS example.

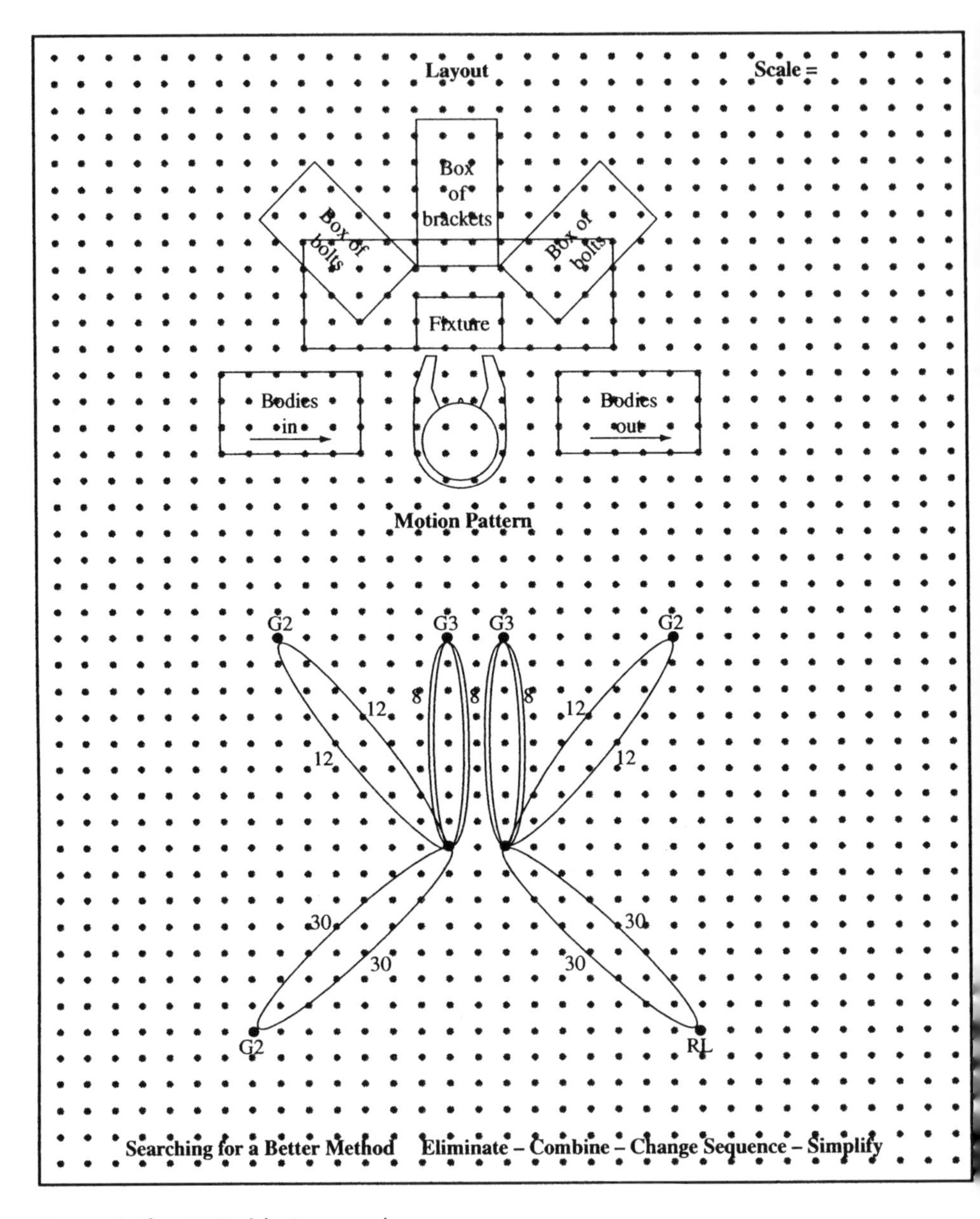

Figure 3–5b PTSS side B example.

Fred Meyers & Associates **Time Study Worksheet** ☐ Snapback ☒ Continuous

Operation Description: Assemble Parts 2 & 4, Machine Screw & Stake, Inspect

Part Number	Operation No.	Drawing No.	Machine Name	Machine Number	
4650-0950	1515	4650-0950	Press	21	☒ Quality ok?
Operator Name	**Months on Job**	**Department**	**Tool Number**	Feeds & Speeds: None	☒ Safety checked?
Meyers	5	Assembly	M61	Machine Cycle: 0.030	☒ Setup proper?
Part Description: Golf Club Sole Assembly	Material Specificatons: Woco & Steel			Time: 8:30 AM	Notes:

Element No.	Element Description		1	2	3	4	5	6	7	8	9	10	Total / Cycles	Average Time	% R	Normal Time	Frequency	Unit Normal Time	Range	R/X	Highest
1	Assembly	R	9	41	71	1.07	38	77	2.08	48	77	3.07	.76 / 9	.084	90	.076	1 / 1	.076	.03		✓
		E	.09	.09	.09	(15)	.08	.08	.10	.07	.08	.08									
2	Drive screw	R	15	46	79	13	43	82	14	53	82	93	.51 / 9	.057	100	.057	1 / 1	.057	.03	.53	✓
		E	06	05	08	06	05	05	06	05	05	(10)									
3	Press	R	28	59	94	27	66	95	28	66	96	4.06	1.22 / 9	.136	110	.150	1 / 1	.150	.02		
		E	13	13	15	14	(23)	13	14	13	14	13									
4	Inspect	R	32	62	92	30	69	98	41	69	99	4.09	.25 / 8	.031	100	.031	1 / 1	.031	.01		
		E	.04	.03	(.02)	.03	.03	.03	(13)	.03	.03	.03									
5	Load screws	R										3.83	.76 / 1	.76	125	.950	1 / 10	.095	—		
		E										.76									
		R																			
		E					*1		*2			*3									
		R																			
		E																			
		R																			
		E																			

Foreign elements:
* 1.23 Part jammed
* 2.13 Tried to rework part
* 3.10 Restart from loading screws

Engineer: Fred Meyers Date: 2 / 25 / xx

Approved by: Fred Meyers Date: 2 / 26 / xx

Notes:
Load screws could be improved to eliminate .095 minute (save)

.409
-.095
.314
+.031
.345
.00575 hr
174 pieces/hr

.00750
.00575
.00175 hr/unit
x $10.00 /hr
.0175 $/unit
500,000 hrs
$8,750

R/X	No. Cycles
.1	2
.2	7
.3	15
.4	27
.5	42
.6	61
.7	83
.8	108
.9	138
.10	169

(48 — boxed between .5 and .6)

Total normal minutes	.409
Allowance + 10 %	.041
Standard minutes	.450
Hours per unit	0 0 7 5 0
Units per hour	133

On back
Work station layout
Product sketch

Figure 3–6 Time study example: Continuous form.

Time study is defined as the process of determining the time required by a skilled, well-trained operator working at a normal pace doing a specific task. Several types of stopwatches could be used:

1. *Snapback:* in one hundredths of a minute
2. *Continuous:* in one hundredths of a minute
3. *Three watches:* continuous watches
4. *Digital:* in one thousandths of a minute
5. *TMU (time-measured unit):* in one hundred thousandths of an hour
6. *Computer:* in one thousandths of a minute

All but the TMU watch read in decimal minutes. The TMU watch reads in decimal hours. Digital watches and computers are much more accurate, and many have memory functions that improve accuracy.

TIME STUDY PROCEDURE AND THE STEP-BY-STEP FORM

The time study procedure has been reduced to 10 steps, and the time study form has been designed to help the time study technologist perform the 10 steps in the proper sequence. (Figure 3–7 shows a blank time study form with circled numbers.) This section is organized according to the following 10 sequential steps:

Step 1. Select the job to study.
Step 2. Collect information about the job.
Step 3. Divide the job into elements.
Step 4. Do the actual time study.
Step 5. Extend the time study.
Step 6. Determine the number of cycles to be timed.
Step 7. Rate, level, and normalize the operator's performance.
Step 8. Apply allowances.
Step 9. Check for logic.
Step 10. Publish the time standard.

Within each step, the blocks of the time study form involved are defined. The circled numbers refer to the blocks on the time study form. The form is designed for both continuous and snapback time study techniques. Everything except block 16 is exactly the same.

Step 1. Select the job to study.

Requests for time study can come from every direction:

1. Unions can question time standards and request a restudy.
2. Supervisors, who are judged partly on the performance of their subordinates, can request a restudy.
3. The job could change, requiring a new standard.

Fred Meyers & Associates **Time Study Worksheet** ☐ Snapback ☐ Continuous

Operation Description (1)

Part Number (2)	Operation No. (3)	Drawing No. (4)	Machine Name (5)	Machine Number (6)	☐ Quality ok?
Operator Name (7)	Months on Job (8)	Department (9)	Tool Number (10)	Feeds & Speeds	☐ Safety checked?
Part Description: (11)	Material Specificatons:			Machine Cycle (12) Time	☐ Setup proper? (13) Notes:

Element No.	Element Description		Readings 1	2	3	4	5	6	7	8	9	10	Total Cycles	Average Time	% R	Normal Time	Frequency	Unit Normal Time	Range	$\frac{R}{X}$	Highest ✓
		R																			
		E																			
		R																			
		E																			
(14)	(15)	R						(16)					(17)	(18)	(19)	(20)	(21)	(22)	(23)	(24)	(25)
		E																			
		R																			
		E																			
		R																			
		E																			
		R																			
		E																			
		R																			
		E																			
		R																			
		E																			

Foreign elements: (27)

Engineer: (33) Date: _/_/_ (34)

Approved by: (35) Date: _/_/_

Notes:

(26)

$\frac{R}{X}$	No. Cycles
.1	2
.2	7
.3	15
.4	27
.5	42
.6	61
.7	83
.8	108
.9	138
.10	169

Total normal minutes __(28)__
Allowance +__(29)__ % __(30)__
Standard minutes __(30)__
Hours per unit __(31)__
Units per hour __(32)__

On back
Work station allowance (36)
Product sketch (37)

Figure 3–7 Time study example: Step-by-step form.

4. New jobs may have been added to the plant.
5. New products can be added, requiring many new time standards.
6. Industrial technologists can improve methods, requiring a new time standard.
7. Cost reduction programs can require new standards—new machinery, tools, materials, methods, and so on.

Once a reason for studying a job has been determined, the time study technician may have several people doing the same job. Which person do you time study? The best answer is two or three, but those people you do *not* want to time study are

1. *The fastest person on the job.* The other employees may think you are going to require them to keep up. Even though you can do a good job of setting a time standard on this person, you do not want to create employee relations problems.
2. *The slowest person on the job.* No matter how you rate the job and no matter how good the time standard is, the employees will wonder how you came up with that standard.
3. *Employees with negative attitudes that will affect their performance while being studied.* If you can sidestep a potential problem, you should.

The person or persons to be time studied should have sufficient time on the job to be qualified, well-trained operators. For this reason, blocks ⑦ and ⑧ have been included on the time study form:

⑦ Operator's name
⑧ Months on the job

The employee should have been on the job for at least 2 weeks.

Once the job has been selected to study, the following information has been determined:

② Part number
③ Operation number
④ Drawing number
⑤ Machine name: a generic name like *press, welder, lathe, drill,* and so on
⑥ Machine number: a specific machine with specific speeds and feeds
⑨ Department: where the machine is located (this can be a number or name)

Step 2. Collect information about the job.

Now that the job has been identified, the technologist must collect information for the purpose of understanding what must be accomplished. The information required is as follows:

① Operation description: a complete description of what needs to be accomplished.

④ Drawing number: will lead to a blueprint to show items like the following:

a. ⑪ and ㊲: part description and material specification (a place on the back of the time study form has been set aside for a product sketch, if needed).

b. ⑩: tool numbers, and sizes of tools such as fixtures, drill sizes, and so on.

c. ⑫: feeds and speeds of equipment; this depends on the sizes of parts and material specifications found on the blueprint; they must be recorded.

⑬ When reviewing the workstation and before starting the time study, the technologist must check the following:

✓ Is quality OK? Quality control must confirm that the quality of the product being produced is high. Is the operator checking parts on the proper schedule? Time standards from producing scrap are worthless.

✓ Has safety been checked? If all the safety devices are not in place, then the technologist would be wasting time setting a standard for the wrong method.

✓ Is the setup properly done? This is the time to see that the proper method, tools, and equipment are in place. Are the materials and tools correctly positioned? Are there unnecessary moves or elements being performed?

If anything is wrong, it must be corrected before a time study can be performed. If the operator must be retrained, the time study should be postponed until retraining is complete.

㊱ A big part of collecting the information is the workstation layout. The back of the time study form allows for a workstation layout, but this may not be needed if done on another of the previous forms (multiactivity form). The workstation layout is one of the best ways to describe the operation. Chapter 7 shows what must be included on a workstation layout.

Step 3. Divide the job into elements.

Elements are units of work that are indivisible. Time study elements should be as small as possible, but not less than .030 minute.

The elements should be as descriptive as possible and must be in the sequence that the methods call for and be made as small as is practical.

Principles of Elemental Breakdown

1. It is better to have too many elements than too few.
2. Elements should be as short as possible, but not less than .030 minute. Elements over .200 minute should be examined for further subdivision.
3. Elements that end in sound are easier to time because the eyes can be looking at the watch while the ears are anticipating the sound.

4. Constant elements should be segregated from variable elements to show a truer time.
5. Separate the machine-controlled elements from the operator-controlled elements so work pace can be differentiated.
6. Natural breaking points are best. The beginning and ending points must be recognizable and easily described. If the element description is not clear, the description or breakdown must be rethought.
7. The element description describes the complete job, and the ending points are clearly marked.
8. Foreign elements should be listed in the order of occurrence. Foreign elements are not listed until they occur during the study.

The reasons for breaking down a job into elements are as follows:

1. It makes the job easier to describe.
2. Different parts of the job have different tempos. The time study technician will be able to rate the operator better. Machine-controlled elements will be constant and normally 100 percent, whereas the operator may be more or less proficient at different parts of the job.
3. Breaking down the job into elements allows for moving a part of the job from operator to operator. This is called *line balancing*.
4. Standard data can be more accurate and more universally applied with smaller elements. All work is made up of common elements. After a number of time studies, the technologist can develop formulas or graphs to eliminate the need for time study. Standard data should be the goal of all time study departments.

On the time study form shown in Figures 3–7 and 3–8, two columns have been assigned to elements:

⑭ Element number: The element number is just a sequential number and is useful when more than 10 cycles are timed. Instead of describing each element over and over again, just reference the element number.

⑮ Element description: Be as complete as possible. The ending points should be clear.

㉗ Foreign elements: These foreign elements will be eliminated from the study, but you don't want to hide anything. Therefore, a reason for throwing out the time is required. Foreign elements marked with an asterisk (*) in the body of the study are referred to this box.

Step 4. Do the actual time study.

This is the essence of the stopwatch time study. Block ⑯ on the step-by-step form is for recording the time for each element. The form has room for 8 elements (8 lines) and 10 cycles (columns) for 80 readings. Most studies will have only 3 or 4 elements, so there is room on one sheet for 20 cycles. This form can be used for either snapback or continuous time study.

Fred Meyers & Associates **Time Study Worksheet** ☐ Snapback ☐ Continuous

Operation Description: Assemble Parts 2 &4, Machine Screw & Stake, Inspect

Part Number 4650-0950	Operation No. 1515	Drawing No. 4650-0950	Machine Name	Machine Number 21	☐ Quality ok?
Operator Name Meyers	Months on Job 5	Department Assembly	Tool Number M61	Feeds & Speeds	☐ Safety checked?
Part Description:	Material Specificatons:			Machine Cycle	☐ Setup proper?
				Time	Notes:

Element No.	Element description		Readings 1	2	3	4	5	6	7	8	9	10	Total Cycles	Average Time	% R	Normal Time	Frequency	Unit Normal Time	Range	R/X	Highest ✓
	Assembly	R	9	41	71	1.07	38	77	2.08	48	77	3.07			90		1/1				
		E																			
	Drive screw	R	15	46	79	13	43	82	14	53	82	93			100		1/1				
		E																			
	Press	R	28	58	94	27	66	95	28	66	96	4.06			110		1/1				
		E																			
	Inspect	R	32	62	92	30	69	98	41	69	99	4.09			100		1/1				
		E																			
	Load screws	R										3.83			125		1/10				
		E																			
		R															1/1				
		E																			
		R															1/1				
		E																			
		R															1/10				
		E																			

Foreign elements:

Notes:

$R/\bar{X}$	No. Cycles
.1	2
.2	7
.3	15
.4	27
.5	42
.6	61
.7	83
.8	108
.9	138
.10	169

Total normal minutes ______
Allowance +_ _ _ _ _10 % ______
Standard minutes ______
Hours per unit - - - - -
Units per hour ______

On back
Work station layout
Product sketch

Engineer: Date: _/_/_

Approved by: Date: _/_/_

Figure 3–8 Time study problem: Continuous technique.

Continuous time study is the most desirable time study technique. The stopwatch remains running through the duration of the study, and element ending times are recorded behind the *R* for *reading*.

Continuous Example

		1	*2*	*3*	*4*	*5*
	R	.16	.83	1.50	2.17	2.83
Load and clamp	E					
	R	.55	1.23	1.90	2.57	3.23
Run machine	E					
	R	.66	1.33	2.01	2.67	3.32
Unload and place aside	E					

Note that each time is getting larger and that five parts were run in a total time of 3.32 minutes. In Step 5 the elemental times are calculated, but at this time you are still out in the plant collecting data.

Snapback studies allow the technician to read the watch and reset it immediately to time the next element. The same study is shown below using the snapback technique.

Snapback Example

		1	*2*	*3*	*4*	*5*
	R					
Load and clamp	E	.16	.17	.17	.16	.16
	R					
Run machine	E	.40	.40	.40	.40	.40
	R					
Unload and put aside	E	.10	.10	.1	.10	.09

Note that the elemental time (*E*) is already calculated. Look at the load and clamp time; the times look consistent—.16, .17, .17, .16, and .16. The time for loading and clamping is immediately obvious. This same information will be available in a continuous time study, but a lot of arithmetic is required first. In the snapback time study technique, the *R* row can be used for rating the operator on each element of work. (We discuss this in more detail later when we consider rating, leveling, and normalizing.)

Step 5. Extend the time study.

Now that the time study has been done, a bigger job is at hand. The continuous method has one more step than the snapback method, so we will concentrate on the continuous method.

⑯ Subtract the previous reading from each reading. The previous element reading was its ending time and the beginning of this element. Subtracting the beginning time from the ending time gives elemental time.

⑰ Total/cycles: The total refers to the total time of the appropriate cycles timed. Some cycles may be eliminated because they include something that does not reflect the elemental time.

Foreign elements are eliminated from further consideration. Cycles are the number of applicable elemental times included in the total time.

⑱ Average time: Average time is the result of dividing total time by the number of cycles. On the average, it took .40 minute of machine time on the last example.

⑲ Percent *R:* Percent rating refers to your opinion of how fast the operator was performing. The rating divided by 100, multiplied by the average time, equals normal time.

$$\text{Average time} \times \frac{\text{rating percent}}{100} = \text{normal time}$$

Later in this chapter we discuss rating in detail.

⑳ Normal time: As stated earlier, *normal time* is defined as the amount of time a normal operator working at a comfortable pace would take to produce a part. Normal time is calculated above and is explained further below for block ㉒.

㉑ Frequency: Frequency indicates how often a task is performed. For example, moving 1,000 parts out of the workstation, moving the empty tub to the other side of the workstation, and bringing in a full tub of 1,000 new parts to the workstation would occur only once in 1,000 cycles (1 per 1,000). If quality control asked the operator to inspect one part out of every 10, *1/10* would be placed in this column. The biggest use of this column is when the operator is producing two parts at a time; then *1/2* is placed in this column. If *1/1* goes in the column, it can be left blank.

㉒ Unit normal time: Unit normal time is calculated by multiplying the frequency by the normal time.

Examples:

Normal Time		*Frequency*		*Unit Normal Time*
1.160	×	1/1,000	=	.001 minute
.400	×	1/10	=	.040 minute
.100	×	1/2	=	.050 minute
.050	×	1/1	=	.050 minute

Every element must reflect the time to produce one unit of production. No one wants a standard for pairs, and mixing frequency of units leads to bad time standards. Be very careful here.

Step 6. Determine the number of cycles to be timed.

The accuracy of time study depends on the number of cycles timed: The more cycles that are studied, the more accurate is the study. Almost all time study work is aimed at an accuracy of ±5 percent with a 95 percent confidence level, so the question is, How many cycles should be studied to achieve this accuracy? Blocks ㉓ through ㉖ will help determine the number of cycles needed.

A detailed discussion of the statistical aspect of the subject and the mathematical approach in determining the number of cycles is not within the scope of this book. As a rule of thumb, 20 to 25 observations should provide sufficient accuracy for our purposes.

Step 7. Rate, level, and normalize the operator's performance.

⑲ Percent rating is the technologist's opinion of the operator's performance. Rating, leveling, and normalizing all mean the same thing, and the term *rating* is used from this point on. Rating is the most challenged aspect of time study; for that reason, it is the most important subject of this chapter (it is discussed in detail later in this chapter).

$$\text{Average time} \times \frac{\text{rating}}{100} = \text{normal time}$$

⑱ → ⑲ → ⑳

Step 8. Apply allowances ㉙.

Allowances are added to a time study to make the time standard practical.

Total normal time + allowances = standard time

㉘ ㉙ ㉚

There are several methods of applying allowances, and there are several types of allowances. We will discuss allowances in detail later in this chapter.

Step 9. Check for logic.

Once the time study has been extended, the test for logic should be applied in two ways:

1. The average time ⑱ should look like the elemental times. If an error in adding was made, a test for logic will prevent a mistake. The easiest mistake to make is decimal error. Be careful not to make decimal errors because they look bad—1,000 percent of errors result from misplacing a decimal only one place, which is why being consistent with decimal placement is so important.
   a. Read stopwatches in two places: .01.
   b. From average time on, use three places: .001.
   c. Hours per unit are five places: .00001.

2. The second test for logic is the total normal time for one unit. During your study, you timed a specific number of parts in a certain amount of time. For example, 10 cycles were timed in 7.5 minutes (7.5 was the last reading in the 10th column). The average time should be somewhere around .75 minute each. Are you close to the total normal time? If not, there is a major error. *Warning:* Do not forget that if the operator is producing two at a time, twice as many parts are being produced.

Step 10. Publish the time standard.

Three numbers are required to communicate a time standard:

1. Decimal minutes ㉚
2. Hours per unit ㉛
3. Pieces per hour ㉜

Starting with standard minutes, dividing ㉚ by 60 minutes per hour equals hours per unit ㉛, and pieces per hour ㉜ is $1/x$ of ㉛ (or divide hours per unit into one hour).

Every company has a method of recording time standard information. Earlier in this chapter, an operations chart for a water valve factory was shown. The time standards could be placed on that operations sheet (review Figure 3–3). The production route sheet is another common tool for communicating the time standard. The computer is the most common method of storing and communicating to everyone what the time standard is for each job.

A few more pieces of information remain to be discussed on the step-by-step time study form:

㉝ Engineer: The time study technologist puts her name here.

㉞ Date: A time study with an incomplete date is worthless.

㉟ Approved by: This is where the chief engineer or manager signs, approving your work. You never fill this in.

An example problem has been included in Figure 3–8. The data have been collected, the job has been broken down into elements, and the time study has been made. You need to extend the study and develop a time standard. This was a continuous time study, and that should be obvious because the times are always getting larger. The extension will start with the subtraction of element readings to find elemental time.

Rating, Leveling, and Normalizing

Rating is the process of adjusting the time taken by an individual operator to what could be expected from a normal operator. The industrial engineer must understand the industry standards of *normal.* Rating an operator includes four factors: (1) skill, (2) consistency, (3) working conditions, and (4) effort (which is most important).

Three of these four factors are accounted for in other ways and have little effect on rating. Effort will be the primary concern.

1. *Skill.* The effect of skill is minimized by timing only people who are skilled. Operators must be fully trained in their work classification before being time studied. A welder must a be qualified welder before being considered as a subject for time study. Two years of training may be required to become a welder, and in addition, this welder must be on this job for at least 2 weeks before performing the job sufficiently. Habits of motion patterns must be routine enough so that the operator does not have to think about what comes next and where everything is located. Very skilled operators make a job look easy, and the industrial technologist must not let this skill affect the rating. On the other hand, if an operator shows lack of skill, such as dropping, fumbling, inconsistent timing, stopping and starting, and so on, the technologist should postpone the study or find someone else to time study.

2. *Consistency.* Consistency is the greatest indication of skill. Operators are consistent when they run the elements of the job in the same time, cycle after cycle. The time study technician begins to anticipate the ending point while looking at the watch and listening for the ending point. The operator is said to be like a machine: Consistency is used to determine the number of cycles. A consistent operator needs to run only a few parts before the cycle time is known with accuracy. The skill of the operator should be evident to the time study technician, and the technician's rating of the operator should be high. When inconsistency is present, the technologist must take many more cycles to be acceptably accurate in the time study. This inconsistency tends to affect the technologist's attitude and rating of the operator in a negative way, and the best thing to do is to find someone else to study. It is more fun rating and working with operators with great skill.

3. *Working conditions.* Working conditions can affect the performance of an operator. In the early twentieth century, this was much more of a problem than it is today. But if employees are asked to work in hot, cold, dusty, dirty, noisy environments, their performance will suffer. These poor working conditions can be eliminated if the true cost is shown. The way to account for poor working conditions today is to increase the allowances (discussed later in this chapter). If operators are required to lift heavy materials in the performance of their duties, 25 percent more time can be added to the time standard as an allowance. Working conditions are not part of modern rating.

4. *Effort.* Effort is the most important factor in rating. Effort is the operator's speed or tempo and is measured based on the normal operator working at 100 percent. As discussed earlier, 100 percent performance rating is defined as

a. Walking 264 feet in 1.000 minute or 3 miles per hour
b. Dealing 52 cards into four hands around a 30 × 30-inch cardtable in .500 minute
c. Assembling 30 3/8 × 2-inch pins into a pinboard in .435 minute

Effort can be seen easily in walking. Walking at speeds less than 100 percent is uncomfortable for most people, and walking at 120 percent requires a sense of urgency that indicates increased effort.

Psychology has been good to the time study technician. The normal tendency of people being watched is to speed up. Being watched makes people nervous, and nervous energy is converted by the body into a faster tempo. The time study technician then gets a frequent chance to rate over 100 percent. When an operator works at 120 percent, the technologist has the pleasant experience of telling the operator, "You are fast. I'm going to have to give you 20 percent more times so that an average person can do the job." That is fun to say, and it happens often.

When rating, the technician must keep tuned into normal pace. This requires continued practice on her part—forever. Experiments and videos have been developed to help keep rating accurate.

PTSS has been developed based on the concept of normality according to industry standards, and a synthetic rating is developed by time studying a job that has been proven by PTSS. A good learning technique used at many companies is to have new technologists time study known jobs and compare their time standards to the known time standards. Another good learning experience is to time several people on the same job. Effort and skill are the only differences in time, so proper rating should make all the normal times the same (see Figure 3–9.)

Many companies use time study rating films developed by industrial associations and professional organizations:

Society for the Advancement of Management (SAM)

Tampa Manufacturing Institute

Ralph Barnes and Associates

Faehr Electronic Timers, Inc.

All these groups produce time study rating films. *Industrial Engineering* magazine would also be a good source.

ALLOWANCES

Allowances are extra time added to the normal time to make the time standard practical and attainable. No manager or supervisor expects employees to work every minute of the hour. What should be expected of the employee? This was the question asked by Frederick W. Taylor over 100 years ago. Would you expect the employee to work 60 minutes per hour? How about 40 minutes? 50 minutes? This section will assist the technologist in answering Taylor's question.

Types of Allowances

Allowances fall into three categories:

1. Personal
2. Fatigue
3. Delay

1. Healthy people in the right frame of mind easily turn in 100 percent performance on correctly standardized jobs. For incentive pay, good performers usually work at paces from 115 percent to 135 percent, depending on the jobs and individuals.
2. For most individuals, *it is uncomfortable to work at a tempo much below 100 percent and extremely tiring to operate for sustained periods at paces lower than 75 percent;* our reflexes are naturally geared to move faster.
3. Poor efficiency on a correctly standardized job usually results from stopping work frequently—"goofing off" for a variety of reasons. Specifically, *substandard production seldomly results from inability to work at a normal pace.*
4. Some standards of 100 percent
   1. Walking 3 miles per hour or 264 feet per minute
   2. Dealing cards into four stacks in .5 minute
   3. Filling the pinboard in .435 minute
5. Very seldomly can *true* performance of over 140 percent be found in industry.
6. When an operator consistently comes up with extremely high efficiencies, it is usually a sign that the method has been changed or the standard was wrong in the first place.
7. Operators work pace during a time study does not affect the final standard. Their *actual* time is multiplied by the performance rule to give a job standard that is fair for all employees.
8. Inasmuch as healthy employees can *easily* vary work pace from approximately 80 percent to around 130 percent—through a range of 50 percent—reasonable inaccuracies in the setting of standards should be sensibly accepted.
9. Ineffective foremen usually fight job standards. *Good supervisors, however, sincerely help in the standard-setting effort, clearly realizing that such information is their best planning and control tool.*
10. *Methods usually influence production more than work pace.* Don't ever get so absorbed in how quickly or how slowly operators "seem" to be moving that you fail to consider whether or not they are using the right method.

Figure 3–9 Ten fundamentals of pace rating (courtesy of Tampa Manufacturing Institute).

Personal Allowance

Personal allowance is that time an employee is allowed for personal activities such as

1. Talking to friends about nonwork subjects
2. Going to the bathroom
3. Getting a drink
4. Any other operator-controlled reason for not working

People need personal time, and no manager would begrudge an appropriate amount of time spent on these activities. An appropriate amount of time has been defined as about 5 percent of the workday, or 24 minutes per day.

Fatigue Allowance

Fatigue allowance is the time an employee is allowed for recuperation from fatigue. Fatigue allowance time is given to employees in the form of work breaks, more commonly known as coffee breaks. Breaks occur at varying intervals and are of varying duration, but all breaks are designed to allow employees to recuperate from on-the-job fatigue. Most employees today have very little physical drudgery

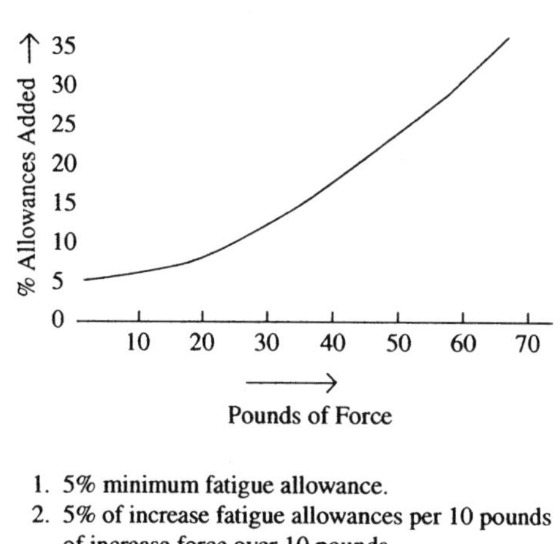

Figure 3–10 Fatigue allowance curve: Percent allowances per pound of force.

involved with their jobs, but mental fatigue is just as tiring. If employees use less than 10 pounds of effort during the operation of their jobs, then 5 percent fatigue allowance is normal. A 5 percent increase in fatigue allowance is given for every 10-pound increase in exertion required of the employee (see Figure 3–10).

Example: An employee must pick up a 50-pound part. Fatigue allowance is $(50 - 10) \div 10 = 4.0$ units of 10 pounds.

$$5 \text{ percent} + (4 \times 5) = 25 \text{ percent allowance}$$

Explanation of Example: The basic fatigue allowance is 5 percent and an additional 5 percent fatigue allowance is added for each 10 pounds of force required over 10 pounds. Fifty pounds is 40 pounds more than basic; 40 pounds is four units of excess weight (10 pounds is one unit); four units times 5 percent percent of excess fatigue; 20 percent plus 5 percent basic equals 25 percent fatigue allowance.

The weight has to be picked up every minute. If the frequency were once every 5 minutes, the 50 pounds would be divided by 5.

$$5 \text{ percent} + (\frac{4}{6} \times 5) = 9 \text{ percent}$$

The basic fatigue allowance is still 5 percent. When lifting only one 50-pound object every 5 minutes, only one-fifth of the excess weight is considered. Forty pounds is four units of weight, so

$$\frac{\text{Four units} \times 5 \text{ percent}}{5} = 4 \text{ percent excess fatigue allowance}$$

Five percent basic plus 4 percent excess equals 9 percent fatigue allowance.

The duration of breaks must now be calculated. The normal 5 percent fatigue allowance is commonly interpreted as two 12-minute breaks, one in midmorning and one in midafternoon, or a combination of the two, adding up to 24 minutes. Five percent of the 480 minutes in an 8-hour day is 24 minutes.

Seventeen percent allowances would equal 82 minutes per day. How will this 82 minutes be split for frequency and duration of breaks? We suggest that 11 minutes be given every hour except the hour before lunch. Seven 11-minute breaks equals 77 minutes, plus a 5-minute cleanup at the end of the shift. Note that a heavy job such as the one we are discussing here will tire out the employee faster than light or mental work, and the increased breaks are not only justified but they will also increase production. Breaks from work allow employees to recuperate, so when they return to work their production rate is higher than it would have been without a work break. The break more than pays for itself.

1. Five percent is the minimum fatigue allowance.
2. Five percent increased fatigue allowances per 10 pounds of increased force over 10 pounds.
3. Force is the weight of the part if lifted.

Delay Allowances

Delay allowances are unavoidable because they are out of the operator's control. Something happens to prevent the operator from working. The reason must be known and the cost accounted for to develop the cost justification. Examples of unavoidable delays include

1. Waiting for instructions or assignments
2. Waiting for material or material handling equipment
3. Machine breakdown or maintenance
4. Instructing others (training new employees)
5. Attending meetings, if authorized
6. Waiting for setup; operators should be encouraged to set up their own machines—a setup is complete when quality control approves
7. Injury or assisting with first aid
8. Union work
9. Reworking quality problems (not operator's fault)
10. Nonstandard work—wrong machine or other problem
11. Sharpening tools
12. New jobs that have not been time studied yet

The operator's performance must not be penalized for problems out of her control. (Delays that are controlled by the operator are called personal time and are not considered here.)

Three methods are available to account for and control unavoidable delays:

1. Add delay allowances to the standard.
2. Time study the allowances and add them to the time standard.
3. Charge the time to an indirect charge.

The goal of time study is to eliminate delay allowances. This is best done by time studying the delay and adding that time to the time standard. However, some delays are so complicated that negotiating an allowance with the operator will save time and money for the company. For example, suppose you ask the question, "How much time do you spend a day cleaning the machine?" The operator will always say, "Well, it depends," and the technologist must ask something such as

What is the longest time?

What is the shortest time?

Do you think 15 minutes is a good average?

If the operator agrees that 15 minutes per day is a good figure, the technologist will calculate a delay allowance as follows:

$$\frac{\text{15 minutes cleanup}}{\text{480 minutes per shift}} = 3 \text{ percent}$$

A 3 percent allowance will be added to the personal allowance of 5 percent plus a fatigue allowance of 5 percent to produce a 13 percent total allowance.

Generally, unavoidable delays can be eliminated or anticipated. Time standards in the form of standard data can be established and added to the time study to compensate the operator. An unavoidable delay is a foreign element. Those unavoidable delays that cannot be anticipated will require operators to charge their time to an indirect account—meeting, injury, machine breakdown, and rework are examples. Supervisors will be required to approve all indirect charges, and the time should be more than 6 minutes to be statistically significant. Employees must not be penalized for management's lack of planning, but the supervisor must be given as much advance notice as possible. Reassignment may be in order.

One last warning about delay allowances: Don't put anything in the time standard that you cannot live with. It is difficult to get this out of the standard once included. Most companies have eliminated delay allowances but have allowed operators to punch out for anything not covered by the time standard.

Personal fatigue and delay allowances are added together, and the total allowance is added to the normal time:

$$\text{Normal time} + \text{allowance} = \text{standard time}$$

Methods of Applying Allowances

Allowances are added in four different ways. The forms in this text use just one of these methods, but there are good reasons for using the other methods. Each company has its own time study form and procedure. The form tells you which method of applying allowances to use. The four methods are presented here in order of ease of application.

Method 1: 18.5 Hours per 1,000

This method is the simplest of all and reduces the mathematical steps. It is also based on a constant allowance—in this case, 10 percent.

If a job takes 1.000 minute normal time, how many pieces per hour could be produced? At the rate of one per minute, 60 could be produced per hour, but you want to be practical and add 10 percent allowances. Ten percent of 60 is 6, so 54 pieces per hour would be an appropriate time standard. How many hours would it take to produce 1,000 units at the rate of 54 per hour? One thousand divided by 54 equals 18.5 hours per 1,000—the name of this method. Three numbers are required to communicate a time standard:

Decimal minute = 1.000
Hours per 1,000 = 18.5
Pieces per hour = 54

All time standards start with a decimal minute, so if the next standard is .5 minute, the hours per 1,000 equals .5 × 18.5 = 9.25 hours per 1,000, and the pieces per hour is $1/x$ or 108 pieces per hour. Try these examples:

Normal Minutes	*Hours per 1,000 (18.5)*	*1/X Pieces per Hour*
.250	4.625	216
.333		
.750		
1.459		
2.015		

Notice that no calculations are made to add allowances; it is all in the 18.5.

What would the hours per 1,000 be with 15 percent allowances?

Method 2: Constant Allowance Added to Total Normal Time

Method 2 is used in this text and is the most common technique used in industry. Each department or plant has only one allowance rate. The average allowance is between 10 and 15 percent. An explanation of what makes up the allowance, such as the one below, must be included:

Personal time	= 24 minutes
Two breaks at 10 minutes	= 20 minutes
Cleanup time	= 4 minutes
Total allowances	= 48 minutes

$$\frac{48 \text{ minutes}}{480 - 48 \text{ minutes}} = 11 \text{ percent}$$

Eleven percent is added to normal time to get standard time, or 111 percent times normal times equals standard time.

$$1.000 + .11 = 1.110 \text{ minutes}$$
$$1.000 \times 111 \text{ percent} = 1.110 \text{ minutes}$$

The time study form will tell you which calculation to make.

Method 3: Elemental Allowances Technique

The theory behind this technique is that each element of a job can have different allowances, as in the following example:

Element Description	*Unit Normal Time*	*Allowance %*	*Standard Time*
1. Load machine	.250	15	.288
2. Time machine	.400	5	.420
3. Unload machine	.175	10	.193

Note that each element allowance is different. Element 1 is operator controlled, and a heavy part is involved. Therefore, more allowances were included. Element 2 is a machine element, and the operator just stands there—no fatigue was given. Element 3 is a normal 5 percent fatigue, plus 5 percent personal allowance.

The obvious advantage of this method is improved elemental time standards. The disadvantage is the increased mathematical effort required. The time study form would have to be redesigned to accommodate this method and, as with all allowances, the form would show you which technique to use.

Method 4: The PF&D Elemental Allowance Technique

As in Method 3, the allowance is placed on each element, and the personal, fatigue, and delay (PF&D) method shows exactly how the allowance was developed. This technique is the most complete of all the techniques.

Example:

Element Description	*Unit Normal Time*	*Allowances %*				
		P	*F*	*D*	*Total*	*Standard Time*
1. Load machine	.250	5	10	0	15	.288
2. Run machine	.400	0	0	5	5	.420
3. Unload machine	.175	5	5	0	10	.193

This allowance technique takes a lot of time and effort. It is very descriptive, but the cost is too high for most companies.

Allowances are an important part of the time standard, and properly established allowances will assist in the continued improvement of the quality of work life. If a job has undesirable aspects that do not reflect on the individual cycle, the allowances must reflect this undesirability. In that way, the money exists to justify a needed change. A plantwide base rate of 10 percent is still very desirable, but additional allowances can be added as needed. The forms used in this text allow for a range of allowances.

Work Sampling

Work sampling is the same scientific process used in Nielsen ratings, Gallup polls, attitude surveys, and federal unemployment statistics. Technicians observe people working and draw conclusions. In fact, everyone who has ever worked with someone else has done work sampling; you have an opinion of how hard the other person works:

"Every time I look at her, she's working."
"He's never working."
Somewhere in between

Supervisors, using informal work sampling, are forming attitudes about employees all the time.

Industrial engineers can walk through a plant and state, "This plant is working at 75 percent performance." They should add "10 percent or so," depending on how many people they observed (number of samples). You could walk through a plant of 250 people one time and count people who are working and those who are not working and calculate the performance of that plant within percent. Industrial engineering consultants often start their consulting proposal with such statistics. Consultants expect to find 60 percent performance in plants without standards, but that is an average. A specific plant may have better management and be averaging between 70 and 75 percent. They could not save as much in this plant.

Setting standards using work sampling is not difficult. The industrial engineer samples a department and finds the following statistics:

Task	*No. of Observations*	*% Total*	*Hours Worked*	*Pieces Produced*	*Pieces per Hour*[a]
Assemble	2,500	62.5	625	5,000[b]	8
Idle	1,500	37.5	375	—	
Total	4,000	100.0	1,000[c]		

[a] Pieces per hour $= \frac{\text{pieces} = 5{,}000}{\text{hours} = 625} =$ 8 pieces per hour.

[b]From supervisor (number of finished products put in the warehouse).

[c]From payroll (hours paid during the study).

Eight pieces per hour is not quite the time standard, but you haven't added allowances. How much time is in the 625 hours for breaks, scheduled or unscheduled? How much time is in there for delays? None. Actual hours worked is 625. All other nonwork time is part of the 375 hours that are thrown away. You could add an appropriate amount of extra time to cover personal time, fatigue time, or delay time (allowances). Ten percent extra time is considered normal. A time standard of 7.3 pieces per hour would be appropriate.

Standard Data

Standard data should be the objective of every motion and time study department. Standard data are the fastest and cheapest technique of setting time standards, and standard data can be more accurate and consistent than any other technique of time study. Starting with many previously set time standards, the industrial engineer tries to figure out what causes the time to vary from one job to another on a specific machine or class of machine. For example, walking time would be directly proportional to the number of feet, paces, yards, or meters walked. There might be two curves on the graph: obstructed and unobstructed.

A second example is counting playing cards. Time for counting cards would be directly proportional to the number of cards counted. Can you think of any other reasons for the time to vary?

There are several ways of communicating the time standard to future generations of factory workers, supervisors, and engineers: (1) graph (see Figure 3–11), (2) table, (3) worksheet (again, see Figure 3–11), and (4) formula.

Metal-cutting machines are examples of the need for and use of formulas. A blueprint calls for the drilling of a hole through the steel plate. Three pieces of information are required:

1. What are the diameter and depth of the hole?
2. What is the material?
3. What tool do we use?

With this information, look up the feeds and speeds in the *Machinery's Handbook.*[1] Feeds and speeds are communicated as follows:

Speed 500 feet per minute
Feed .002 inch per revolution

By substituting this information into three simple formulas, you determine the time standard.

Other machines, like welders, have simpler formulas, such as 12 inches per minute. The machine manufacturers are a good source of standard data.

[1] *Machinery's Handbook,* New York: The Industrial Press.

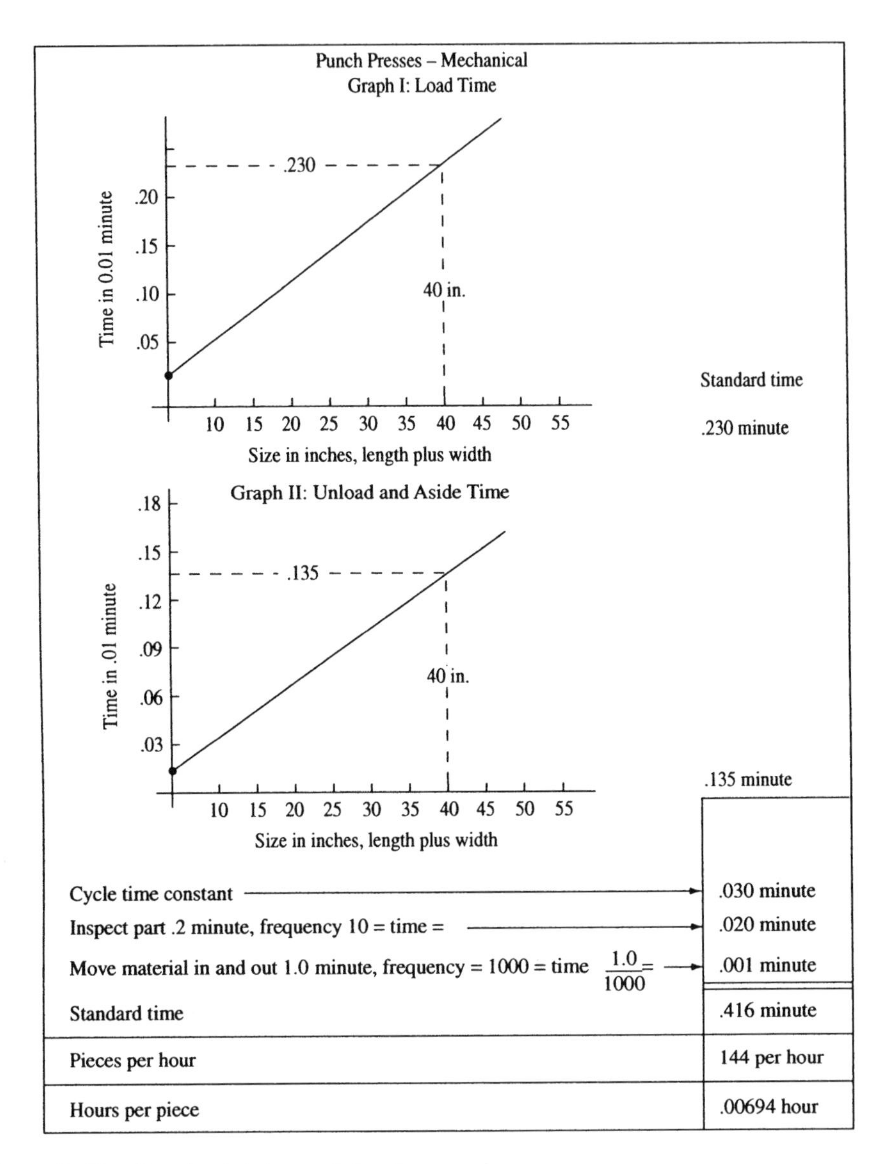

Figure 3–11 Standard data worksheet example.

Expert Opinion Time Standards and Historical Data

An expert opinion time standard is an estimation of the time required to do a specific job. This estimate is made by a person with a great experience base. Many people say, "You can't set time standards on my work." A good industrial engineer's response would be, "You are right, but I know someone who can—you can!" The individual nature of many staff and service workers makes setting time standards

Cycle Time	Volume of Production		
	High 1,000s	*Medium 100s*	*Low 10s*
Long	Work sampling	Work sampling Stopwatch	Expert opinion Work sampling History
Medium	Work sampling Stopwatch PTSS	Stopwatch Work sampling	Expert opinion History Stopwatch
Short	PTSS	PTSS Stopwatch	Stopwatch Expert opinion

Note: Standard data are the ultimate time standard technique and can be used in all situations. Standard data should be the goal of all time study departments.

Figure 3–12 Which time standards technique do we use?

with the more traditional techniques unprofitable. Engineering maintenance and some office workers never do the same thing twice, but goals are still needed (time standards). Maintenance work is controlled by work order. Why not ask an expert how long this requested work will take? In well-managed companies, new maintenance projects will not be approved until the job is estimated. These time standards would be used to schedule and control maintenance work, just as you would schedule and control the work performed by a machine operator.

The expert in an expert opinion time standards system is usually a supervisor. In larger departments, a specialist may be used. For example, in the maintenance department, the person would be called a maintenance planner. The expert would estimate every job and maintain a backlog of work. The backlog of work would give the department time to plan the job, thereby performing that job more effectively.

Historical data are an accounting approach to expert opinion time standard systems. A record is kept of how much time was used on each job. When a new job comes along, it is compared to a previous job standard. These standards are then used in a labor performance control system. The problem with historical time standards is that they do not reflect the time the job should have taken. Inefficiency is built into such a system, but a bad standard is better than no standard at all.

Figure 3–12 will help you to choose the correct technique for setting time standards.

TIME STANDARDS FOR MANUFACTURING FACILITIES DESIGN

Time study does not always mean stopwatch time study. The stopwatch method of setting time standards is not very useful in manufacturing facilities design because you need to know the required time for each element of work before production begins. For this reason, predetermined time standards or standard data systems are used. Time standards are used for five main purposes in facilities design:

1. Determining the number of workstations and machines
2. Determining the number of people
3. Determining conveyor line speeds
4. Balancing assembly and packout lines
5. Loading work cells

We will discuss each of these topics in the next chapter.

QUESTIONS

1. What are time standards used for?
2. What is the definition of a time standard?
3. What three numbers make up a time standard?
4. What is productivity? How is it measured for individuals?
5. What are the five techniques of setting time standards?
6. Which technique would be used when no workstation is available?
7. Which technique is the most popular?
8. Which technique would be used for maintenance work?
9. Which is the best technique for setting time standards?
10. Which technique is both a method and a time study technique?
11. What are the three basic levels of productivity?
12. Define *takt time.*
13. How many machines should you buy and how many people should you hire if 3,000 units are needed per shift in a 75 percent efficient plant that has 10 percent downtime? The machine time standard is .284 minute. How much will a unit cost to produce if the operator earns $15 per hour? How many units will be produced per shift? What is the takt time?
14. Calculate the pieces per hour and the hours per unit for the following:

Time Standard Minute	*Pieces per Hour*	*Hours per Piece*
.300	—	—
2.000	—	—
.450	—	—
.050	—	—

15. Define *measured* or *observed time, normal time,* and *standard time.*
16. What is meant by allowances and what are the most common types of allowances?
17. How does standard time differ from normal time?

18. Given a 10-hour shift, if a total of 20 minutes is allowed for breaks and a 10-minute cleanup period is given at the end of the shift, determine the takt time. Assume 85 percent efficiency. During this shift, 2,500 units are to be built.

19. The average observed (measured) time for an operation is given as .570 minute. The operator is rated at 95 percent. Allow a total of 15 percent for PF&D. If the direct labor wages are $12 per hour, determine (a) normal time, (b) standard time, (c) pieces per hour, (d) hours per 1,000 pieces, (e) standard direct labor cost per unit, and (f) direct labor cost per unit as produced by this operator.

20. Three operations are performed on a part. If these operations produce 1, 3.5, and .5 percent scrap, respectively, determine the number of blanks required to begin the process if 5,000 finished units are required.

CHAPTER 

Process Design

OBJECTIVES:

Upon the completion of this chapter, the reader should:

- Understand the role of process design in facilities planning
- Be able to define and construct routing sheets
- Be able to calculate conveyor speed
- Understand the concept of and be able to calculate "fraction" equipment
- Understand the concept of line balancing and be able to balance an assembly line

The process engineer may be the same person as the facilities designer, but the larger the firm is, the less this is true. In larger firms, the manufacturing facilities designer is a collector of information used in the facility design. Larger firms may have departments called processing, tool design, time standards, ergonomics, production packaging, and others. Basically, the process engineer or designer, whether a person or a department, is in charge of all these tasks. The process designer determines how the product and all its components will be made. The information provided by the process designer would include the following:

1. The sequence of operations to manufacture every part in the product (the company "makes" parts only because the "buy" parts will not be our company's problem)
2. The needed machinery, equipment, tools, fixtures, and so on
3. The sequence of operations in assembly and packout
4. The time standard for each element of work (this may be another department in many companies)
5. The determination of the conveyor speeds for cells, assembly and packout lines, and paint or other finishing systems
6. The balance of the workloads of assembly and packout lines
7. Load work cells

8. The development of a workstation drawing for each operation using all the principles of motion economy and ergonomics.

This book will consider the process designer and the facilities designer to be one and the same person, and all the information produced in the manufacturing department will be done by the same person—you. This would be a good job description and one of the best work experiences you will ever have. Then, you will really understand how your plant operates.

Process design can be divided into two broad categories, *fabrication* and *assembly*. Fabrication process design is initially planned on a route sheet. Assembly and packout process design uses the techniques of assembly charts and assembly line balancing.

FABRICATION: MAKING THE INDIVIDUAL PARTS

The sequence of steps required to produce (manufacture) a single part is referred to as the *routing*. The part is routed from the first machine to the second machine and so on until you have a finished part that will be united with other parts. The form used to describe this routing is called the *route sheet.*

Route Sheets

A route sheet (see Figures 4–1 and 4–2) is required for each individual fabricated part of the product (for make parts). If the finished product that is to be manufactured has 30 different parts and you buy 10 from outside the company (buyouts), and make 20 parts in-house, you will need 20 route sheets. The route sheet lists the operations required to make that part in the proper sequence. The route sheet gets its name from the way it is used. For example, you need to produce an order of 2,500 axles for a wagon. A copy of the route sheet would be issued by the production and inventory control department showing the order quantity. This order would then be given to the stores department where the raw material for 2,500 axles would be pulled and transported to the first operation (according to the route sheet). The route sheet would accompany the material from operation to operation telling the operators what to do. The route sheet will tell the plant personnel about the part number, part name, quantity to produce (left blank until needed), operation number, operation description, machine number (if available), machine name, tooling needed, and time standard.

The route sheet ends with the last operation prior to being assembled with other parts. For example, if three parts are going to be welded together, the individual parts lose their identity once joined with other parts, so that the route sheet would end before welding. If an individual part goes through a clean, paint, and bake operation before being assembled, then the clean, paint, and bake procedure would be included on the route sheet.

Part Name	Box Body	Part Number 1600		Date 10/22/XX	Quantity 1,000
A	*B*	*C*	*D*	*E*	*F*
Operation number	Operation and tooling	Machine name	Machine number	Pieces per hour	Hours per piece
	Description				
5	Cut to length	Shear	12	1,200	.00083
10	Cut to width	Shear	12	400	.0025
15	Notch corners	Press	65	300	.00333
20	Punch four holes	Press	65	300	.00333
25	Form two short legs	Press break	55	250	.004
30	Form two long legs	Press break	55	250	.004
35					
40					

Figure 4–1 Route sheet.

Route Sheet 1

Part Number 7440 Part Name Body

Raw Material 1,020 cold rolled steel 18 × 24 inches 20 gal.

Order Quantity ____________

Operation No.	*Machine Name*	*Operation*	*Pieces per Hour Time Standard*
5	Strip shear	Cut to width	1,400
10	Chop shear	Cut to length	1,175
15	Punch press	Punch catch holes	650
20	Press brake	Form two legs	475

Route Sheet 2

Part Number 7420 Part Name End

Raw Material 1,020 cold rolled steel 6 1/2 × 6 1/4 inches 20 gal.

Order Quantity Twice the number of bodies

Operation No.	*Machine Name*	*Operation*	*Pieces per Hour Time Standard*
5	Strip shear	Cut to width	1,850
10	Chop shear	Cut to length	950
15	Punch press	Punch hinge holes	825
20	Press brake	Form three sides	595

Figure 4–2 Sample route sheets.

The sequence of operations as shown on the route sheet affects the proper layout of the equipment on the production floor. You want the material to flow smoothly through the plant from the raw material stores to the first operation, to the second operation whose machine is right next to the first machine. This will ensure that the part travels as short a distance as possible. Process-oriented layouts are where you group all like machines together and bring all parts to them, whereas product-oriented layouts place machines where they are needed to eliminate excessive moving. Skipping over machines and backtracking will result from process layouts and must be discouraged because it adds costs without adding to the value (muda). When many parts are fabricated in one group of machines (called a process layout), jumping around may be necessary, but you want to minimize this jumping, skipping, and backtracking. There are two ways to change the sequence in order to make the flow through the plant smoother:

1. Change the route sheet (paper change) if possible so that the sequence of operation agrees with the other parts or the existing (or proposed) plant layout.
2. Change the physical layout of the machines so that the machines are in the correct sequence.

Changing the paperwork is the first choice because it is the least expensive way.

Time standards are an important part of the route sheets. Time standards are used to determine how many machines are needed in the layout. They are another piece of information that may come from another group within the manufacturing engineering department, but in many companies, time standards are developed by the manufacturing facilities designer.

In Chapter 2, Figure 2–3a showed an exploded drawing of a toolbox. This toolbox has nine different manufactured parts, thus nine route sheets will be needed.

Part No.	*Part Name*	*Quantity per Box*
1	Body ends	2
2	Shelf brackets	2
3	Body	1
4	Cover ends	2
5	Cover body	1
6	Hinge	1
7	Tray ends	2
8	Tray body	1
9	Tray handle	1

Figure 4–2 shows two of these route sheets. Figure 4–3 is a summary of all nine route sheets.

Part Name	*Body Ends*	*Tray Brackets*	*Body*	*Cover Ends*	*Cover*	*Hinge*	*Tray Ends*	*Tray*	*Handle*
Time Standards in Pieces per Hour									
Parts per unit	2	2	1	2	1	1	2	1	1
Operations									
Strip shear	1,850	2,750	1,400	2,100	1,750	—	2,250	1,850	—
Chop shear	950	1,400	1,175	1,050	1,320	935	1,220	1,410	—
Punch	825	—	650	870	759	—	—	—	—
Form	595	841	475	616	528	—	629	567	—
Roll form	—	—	—	—	—	—	—	—	375
Time Standards in Decimal Minutes per Unit* (Divide the Above Pieces per Hour into 54 Minutes)									
Strip shear	.029	.020	.039	.026	.031	—	.024	.029	—
Chop shear	.057	.039	.046	.051	.041	.058	.044	.038	—
Punch	.065	—	.083	.062	.071	—	—	—	—
Form	.091	.064	.114	.088	.102	—	.086	.095	—
Roll form	—	—	—	—	—	—	—	—	.144

*The time standards in decimal minutes will be used to determine the number of machines needed.

Figure 4–3 Summary of route sheets.

The Number of Machines Needed

The question of how many machines you should buy can be answered only when you know

1. How many finished units are needed per day
2. Which machine runs what parts
3. What is the time standard for each operation

How many finished units are needed per day? The marketing department decides how many products to produce (manufacture) per day. For example, let's say you are going to build 2,000 toolboxes per 8-hour shift. From the figure, you calculate the plant rate (takt time) as follows:

60 minutes per hour × 8 hours	= 480 minutes
Less downtime 10 percent	48 minutes
Available minutes per shift	432 minutes
Expected efficiency approx.	80 minutes
Effective minutes	= 345.6 minutes per 8-hr shift

You have 345.6 minutes to produce 2,000 units.

$$\text{Takt time} = \frac{345.6 \text{ minutes}}{2{,}000 \text{ units}} = .173 \text{ minute per unit}$$

Which machine runs what parts? The route sheets discussed previously tell which machines are needed to produce each part. Figure 4–3 summarizes the nine route sheets needed to manufacture the toolbox. Note that seven different parts are run on the strip shear, eight different parts on the chop shear, four different parts on punches, seven parts on the press brakes, and only one part on the roll former.

What is the time standard for each operation for each part? The time standard for every operation on every part is in both pieces per hour and decimal minute (see Figure 4–3). You need the decimal minute time standards to compare with the takt time, calculated in question 1 of this section.

Once you know the plant rate (*R* value), the machines to be used, and the time standards, divide the time standard (decimal minute) by the *R* value. The resultant number of machines should be in two decimal places (i.e., .34 machine). Once all the machine requirements for each operation have been calculated, total similar machine requirements and round up recommending the purchase of enough machines. The numbers shown in Figure 4–4 are the result of dividing those in Figure 4–3 by the *R* value of .173 (to produce 2,000 toolboxes per shift). This information on the number of machines required will be used later to determine the number of square feet of floor space needed in the fabrication department. Chapter 7 discusses workplace design and space determination, and that is based on the above fabrication equipment requirements.

You will need to acquire two strip shears, four chop shears, three punch presses, six press brakes, and one roll former to produce 2,000 toolboxes per day. Always round up on the total machines; otherwise, a bottleneck will be created and the

Part Name	2 *Body Ends*	2 *Tray Brackets*	1 *Body*	1 *Cover Ends*	1 *Cover Body*	1 *Hinge*	2 *Tray Ends*	1 *Tray Body*	1 *Tray Handle*	*Total Machines*
Machines										
Strip shear	.34*	.24	.23	.30	.18	—	.28	.17	—	1.74
Chop shear	.66	.45	.27	.59	.24	.33	.51	.22	—	3.27
Punch press	.75	—	.48	.72	.41	—	—	—	—	2.36
Press brake	1.05	.74	.65	1.02	.59	—	.99	.55	—	5.59
Roll former	—	—	—	—	—	—	—	—	.83	.83

*.029 ÷ .173 × 2 parts per unit = .34 machine.

Figure 4–4 Machine requirements spreadsheet.

plant will not produce 2,000 toolboxes per day—unless the employees work overtime. If, due to economic considerations, rounding up cannot be justified, overtime may need to be planned for these operations in order to meet production requirements and to alleviate bottlenecks. If investment can be justified, and the production volume is warranted, then rounding up is recommended.

WORK CELL LOAD CHART

Earlier in this chapter, process-oriented and product-oriented layouts were explained. To better help you understand the concept of work cell, we need to define the concept of group technology.

Group technology takes advantage of similarity in parts or features in a group or family of parts so that these parts can be processed as a group. Group technology requires that engineering drawings include a certain coding scheme that specifies the *type* and *parameters* of the required processing. The type of the process specified by the code, for example, may be a drilling operation. The parameters that are included would specify the hole dimensions. Those parts with similar codes, regardless of their end products or destination, can then be grouped and processed together, taking advantage of one setup and minimizing the setup cost.

Whereas the idea of work cells has been around since the 1920s, the use of group technology has added special significance and increased the use of work cells. Where volume warrants, a small group of machinery and equipment can be arranged to process not only several units of the same product but also a family of batches from several different products that have been identified through group technology as requiring the same series of operations.

A *work cell* is a collection of equipment required to make a single part or a family of parts with similar characteristics. This equipment is placed in a circle around an operator or operators. (Figure 4–5 shows a typical work cell layout.) The operator (most often a single operator) then takes a part from the in-basket and moves that part around the circle of equipment. Equipment usually includes automatic machines that only need to be loaded, activated, and then unloaded. Once the machine is loaded and activated, the operator moves the just-completed part from the first machine to

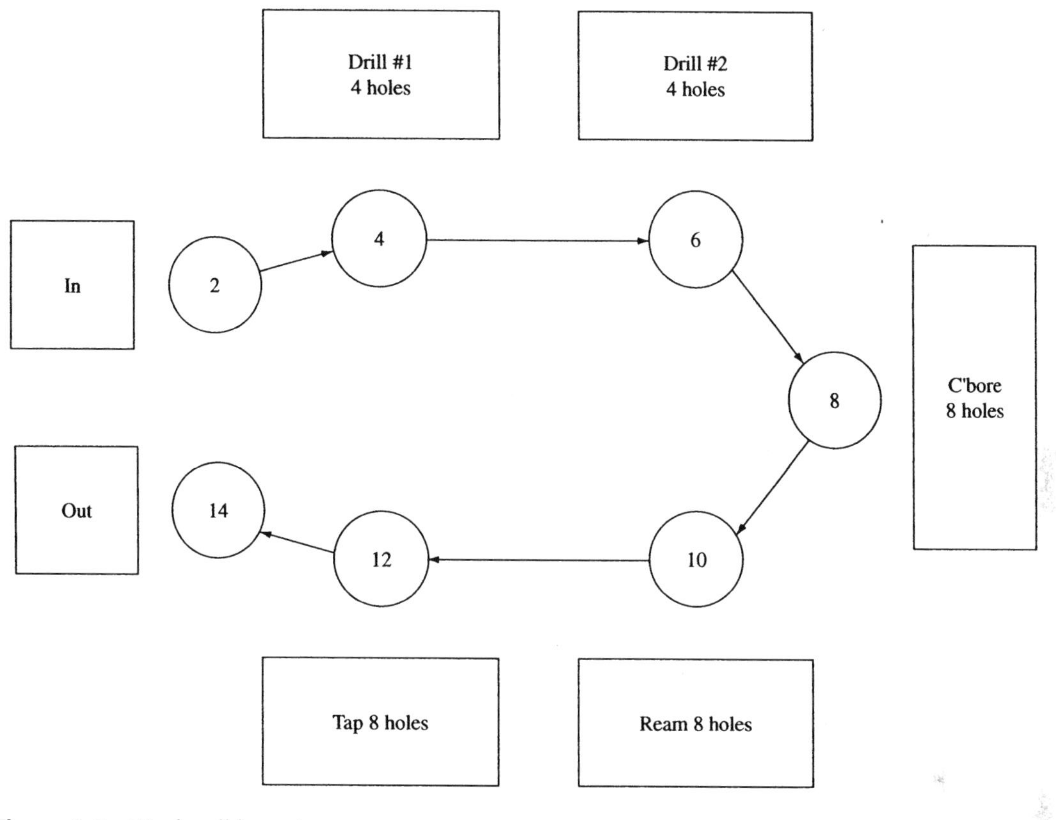

Figure 4–5 Work cell layout.

the second, where the operator removes the previous part and loads the next part. This proccss continues around the cell: taking parts out of one machine, putting new parts back into this machine, then activating that machine until arriving at the last machine, where the part is removed, inspected, and placed in the finished parts basket. Work cells are being developed at a very fast rate because they

1. Significantly reduce setup time
2. Eliminate all storage between operations
3. Eliminate most of the moving time between operations
4. Eliminate delays spent waiting for the next machine
5. Reduce costs
6. Reduce inventory (work-in-process reductions)
7. Reduce manufacturing in-process time

These are the goals of lean manufacturing and a good description of eliminating muda. The work cell concept considers operator utilization to be more important than machine utilization.

The work cell load chart is different from the previous techniques in that the end result does not have to be a complete part or product; it could be for only a few operations. You could end up with a complete part; however, that is not the goal of a work cell. For example, you may cast or forge a part elsewhere, machine the part in a cell, and chrome plate it in a third area. Once we determine what the cell is going to do, we need to micromotion-study all the operations involved. This was discussed in Chapter 3.

Work cell load charts are special operations charts used for multimachine situations (see Figures 4–6 and 4–7, for examples). The charts list the operator time, machine time, and walking time required to run a work cell to produce one part per cycle using many machines. Consequently, the work cell load charts show the total cycle time, operator utilization, and machine utilization. Because they are visual, work cell load charts help people to see problems and to make improvements on the operation by properly loading the operators or machines and so on.

Step-by-Step Procedure for Preparing a Work Cell Load Chart

The circled numbers in this discussion correspond to the categories shown in Figure 4–7.

① *Operation No.:* This is just a numerical sequence of steps. Good procedure uses numbers that leave room for expansion, like 2, 4, 6, 8 or 5, 10, 15, 20. This will allow the insertion of new operations between existing operations without having to renumber everything.

② *Operation Description:* This will include machine names and operation descriptions being performed. It should be descriptive enough to communicate to others what is being accomplished, so that they can follow the sequence of operations.

For 3 to 5: Time standards in decimal minutes. These times were developed by time study techniques discussed in Chapter 3.

③ *Manual:* The time it takes the operator to load, unload, inspect, and do anything else the operator is required to do. This time is totally under the control of the operator.

④ *Machine:* Once the operator activates the machine, the machine does its job automatically, and the operator goes on to the next operation. This machine time is usually calculated using feeds and speed formulas, discussed in Chapter 3. Machine times are usually out of the control of the operator.

⑤ *Walk:* Walk time is the time it takes an operator to move from one machine or operation to the next. The time standard for walking has already been determined to be .005 minute per foot, and it can be easily calculated from a workstation layout. For example, it would take .050 minute for the average person to move 10 feet. This statistic is based on the basic time standard of

Part Number 1675 **Part Name** crank **Date** 5/13/xx **Engineer** Justin M.

Operation Number	Operaton Description	Time in .001 Minute			Time in .025 Minute
		Manual	Machine	Walk	
2	Get new part	54	0	45	
4	Drill #1	49	455	25	
6	Drill #2	55	470	35	
8	Counterbore	75	289	35	
10	Ream	111	115	35	
12	Tap	175	300	25	
14	Inspect & aside	55	0	25	
	Total	.574	1.629	225	

Total cycle time = .8 minute + 10% allowance = .88 minute

Pieces per hour = 68

Hours per piece = .01467

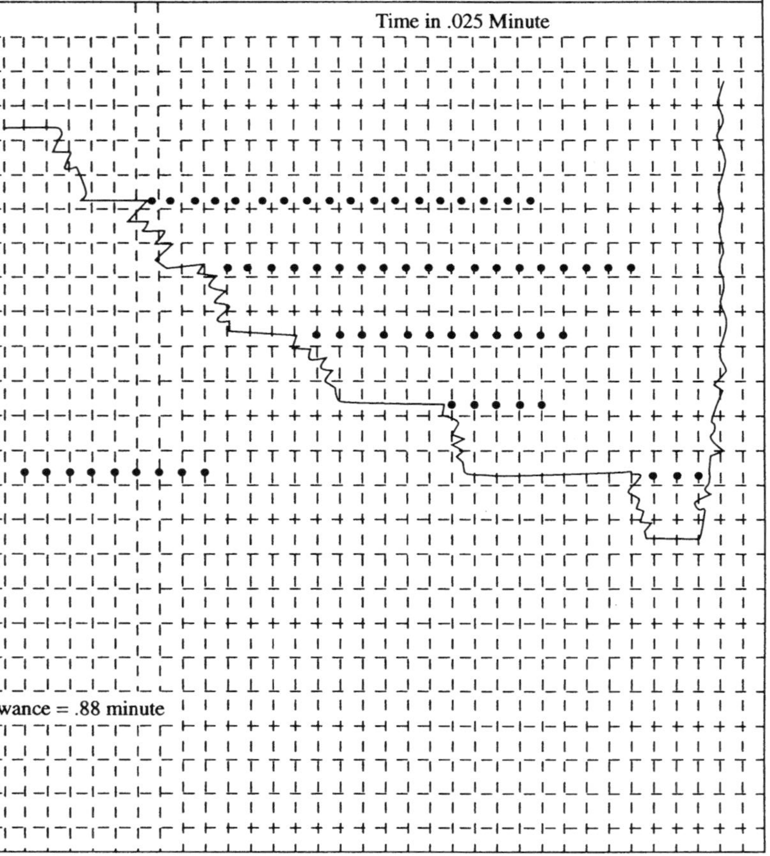

Figure 4–6 Work cell load chart.

Operation Number	Operation Description	Time in Minutes			Time in ___ Minutes
		Manual	Machine	Walk	
(1)	(2)	(3)	(4)	(5)	(6)

Figure 4–7 Work cell load chart: Step-by-step form.

walking 3 miles per hour, plus a 25 percent allowance for obstruction and turning.

⑥ *Operation accumulation time graph:* This is the visual "meat" of the form. The time data are plotted on the chart using three standard symbols:

______ : Solid line represents manual or operator time

_ _ _ _ : Dotted line represents machine or automatic time

/\/\/\/\ : Zigzag line represents walk time to next operation

The resulting graph visually shows the workload on a time scale and can be used to balance the operator and machines better. Also, the total cycle time results in a time standard. With analysis and a little imagination, an improvement can be attained.

ASSEMBLY AND PACKOUT PROCESS ANALYSIS

Once all parts are produced by the fabrication departments or received from the suppliers and available for assembly, new analytical tools are needed. Subassembly, welding, painting final assembly, and packout are all functions included in this area of the plant.

The Assembly Chart

The *assembly chart* (Figure 4–8) shows the sequence of operations in putting together the product. Using the exploded drawing (Figure 2–3a) and the parts list (Figure 2–4) shown in Chapter 2, the layout designer will diagram the assembly process. The sequence of assembly may have several alternatives. Time standards are required to decide which sequence is best. This process is known as *assembly line balancing.*

Time Standards for Every Task

The tasks should be as small as possible, so that the layout designer has the flexibility of giving that small task to several different assemblers (see Figure 4–9). The time standard setting techniques used for assembly line design are from either a predetermined time standard system or standard data. If you had 10 bolts to assemble, you would want a time standard (decimal minute) per bolt and a separate time for running the bolt down and tightening because this process gives the most flexibility.

Plant Rate and Conveyor Speed

Conveyor speed is dependent on the number and units needed per minute, the size of the unit, the space between units, and, sometimes, the hook spacing. *Conveyor belt speed* is recorded in feet per minute. Therefore, the size of the part plus the space

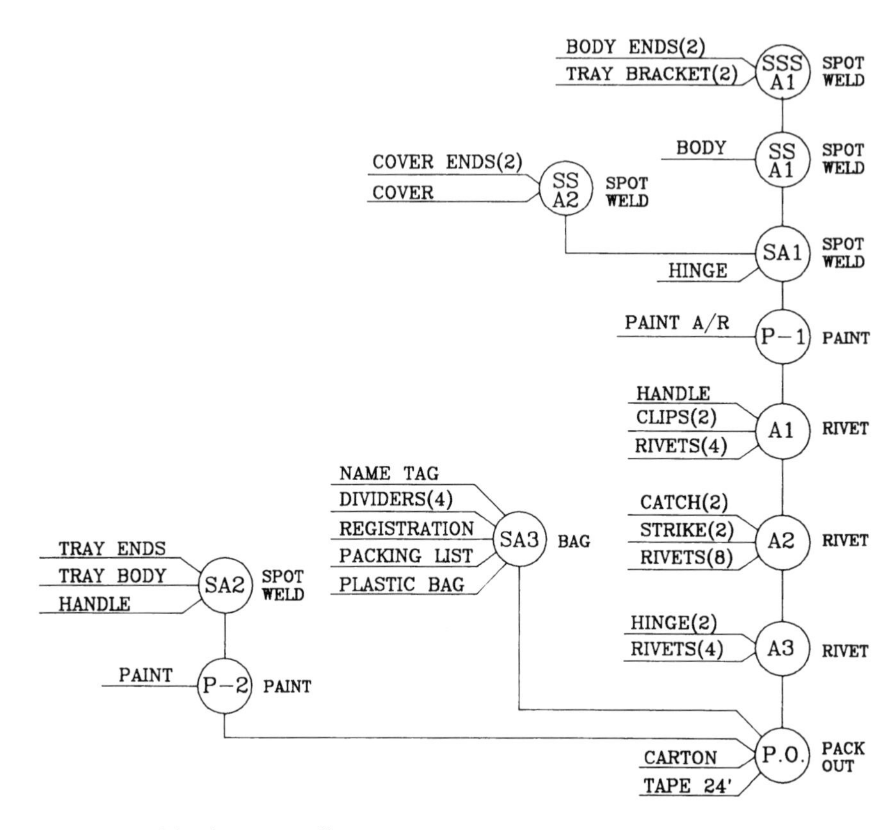

Figure 4–8 Assembly chart—Toolbox.

Operation No.	*Operation Description*	*Minute*
SSSA1	Assemble and spot/weld two tray brackets to two body ends	.153 each
SSA1	Weld two body ends to body	.291
SSA2	Weld two cover ends to cover body	.260
SA1	Weld hinge to body, weld cover to hinge, and hang an overhead conveyor	.356
P1	Clean, paint, and bake	automatic
A1	Rivet handle and two clips to cover	.310
A2	Rivet two catches to cover two strikes to body	.555
A3	Rivet two hinges to body and cover	.250
SA2	Weld tray ends to tray and add handle. Hang on overhead conveyor	.415
P2	Clean, paint, and bake tray	automatic
SA3	Bag loose parts	.250
P.O.	Pack tray into toolbox, place plastic bag into toolbox and close. Form carton and pack toolbox	.501

Figure 4–9 Assembly and packout time standards.

between parts (measured in feet) times the number of parts needed in one minute equals feet per minute.

Example: Charcoal grills are in cartons 30 × 30 × 24 inches high. A total of 2,400 grills are required every day.

	480	minutes per 8-hour shift
less	50	minutes schedulted downtime (breaks, etc.)
	430	available minutes
	80%	anticipated performance
	344	effective minutes of work per day
÷	2,400	grills per day
=	.143	minute per grill

$$\frac{1 \text{ minute}}{.143 \text{ minute per grill}} = 7 \text{ grills per minute}$$

7 grills per minute × 2-1/2 feet per grill (30 inches) = 17.5 feet per minute

The conveyor must run at 17.5 feet per minute, or the plant will not produce its 2,400 grills. Check the production rate:

	7	grills per minute	×	430 minutes per shift
×	80%	performance	=	2,408 grills per shift

Paint Conveyor Speed

The overhead paint conveyor speed is complicated additionally by multiple parts per hook and hook spacing.

Example: You are going to paint the following parts on one overhead conveyor system:

Part No.	*Parts per Hook*	*Quantity to Paint (per shift)*	*Needed Hooks per Day*	*Hooks per Minute **
15	1	500	500	1.45
263	4	300	75	.22
44	2	1,000	500	1.45
14	8	2,000	250	.73
21	2	100	50	.15
03	1	125	125	.36
				4.36

*Based on 430 minutes @ 80 percent = 344.

A total of 4.36 hooks need to pass any one point in one minute if the hooks are one foot apart. So, 4.36 feet per minute would be the conveyor speed. If the hooks were 1-1/2 feet apart, the conveyor speed would be 6.54 feet per minute (4.36 × 1.5) Figure 4–10 shows some different kinds of paint hooks.

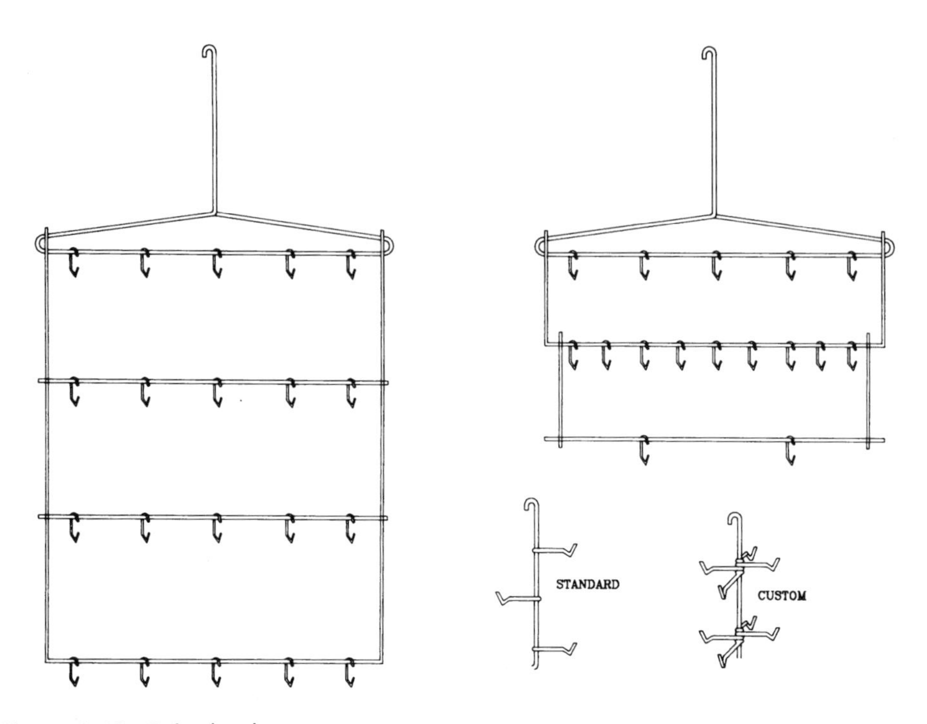

Figure 4–10 Paint hooks.

This conveyor speed also determines the size of the drying oven and baking oven. Let us say that your parts need 10 minutes at 400° to dry. Ten minutes at 6.54 feet per minute equals 65.4 feet of conveyor in an oven. Most drying ovens take the parts into and out of the oven from the same end, so the oven will be about 33 feet long and 4-1/2 feet wide.

Looking at the toolbox example, you will use hook spacings of 18 inches because of the part size. Which parts need painting? The box and tray assembly only. Each is placed on a hook by itself and sent through an electrostatic spray system. But first, it must be cleaned and dried. After painting, it will be baked and cooled. The drying needs 450° for 10 minutes. Fifteen minutes is required for cooldown before an assembler can take it off the line.

	Parts per Hook	*Quantity per Shift*	*Needed Hooks per Day*	*Hooks per Minute**
1. Box assembly	1	2,000	2,000	5.78
2. Tray assembly	1	2,000	2,000	5.78
				11.56

*Based on a takt rate or *R* value of .173 or 5.78 per minute.

A total of 11.56 hooks are needed every minute and each hook is 18 inches apart (1-1/2 feet), so 11.56 times 1.5 equals 17.34 feet per minute conveyor speed.

You need 10 minutes of drying and baking time, so 17.34 feet per minute times 10 minutes equals 173 feet of drying and baking. Use a double-decker dryer 90 feet long with drying on the top (hottest) and baking on the bottom. Cooling time of 15 minutes times 17.34 feet per minute means 260 feet from baking to the first assembler.

Like magic, the data for your layout is coming together. In reality, it isn't magic at all but a systematic approach that provides the answers.

When the toolboxes are taken off the overhead conveyor line after painting to rivet on the handle, the finished toolbox can be placed on a flat belt conveyor. That conveyor speed needs to be 5.78 toolboxes per minute times the spacing. Because this toolbox is 18 inches long, 24 inches between centers should be adequate spacing. Two feet per box times 5.78 boxes per minute equals 11.56 feet per minute. The tray will stay on the hook until it reaches the packout station. Trays and toolboxes will be on every other hook.

Although there is enough information to lay out the clean, paint, and bake area now, we will wait until Chapter 7 and calculate all production space requirements then.

Assembly Line Balancing

The purposes of the assembly line balancing technique are

1. To equalize the workload among the assemblers
2. To identify the bottleneck operation
3. To establish the speed of the assembly line
4. To determine the number of workstations
5. To determine the labor cost of assembly and packout
6. To establish the percentage workload of each operator
7. To assist in plant layout
8. To reduce production cost

The assembly line balancing technique builds on the assembly chart (Figure 4–8) time standards (Figure 4–9), and the plant rate (takt rate or *R* value) is calculated in the last section.

The objective of assembly line balancing is to give each operator as close to the same amount of work as possible. This can be accomplished only by breaking down the tasks into the basic motions required to do each piece of work and reassembling the tasks into jobs of near equal time value. The workstation (or stations) with the largest time requirement is designated as the *100 percent station* and limits the output of the assembly line. If industrial engineers want to improve the assembly line (reduce costs), they concentrate on the 100 percent station. Reducing the 100 percent station in the example below by 1 percent would save the equivalent of .25 people, a multiplying factor of 25 to 1.

Figure 4–11 shows an example of an assembly balance problem for the toolbox production.

SA3 could be taken off the assembly line and handled separately from the main line to save money. Given the standard time of .250 minutes for SA3, one SA3 could be assembled in .0047 hours, or 240 SA3s per hour if the operation were performed

	Time Standard	*Number of Stations*	*Rounded Up*	*Average Time*	*Percent Loaded*	*Hours per Unit*	*Pieces per Hour*
SSSA1	.306	1.77	2	.153	92	.00557	180
SSA1	.291	1.68	2	.146	87	.00557	180
SSA2	.260	1.50	2	.130	78	.00557	180
SA1	.356	2.06	3	.119	71	.00835	120
A1	.310	1.79	2	.155	93	.00557	180
A2	.555	3.20	4	.139	83	.01113	90
A3	.250	1.44	2	.125	75	.00557	180
SA2	.415	2.40	3	.138	83	.00835	120
SA3	.250	1.44	2	.125	75	.00557	180
P.O.	.501	2.90	3	.167	100	.00835	120
Total	3.494		25			.06960	

Figure 4–11 Initial assembly line balance.

in a batch mode. However, in an assembly line mode, operations are slowed to the rate of the slowest activity; hence, the rate of production for SA3s is dropped to only 180 pieces per hour, or .00557 hours per unit.

	.0057	balanced cost
−	.00417	individual cost
	.00140	hour per unit savings
×	500,000	units per year
	700	hours per year
@	$15.00	per hour
=	$10,500.00	per year savings

This is called the *cost of balancing*. In this case it is too high.

Subassemblies that can be taken off the line must be

1. Poorly loaded. The smaller the percentage that is loaded, the more desirable it is to subassemble. For example, a 60 percent load on the assembly line balance would indicate 40 percent lost time. If you take this job off the assembly line (not tied to the other operators), you could save 40 percent of the cost.
2. Small parts that are easily stacked and stored.
3. Easily moved. The cost of transportation and inventory will go up, but because of better labor utilization, total cost must go down.

A subassembly on the assembly line balance form would look like this:

Question No.	*Time Standard*	*No. of Stations*	*Rounded Up*	*Average Time*	*Percent Loaded*	*Hours per Unit*	*Units per Hour*
SA3	.250	1.44	1.44	2.50	Sub	.00417	240

Look back at Figure 4–11 and SA3. You have saved plenty, but can you do this to SA1? No, because it is a large part that is not easily stacked, stored, or moved.

The assembly line balance in Figure 4–11 is not a good balance because of the low percentage loads. An improvement is possible (look at the 100 percent station). If you add a fourth packer, you will eliminate the 100 percent station at packout (P.O.). Now the new 100 percent (bottleneck station) is A1 (93 percent). By adding this person, you will save 7 percent of 25 people or 1.75 people and increase the percent load of everyone on the assembly line (except P.O.). You might now combine A1 and A2, and further reduce the 100 percent. The best answer to an assembly line balance problem is the lowest total number of hours per unit. If you add an additional person, that person's time is in the total hours. Try to improve the toolbox assembly line balance, then see how that affects the assembly line in Figures 4–12 and 4–13.

Notes on the assembly line balance (Figure 4–11):

1. The busiest workstation is P.O. It has .167 minute of work to do per packer. The next closest station is A1 with .155 minute of work. As soon as you identify the busiest workstation, you identify it as the 100 percent station, and communicate

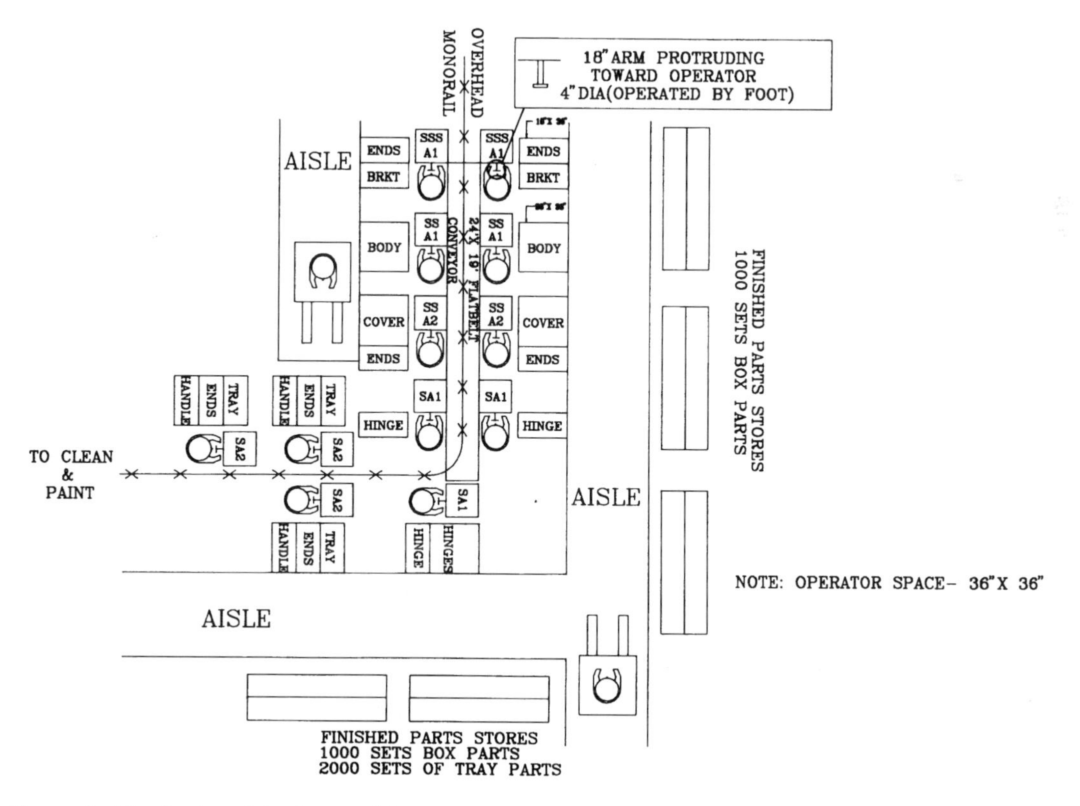

Figure 4–12 Spot weld subassembly.

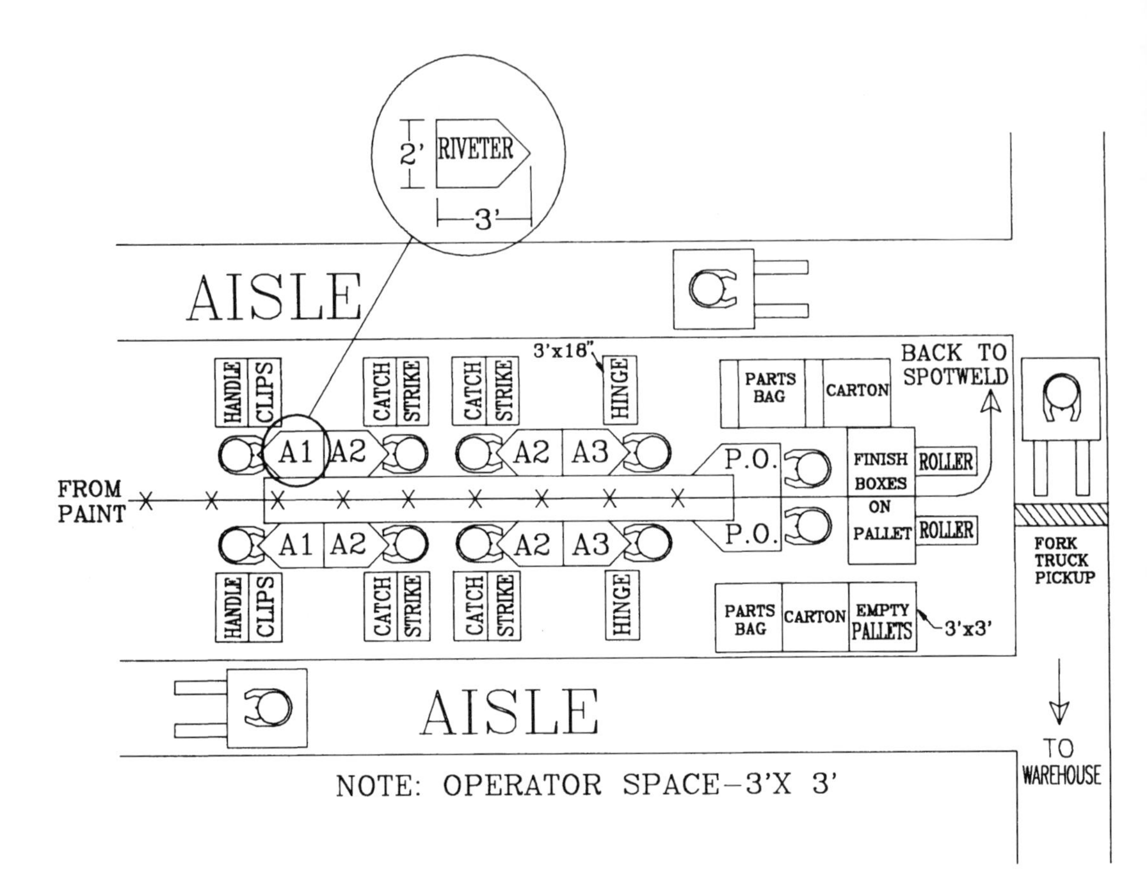

Figure 4–13 Assembly and P.O. line.

that this time standard is the only time standard used on this line from now on. Every other workstation is limited to 360 pieces per hour. Even though other workstations could work faster, the 100 percent station limits the output of the whole assembly line.

2. The total hours required to assemble one finished toolbox is .06960 hour. The average hourly wage rate times .06960 hour per unit gives you the assembly and packout labor cost. Again, the lower this cost is the better the line balance is.

Line balancing is an important tool for many aspects of industrial engineering, and one of the most important used is the assembly line layout. The back of the assembly line balancing form is designed for an assembly line layout sketch. Look at the examples in Figures 4–12 and 4–13.

Packout work is considered the same as assembly work as far as assembly line balancing is concerned. Many other jobs may be performed on or near the assembly line, but they are considered subassemblies and are not directly balanced to the line because subassemblies can be stockpiled. Their time standards stand on their own merit.

STEP-BY-STEP PROCEDURE FOR COMPLETING THE ASSEMBLY LINE BALANCING FORM

The assembly line balancing form shown in Figure 4–14 includes the following categories:

① *Product No.:* The product drawing or product part number.

② *Date:* The complete date of development of this solution.

③ *By I.E.:* The name of the engineer doing the assembly line balance.

④ *Product Description:* The name of the product being assembled.

⑤ *Number of Units Required per Shift:* The quantity of production required per shift—given to the engineer by the sales department. The engineer's objective is to get as close to this quantity as possible without going below.

⑥ *Takt time:* The plant rate or takt rate has been discussed previously in this chapter, but this block is designed for a specific plant with the following past experience.

a. Existing products have run at 85 percent efficiency.

b. New products average 70 percent efficiency during the first year.

c. Eleven percent allowances are added to each standard. The *R* value in this plant is calculated by dividing 300 or 365 minutes by the number of units per shift (Step ⑤). The result is the plant or takt rate—the *R* value.

⑦ *No.:* This is a sequential operation number. Operation numbers provide a simple, useful method of referring to a specific job.

⑧ *Operation/Description:* A few well-chosen words can communicate what is being done at this workstation. Parts' names and job functions are the key words.

⑨ *Takt time:* The takt time calculated in block ⑥ above goes behind each operation. The plant or takt rate is the goal of each workstation, and putting the takt time on each line keeps that goal clearly in focus.

⑩ *Cycle Time:* The cycle time is the time standard set by combining elements of work together into jobs. The goal is the takt time, but that specific number can seldom be achieved. Cycle time can be changed by moving an element of work from one job to another; however, elements of work are a large proportion of most jobs. Faster equipment or smarter methods may reduce the cycle time, and this is a good cost reduction tool to be discussed later.

⑪ *No. Stations:* The number of stations is calculated by dividing the takt time ⑨ into the cycle time ⑩ and rounding up. If the number of stations is rounded down, the goal (number of units per shift ⑤) will not be achieved. Management may round down the number of workstations because of cost, but if it does, it knows the goal will not be achieved without overtime, and so on. But that is management's decision, not the engineer's. If the number of workstations is rounded down, the workstation will be the bottleneck, the restriction, the slowest station, or the 100 percent station.

Fred Meyers & Associates **Assembly Line Balancing**

Product No. ①	Product Description: ④	takt Time ⑥ Existing Product = 365 Minutes / Units Req'd/Shift = R
Date: ②		
By I.E.: ③	Number Units Required per Shift ⑤	Calculations New Product = 300 Minutes / Units Req'd/Shift = R

No.	Operation Description	takt Time	Cycle Time	No. Stations	Avg. Cycle Time	% Load	Hrs./1000 Line Balance	Pcs.Hr Line Balance
⑦	⑧	⑨	⑩	⑪	⑫	⑬	⑭	⑮

Figure 4–14 Assembly line balancing step-by-step form.

⑫ *Avg. Cycle Time:* The average cycle time is calculated by dividing the cycle time ⑩ by the number of workstations ⑪. It is the speed at which this workstation produces parts. If the cycle time of a job is one minute, and four machines are required, the average cycle time is .250 minute (1.000 divided by 4 equals .250) or a part would come out of those four machines every .250 minute. The very best line balance would be for every station to have the same average cycle time, but this never happens. A more realistic goal is to work at getting them as close as possible. The average cycle time will be used to determine the percentage workload of each workstation, the next step.

⑬ *% Load:* The percentage load tells how busy each workstation is compared to the busiest workstation. The highest number in the average cycle time column ⑫ is the busiest workstation and, therefore, is called the 100 percent station (100 percent is written in the percent load column). Now every other station is compared to this 100 percent station by dividing the 100 percent average station time into every other average station time and multiplying the result times by 100. The result will equal the percent load of each station. The percent load is an indication of where more work is needed or where cost reduction efforts will be most fruitful. If the 100 percent station can be reduced by 1 percent, then you will save 1 percent for every workstation on the line.

Example: To calculate percent load, look back at the example in Figure 4–11 of this chapter. The average cycle times were .153, .146, .130, .119, .155, .139, .125, .138, .125, and .167. Reviewing these average cycle times reveals that .167 is the largest number and is designated the 100 percent workstation. A good practice is to circle the .167 and the 100 percent to remind you that this is the most important workstation on the line, and no other time standard has any further meaning. Now that the 100 percent station is determined, the percent load of every other workstation is determined by dividing .167 into every other average cycle time:

Operation SSSA1 = .153 divided by .167 = 92 percent
SSA1 = .146 divided by .167 = 87 percent
SSA2 = .130 divided by .167 = 78 percent
and so on

Where will the supervisor put the fastest person? The P.O. operation! Where will the industrial engineer look for improvement or cost reduction? The P.O. operation, the 100 percent loaded station.

Electronic copies of the routing sheet, line balancing, and fabrication equipment spreadsheet are provided for your use. These forms may be downloaded, copied, and utilized for your project.

A good balance would have all workstations in the 90 to 100 percent range. One workstation below 90 percent can be used for absenteeism. A new person can be put on this station without slowing down the whole line.

⑭ *Hrs./1,000 Line Balance:* The hours per unit produced can most easily be calculated by dividing the 100 percent average cycle time (which is circled on the line balance) by 60 minutes per hour:

$$\frac{.167 \text{ minute per unit}}{60 \text{ minutes per hour}} = .00278 \text{ hour per unit}$$

The .167 time standard is for one person, so that if two people are required on an operation, two times .00278 hour per unit will be required.

Two people = .00557 hour per unit
Three people = .00835 hour per unit
Four people = .01113 hour per unit

Also, as a quality check, if you multiply the number of operators on the assembly line times .00278 hour per unit, you get the total hours to make one unit. In the toolbox problem .06958 (25 × .0027833 = .06958) hour of labor will be required to assemble each toolbox. Another piece of logic is that everyone on an assembly line must work at the same rate. In other words, the person with the least amount of work to do cannot do more than the operator ahead of him.

⑮ *Pcs./Hr. Line Balance:* Pieces per hour is $1/x$ of hours per unit, or divide the hours per unit into one. Notice in Figure 4–11 that all the stations produce 360 pieces. Station A1 has two operators, each producing 180 pieces per hour for 360 pieces per hour total.

⑯ *Total Hours per Unit:* To get total hours per unit, add the number of hours from all the operations. The hours per unit for one operator times the total number of operators on the line also equals the total hours per unit. The total of column 11 is the total operators.

⑰ *Average Hourly Wage Rate:* This would come from the payroll department, but, for example, let us say $15.00 per hour is the average hourly wage rate.

⑱ *Labor Cost per Unit:* In the example, .06960 hour times $15.00 per hour = $1.044 per unit labor cost. The lower the cost is, the better the line balance.

⑲ *Total Cycle Time:* The total cycle time tells the exact work content of the whole assembly, and if treated like any other time standard, it can show what a perfect line balance would be.

In the example, 3.494 minutes divided by 60 minutes per hour equals .05823 hour per unit. The line balance came out to .06960, or .01137 hour more. This .01137 hour is potential cost reduction, and what cannot be removed by cost reduction is called the cost of line balance.

CALCULATING THE EFFICIENCY OF THE ASSEMBLY LINE

The efficiency of an assembly line can be determined in a number of ways. The direct labor cost per unit can be easily calculated from the last two columns of the line balancing chart. The last column in Figure 4–14 shows the total number of pieces or units that are produced per *person-hour* on the line. The inverse of this figure is the required person-hours per piece. Therefore, multiplying this number by the hourly wages yields the direct labor cost per unit. As you attempt to balance the line further, any increase or decrease in this cost is a clear indication of how successful you are in improving the line.

You can also calculate the numerical or percentage efficiency for the line using either the hours per 1,000 columns or the cycle time column. Figure 4–14 shows calculations for hours per 1,000 and hours per 1,000 line. The hours per 1,000 column uses the cycle time as if this were a stand-alone operation. That is, the operator is not hindered by the flow of the line. The hours per 1,000 line column uses the adjusted cycle time as determined by the 100 percent station. The sum of the first column tells us how many units could be assembled if each step could be performed at its standard cycle time. The second column shows what the actual production is because not all stations are performing at their maximum capacity. Keep in mind that an assembly line or any sequential series of operations can operate only as fast as the *slowest* member of the team.

$$\text{Line efficiency} = \frac{\text{sum of hours per 1,000}}{\text{sum of the hours per 1,000 line balance}} \times 100$$

Similarly, line efficiency can be calculated as follows. The average cycle time for the 100 percent station is the fastest speed at which any operator on this line can work. This time multiplied by the total number of operators on the line is the total cycle time per unit. Why? This figure divided by the total cycle time ⑲ in Figure 4–14 will show the efficiency of the line:

$$\text{Line efficiency} = \frac{\text{sum of 1 (19) cycle time}}{\text{total cycle time}} \times 100$$

Both methods of calculation will yield the same answer. Once again, any change in the line is reflected in the line efficiency.

Management often equates the addition of workstations or operators to an increase in the labor cost per unit. The above calculations, especially those dealing with the cost per unit, clearly demonstrate that adding workers to the line by virtue of increasing the line efficiency often decreases the cost per unit.

Use of Computer Simulation

Computer simulation and modeling are powerful tools in designing work cells and aiding with balancing lines and work cell loads. Various software packages allow the user to design an entire facility or just a portion of a facility, such as a work cell. One can then simulate the working of the cell under various conditions and examine various scenarios. Figures 4–15 through 4–18 show a manufacturing work cell that was designed by using the computer simulation package ProModel. The use of simulation and simulation software will be discussed in greater detail in Chapter 15. The cell was set up by using a library of icons available in the software. The decision regarding equipment selection and the arrangement, as well as the number of resources—that is, the number of equipment and operators—is made by the operator. This cell was designed for the production of cogs. System parameters such as machine time, cycle time, walking distances, and so on are obtained in the same manner as the facilities planner would obtain them under any conditions, and they are entered into the simulation run. Once the data are entered, the designer has the opportunity to play various scenarios in order to optimize the cell.

As shown in Figure 4–15, this cell consists of a receiving area, two NC lathes, a degreasing station, and an inspection area. The purpose in this example is to determine the number of workers that would minimize the cost per unit of production in the cell.

Three simulation runs were performed. The first run utilized only one operator as shown in Figure 4–16. As illustrated in Figures 4–17 and 4–18, the second

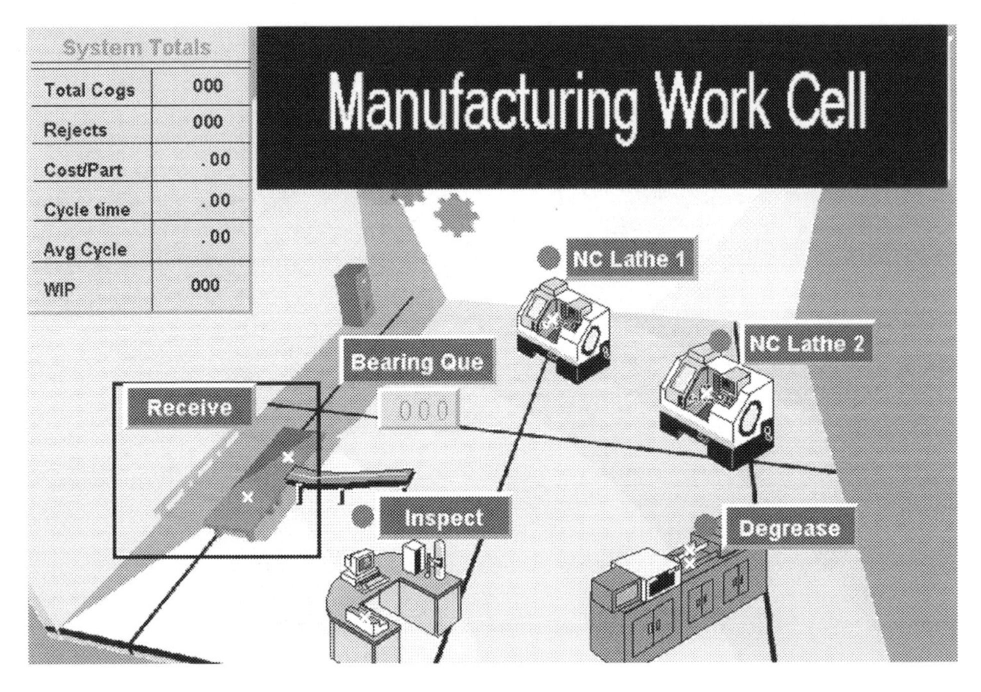

Figure 4–15 A simulated manufacturing cell (courtesy of ProModel Corporation).

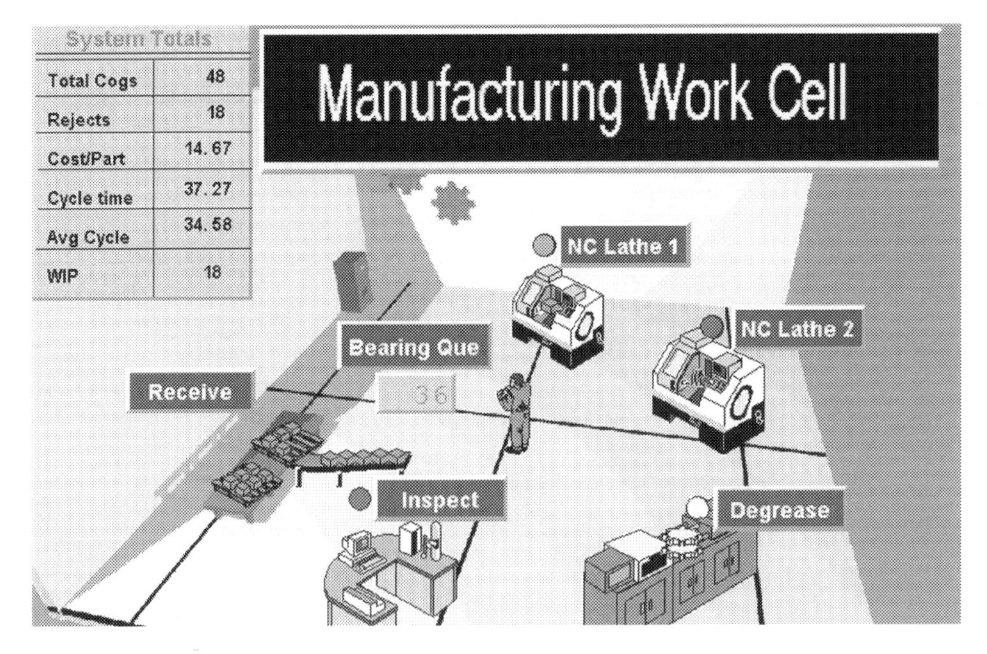

Figure 4–16 The simulated status of the manufacturing cell utilizing one operator after a fixed period of time (courtesy of ProModel Corporation).

and third runs employed two and three operators, respectively. Each simulation was run for a total of 10 hours of actual production time. Simulation runs produce detail statistics pertaining to production volume, machine usage, equipment and operator idle time, and an array of other useful information. Some of these statistics and reports will be presented in Chapter 15. Summary data are given in Figures 4–16 through 4–18 regarding each run. Summary statistics in Figure 4–16 show that when using only one operator in the cell, during the 10-hour production run, 48 cogs were produced. The average cycle time was approximately 34.58 minutes per part and the cost per unit is estimated at $14.67. When the number of operators is increased to two, during the same elapsed time 82 cogs are machined with an average cycle time of 25.32 minutes per unit. The production cost is reduced to $11.56 per cog (see Figure 4–17). Increasing the number of workers to three does not seem to be beneficial. As shown in Figure 4–18, although the average cycle time is not significantly changed, the cost per unit is increased by approximately $.28 per unit, presumably due to an increase in direct labor cost. Furthermore, the total production has remained the same, indicating that the machine capacity has been reached. This rather simple simulation allows the process engineer to determine that the optimum number of operators for this cell is two. Of course, other scenarios that manipulate the number and the type of equipment can also be carried out.

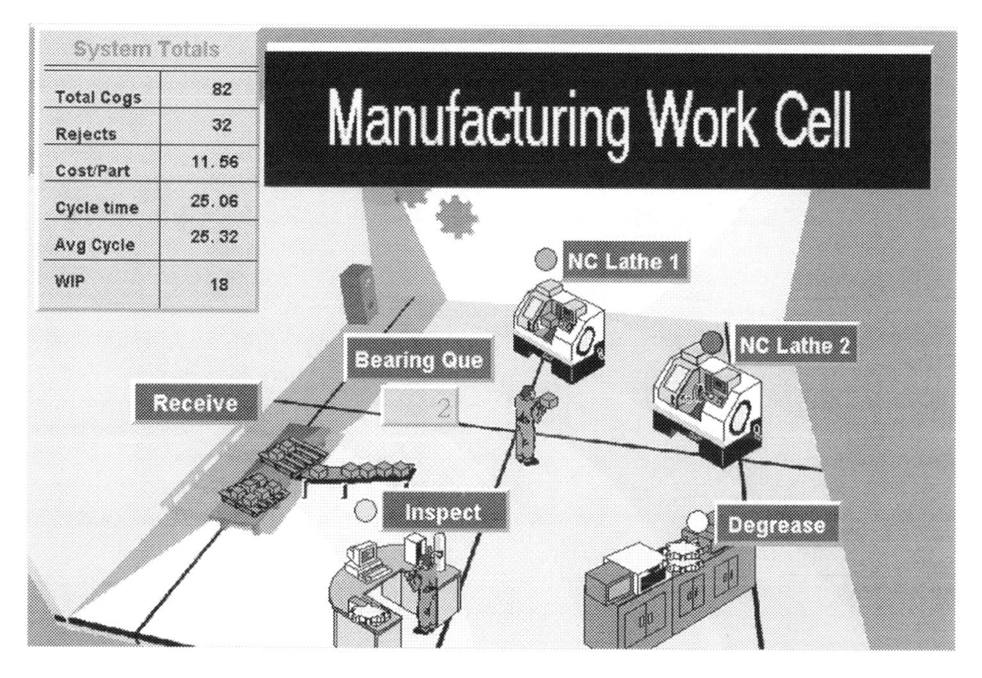

Figure 4–17 The simulated status of the cell utilizing two operators after the same period of elapsed time (courtesy of ProModel Corporation).

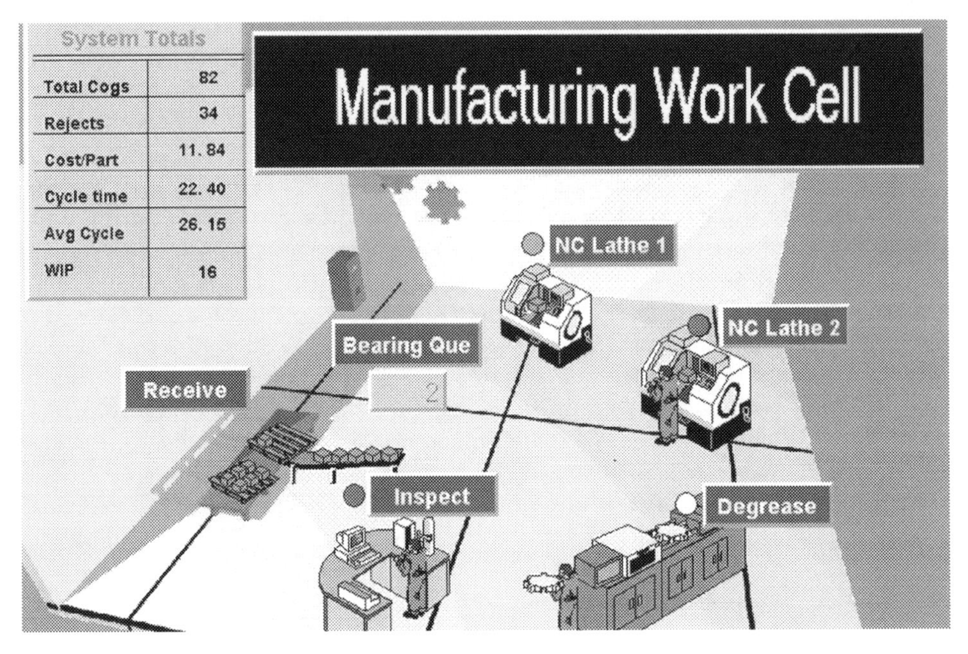

Figure 4–18 The simulated status of the cell utilizing three operators after the same period of elapsed time (courtesy of Promodel Corporation).

LAYOUT ORIENTATION

Mass production and job shop are the two basic layout orientations. *Mass production* is product oriented and follows a fixed path through the plant. The assembly line best illustrates the mass production orientation. Mass production orientation is preferred over job shop because the unit cost is lower, but not all products follow a fixed path.

The *job shop* orientation layout is process oriented (built around machine centers). Fabrication departments are usually laid out this way because the paths taken by the parts are not consistent. This is called a *variable path flow.*

Because mass production is preferred over job shop, several new techniques have been developed to move job shop orientation closer to mass production:

1. Group technology tries to classify parts into groups with similar process sequences. The equipment then can be placed in a straight line or a work cell approaching a fixed path. The plant could have a sheet metal line, plastics line, bar stock line, casting line, and so on. Reducing cross traffic, backtracking, and feet of travel are the objective.

2. A work cell is a group of machines dedicated to making one complicated part. One or two operators may run 6 to 10 machines. The machines remain set up for this part indefinitely. Some machines may not be fully utilized, but the lost

time is counterbalanced by less inventory required, less material handling, and much shortened throughput time (the time a part spends in production).

Most plants use job shop orientation for the fabrication end of the plant and mass production orientation for assembly lines and packout. As opportunities arise, group technology and work cells are created.

In the next chapter, flow analysis techniques will be discussed to optimize layouts of fabrication and assembly areas.

QUESTIONS

1. What is process design?
2. What are the two categories of process design?
3. What is a route sheet?
4. What information is included on a route sheet?
5. What determines how many machines to buy?
6. Which time standard (decimal minute, pieces per hour, or hours per unit) compares to the takt time?
7. What is an assembly chart (see Figure 4–8)?
8. What information is needed to calculate an assembly line conveyor speed?
9. What additional information is needed to calculate paint conveyor speed?
10. What are the eight purposes of assembly line balancing?
11. Rebalance Figure 4–11 by adding a fourth packout person and make a subassembly out of SA3. Then answer the following:
    a. What is the total hours per unit? ____________
    b. How many units per shift will be made at 100 percent? ____________
    c. How many people are now used? ____________
    d. What is the new 100 percent station? ____________
    e. Is this a better balance? ____________
    f. How much money is saved if you produce 700,000 units per year and the employees are paid $10.00 per hour? ____________
12. What are the two primary layout orientations?
13. Balance the following assembly line to produce 1,500 units per 8-hour shift at 85 percent and 30 minutes personal time:

Operation No.	*Time Standard*
1	.390
2	.235
3	.700
4	1.000
5	.240
6	.490

14. Calculate the efficiency of the line in question 13.
15. Explain how adding personnel to a line can reduce cost per unit.
16. Define and contrast process-oriented and product-oriented layouts.
17. Define *group technology* and explain how it is implemented.
18. Explain the concept of work cell.
19. Given a takt time or R value of .452 minute, balance the following line. What is the efficiency of the line?

Operation No.	*Time Standard*
1	.455
2	.813
3	.233
4	.081
5	.945

CHAPTER

Flow Analysis Techniques

OBJECTIVES:

Upon the completion of this chapter, the reader should:

- Understand the importance of material flow and flow analysis
- Be able to identify and construct various flow analysis tools
- Be able to calculate flow efficiency by the use of a from-to chart

Flow analysis is the heart of plant layout and the beginning of the material handling plan. The flow of a part is the path that the part takes while moving through the plant. Flow analysis not only considers the path that every part takes through the plant but it also tries to minimize the (1) distance traveled (measured in feet), (2) backtracking, (3) cross traffic, and (4) cost of production.

Flow analysis will assist the manufacturing facilities designer in the selection of the most effective arrangement of machines, facilities, workstations, and departments. It is said that if you improve the product flow, you will automatically increase profitability. You can improve flow by developing product or part classes or families (parts with similar process steps) and implementing the concept of group technology. You can try to get each part to take a similar path and to move the parts automatically. The flow of parts and, therefore, the layouts of the plant will differ greatly with the two basic types of layout orientation—the process-oriented facility and the product-oriented facility. Product-oriented layouts will have less of everything (part of the definition of lean manufacturing) than the process-oriented layouts, but a large number of different parts or products with varying process steps may dictate a process-oriented layout. The flow analysis tools and techniques that are appropriate for a process-oriented layout are discussed in the next section.

It is equally important to consider and analyze the traffic patterns and paths that employees take throughout the facility in the course of the day. For example, most employees drive to work, park their cars, enter through the employee entrance, punch their time cards, go to their lockers, go to the cafeteria, and then go to their workstation. Use this flow to place these service facilities conveniently for the employees. In the course of the day, employees may also have to leave their workstation to walk to the tool room, to use the bathroom, or to take a drink of water from the water fountain. When designing the flow pattern, keeping in mind that the employees' walking time is a nonproductive time may help to focus on the importance of flow analysis.

A core principle of lean manufacturing is the product-oriented flow layout. It establishes the basis for high factory performance and has many advantages, as compared to process-oriented flow layouts that are planned around a group of similar equipment. Let us look at the advantages of product-oriented flow and think about cells and assembly lines when reviewing this list:

1. There is simplified coordination and production scheduling—first-in-first-out and no setting aside (no work in process).
2. Users and makers of parts can see and talk to each other, thereby seeing and solving problems more quickly.
3. There is less work-in-process inventory. A golf club manufacturer had 6 months of in-process inventory when it had a process-oriented layout. The manufacturer changed to a product-oriented layout and reduced in-process inventory to less than 2 days.
4. This layout eliminates excessive handling of materials. Work cells combine several machines in a U-shaped layout around an operator moving one part around the circle until it is complete. The old way would have moved tubs of parts from machine to machine.
5. Quality problems are easier to identify and quicker to fix because there are so few parts in the system.
6. Flow of material and operators' work can be standardized (written up as a standard practice) way to do the job, and it is used as an instructional plan for new employees.
7. Less floor space is required for all the above reasons.
8. This layout provides the foundation for continuous improvement, which is another basic requirement of lean manufacturing.

The following list summarizes the differences between the material flow in a process-oriented layout and a product-oriented layout:

1. Material flow will be much smoother in product-oriented flow plans, but material still travels according to the sequence of operations specified on the route sheet.
2. The distance material has to travel through the production process will be much shorter.
3. There is less confusion about which process sequence to use, or when and where finished material should be transferred.
4. One concern with the product-oriented layout is the restriction of the machine capacity. Instead of using one fast, flexible, expensive machine that can produce many different parts, cheaper machines that can be used solely for one part are acquired.

Once a sophisticated, expensive machine is purchased, it may become an obstacle in converting to different products and a product-oriented flow layout.

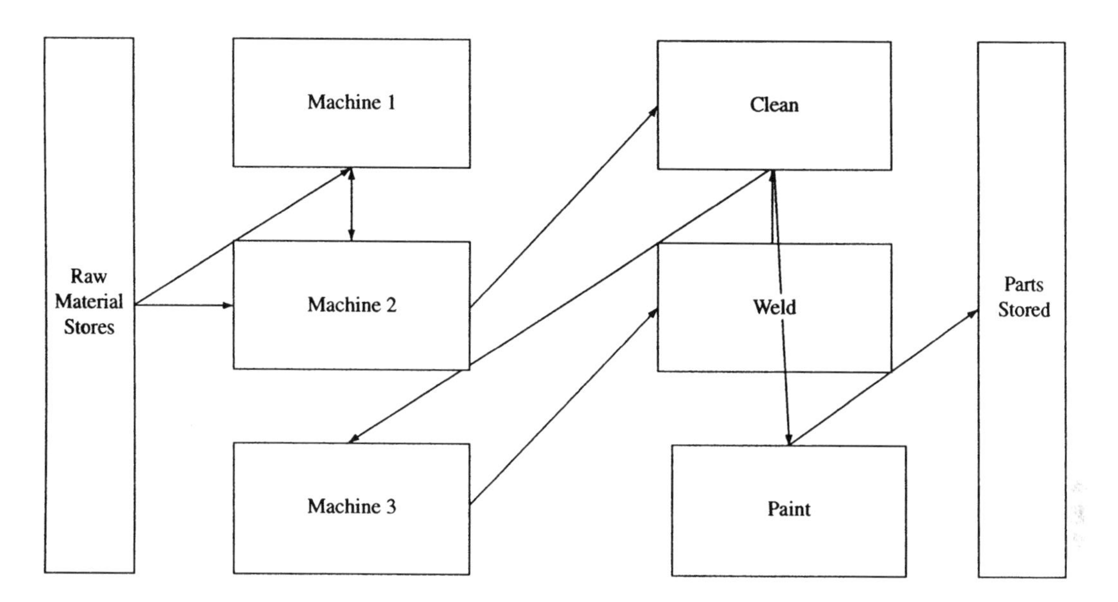

Figure 5–1a Process flow layout.

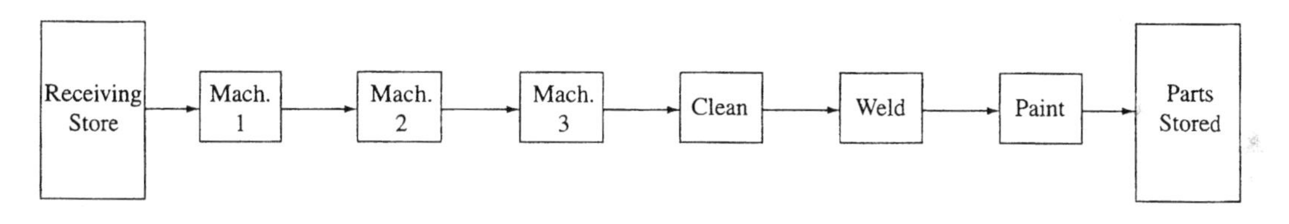

Figure 5–1b Product flow layout.

Sophisticated equipment is often a compromise of features. The machine may perform many functions, but none may perform as well as a single, special purpose machine. Product flow with multiple steps yields the opportunity to return to simpler, less costly equipment that will perform specialized tasks. Tools such as production balancing, kanban, work cell development, employee involvement, and quality are greatly enhanced by working with a product-oriented flow layout.

Existing plants will realize improvements by changing to product flow facilities design. The first step is to identify the existing flow path on a layout (called a *flow diagram*) for the products made by your company (Figures 5–1a and 5–1b, 5–13, and 5–14). In some plants, this may not be easy. Different jobs require different machines, so jobs take various paths around the plant. There are always some good product flow paths, and they must be carefully identified. In many plants, presses and press operators are in one location, whereas shears and their operators are in another location. Figures 5–1a, 5–1b, and 5–14 show how the flow in such a layout would look.

Where products are more standardized (similar), it is much easier to identify the flow paths. Many plants may have started out with an efficient product flow but have lost their efficiency due to expansion and growth. Machines and people are located and organized without consideration of the product flow. In some cases, it may be important to improve the operations without moving the people and equipment. For example, the equipment may be in the wrong location but would be too costly to move. Small plants may also tolerate their present organization because the operation is fairly compact, transport distances are short, and operators can see and talk to one another. This layout assumes that great efficiencies can be gained by having all the press operations grouped together under one supervisor, all the machining operations grouped together under another supervisor, and so on. However, this is not good for product flow. It is hoped that workers will become skilled at performing just one operation and perform it with high efficiency. However, losses in poor flow, increased material handling, and increased inventory requirements make process layouts very expensive. Scheduling and movement of material through the process operations will frequently become complex. Often it is found that a given part travels several times the length of the factory in going through all the required operations. A hand tool manufacturing company made many different families of tools such as sockets, wrenches, screw drivers, and so on. The plant was laid out in process order where all tools traveled through most of the departments. The average 3/8-inch drive socket traveled over 6,000 feet in the process of being manufactured (22 operations). A product layout was proposed and implemented where the socket traveled only 300 feet. It does not take much imagination to figure out which layout was the most economical.

In the product-oriented layout (Figure 5–1b), machines are moved and grouped according to part or product families. The production flow is greatly simplified. To function properly, operators within a manufacturing cell are cross-trained on all the operations performed in the cell. This increases the flexibility to respond to special lots within the cell and has a direct impact on product quality because the operators can see the whole effect of the operations and can quickly identify the root cause of any problems generated within the cell.

An over-the-road flatbed trailer manufacturer set up a plant with 17 cells. The subassembly operations could get as far as eight trailers ahead, but the main production line (13 cells) was connected together on a track where a pull system was used. A pull system has a completed part in each position (cell), and when it is needed, the next operation takes the part and this cell goes back to work making another one. A properly balanced cell will complete the next unit just moments before it is needed.

Flow analysis techniques will assist the manufacturing facilities designer to choose the best arrangement of machines, workstations, employee services, support services, and departments. There are three groups of flow analysis tools—techniques for (1) the fabrication of individual parts; (2) the total plant flow; and (3) people and information flow, which will be studied in the next chapter and in Chapter 12 on office layout.

FABRICATION OF INDIVIDUAL PARTS

These techniques are used mostly in process layouts. The study of individual parts flow results in the arrangement of machines and workstations. Route sheets are the primary source of information. To establish this best arrangement of equipment, facilities designers use four techniques:

1. String diagram
2. Multicolumn process chart
3. From-to chart
4. Process chart

You may not use each technique every time, but using more than one technique is good practice. To show how these techniques work, consider a small group of parts (see Figure 5–2) with the following routing (flow). This flow routing will be considered inflexible so that you must lay out (or lay out again) the workstations. You need 2,000 units per day of all parts and the parts weigh .5, 9, .5, 15, and 3.75 pounds, respectively. Each machine is identified with a letter (R, A, B, C, D, E, F, S). R is the incoming material location (called receiving), and S is the shipping end of the line. Using some creative ability, the machines will be laid out in alphabetical order first, then checked for efficiency.

String Diagram

In a *string diagram,* circles represent the equipment and the lines between circles indicate flow (see Figures 5–3 and 5–4). Flow lines between adjacent circles are from middle of circle to middle of circle. If you jump a department, you will place the line above the circles. If the flow is backward, called *backtracking* (going toward R), the flow line is drawn under the circles (Figure 5–3).

Look at the important relationships (the two circles with several lines between them). What is clear in the string diagram in Figure 5–3 is that this arrangement of machines produces a lot of travel. To improve this layout, look for important relationships (more than one part taking the same route). Look at the C to D relationship.

Part No.	*Routing (Operation Sequence)*
1	R A B D C F S
2	R B D C A S
3	R E F B A C D S
4	R F A C D S
5	R C A D S

Figure 5–2 Routing for five parts.

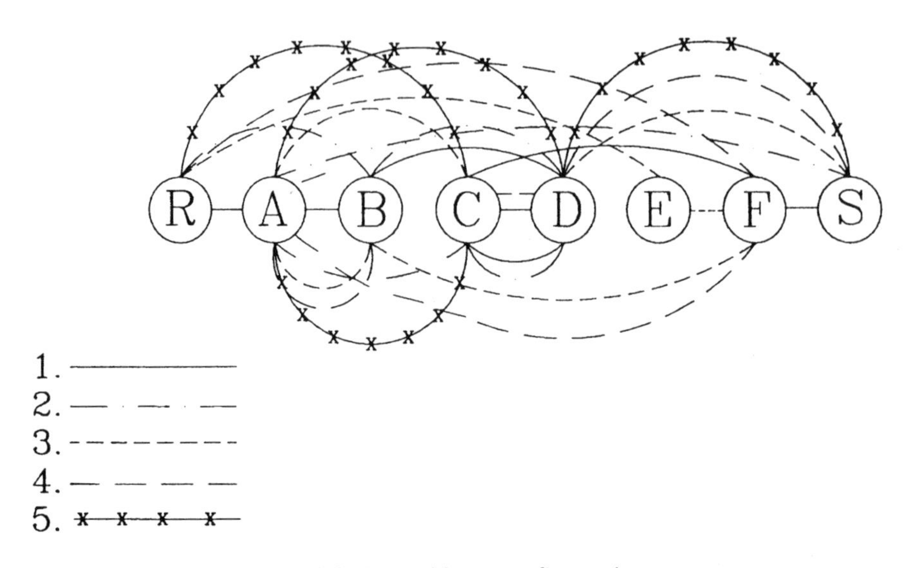

Figure 5–3 String diagram (alphabetical layout—first try).

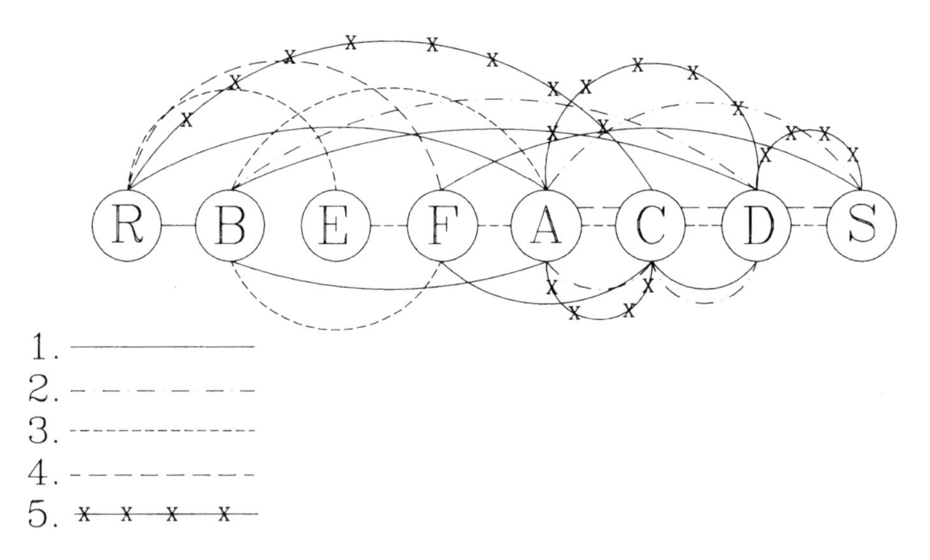

Figure 5–4 String diagram—Improved method.

Four of the five parts make this trip, so this relationship is important and C must stay close to D. Here are other important relationships:

1. B–D has two lines.
2. A–C has four lines.
3. D–S has three lines.

So, let us rearrange the layout (see Figure 5–4). Which is best? You could move each part seven steps from R to S, so that a perfect layout would require moving only 7 steps times 5 parts equals 35 steps. A step is the distance between the center of one circle to the center of the adjacent circle. If you jump one circle, two steps would be required.

In the first alphabetical layout, part 1 went from R to A to B to D to C to F to S for a total of 9 steps. Part 2 traveled 13 steps; part 3 traveled 17 steps; part 4 traveled 17 steps; and part 5 traveled 11 steps.

Part No.	*No. of Steps Traveled*
1	9
2	13
3	17
4	17
5	11
Total	67

$$\text{Efficiency} = \frac{35}{67} = 52 \text{ percent}$$

The second layout produced fewer steps:

Part No.	*No. of Steps Traveled*
1	19
2	11
3	11
4	7
5	9
Total	57

$$\text{Efficiency} = \frac{35}{57} = 61 \text{ percent}$$

How efficient can you make this layout?

Multicolumn Process Chart

Using the same routing information used in the string diagram for the five parts, a *multicolumn process chart* shows the flow for each part right next to but separate from each other (see Figures 5–5 and 5–6). First of all, you list the operations down the left-hand side of the page, then set up a column next to the list of operations—one for each part as follows (see Figure 5–5):

$$\text{Efficiency} = \frac{35}{67} = 52 \text{ percent}$$

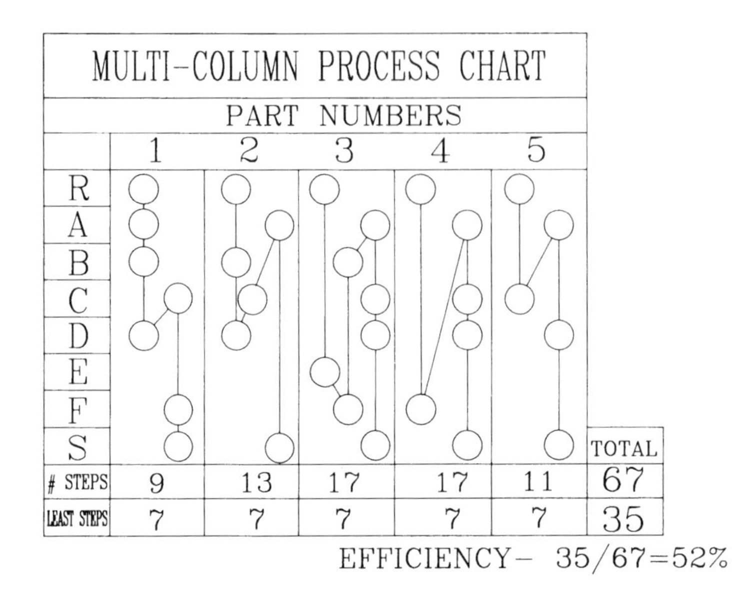

Figure 5–5 Multicolumn process chart.

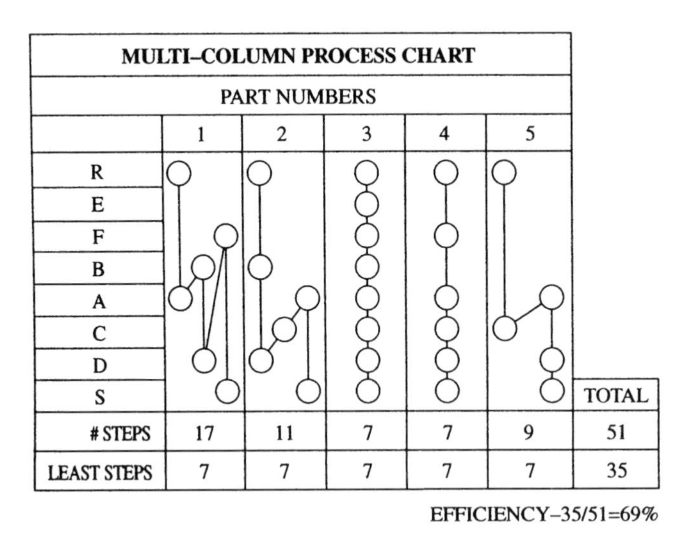

Figure 5–6 Multicolumn process chart—Improved layout.

This is the same efficiency established in the string diagram. Let us try to improve it again, but make it different from (better than) the second string diagram. Look for clues to improvements. Figure 5–6 is an improved layout.

You have come a long way toward a perfect layout. The alphabetical layout produced only a 52 percent efficiency; now it is 69 percent efficient. You can still improve efficiency. Before you try another improvement, study the third technique of flow analysis, the from-to chart.

From-To Chart

The *from-to chart* is the most exact technique of the three. Designers can develop an efficiency that considers the importance of the parts. Up until now, we have considered each part as equal in importance, but at the beginning of this chapter, the quantity and weight of each part was recorded. Figure 5–7 shows a chart of those data given earlier.

*Routing for Five Parts**

Part No.	*Routing (Operation Sequence)*
1	R A B D C F S
2	R B D C A S
3	R E F B A C D S
4	R F A C D S
5	R C A D S

*From Figure 5–2.

The relative importance of part 4 is 30 times more important than parts 1 and 3, so it should have 30 times more effect on the layout.

The from-to chart is a matrix. The sequence of operations is written down the left-hand side of the form and across the top. The vertical sequence of machines is the "from" side of the matrix. The horizontal sequence of machines is the "to" matrix. Everything moves *from* some place *to* some place. Each time a move is required, a weighted value is placed in that coordinate (see Figure 5–8). For an example of all five parts, see Figure 5–9.

To evaluate this alternative, you assign penalty points to each move depending on how far the move is away from the present location. For example, the move R to A is right next door, so you multiply that weight times 1 (one block). R to B is two blocks

Part No.	*Quantity per Day*	*Weight in Pounds*	*Total Weight (lbs)*	*Relative Importance**
1	2,000	0.5	1,000	1.0
2	2,000	9.0	18,000	18.0
3	2,000	0.5	1,000	1.0
4	2,000	15.0	30,000	30.0
5	2,000	3.75	7,500	7.5

*These numbers and the routing in Figure 5-2 create the weighted value of each move.

Figure 5–7 Part quantity and weight data.

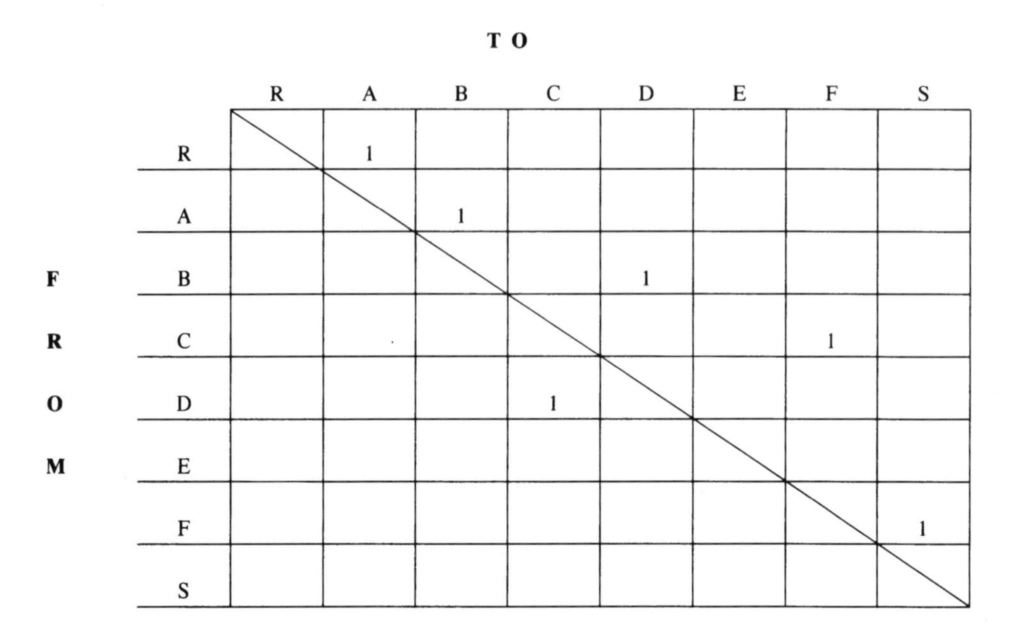

FROM \ TO	R	A	B	C	D	E	F	S
R		1						
A			1					
B					1			
C							1	
D				1				
E								
F								1
S								

Figure 5–8 From-to chart—Example of Part 1 with a relative value of 1.

away, so you multiply the 18 in that block times 2, three blocks away times 3, and so on. In Figure 5–10, the circled numbers are the penalty points (p.p.). Below and to the left of the diagonal line indicates backtracking, so that the penalty points are doubled.

The efficiency of the alphabetical layout is

$$\frac{283}{1{,}077.5} = 52 \text{ percent}$$

Now look for clues to improvement. The highest penalty points are the best clues. For example, from F to A has a penalty point of 300. This means that F wants to be closer to A. The move of R and F has a penalty point of 180. This means F wants to be closer to R. A new layout change of sequence will change both the vertical and horizontal sequences. Figure 5–11 illustrates an improved layout:

$$\text{Efficiency} = \frac{283}{548} = 52 \text{ percent}$$

This still can be improved. Find the best layout. A 56 percent is the best layout possible. A perfect layout (straight-through flow—no backtracking) is not possible in this case because of the different routing for each part, assuming there is no other routing possible. If you could change the routing of even one part, you could improve the efficiency. Practical limitations may dictate the routing, so you are left with the need to arrange the machines and equipment the best possible way. The string diagram, the multicolumn process chart, and the from-to chart are techniques to help find that best layout.

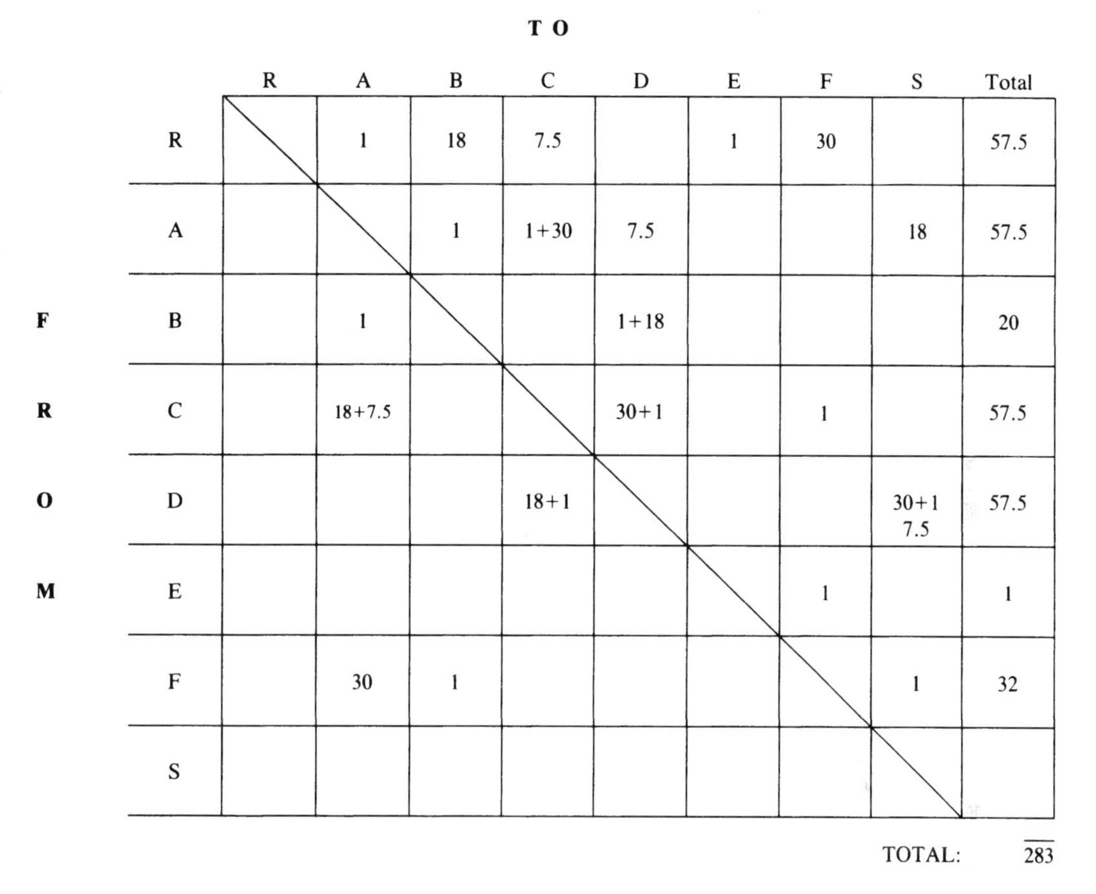

TO

		R	A	B	C	D	E	F	S	Total
	R		1	18	7.5		1	30		57.5
	A			1	1+30	7.5			18	57.5
F	B		1			1+18				20
R	C		18+7.5			30+1		1		57.5
O	D				18+1				30+1 7.5	57.5
M	E							1		1
	F		30	1					1	32
	S									
									TOTAL:	283

Figure 5–9 From-to chart—Alphabetical layout.

The first part of this chapter listed four goals of flow analysis. Minimizing distance traveled and backtracking are the first two goals. These first three techniques (string diagram, multicolumn process chart, and from-to chart) address these goals. Discouraging backtracking was best addressed by these techniques, but distances were considered only in relative terms. Future techniques will allow you to calculate the exact distance measured in feet. Minimizing the cost of production is the ultimate goal of flow analysis. The final fabrication flow analysis technique addresses this point.

Process Chart

The *process chart* (see Figure 5–12) is used for just one part, recording everything that happens to that part from the time it arrives in the plant until it joins the other parts. Symbols are used to describe what happens:

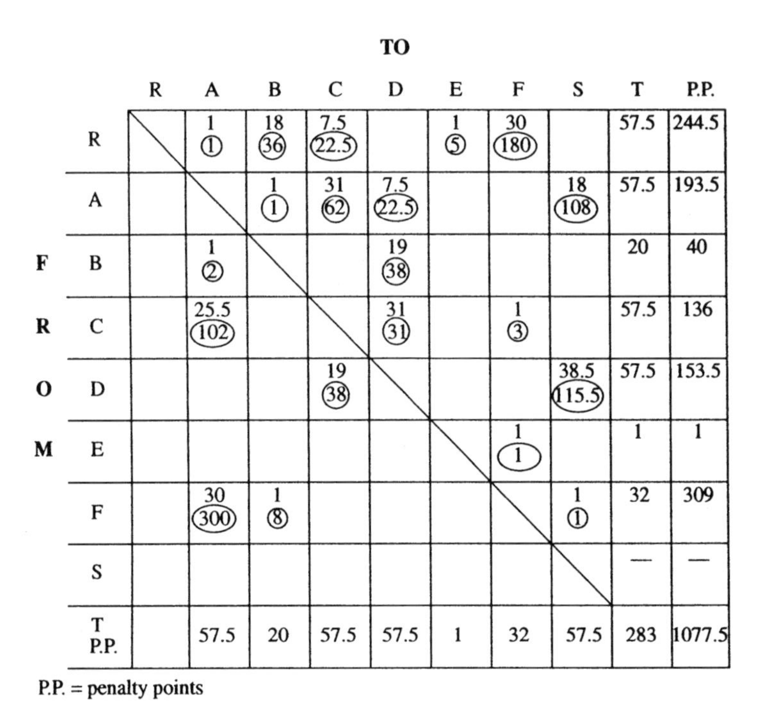

TO (columns) / FROM (rows)

	R	A	B	C	D	E	F	S	T	P.P.
R		1 (1)	18 (36)	7.5 (22.5)		1 (5)	30 (180)		57.5	244.5
A			1 (1)	31 (62)	7.5 (22.5)			18 (108)	57.5	193.5
B		1 (2)			19 (38)				20	40
C		25.5 (102)			31 (31)		1 (3)		57.5	136
D				19 (38)				38.5 (115.5)	57.5	153.5
E							1 (1)		1	1
F		30 (300)	1 (8)					1 (1)	32	309
S									—	—
T P.P.		57.5	20	57.5	57.5	1	32	57.5	283	1077.5

P.P. = penalty points

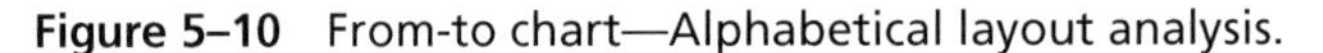

Figure 5–10 From-to chart—Alphabetical layout analysis.

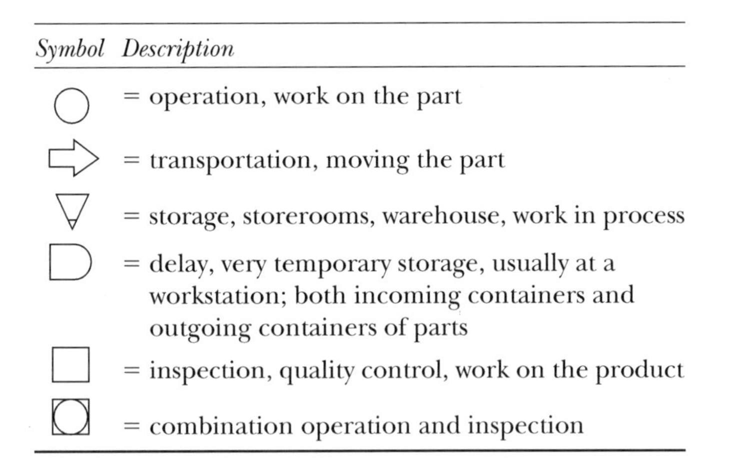

Symbol	*Description*
○	= operation, work on the part
⇨	= transportation, moving the part
▽	= storage, storerooms, warehouse, work in process
D	= delay, very temporary storage, usually at a workstation; both incoming containers and outgoing containers of parts
□	= inspection, quality control, work on the product
⊡	= combination operation and inspection

Process charting lends itself to a standard form. A properly designed form will lead designers to ask questions of each step. Designers want to know the why, who, what, where, when, and how of every operation, transportation, inspection, storage, and delay. Once designers understand the answers to these questions, they can ask

TO

FROM		R	E	F	B	A	C	D	S	T	P.P.
	R		1 (1)	30 (60)	18 (54)	1 (4)	7.5 (37.5)			57.5	156.5
	E			1 (1)						1	1
F	F				1 (1)	30 (60)			1 (5)	32	66
R	B					1 (1)		1+18 (57)		20	58
O	A				1 (2)		1+30 (31)	7.5 (15)	18 (54)	57.5	102
M	C			1 (6)		18+7.5 (51)		1+30 (31)		57.5	88
	D						1+18 (38)		1+30+7.5 (38.5)	57.5	76.5
	S									—	—
	T		1	32	20	57.5	57.5	57.5	57.5	283	
	P.P.		1	67	57	116	106.5	103	97.5		548.0

Figure 5–11 From-to chart—Improved alphabetical layout analysis.

1. Can I eliminate this step?
2. Can I automate this step?
3. Can I combine this step with another one?
4. Can I change the routing to reduce distances traveled?
5. Can I move workstations closer together?
6. Can I justify production aids to increase effectiveness?
7. How much does this part cost to produce?

Step-by-Step Description for the Process Chart

This step-by-step procedure accompanies Figure 5–12.

FRED MEYERS & ASSOCIATES PROCESS CHART

☐ PRESENT METHOD (1) ☐ PROPOSED METHOD DATE: (2) PAGE __ OF __ .

PART DESCRIPTION: (3)

OPERATION DESCRIPTION: (4)

SUMMARY	PRESENT		PROPOSED		DIFF.	
	NO.	TIME	NO.	TIME	NO.	TIME
○ OPERATIONS						
⇨ TRANSPORT.				(5)		
☐ INSPECTIONS						
D DELAYS						
▽ STORAGES						
DIST. TRAVELED		FT.		FT.		FT.

ANALYSIS: (6)
WHY WHEN
WHAT WHO
WHERE HOW

FLOW DIAGRAM ATTACHED (IMPORTANT) (7)

STUDIED BY:

STEP	DETAILS OF PROCESS	METHOD	OPERATION	TRANSPORT	INSPECTION	DELAY	STORAGE	DISTANCE IN FEET	QUANTITY	TIME HRS/UNIT .00001	COST PER UNIT	TIME/COST CALCULATIONS
1			○	⇨	☐	D	▽					
2			○	⇨	☐	D	▽					
3	(8)	(9)	○	⇨	(10)	D	▽	(11)	(12)	(13)	(14)	(15)
4			○	⇨	☐	D	▽					
5			○	⇨	☐	D	▽					
6			○	⇨	☐	D	▽					
7			○	⇨	☐	D	▽					
8			○	⇨	☐	D	▽					
9			○	⇨	☐	D	▽					
10			○	⇨	☐	D	▽					
11			○	⇨	☐	D	▽					
12			○	⇨	☐	D	▽					
13			○	⇨	☐	D	▽					
14			○	⇨	☐	D	▽					
15			○	⇨	☐	D	▽					
16			○	⇨	☐	D	▽					
17			○	⇨	☐	D	▽					

Figure 5–12 Sample process chart.

1. *Present Method or Proposed Method:* A checkmark in one of the two boxes is required. A good industrial engineering practice is always to record the present method so that the improved (proposed) method can be compared. Setting the cost with the present and proposed methods will require justification of the proposal, especially if any costs are involved. Recording and advertising cost reduction dollars saved is a smart idea for any engineer.
2. *Date* __________ *Page* __________ *of* __________*:* Always date your work. Work tends to stay around for years, and someday you will want to know when you did this great work. Page numbers are important on big jobs to help keep the proper order.
3. *Part Description:* This is probably the most important information on the form. Everything else would be useless if you did not record the part number. Each process chart is for one part, so be specific. The part description also includes the name and specifications of the part. Attaching a blueprint to the process chart would be useful.
4. *Operation Description:* In this block, you record the limits of the study; *for example,* from the receiving department to the assembly department. Also, any miscellaneous information can be placed here.
5. *Summary:* The summary is used only for the proposed solution. A count of the operations, transportation, inspection, delays, and storage for the present and proposed methods are recorded and the difference (savings) is calculated.

   The distance traveled is calculated for both methods and then the difference is determined. The time standards in minutes or hours are summarized and the difference is calculated. Thus, the result of all the work of present and proposed process charting is cost reduction information.
6. *Analysis:* Why, what, where, when, how, and who—these questions are asked of each step (line) in the process chart. "Why" is first. If you do not have a good reason for a step, you can eliminate it and save 100 percent of the cost. The questioning of each step is how you come up with the proposed method. With these questions, designers try to

   a. Eliminate every step possible because this produces the greatest savings.
   b. Combine steps when they cannot be eliminated to spread the cost and possibly to eliminate steps in between. For example, if two operations are combined, delays and transportation can be eliminated. If transportation is combined, many parts will be handled as one.
   c. Change the sequence of operations to improve and reduce flow and to save many feet of travel when steps cannot be combined or eliminated.

   As you can see, the analysis phase of process charting gives the process meaning and purpose. We will come back to Step 6 after Step 15 for cost reduction.
7. *Flow Diagram Attached (Important):* Process charting is used in conjunction with flow diagramming. The same symbols can be used in both techniques. The process chart is the words and numbers, whereas the flow diagram is

the picture. (The flow diagram is the technique described in the next section.) The present and proposed methods of both techniques must tell the same story; they must agree.

Studied By: is where the facilities designer's name goes.

8. *Details of Process:* Each line in the flow process chart is numbered, front and back. One chart can be used for 42 steps. Each step is totally independent and stands alone. A description of what happens in each step aids the analyst's questions. Using as few words as possible, describe what is happening. This column is never left blank.
9. *Method:* Method usually refers to how the material was transported—by fork truck, hand cart, conveyor, by hand—but methods of storage could also be placed here.
10. *Symbols:* The process chart symbols are all listed here. The analyst should classify each step and shade the proper symbol to indicate to everyone what this step is.
11. *Distance in Feet:* This step is used only with the transportation symbol. The sum of this column is the distance traveled in this method. This column is one of the best indications of productivity.
12. *Quantity:* Quantity refers to many things:
    a. Operation: How many pieces per hour were produced.
    b. Transportation: How many pieces were moved at a time.
    c. Inspection: How many pieces per hour were inspected if under time standard or frequency of inspection.
    d. Delay: How many pieces are in a container. This will tell us how long the delay is.
    e. Storage: How many pieces there are per storage unit.

    All costs will be reduced to a unit cost or cost per unit, so knowing how many pieces are moved at one time is important.
13. *Time in Hours per Unit (.00001):* This step is for labor cost. Cost for storage and delays will be set in another way—inventory carrying cost. This column will be used only for operations, transportation, and inspection. Time per unit is calculated in two ways:
    a. Starting with pieces per hour time standards, say, 250 pieces per hour, divide 250 pieces per hour into one hour, and you get .00400 hour per unit. On the process chart, place 400 in the time column, knowing that the decimal is always in the fifth place.
    b. Starting with a material handling time of 1.000 minute to change a tub of parts at a workstation with a hand truck, you have 200 parts in that tub. How many hours per unit is your time standard?

$$\frac{1.000 \text{ minute per container}}{200 \text{ parts per container}} = .005 \text{ minute per part}$$

$$\frac{.005 \text{ minute per part}}{60 \text{ minutes per hour}} = .00008 \text{ hour per part}$$

14. *Cost per Unit:* Hours per unit multiplied by a labor rate per hour equals a cost per unit. For example, in the above two problems using a labor rate of $15.00 per hour cost per unit would be
    a. .00400 × $15.00 = $.06 per unit
    b. .00008 × $15.00 = $.0012 per unit

    The cost per unit is the backbone of process charting. Because you are always looking for a better way, the least expensive method overall is the best method.
15. *Time/Cost Calculations:* Industrial engineers are required to calculate costs based on many different things, and, therefore, how costs were calculated tends to get lost. This space is provided to record the formulas developed to determine the costs so that they do not have to be redeveloped over and over again.

TOTAL PLANT FLOW

The three techniques studied in this section include the following:

1. Flow diagram
2. Operations chart
3. Flow process chart

We will consider every step in the process for fabrication, assembly, and packout of the product. The techniques use the same symbols as those used in the process chart, but in a different way. All parts are considered, not just one.

Flow Diagrams

Flow diagrams show the path traveled by each part from receiving to stores to fabrication of each part to subassembly to final assembly to packout to warehousing to shipping. These paths are drawn on a layout of the plant.

The flow diagram will point out problems with such factors as cross traffic, backtracking, and distance traveled.

When the paths traveled by all the parts are drawn on the same layout, the flow diagram will also bring attention to the heavy traffic paths and the activity centers between which the most traffic occurs. More congested paths may require wider aisles or an alternative material handling solution. Production centers between which heavy traffic occurs may need to be moved closer to each other to reduce the overall distances traveled.

Figures 5–13 and 5–14 each show the flow path for a single item through a facility for illustration purposes. Whereas these analyses are performed in isolation from the paths of other parts, one can still investigate as to whether these flow patterns can be improved. For instance, can the distance traveled in Figure 5–13 be shortened and one of the temporary storage skids be eliminated? Can you envision a solution for the backtracking problem in Figure 5–14?

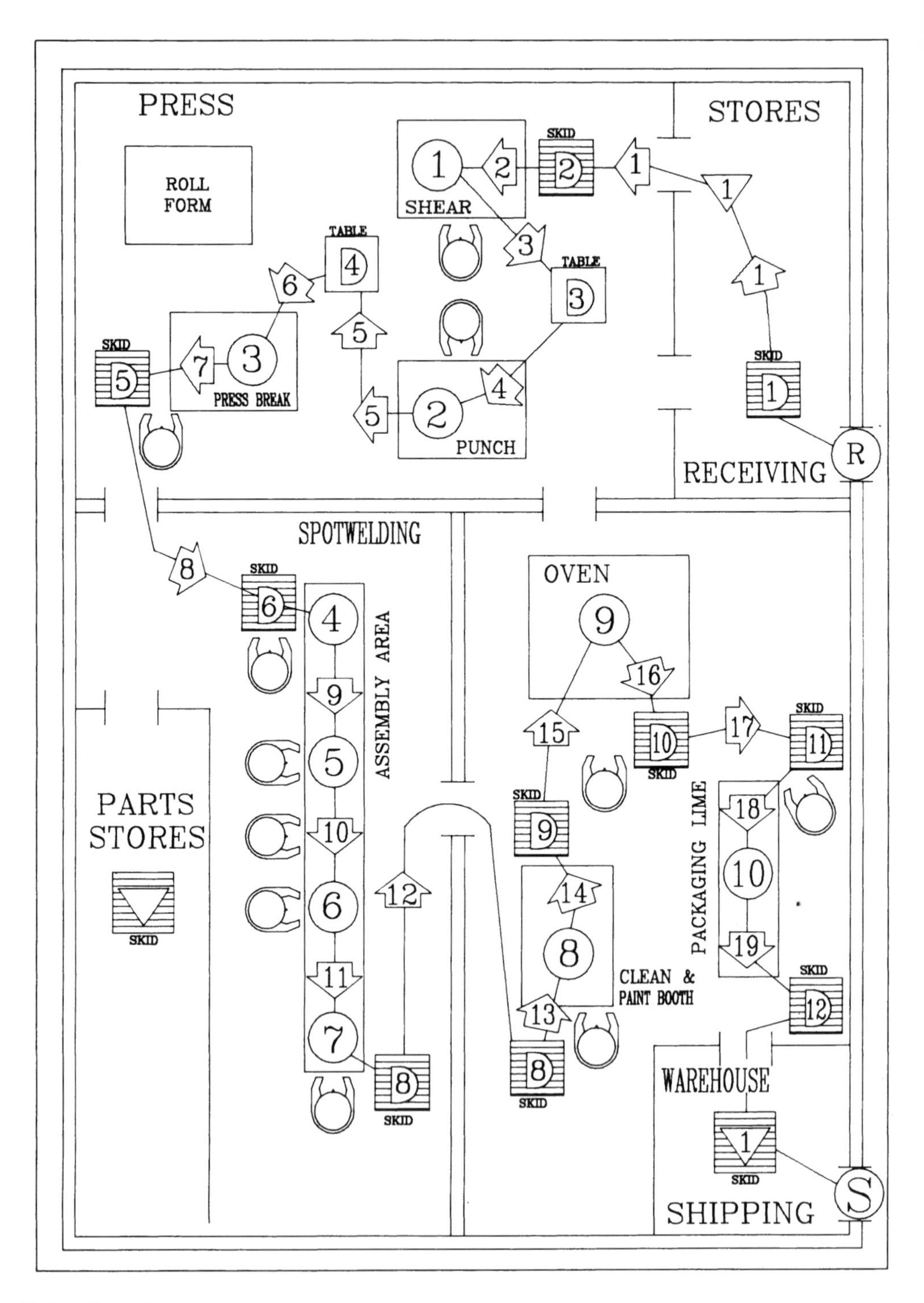

Figure 5–13 Flow diagram.

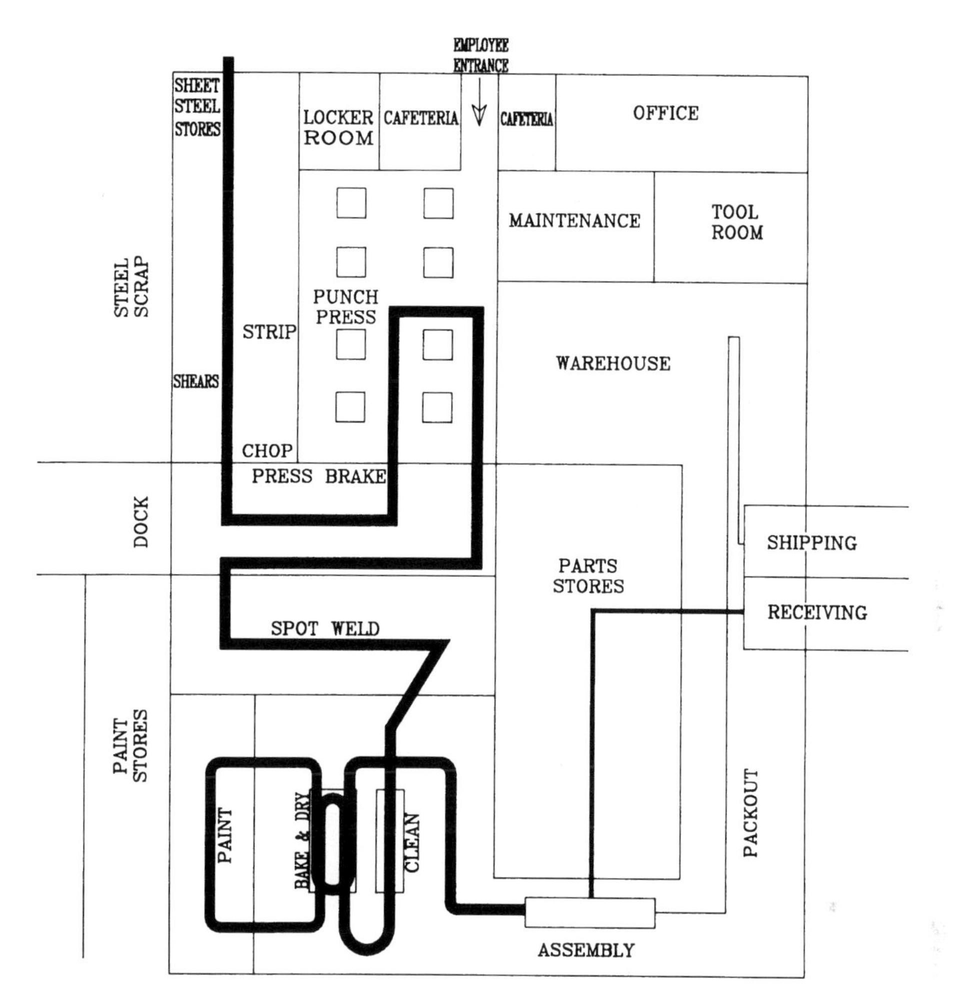

Figure 5–14 Flow diagram—Toolbox plant.

Cross Traffic

Cross traffic is where flow lines cross. Cross traffic is undesirable and a better layout would have fewer intersecting paths. Anywhere traffic crosses is a problem because of congestion and safety considerations. Proper placement of equipment, services, and departments will eliminate most cross traffic.

Backtracking

Backtracking is material moving backward in the plant. Material should always move toward the shipping end of the plant. If it is moving toward receiving, it is moving backward. Backtracking costs three times as much as flowing correctly. For example, five departments have a flow like this:

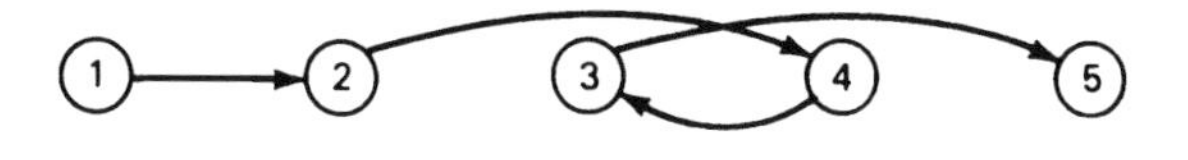

How many times did material move between departments 3 and 4? Three times! Twice forward and once backward. If you arranged this plant and changed around departments 3 and 4, you would have straight-through flow:

This arrangement has no backtracking. Efficiency-wise, material travels less distance. In the first example, it traveled six blocks (a block is one step between departments next to each other). In the straight line flow, it traveled only four blocks—a 33 percent increase in productivity.

Distance Traveled

Distance costs money to travel. The less distance traveled, the better it is. The flow diagram is developed on a layout, and the layout can be easily scaled and the distance of travel calculated. By rearranging machines or departments, you may be able to reduce the distances traveled.

Because flow diagrams are created on plant layouts, no standard form is used and there are few conventions to restrict the designer. The objective is to show all the distances traveled by each part and to find ways of reducing the overall distance.

The flow diagram is developed from route sheet information, assembly line balance, and blueprints. The route sheet specifies the fabrication sequence for each part of a product. This sequence of steps required to make a part is very practical and has some room for flexibility. One step may come before or after another step, depending on conditions. The sequence of steps should be changed to meet the layout if possible because that requires only a paperwork change. But if the sequence of operations cannot be changed and the flow diagram shows backtracking, moving equipment may be necessary. The objective will always be to "make a quality part the cheapest, most efficient way possible."

Step-by-Step Procedure for Developing a Flow Diagram

Step 1. The flow diagram starts with an existing or proposed scaled layout.

Step 2. From the route sheet, each step in the fabrication of each part is plotted and connected with a line, and color codes or other methods of distinguishing between parts are used.

Step 3. Once all the parts are fabricated, they will meet, in a specific sequence, at the assembly line. The position of the assembly line will be determined by where the individual parts came from. At the assembly line, all flow lines join together and travel as one to packout, warehouse, and shipping. A well-thought-out flow diagram will be the best technique for developing a plant layout.

Plastic overlays on plant layouts are often used to develop flow lines for flow diagrams. The flow lines can be drawn with a grease pencil and grouped by classes for plants with several different parts. It does not take a large product to make the fabrication departments of a plant layout look like a bowl of spaghetti. Using several plastic overlays will simplify the analysis.

A new industrial engineer will learn much from the creation of a flow diagram, and an experienced engineer will always find ways to improve the flow of material.

Refer back to Figure 5–13, which shows the flow of one part through a plant. Can you recommend improvements? Figure 5–14 shows the flow in the toolbox example.

The Operations Chart

The *operations chart* (see Figures 5–15 and 5–16) has a circle for each operation required to fabricate each part, to assemble each part to the final assembly, and to pack out the finished product. On one piece of paper, every production operation, every job, and every part are included.

Operations charts show the introduction of raw materials at the top of the page, on a horizontal line. (See Figure 5–15.) The number of parts will determine the size and complexity of the operations chart.

Below the raw material line, a vertical line will be drawn connecting the circles (steps in the fabrication of that raw material into finished parts). Figure 5–15 illustrates these points. Once the fabrication steps of each part are plotted, the parts flow together in assembly. Usually, the first part to start the assembly is shown at the far right of the chart. The second part is shown to the left of that, and so forth working from right to left (see Figure 5–16). Some parts require no fabrication steps. As described in an earlier chapter, these parts are called buyouts. Buyout parts are introduced above the operation at which they will be used (shown on the bottom of Figure 5–16—packout operation). In the packout operation, six products will be placed in a master carton and taped closed.

The operations chart shows a lot of information on one page. The raw material, the buyouts, the fabrication sequence, the assembly sequence, the equipment needs, the time standards, and even a glimpse of the plant layout, labor costs, and plant schedule can all be derived from the operations chart. Is it any wonder that plant layout designers consider this one of their favorite tools?

The operations chart is different for every product, so a standard form is not practical. The circle is universally accepted as the symbol for operations. There is more convention in operations charting than in flow diagramming, but designers should not be too rigid in their thinking.

Step-by-Step Procedure for Preparing an Operations Chart

Step 1. Identify the parts that are going to be manufactured and those that are going to be purchased complete.

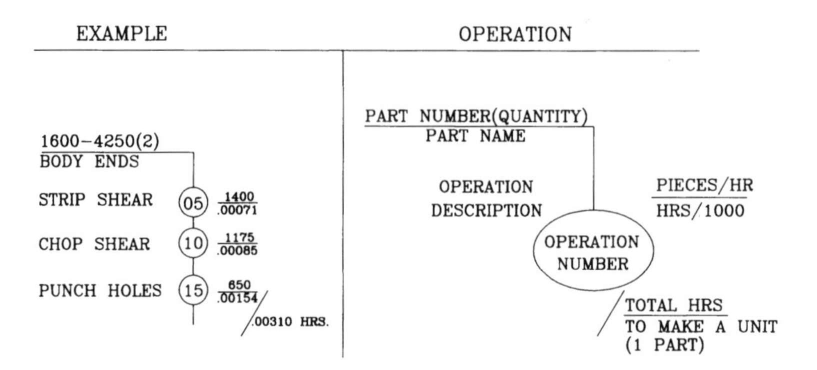

Figure 5–15 Sample operations chart.

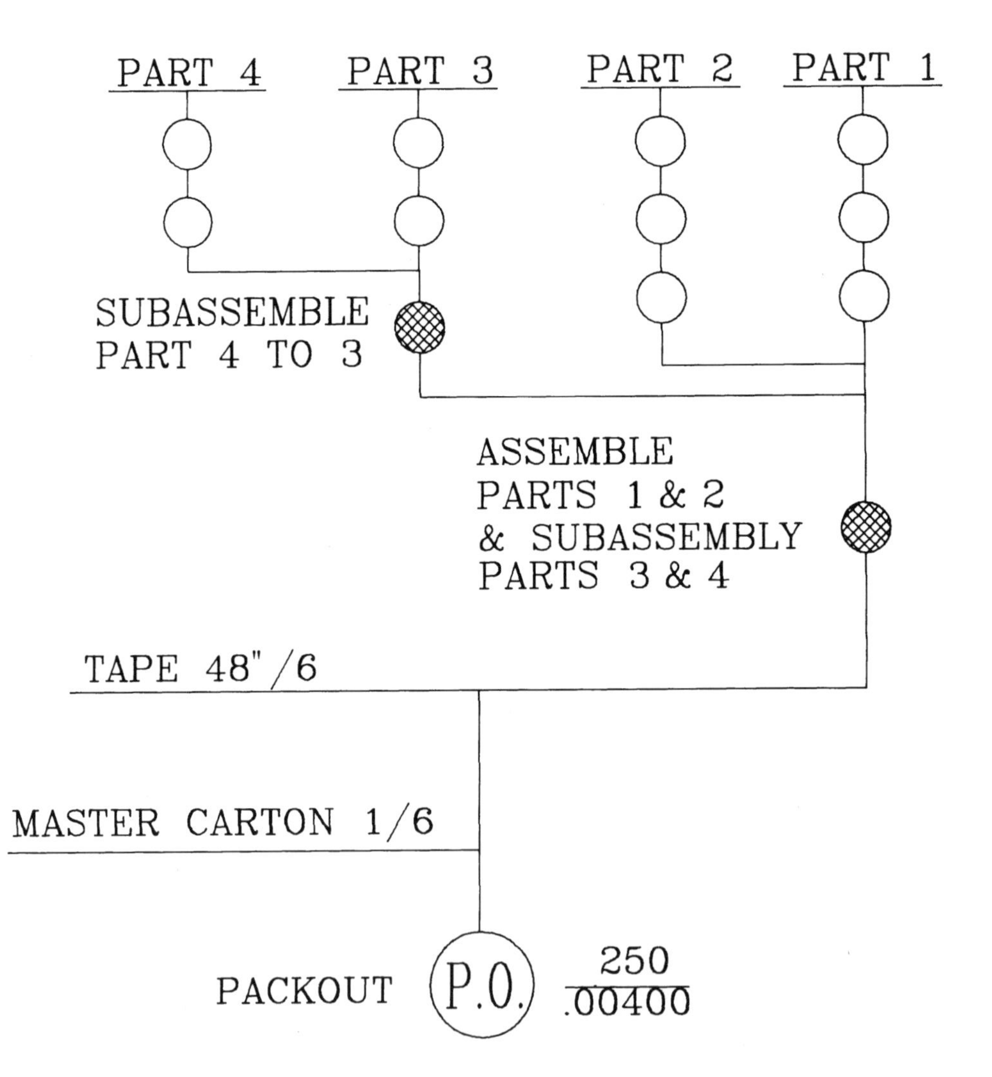

Figure 5–16 Operations chart design.

Step 2. Determine the operations required to fabricate each part and the sequence of these operations.

Step 3. Determine the sequence of assembly of both buyout and fabricated parts.

Step 4. Find the base part. This is the first part that starts the assembly process. Put that part name on a horizontal line in the upper right of the chart. On a vertical line extending down from the right side of the horizontal line, place a circle for each operation. Beginning with the first operation, list all operations down to the last operation.

Step 5. Place the second part to the left of the first part and the third part to the left of the second part, and so forth until all manufactured parts are listed across the top of the page in reverse order of assembly. All the fabrication steps are listed below the parts with a circle representing each operation.

Step 6. Draw a horizontal line from the bottom of the last operation of the second part to the first part just below its final fabrication operation and just above the first assembly operation. Depending on how many parts the first assembler puts together, the third, fourth, and so on, parts will flow into the first part's vertical line, but always above the assembly circle for that assembly operation.

Step 7. Introduce all buyout parts on horizontal lines above the assembly operation circle where they are placed on the assembly.

Step 8. Put time standards, operation numbers, and operation descriptions next to and in the circle as explained earlier.

Step 9. Sum total the hours per unit and place these total hours at the bottom right under the last assembly or packout operation.

Figure 5–17 is a good example of an operations chart showing subassembly. Some parts will flow together before they reach the assembly line. This could be welding parts together or assembling a bag of parts. This is called subassembly and is treated just like the main assembly, except it is done before the parts reach the far right side vertical line. Bag packing is a good example. All parts are usually buyouts and could be placed at the bottom left of the operations chart, such as bag packing shown in Figure 4–8 SA3 (see Chapter 4) and in Figure 5–17 SA3, bottom left.

Flow Process Chart

The *flow process chart* combines the operations chart with the process chart. The operations chart used only one symbol—the circle or operations symbol. The flow process chart is just five times more, using all five process chart symbols. Another difference is that buyout parts are treated like manufactured parts. No standard form exists to flow process charting (see Figure 5–18).

The flow process chart is the most complete of all the techniques, and when completed, the engineer will know more about the plant's operation than anyone in the plant.

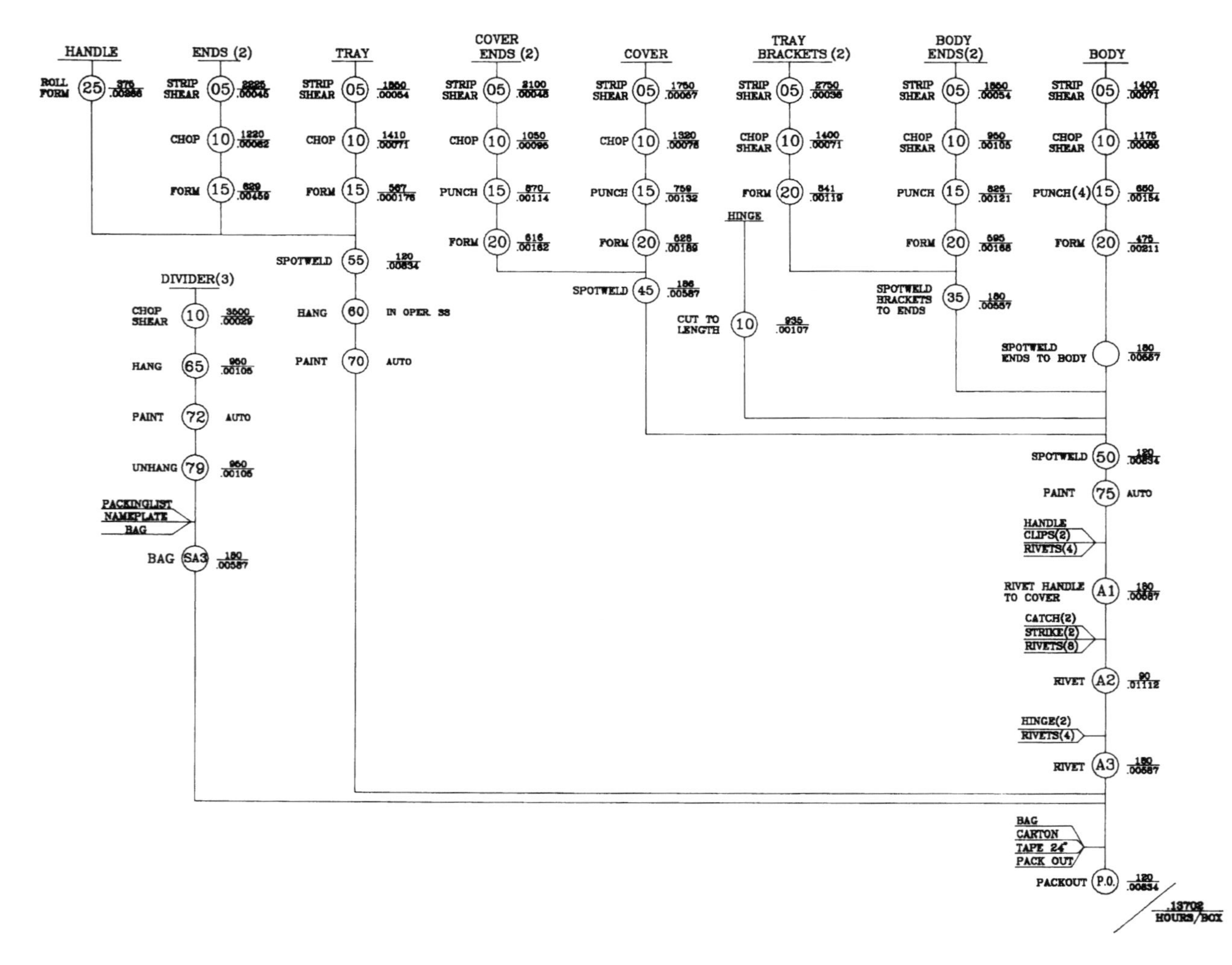

Figure 5–17 Sample operations chart—Subassembly example.

Step-by-Step Procedure for Preparing a Flow Process Chart

Step 1. Start with an operations chart.

Step 2. Complete a process chart for each part.

Step 3. Combine the operations chart and the process chart, working in all the buyouts.

COMPUTER-AIDED FLOW DESIGN AND ANALYSIS

Computers and state-of-the-art software packages can aid with the design and analysis of material flow in the manufacturing facility. The use of this technology allows the user to consider and evaluate many configurations without the expense of physically rearranging the facility to achieve an optimum level of efficiency in material flow. FactoryFLOW is a powerful layout analysis tool capable of integrating the actual facilities drawings with the material flow paths and the production and material handling data. As is the case in the traditional manual method, facilities planners are responsible for developing or obtaining the input data such as time standards, route sheets, and process and equipment requirements. Using a companion program, FactoryCAD, they can prepare a sketch of facilities showing the existing or the proposed location of various activity centers. By integrating the routing data with the layout information and using material flow as the key measure of production design efficiency, the software enables the facilities designers to compare, evaluate, and analyze alternative designs.

The software allows the user to incorporate a large amount of data ranging from production volumes and part routings, to fixed and variable material handling costs. The output of the analysis can then be viewed in a spatial medium, with the designer having the ability to manipulate system parameters in real time to study and compare various scenarios.

The software produces actual path diagrams showing how materials travel among various activity centers. Flow line thickness indicates frequency, hence the cost. Critical paths, bottlenecks, and flow efficiency can be readily determined. In addition to online visual aids, the software produces a wide variety of reports for a detailed analysis of the cost of individual and combined moves.

As with any computerized modeling tool, the system allows the user to investigate various scenarios by easily making changes to the layout, routings, production volumes, material handling systems, and a host of other variables. The results of these changes can be seen immediately and obtained in a report. The ability to manipulate system variables easily allows the user to

- Redesign material flow
- Eliminate or significantly reduce non-value-added handling
- Reduce total part or product travel distance

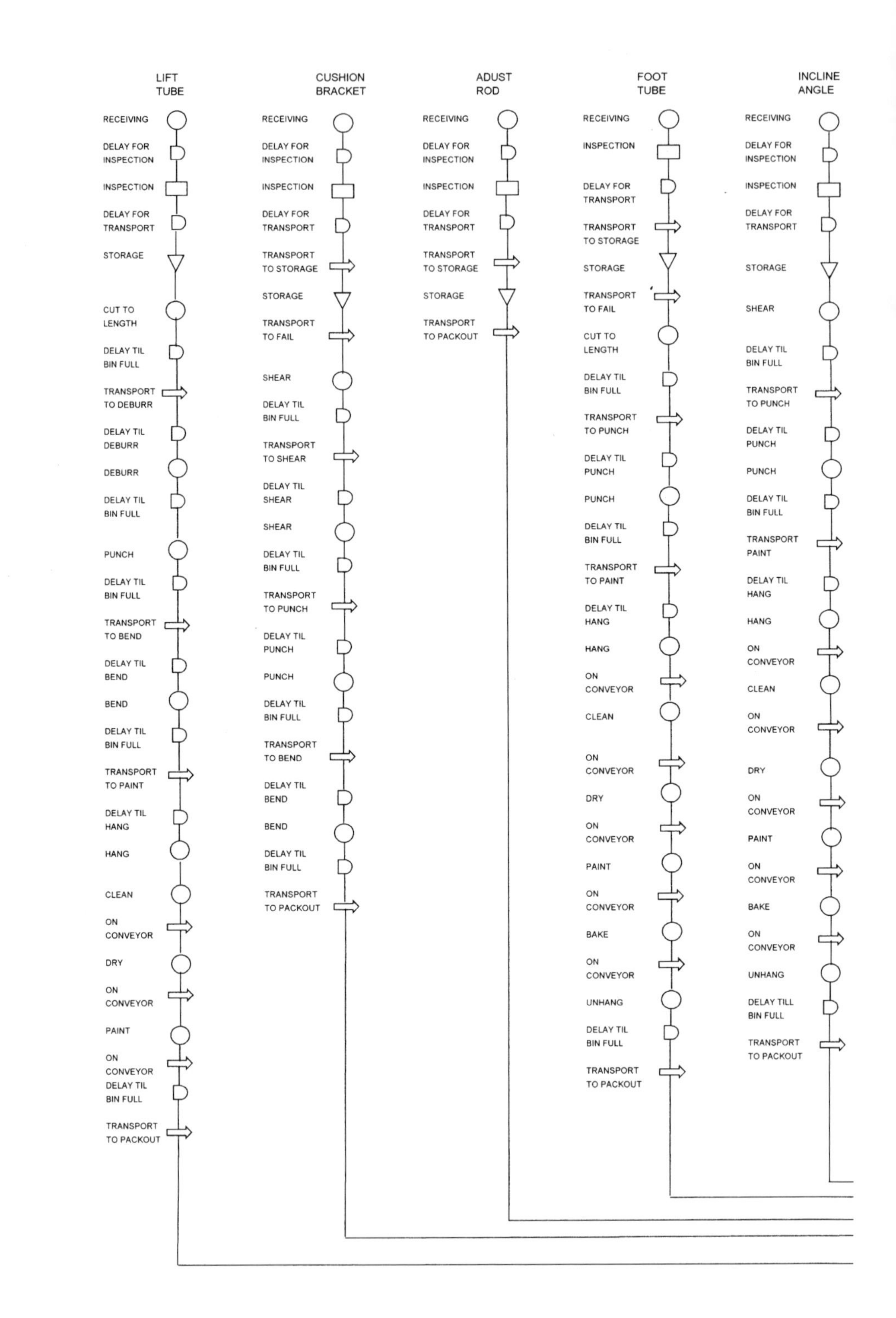

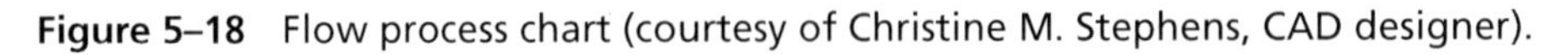

Figure 5–18 Flow process chart (courtesy of Christine M. Stephens, CAD designer).

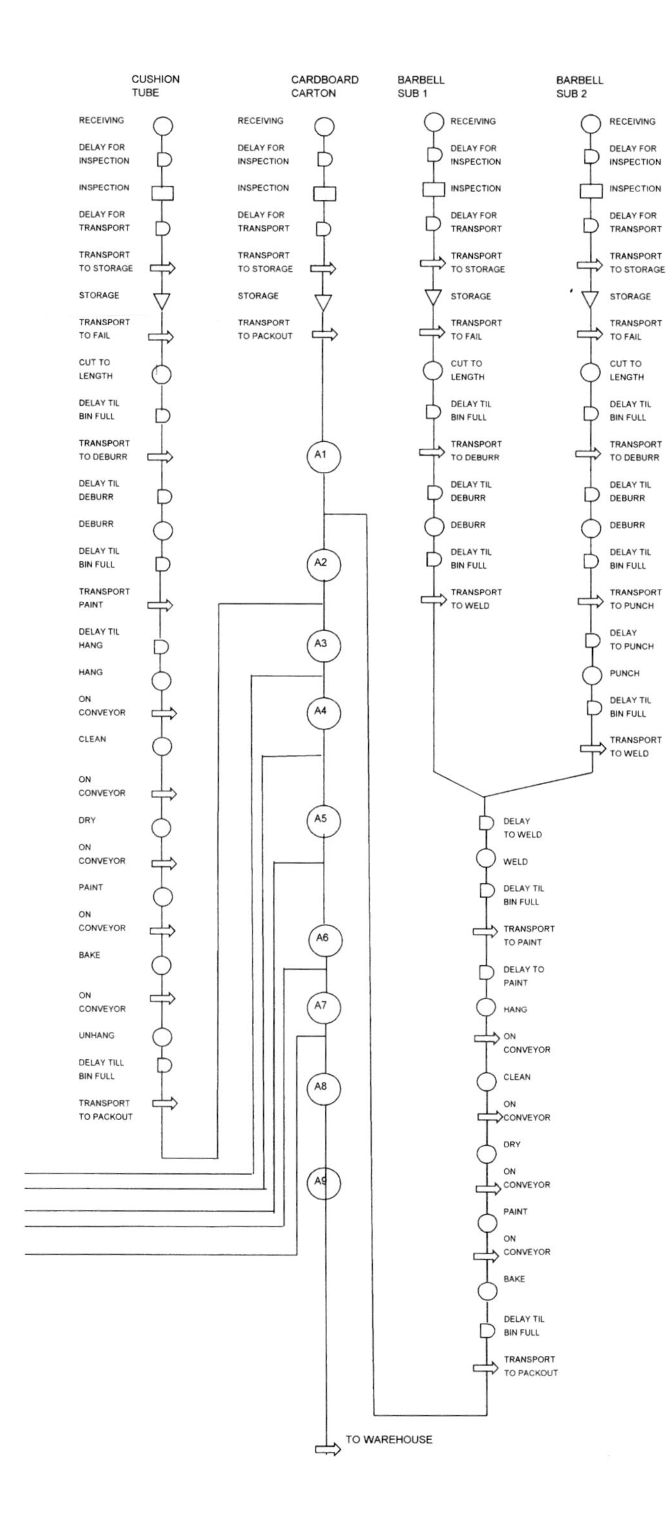
CUSHION TUBE
RECEIVING
DELAY FOR INSPECTION
INSPECTION
DELAY FOR TRANSPORT
TRANSPORT TO STORAGE
STORAGE
TRANSPORT TO FAIL
CUT TO LENGTH
DELAY TIL BIN FULL
TRANSPORT TO DEBURR
DELAY TIL DEBURR
DEBURR
DELAY TIL BIN FULL
TRANSPORT PAINT
DELAY TIL HANG
HANG
ON CONVEYOR
CLEAN
ON CONVEYOR
DRY
ON CONVEYOR
PAINT
ON CONVEYOR
BAKE
ON CONVEYOR
UNHANG
DELAY TILL BIN FULL
TRANSPORT TO PACKOUT
CARDBOARD CARTON
RECEIVING
DELAY FOR INSPECTION
INSPECTION
DELAY FOR TRANSPORT
TRANSPORT TO STORAGE
STORAGE
TRANSPORT TO PACKOUT
A1
A2
A3
A4
A5
A6
A7
A8
A9
TO WAREHOUSE
BARBELL SUB 1
RECEIVING
DELAY FOR INSPECTION
INSPECTION
DELAY FOR TRANSPORT
TRANSPORT TO STORAGE
STORAGE
TRANSPORT TO FAIL
CUT TO LENGTH
DELAY TIL BIN FULL
TRANSPORT TO DEBURR
DELAY TIL DEBURR
DEBURR
DELAY TIL BIN FULL
TRANSPORT TO WELD
BARBELL SUB 2
RECEIVING
DELAY FOR INSPECTION
INSPECTION
DELAY FOR TRANSPORT
TRANSPORT TO STORAGE
STORAGE
TRANSPORT TO FAIL
CUT TO LENGTH
DELAY TIL BIN FULL
TRANSPORT TO DEBURR
DELAY TIL DEBURR
DEBURR
DELAY TIL BIN FULL
TRANSPORT TO PUNCH
DELAY TO PUNCH
PUNCH
DELAY TIL BIN FULL
TRANSPORT TO WELD
DELAY TO WELD
WELD
DELAY TIL BIN FULL
TRANSPORT TO PAINT
DELAY TO PAINT
HANG
ON CONVEYOR
CLEAN
ON CONVEYOR
DRY
ON CONVEYOR
PAINT
ON CONVEYOR
BAKE
DELAY TIL BIN FULL
TRANSPORT TO PACKOUT

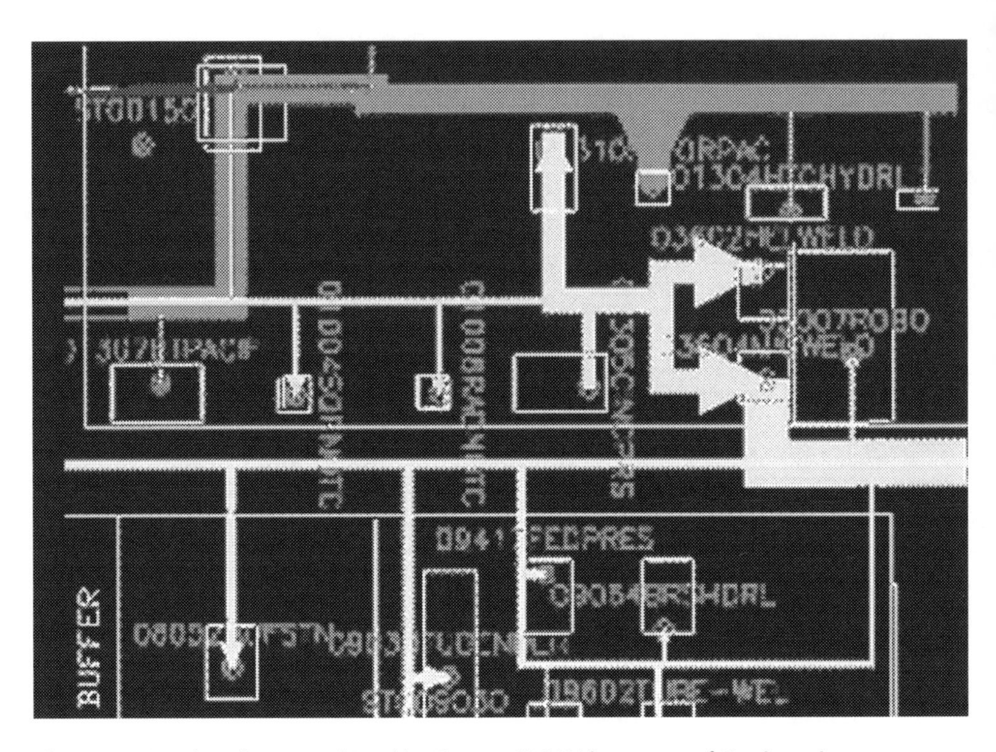

Figure 5–19 Flow lines produced by FactoryFLOW (courtesy of Engineering Animation, Inc.).

- Reduce work-in-process inventories
- Evaluate alternate material handling systems

FactoryFLOW produces flow lines (see Figure 5–19) that illustrate total move distances, move intensities, and costs. Furthermore, the system generates numerical data comparing alternate flow paths and alternate layouts of machines and storage areas. Creating and evaluating layout alternatives are accomplished by moving the equipment with the mouse and having the system recalculate the results. Optimal dock, storage, and equipment locations can be quickly and easily determined.

Figure 5–20 illustrates the proposed layout for an automotive components plant that would have required over $7 million in new machines and $1.2 million in building expansion. Using FactoryFLOW, designers were able to create and evaluate various alternatives. In just 2 weeks, they were able to achieve a more focused material flow to reduce space and tooling requirements and to expand a cost savings of $3.2 million. Figure 5–21 shows the redesigned layout and the enhanced flow lines.

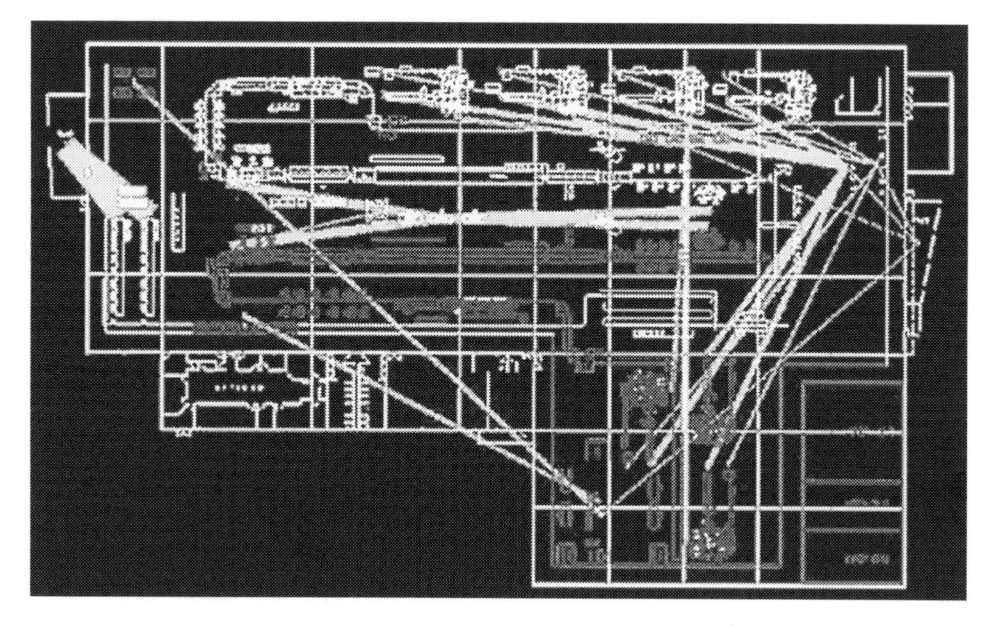

Figure 5–20 Proposed layout for an automotive plant using FactoryFLOW (courtesy of Engineering Animation, Inc.).

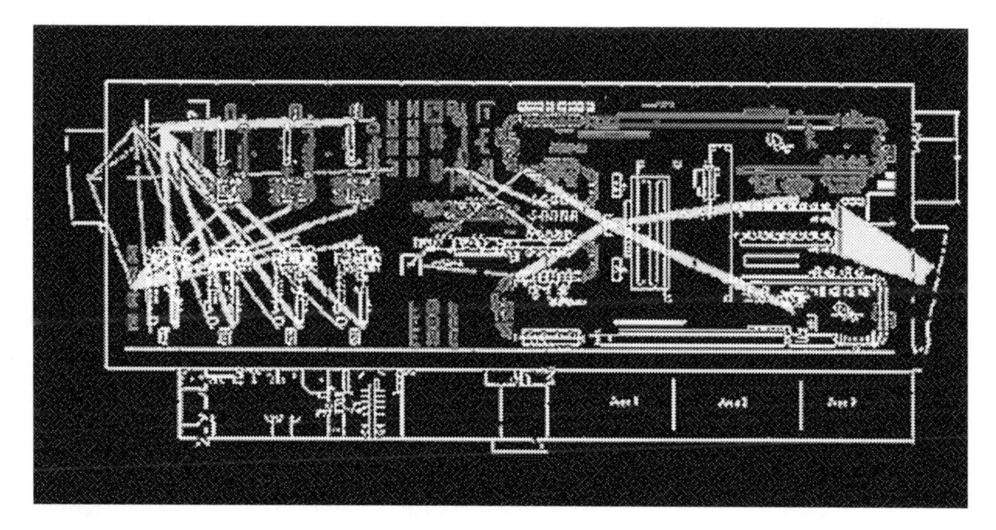

Figure 5–21 Redesigned layout with enhanced flow lines using FactoryFLOW (courtesy of Engineering Animation, Inc.).

CONCLUSION

Flow analysis leads to better plant layouts. Promoting efficiency, effectiveness, and cost reduction are the goals of flow analysis. A detailed analysis of material flow will arm the layout designer with critical information such as (1) operation requirements, (2) material handling needs, (3) storage needs, (4) inspection requirements, and (5) delay reasons.

With this information, the designer is challenged to

1. Eliminate as many steps as possible
2. Combine steps
3. Rearrange equipment
   a. to eliminate cross traffic
   b. to eliminate backtracking
   c. to reduce the distance of travel
4. Reduce production costs in general

Electronic copies of the process chart and from-to chart are provided for your use. These forms may be downloaded and utilized in your project.

QUESTIONS

1. Define *flow line.*
2. What does flow analysis try to do?
3. What are the two basic groups of flow analysis techniques?
4. What are the fabrications of individual parts flow analysis techniques?
5. Draw a string diagram, multicolumn process chart, and a from-to chart for the following four parts:

Part No.	*Weight*	*Sequence*
A	1	1 2 3 4 7
B	2	1 3 2 6 7
C	3	1 3 4 5 6 7
D	4	1 3 4 5 7

   What is the efficiency of the from-to chart?
6. Draw a process chart for the toolbox body shown in Figure 4–1 using the flow diagram in Figure 5–14.
7. What are the three techniques of total plant flow?
8. Draw an operations chart for your project.
9. Flow process charts combine what two techniques?
10. Why is flow analysis and design for human resources as important as that for material?
11. What is FactoryFLOW and what does it attempt to do?
12. What are the advantages of computer-aided flow analysis? What are the possible disadvantages?

CHAPTER

Activity Relationship Analysis

OBJECTIVES:

Upon the completion of this chapter, the reader should:

- Be able to define activity relationship chart and its function
- Be able to identify relationship codes
- Be able to construct an activity relationship chart
- Be able to construct a dimensionless block diagram
- Be able to utilize the block diagram to analyze flow

Manufacturing flow was discussed in Chapter 5, but other departments, services, and facilities must be included to establish good overall flow. Material flows from receiving, to stores, to warehousing, to shipping. Information flows between offices and the rest of the facility, and people move from place to place. Each department, office, and service facility must be placed properly in relationship to each other. The techniques in this chapter will help establish the optimum placement of everything that needs space. Sometimes very little space is necessary; for example, the employee entrance location is important to the employee flow path. From the parking lot to the employee entrance, to the time clock, to the lockers, to the cafeteria, to the work area is a typical operator's flow path when coming to work. The techniques to be studied in this chapter are

1. The activity relationship diagram
2. The worksheet
3. The dimensionless block diagram
4. The flow analysis

These techniques will help the facilities planner place each department, office, and service area in the proper location. The objective is to satisfy as many important relationships as possible in order to create the most efficient layout possible.

The auxiliary services (manufacturing support services), personnel services (bathrooms, cafeterias, etc.), and offices for all those who need them will be part of this chapter, but they will be discussed in much greater detail in Chapters 8, 9, and 12. The four techniques studied in this chapter are sequential. The activity relationship diagram is redrawn into a worksheet, and the worksheet is used to draw the dimensionless block diagram. The flow analysis is then drawn on the dimensionless block diagram.

ACTIVITY RELATIONSHIP DIAGRAM

The *activity relationship diagram,* also called an *affinity analysis diagram,* shows the relationship of every department, office, or service area with every other department and area (see Figure 6–1). It answers the question, How important is it for this department, office, or service facility to be close to another department, office, or service facility? This question needs to be asked about the relationship of every department, office, or service facility with every other department, office, or service facility. Closeness codes are used to reflect the importance of each relationship. As a new person or an outside consultant, you may need to talk with many people to determine these codes, and once they are set, your arrangement of departments, offices, and service facilities is nearly determined for you. The codes are as follows:

Code	*Definition*
A	Absolutely necessary that these two departments be next to each other
E	Especially important
I	Important
O	Ordinary importance
U	Unimportant
X	Closeness undesirable

The A code should be restricted to the movement of massive amounts of material between departments; for example, the raw steel storeroom and the shear department in manufacturing is an A code. For the same reason, the steel receiving department must be close to the steel storeroom. The need for great numbers of people moving could also be classified as A codes; for example, maintenance and the tool and supplies crib are an A code. However, be sparing in the use of this most important code; otherwise, it will become less useful. You will find it difficult to handle more than eight A codes with one department. Sometimes you can combine two departments, offices, or service facilities together on the same line such as on line 4 of Figure 6–1 (assembly and packout). This is like a super A code. Maintenance and the toolroom, and the restroom and lockers, are other examples of departments and service facilities, respectively, that should not be divided.

Use E codes if there is any doubt that it is an A code. Much material or many people move between these two departments, but not everything or everyone moves all the time. For example, everyone needs the restroom, or break room, but not all the time, so

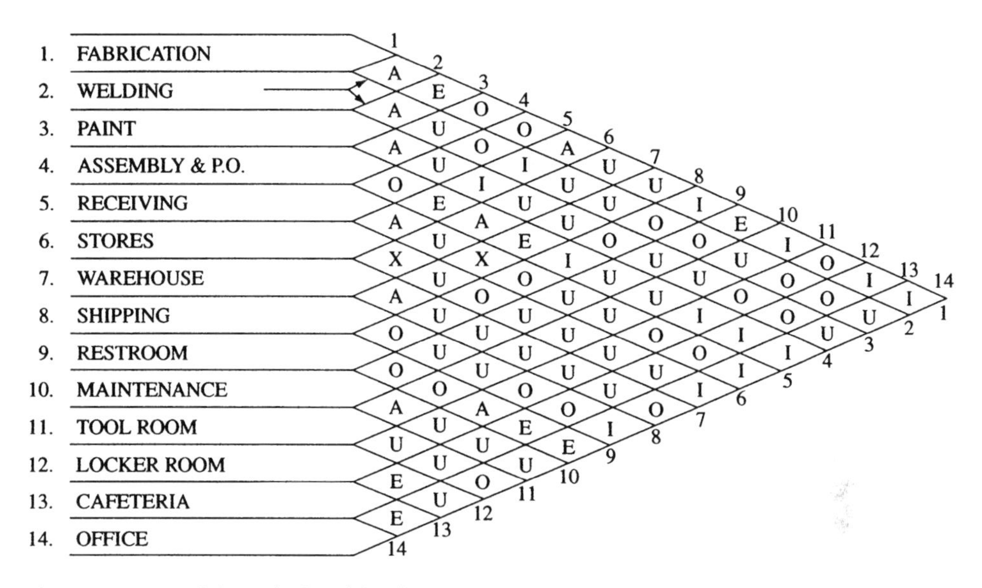

Figure 6–1 Activity relationship diagram.

an E code could be appropriate for departments with several people. Departments with few people may have the same needs, but because of the fewer people, there is less need to place them close to the services. An interesting way to consider the placement of a service facility like a restroom is to think of attaching a rubber band between every employee and the restroom so that each person pulls the restroom toward him or her. If you had only two people, the rubber bands would place the restroom halfway between the two people. With many people, it is just a little more complicated.

Use I and O codes when some level of importance is desired, but these closeness codes are not as useful as the others. It is not a good idea to omit these codes, at least for the first few layout designs.

U codes are useful because they tell you when no activity or interface is needed between two departments. These departments can be placed far away from each other.

X codes are as important as A codes, but for the opposite reason. For example, if the paint department is located next to the welding department, an explosion is possible. Noise, smell, heat, dust, cold, and so on are all good reasons for an X code.

Be sure that you can understand the activity relationship diagram in Figure 6–1. For instance, the closeness code for the relationship between paint and the tool room is a U. Do you see this? Still referring to Figure 6–1, what are the closeness codes for the following departments?

1. Fabrication and stores
2. Paint and fabrication
3. Stores and warehouse
4. Maintenance and office

The answers to these questions are presented at the end of this section.

Here is a step-by-step procedure for developing an activity relationship diagram:

1. List all departments in a vertical column on the left-hand side of the form. For an example, see Figure 6–1.
2. Starting with line 1 (fabrication), establish the relationship code for each following department. Establishing these relationship codes requires an understanding of all the departments, an understanding of management attitudes, and a determination to produce the most efficient layout possible.
3. Reason codes can be used like asterisks. For example, you do not want shipping and receiving close to each other. Why? A *1* could be placed below the X in the 5-8 intersection below the activity relationship code. You would write a reason code key below the diagram:

Reason Code	*Reason*
1	For better flow
2	All material moves between these two departments
3	People movement
and so on	and so on

For example, an A closeness code with a 2 reason code should be on line 5-6 of Figure 6–1.

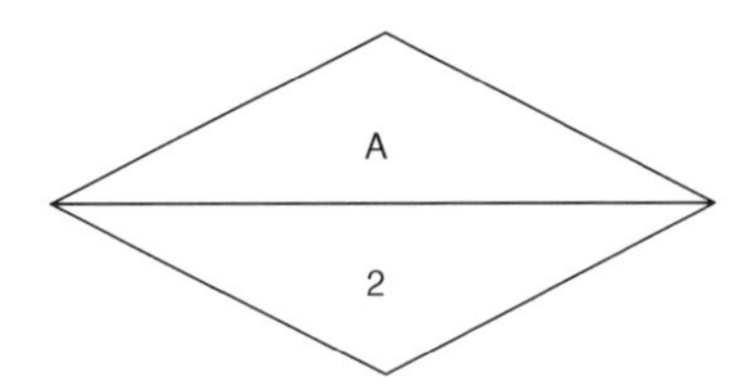

Next week someone may ask, "Why did you code this an A?" Without reason codes, you might not remember why. Reason codes are not used all the time, but often they can be useful. The form shown in Figure 6–11 uses reason codes and is a better example than Figure 6–1. Figure 6–1 is just a teaching example for closeness codes.

Answers to the activity relationship charts from Figure 6–1:

1. A
2. E
3. X
4. U

Determining the Relationship Code

The relationship or affinity codes state the desired degree of closeness between two activity centers. Each code can be broken down into a qualitative and a quantitative component in order to facilitate the assignment of the codes.

The quantitative component of the relationship between two departments or work centers can be based on the actual flow of material. Considerations given to how many parts per day, or how many different parts, or how many tons of material are moved between the two given work centers can be a great aid in determining the proper relationship codes between the two centers. Flow lines can be drawn between the two activity codes to depict the movement of the parts or people. The number of lines or the intensity of the flow will then indicate the desired degree of closeness. One numbering or weighing scheme may assign arbitrary values to the relationship codes as follows: A = 4, E = 3, I = 2, and O = 1. Using the same scale, one can evaluate the intensity of the flow lines between the centers.

The qualitative component in assigning relationship codes can be based on expert opinions and the judgment of individuals as to where two departments or centers should be located in relation to each other and in assigning a relative number to the relationship. To keep matters simple and also to avoid the possibility that the flow-related and the non-flow-related criteria do not overshadow each other, it would be advisable to use the same ranking scale. The average scores from the combined flow and non-flow-related activities can provide a reasonably clear guide in assigning the activity relationship or affinity codes.

For the novice, and often for the experienced planner, it might be tempting to overstate the relationship between the work centers and to overassign code A, in particular. A Pareto analysis approach to assigning the relationship codes may be helpful. A rule-of-thumb approach states that you should not exceed the following percentages for a given code:

Code	*Percentage*
A	5
E	10
I	15
O	25

The remaining relationships will probably be assigned as U, with the exception, of course, of where a code X is deemed necessary.

The total number of relationships, N, between all possible pairs of work centers in any facility can be determined as follows:

$$N = \frac{n(n-1)}{2}$$

where n = number of departments or work centers in the facility. For example, for a facility with 25 different departments or work centers,

$$N = \frac{25(25-1)}{2} = 300 \text{ total relationship codes}$$

Using the rule of thumb described above, the facilities planner in this case should have no more than 15 A relationship codes (300×5 percent = 15). Similarly, it is reasonable to expect that the number of E and I codes should not exceed 30 and 45, respectively.

WORKSHEET

Whereas the development of the *worksheet* is not necessary, it may serve as an interim step between the activity relationship diagram and the dimensionless block diagram. It can also serve as a summary of the activity relationship chart. This step, however, can be skipped and one can move directly to the dimensionless block diagram. The worksheet interprets the activity relationship diagram and becomes the basic data for the dimensionless block diagram.

Here is a step-by-step procedure for the worksheet (see Figure 6–2):

1. List all the activities down the left-hand side of a sheet of paper.
2. Make six columns to the right of the activity column and title them A, E, I, O, U, and X (relationship codes).
3. Taking one activity (department, office, or service facility) at a time, list the activity number(s) under the proper relationship code. Two points will assist the designer here:
   a. Be sure every activity number appears on each line (1–14 must be somewhere on each line).
   b. The relationship codes for an activity center are listed below as well as above the activity name, as shown by the direction arrows in Figure 6–1. For example, line 2's (welding) relationship code with fabrication is an A and is located at the coordinate 1-2.

The activity relationship worksheet shows the same relationships as the activity relationship diagram.

DIMENSIONLESS BLOCK DIAGRAM

The *dimensionless block diagram* is the first layout attempt and the result of the activity relationship chart and the worksheet. Even though this layout is dimensionless, it will

Activities	*A*	*E*	*I*	*O*	*U*	*X*
1. Fabrication	2, 6	3, 10	9, 11, 13, 14	4, 5, 12	7, 8	
2. Welding	1, 3		6	9, 10, 12, 13, 5	7, 8, 4, 11, 14	
3. Paint	2, 4	1	6	12, 13, 9	5, 7, 8, 10, 11, 14	
4. Assembly and P.O.	3, 7	6, 8	9, 12, 13, 14	1, 5	2, 10, 11	
5. Receiving	6		14	4, 2, 1, 9, 12, 13	3, 7, 10, 11	8
6. Stores	5, 1	4	3, 2, 14	9	8, 10, 11, 12, 13	7
7. Warehouse	4, 8			14	5, 3, 2, 1, 9, 10, 11, 12, 13	6
8. Shipping	7	4	14	9, 12, 13	6, 3, 2, 1, 10, 11	5
9. Restrooms	12	13, 14	4, 1	8, 6, 5, 11, 3, 2, 10	7	
10. Maintenance	11	1		9, 2	8, 7, 6, 5, 4, 3, 12, 13, 14	
11. Tool room	10		1	9, 14	8, 7, 6, 5, 4, 3, 2, 12, 13	
12. Locker room	9	13	4	8, 5, 3, 2, I	11, 10, 7, 6, 14	
13. Cafeteria		14, 12, 9	4, 1	8, 5, 3, 2	10, 11, 7, 6	
14. Office		13, 9	8, 6, 5, 4, 1	11, 7	12, 10, 2, 3	

Figure 6–2 Activity relationship worksheet.

be the basis for the master layout and plot plan. Once the size of every department, office, and support facility has been determined, space will be allocated to each activity per the dimensionless block diagram's layout. Area allocation will be discussed in depth in Chapter 13. If you obey the activity codes, a good layout will result. Sometimes it is harder laying out the dimensionless block diagram than when exact sizes are available, because large departments tend to have more A and E relationships than small departments and can have many more departments (activities) on its borders. Here's a step-by-step procedure for a dimensionless block diagram:

1. Cut up a sheet of paper into about 2 × 2-inch squares. (In this example, 14 squares are needed.)
2. Place an activity number in the center of each square (1 to 14 in this example).
3. Taking one square at a time, make a template for that activity by placing the relationship codes in the following positions (see Figure 6–3):
   a. A relationship in the top left corner
   b. E relationship in the top right corner
   c. I relationship in the bottom left corner
   d. O relationships in the bottom right corner
   e. U relationships omitted
   f. X relationships in the center under the activity number
4. Each activity center is represented by one square (see Figure 6–4).
5. Once the 14 templates are ready, you place them in the arrangement that will satisfy as many activity codes as possible.

Start with the activity with the most important closeness codes. For example, in Figure 6–2, activities 1 and 4 have two A and two E codes. Start with either one of these activities. Place the template you chose in the middle of your desk. Look at the A codes, find those A-coded activity templates, and place those templates

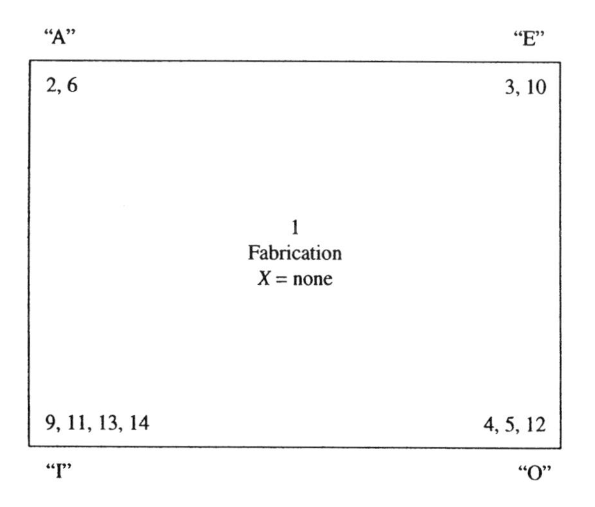

Figure 6–3 One square representing the fabrication department.

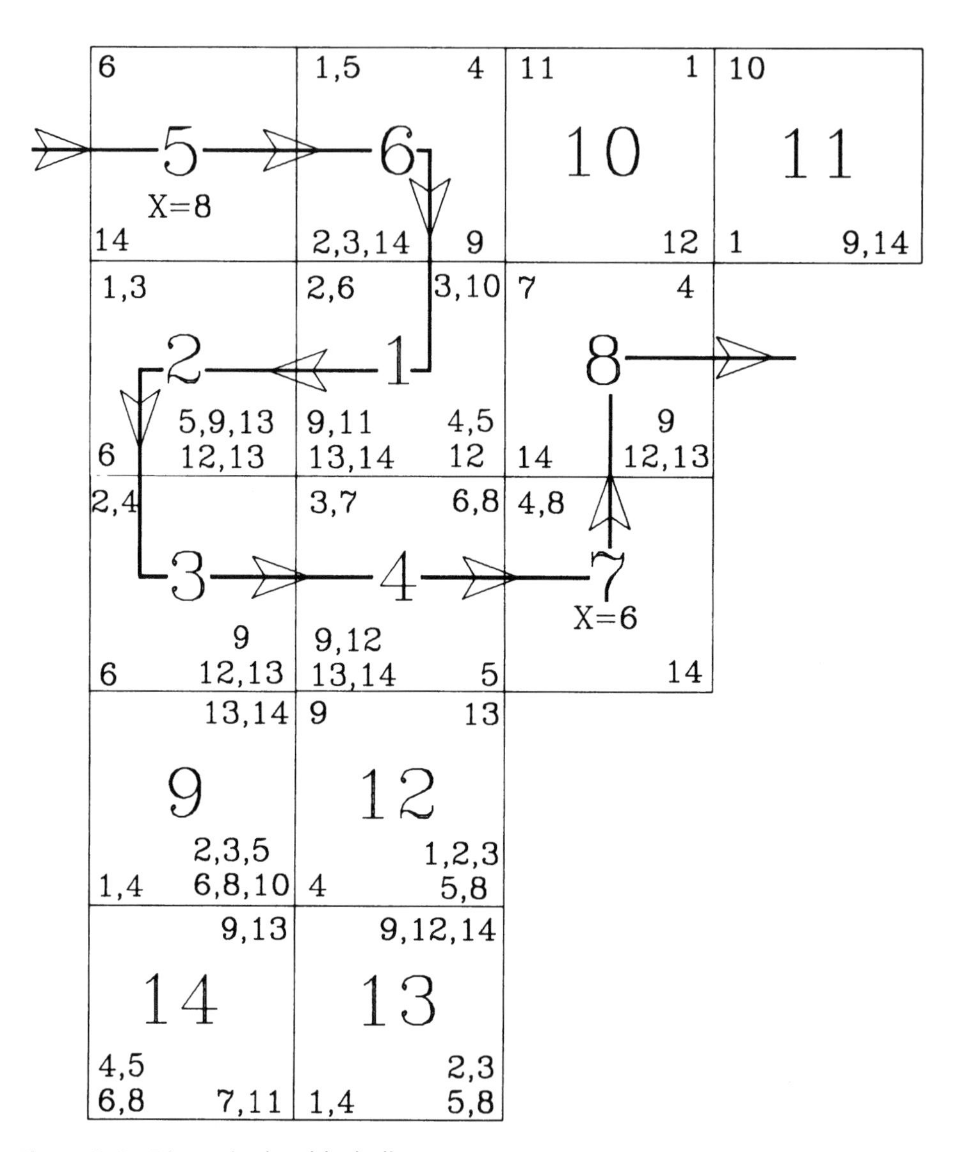

Figure 6–4 Dimensionless block diagram.

adjacent to the first template with a full side against the first template. In the example shown in Figure 6–3, pick up the template for activity 1 (fabrication) and place it in the middle of your desk. Now pick up templates 2 and 6 (because they have an A relationship with 1). Place template 2 on any one of template 1's four sides but ensure full side contact. Place template 6 also where it has a full side with 1, but notice that template 6 needs to be close to (have an I relationship with) template 2 also, so that you allow a corner to touch. You now have three templates (activities) positioned. Continue to pick up additional A relationships and place them where they satisfy the most activity relationships until all departments are accounted for. If you have more than one place to put a template, consider the lesser important (E, I, O codes) relationships (see Figure 6–4). Since this is a dimensionless layout, shape is of no importance; only satisfying the relationship codes is the primary consideration.

All As should have a full side touching. All Es should have at least a corner touching. No X relationship should be touching.

Give two checkmarks for As that are not touching at all or for Xs touching with a full side; one checkmark for As with only a corner touching, with an X touching a corner, or with an E not touching at least one corner. Try to accommodate all the A, E, and X codes. Use the I codes when you can, but there is usually more than enough important codes so that I and O codes are seldom used. You probably noticed that U codes did not even have a place on the template, and if it were not for the ability to check yourself by ensuring that every activity has been included, you could just leave out unimportant relationships.

How many checkmarks do you find? The fewer checkmarks, the better. In this example, 6-4 gets one checkmark, and 4-6 gets one checkmark.

FLOW ANALYSIS

Flow analysis is now performed on the dimensionless block diagram. Starting with receiving, show the movement of material to stores, to fabrication, to welding, to paint, to assembly and packout, to the warehouse, and to shipping. Flow analysis will ensure that important relationships will be maintained and that your layout makes good sense. You would not want material to flow through the corner of a department, nor to jump over one or more departments. Also, you would not want shipping or receiving to be located in the middle of the building.

Electronic copies of the activity relationship chart and the worksheet are provided for your use. You may download, copy and utilize these forms for your project.

COMPUTER-GENERATED ACTIVITY RELATIONSHIP CHART

As discussed in the previous chapters, state-of-the-art software packages are available to aid facilities planners in achieving an optimum or, at least, a near-optimum solution to a layout problem. One area in which planners can enjoy the benefits of the computer-aided layout tools is in the generation of the activity relationship chart and the resulting block diagram.

FactoryPLAN, via a series of interactive menus and on-screen prompts, assists the user in arranging the layout based on the closeness ratings between pairs of activity centers or work areas. The software can be used to lay out a manufacturing facility with discrete product lines using data based on material flow, personnel

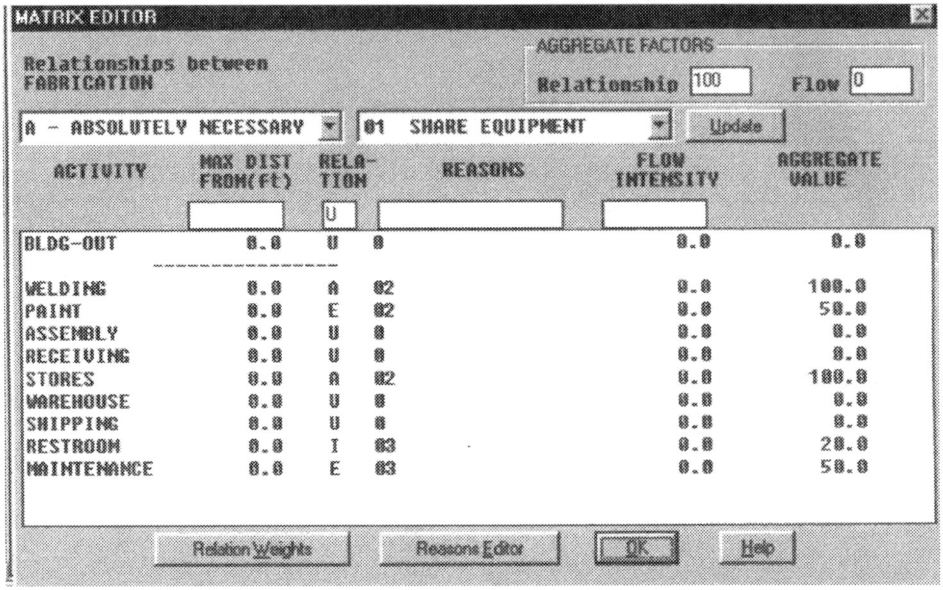

MATRIX EDITOR

Relationships between FABRICATION

AGGREGATE FACTORS — Relationship 100 — Flow 0

A - ABSOLUTELY NECESSARY | 01 SHARE EQUIPMENT | Update

ACTIVITY	MAX DIST FROM(Ft)	RELA-TION	REASONS	FLOW INTENSITY	AGGREGATE VALUE
		U			
BLDG-OUT	0.0	U	0	0.0	0.0
WELDING	0.0	A	02	0.0	100.0
PAINT	0.0	E	02	0.0	50.0
ASSEMBLY	0.0	U	0	0.0	0.0
RECEIVING	0.0	U	0	0.0	0.0
STORES	0.0	A	02	0.0	100.0
WAREHOUSE	0.0	U	0	0.0	0.0
SHIPPING	0.0	U	0	0.0	0.0
RESTROOM	0.0	I	03	0.0	20.0
MAINTENANCE	0.0	E	03	0.0	50.0

Relation Weights | Reasons Editor | OK | Help

Figure 6–5 Relationships between fabrication and welding (courtesy of Engineering Animation, Inc.).

interactions, shared equipment, and a variety of other reasons. The analysis is performed in three steps:

- Create a data file containing activity center names. At this point, the user can also enter the type of activity, such as material handling, inspection, office, and so on that takes place at this center. The user also has the option to specify the required space for the center.
- Once the list is completed, the user is prompted to specify the affinity code and reason code between pairs of work centers. As with the manual method, it is the responsibility of the facilities planner, based on available data, to determine the desired degree of closeness between the work centers and to give a reason code for the required proximity. Figure 6–5 shows the menu for entering the relationship code between fabrication and welding. As shown in the figure, code A and the reason code 2 are stated. Once the relationships between fabrication and other centers are stated, the next center is automatically selected so that the codes between those departments can be filled in.
- The third step of the analysis is the generation of the activity relationship chart and the flow path diagrams.

Figure 6–6 displays the activity relationship chart generated by FactoryPLAN software based on the data supplied by the planner. In this example, the information from Figure 6–1 was used so that a closer comparison could be drawn between a manual and a computer-aided outcome. The tool room, locker room, cafeteria,

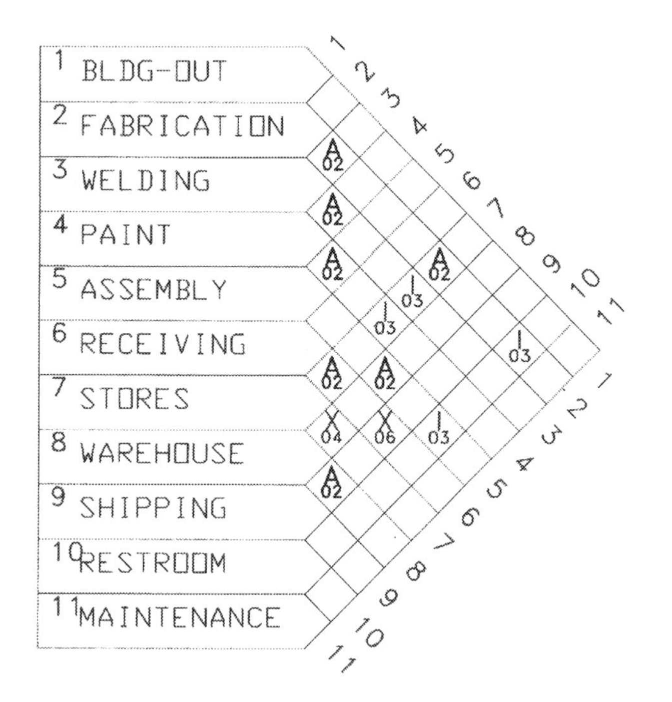

Figure 6–6 FactoryPLAN-generated activity relationship chart (courtesy of Engineering Animation, Inc.).

and office areas were left out to enhance clarity of the flow diagram. You will also notice that in the activity relationship diagram generated by the computer, a center called "BLDG-OUT" is generated by the software. This artificial activity is created to represent the relationships between the activities in the facility and the outside world such as the flow of material into the receiving department and the flow of the material out of the shipping department. Each diamond contains the closeness code and the reason code. The software-generated chart omits the U codes.

The software will generate an optimized layout based on the data that are entered by the user. The layout can then be manipulated and rearranged by simply selecting a work center and moving its location.

The users may also generate their own layout. The software will prompt the users, starting with work centers that have A relationship codes, to place the activity center. The center is placed by a simple click of the mouse. With the placing of each center, flow lines are automatically generated.

The prompting will continue until all work centers have been placed. At any point in the process, the planner may move an activity center to a more desirable location based on the density of the flow lines. Figure 6–7 shows the flow lines that are generated based on the interdepartmental activities and the flow of materials or personnel. Study the number and density of these flow lines. Activity centers can be

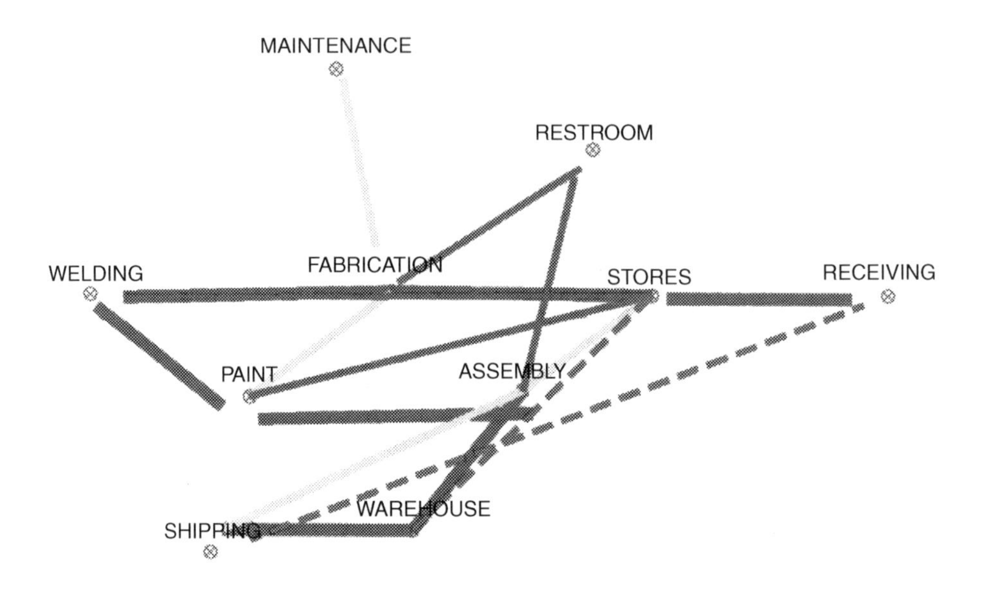

Figure 6–7 Flow lines (courtesy of Engineering Animation, Inc.).

easily moved to shorten the distances of those centers that are connected by the heaviest lines. In Figure 6–7, an attempt was made to duplicate the layout shown in Figure 6–4. Compare the two drawings. One can certainly appreciate the more telling tale of traffic that occurs among all centers as shown in Figure 6–7.

QUESTIONS

1. What are the six activity codes and for what do they stand?
2. From where do these codes come?
3. What are reason codes? Why are they used? How are they used?
4. Why do you need a worksheet?
5. What is a template?
6. Develop a dimensionless block diagram for the activity relationship diagram in Figure 6–8. How many checkmarks did you come up with? Remember, the fewer the better.
7. If a facility contains 15 different departments, how many total affinity codes do you expect?
8. Using the rule of thumb, what would be the maximum number of A, E, and I codes that you would be inclined to use?

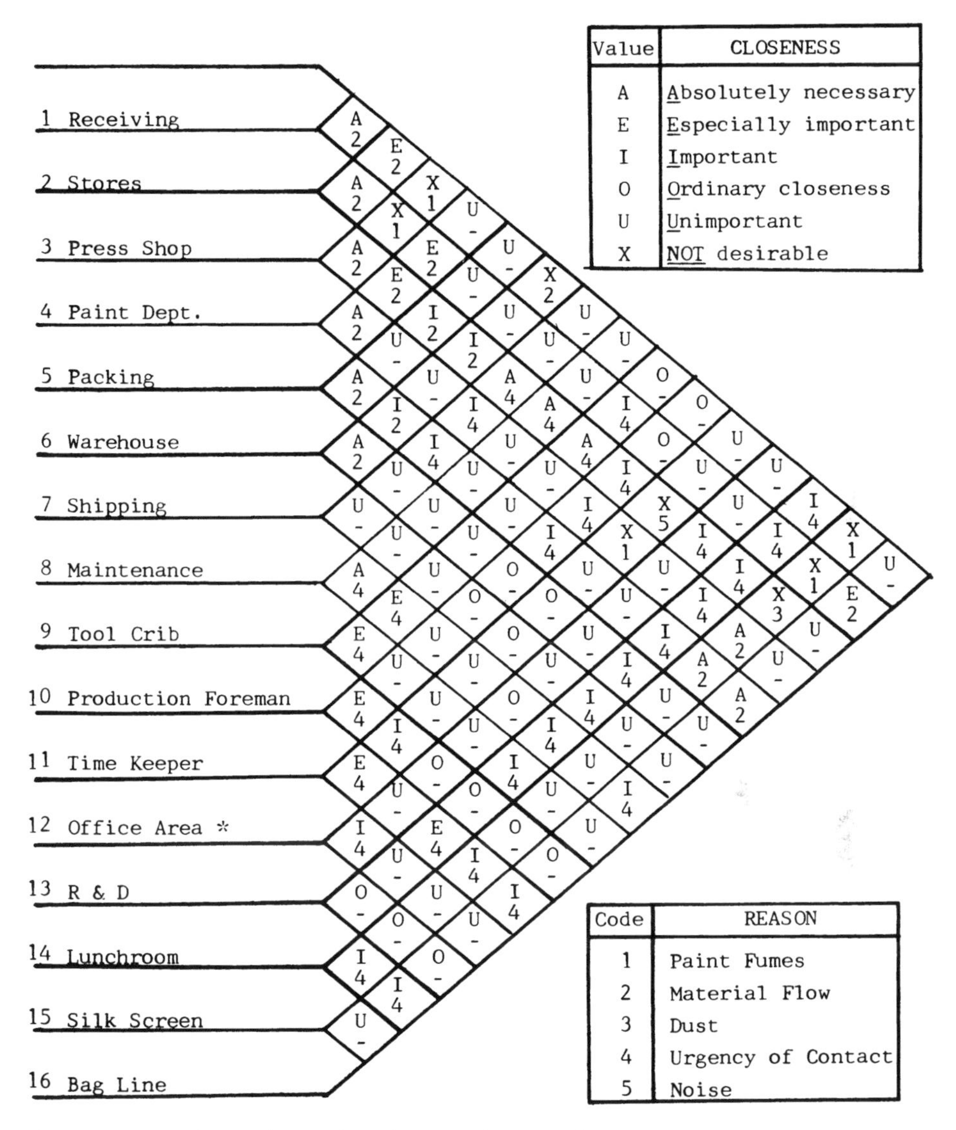

Figure 6–8 Activity relationship diagram for question 6.

CHAPTER 

Ergonomics and Workstation Design Space Requirements

OBJECTIVES:

Upon the completion of this chapter, the reader should:

- Understand ergonomic principles as applied to workstation design
- Understand the concepts of motion economy
- Be able to apply these principles and concepts to work space planning and space determination

WORKSTATION DESIGN

The result of ergonomics and workstation design is a workstation layout, and the workstation layout determines the space requirements. The manufacturing department's total space requirements are just a total of individual space requirements plus a contingency (a little extra) factor.

Ergonomics is the science of preventing musculoskeletal injuries in the workplace. It is the study of workplace design and the integration of workers with their environment. Ergonomic considerations include employee size, strength, reach, vision, cardiovascular capacities, cognition, survivability, and, more recently, cumulative musculoskeletal injuries. Safety and health considerations are now an integral part of workstation design, and workstation designers must continue their education in this area. Ergonomics is an important subject in industry today. The text in Figure 7–1, provided by Aero-Motive Manufacturing Company, describes the importance of ergonomics.

The word "ergonomics" comes from two Greek words, *ergon,* meaning "work" and *nomos,* which means "rules" or "laws." One may loosely translate the word

Ergomation is the successful incorporation of the worker and the process environment. This process has become an integral part of the workplace as studies have proven that thousands of hours are lost due to RMI (Repetitive Motion Injury).

Flexibility plays an important role in the work process; ergomation products have been created to allow for physical differences in operators. As a result, worker movements are deliberate and put the worker in the best possible position.

Modularity is also vital. All aspects of the work area have universal connecting hardware that allows unlimited configurations.

Figure 7–1 Ergomation (adapted from Aero-Motive Manufacturing).

"ergonomics" as "laws or rules of performing or doing work." The discipline of ergonomics is also referred to as human factors or human engineering.

The discussion of ergonomics is best left for a course that deals specifically with that subject. But to the extent that it relates to workstation design, the golden rule may be stated as follows: *Design the work or the workstation so that the task fits the person rather than forcing the human body or psyche to fit the job.* To achieve this seemingly simple, yet extremely important principle, one area of ergonomics, called *anthropometry,* provides insight into the physical measurements of the human body. Using the basic statistical tools, anthropometry defines the range of variations and distribution when dealing with various physical measurements and characteristics of the human body, such as height, strength, and length of reach, among other data.

Anthropometric data, then, help planners design workstations, manual unit loads, or other tools to accommodate the majority of workers. For example, if a workstation is designed for the 5th percentile woman, the 95th percentile man will encounter great difficulty in performing his task at this station. A hand tool requiring the strength or the grip of the 95th percentile man will not accommodate the 5th percentile woman.

Not all aspects of workstation design that are necessary fall under strict measurement and statistical distributions of anthropometrics. Common sense does play a role as well. Understand the natural posture or the comfortable state of the worker. Consider the height of the workbench in relation to the worker's elbows. Are the elbows raised or are they at the 90° angle when doing work? Which one is a more comfortable position? How about the wrists? Do the worker's wrists lay flat, or are they bent in an upward position (the major cause of carpal tunnel syndrome) while doing work?

Improper workstation design costs American industry millions of dollars annually in lost productivity, health, and job-related injuries and accidents.

The resulting workstation design is a drawing, normally a top view, of the workstation, including the equipment, materials, and operator space. Designing workstations has been an activity performed by industrial and manufacturing engineers for nearly a century. During this period, the profession has developed a list of principles of ergonomics and *motion economy* that all new engineers should learn and apply. When the principles of ergonomics and motion economy are properly applied to the design of a workstation, the most efficient and safe motion patterns will result.

"Where to start?" is the first question most often asked by new workstation designers. The answer is very simple—start anywhere! No matter where you start in designing a workstation, another idea will come along making that starting point obsolete. Where to start depends a great deal on what is to be accomplished at that workstation. The cheapest way to get into production is usually the best rule for the starting point. The cheapest way means just that—the simplest machines, equipment, and workstations. Savings must justify any improvement on this most economical method. Therefore, the designer is free to start anywhere, then improve on the first method.

The following information must be included in any workstation design:

1. Worktable, machines, and facilities
2. Incoming materials (material packaging and quantity must be considered)
3. Outgoing material (finished product)
4. Operators' space and access to equipment
5. Location of waste and rejects
6. Fixture and tools
7. Scale of drawing (see Figure 7–2)

A three-dimensional drawing would show an even greater amount of information. Any talented designer could attempt a three-dimensional design. Figure 7–3 is a photo of a well-planned workbench.

The second example of workstation design will be of a machine operation (see Figure 7–4). The needs of this station design are the same as the previous station, but the equipment (machines, jigs, and fixtures) will be added.

Figure 7–5 through 7–10 (pp. 208–211) are workstation designs for the equipment required in the toolbox plant example. Figure 7–11 is a workstation design for the toolbox paint system.

ERGONOMICS AND THE PRINCIPLES OF MOTION ECONOMY

Industrial and manufacturing engineers are continually developing guidelines for safer and more efficient and effective workstation design. Frank and Lillian Gilbreth originally collected these guidelines and titled them "The Principles of Motion Economy." Ralph Barnes has updated and published these principles since 1937. Ergonomics started during World War II and has just recently become an important part of industrial and manufacturing engineering.

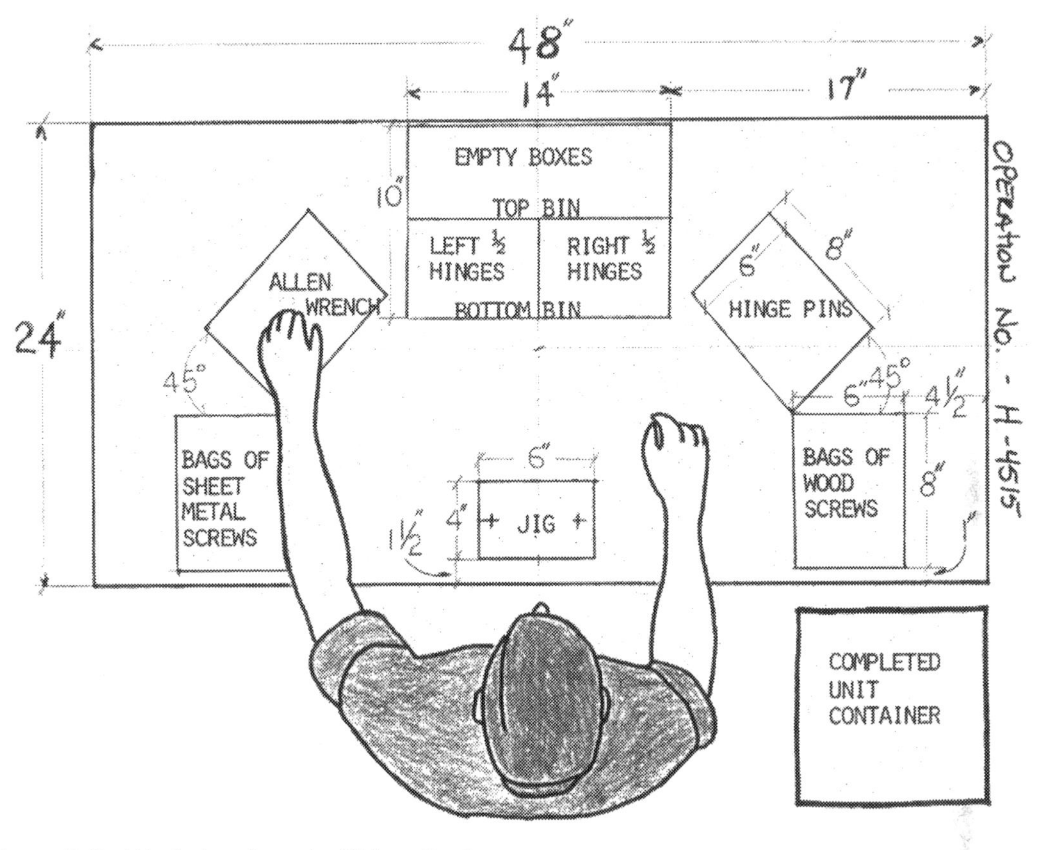

Figure 7–2 Workplace layout—Old method.

Effectiveness is doing the right job. *Efficiency* is using the job right. Effectiveness is important to consider first because doing an unnecessary job is bad, but making a useless job efficient is the worst sin. Safety and efficiency should be the goals of every workstation designer.

Ergonomics and the principles of motion economy should be considered for every job. Sometimes principles will be violated with good reasons. These violations and reasons should be written up for future use. You will have to defend yourself to every new workstation designer, so be prepared—write it up.

The principles are often used together in very creative ways, but knowledge of these principles is the starting point. The only limit to improved workstation design is the designer's creativity.

Principle 1: Hand Motions

First of all, hand motions should be eliminated as much as possible. Let a mechanical device do it, but if needed (and many hand motions are necessary), the hands should operate as mirror images. They should make start and stop motions at the

Figure 7–3 Workbench (courtesy of American Seating Co.).

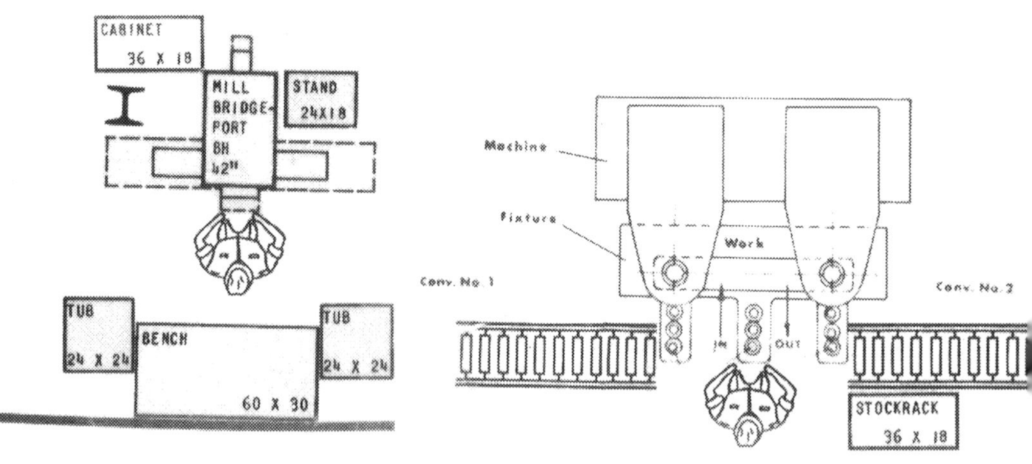

Figure 7–4 Workstation layout.

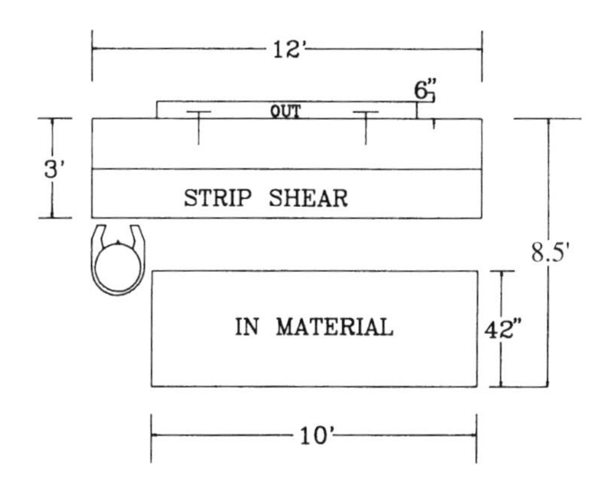

Figure 7–5 Strip shear—Total square feet: 102.

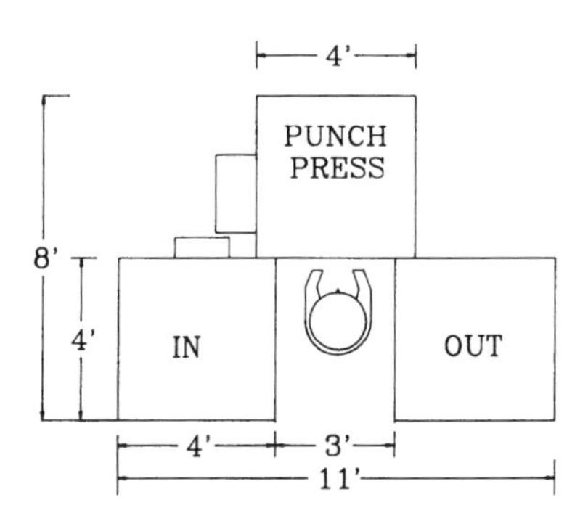

Figure 7-6 Punch press—Total square feet: 88.

same time; they should move in opposite directions; and they should both be working at all times.

If the hands are reaching for two parts at the same time, the bins should be placed the same distance back from the work area and from the centerline of the workstation.

Reaching for only one part with one hand leaves the question of what the other hand is going to do. To keep both hands working at all times is a large challenge and can be most easily accomplished by doing two parts at a time (one complete task with the left hand and one complete task with the right). Holding parts in one hand while assembling other parts to it is a very poor use of the hands. (Think about how you would redesign this task.) This is affectionately called a "one-arm bandit job." It is said that the most expensive fixture in the world is the human hand.

Keep in mind that in workstation design, the issue of workers being right-handed or left-handed is not considered. Furthermore, if hand tools are used they

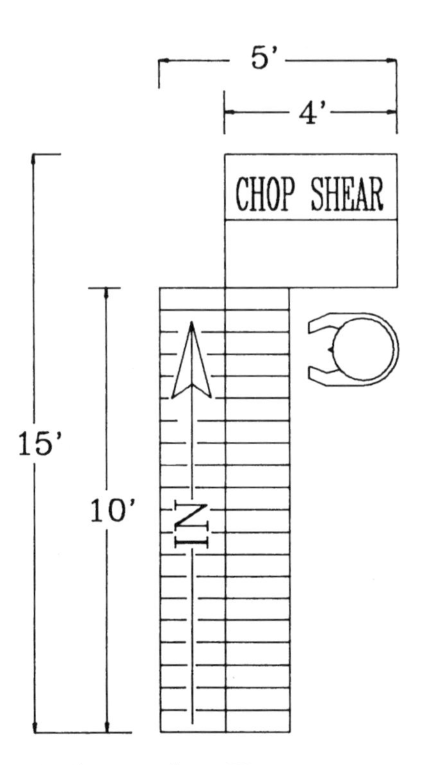

Figure 7–7 Chop shear—Total square feet: 75.

must be designed ergonomically, and they must be easily adapted to both left-handed and right-handed individuals. Considering that over 10 percent of people are left-handed, the probability of having a left-handed person operating at a given station is easily within the realm of possibility.

Principles of motion economy related to hand motions:

1. Eliminate as many hand motions as possible.
2. Combine motions to eliminate other motions.
3. Make motions as short as possible, and discourage leaning because of excessive reaching.
4. Reduce the force required as much as possible.
5. Keep both hands equally busy.
6. Use mirror image moves.
7. Do not use the hand as a holding device.
8. Locate frequently used tools and materials closer to the point of use, and tools and materials used less often farther away. The weight of tools and materials should also influence their closeness to the point of use. Place heavy material closer to the point of use.

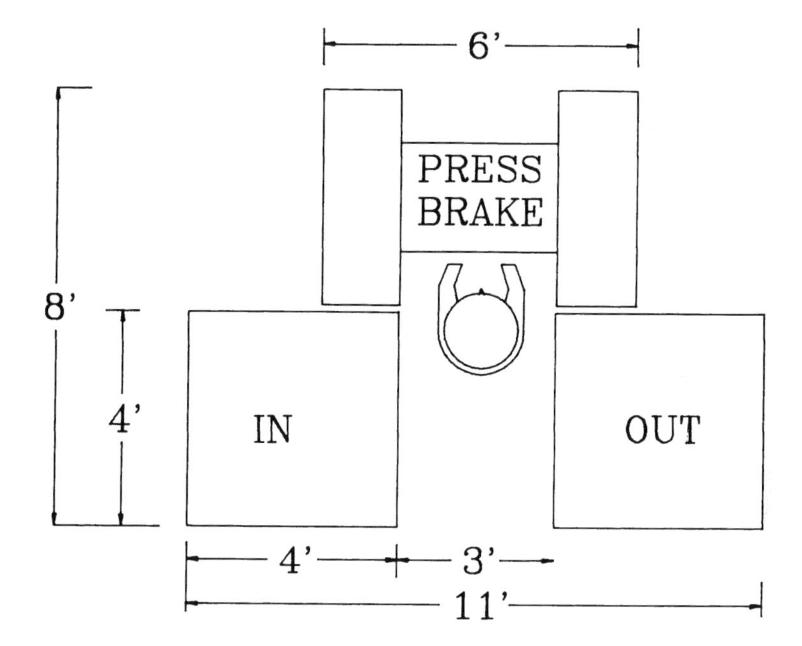

Figure 7–8 Press brake—Total square feet: 88.

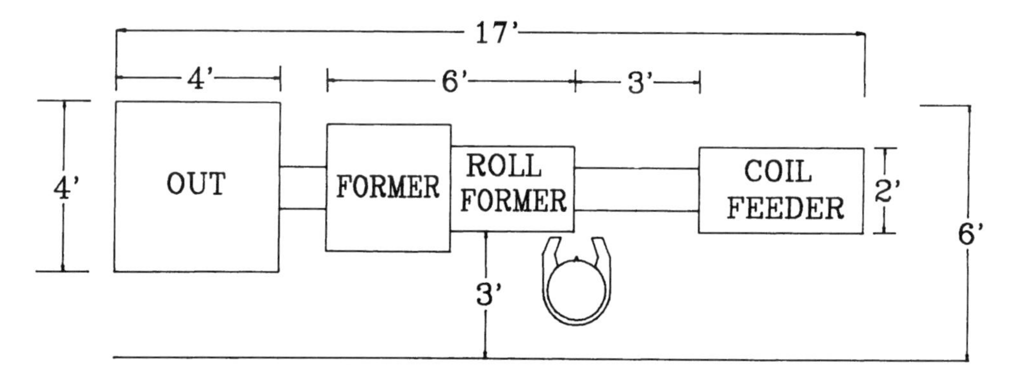

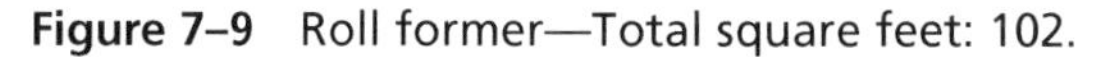

Figure 7–9 Roll former—Total square feet: 102.

Principle 2: Basic Motion Types

Ballistic motions are fast motions created by putting one set of muscles in motion and not trying to stop those motions by using other muscles. Throwing a part in a tub or hitting a panic button on a machine are good examples. Ballistic motions should be encouraged.

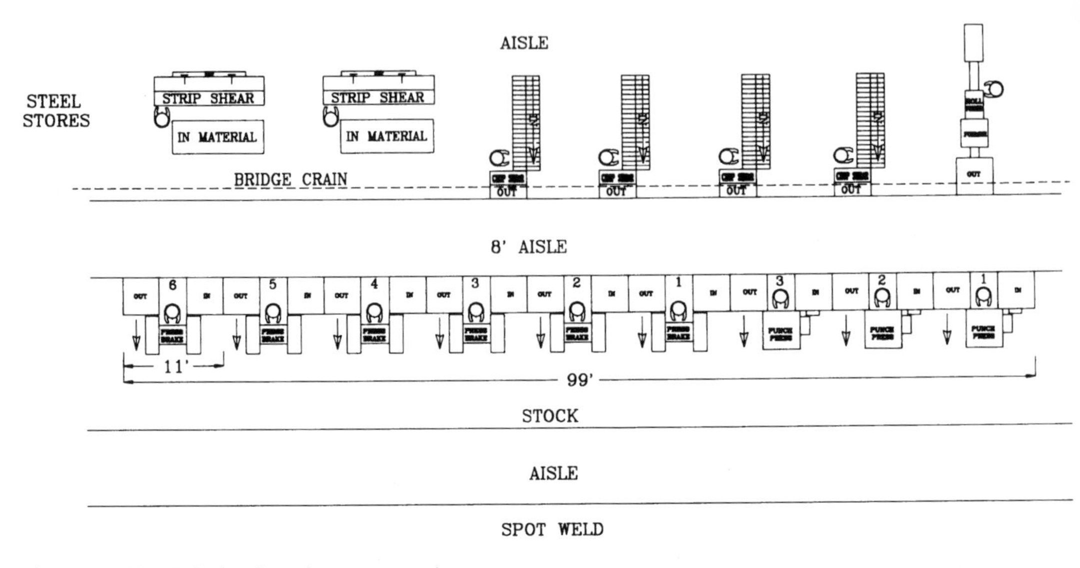

Figure 7–10 Fabrication department layout.

Controlled or *restricted motions* are the opposite of ballistic motions and require more control especially at the end of the motion. Placing parts carefully is an example of a controlled motion. Safety and quality considerations are the best justification for controlled motions, but if there are ways to substitute ballistic motions for controlled motions, cost reduction can result. Controlled motions are to be considered first for elimination—try to design a means for avoiding their use because they are costly, fatiguing, and unsafe.

Continuous motions are curved motions and much more natural than straight-line motions, which tend to be controlled or restricted motions. When the body part has to change direction, speed is reduced and two separate motions result. If direction is changed less than 120°, two motions are required. Reaching into a box of parts lying flat on the table is an example of requiring two motions: one motion to the lip of the box and another down into the box. If the box were placed at an angle, one motion could be used. This principle will be shown in greater detail in the gravity principle section of this chapter.

Principle 3: Location of Parts and Tools

Have a fixed place for all parts and tools and have everything as close to the point of use as possible. Having a fixed place for all parts and tools aids in habit formation and speeds up the learning process. Have you ever needed a pair of scissors, and when you looked where they were supposed to be, they were gone? How efficient were you in the next few minutes? A toolmaker's toolbox is laid out so that the toolmaker knows where every tool is and can retrieve it without looking. This should be a goal in every workstation that planners design.

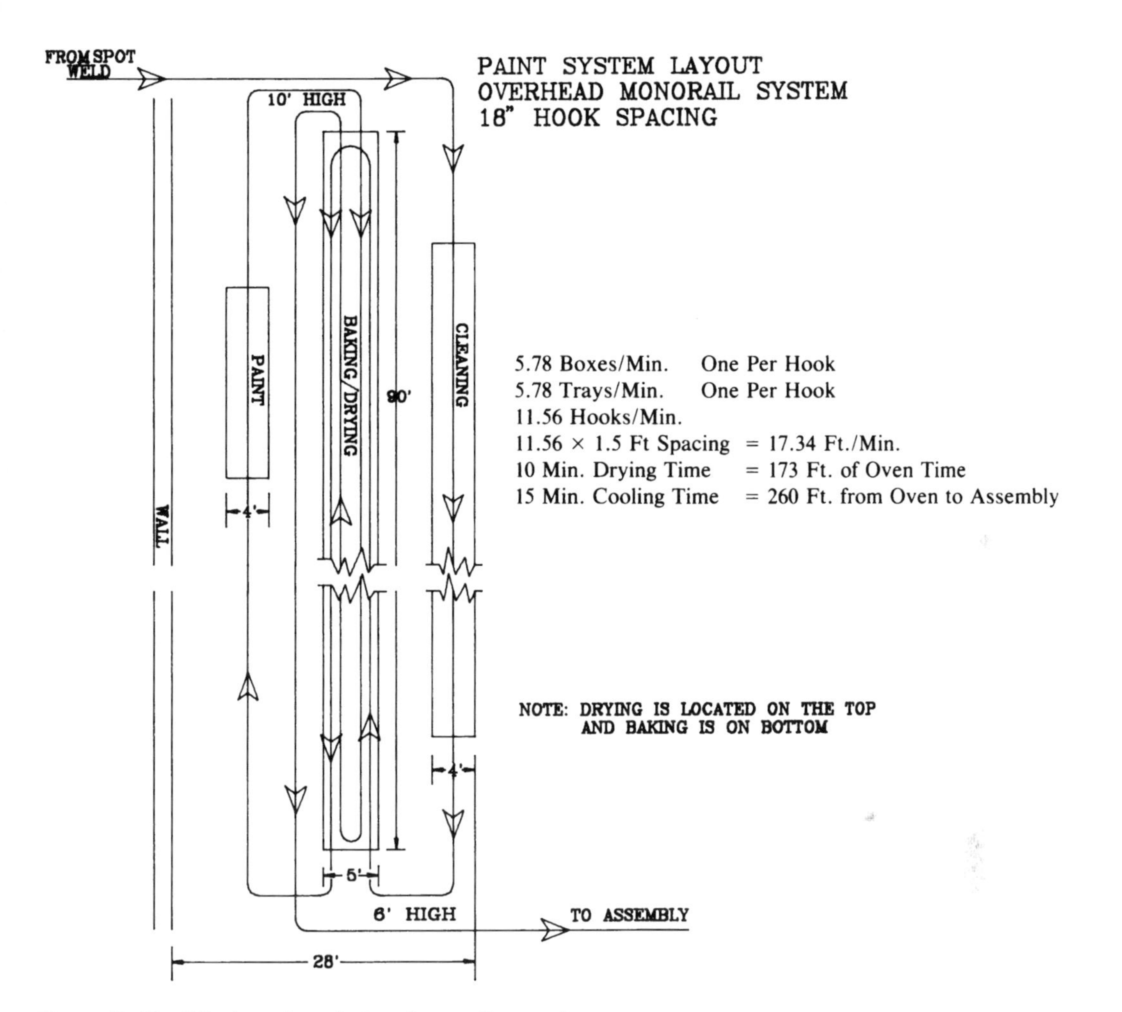

Figure 7–11 Workstation design for toolbox paint system.

The need for having parts located as close as possible to the point of use is quite evident, and it should be no surprise to learn that the farther you reach for something, the more costly and fatiguing that reach will be. Real creativity is required to minimize reaches. You can place parts on two tiers, instead of having one row of parts across the top of the workstation, or maybe three tiers of parts one over the other would be better. You can hang tools from counterbalances over the workstation. Or, you can use conveyors to move parts into and out of the workstation.

Here is a summary of the location of parts and tools:

1. Have a fixed location for everything.
2. Place everything as close as possible to the point of use.

Principle 4: Freeing the Hands from as Much Work as Possible

As stated earlier, the hand is the most expensive fixture that a designer could use. So, you must provide other means of holding parts. Fixtures and jigs are designed to hold parts so that the worker can use both hands. Foot-operated control devices can be designed to activate equipment to relieve the hands for work. Conveyors can move parts past operators so that they don't have to get or set aside the base unit. Powered round tables are also used to move parts past an operator. Fixtures can be electric, air, hydraulic, and manually activated. They can be clamped with little pressure or tons of pressure. Clamping devices can be automatically activated and the hand can be relieved of the task. Clamping devices can have any shape, which will be determined by the shape of the part. A hex nut can be placed in a hex-shaped hole that has no clamping need, but it will be held firm because of the part and fixture shape. Toy manufacturers need to hold toys in a clamping device until the glue dries. The clamp can have the exact shape as the toy bottom and toy top. Fixture design is easy and only your knowledge of the part and needed processes are required to design fixtures. Many tooling vendors would "love" to supply you with fixture building materials and devices. (See Figure 7–12.)

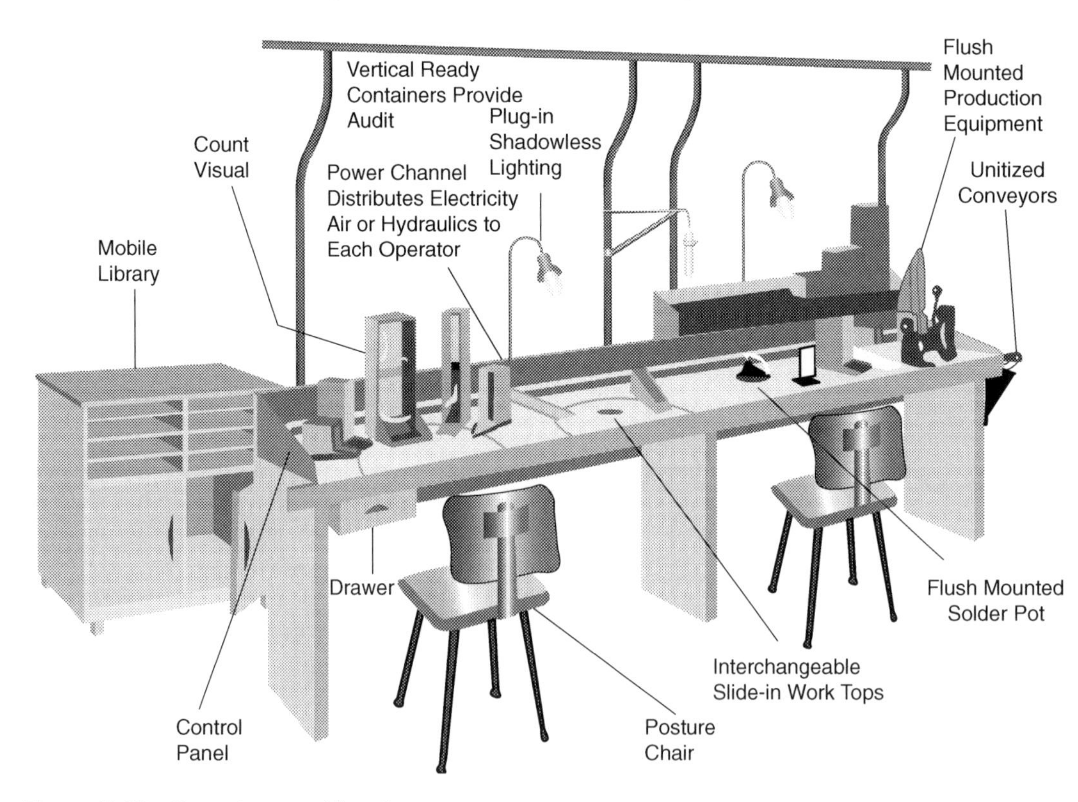

Figure 7–12 Operator considerations.

Principle 5: Gravity

Gravity is free power. Use it! Gravity can move parts closer to the operator. By putting an incline in the bottom of parts hoppers, parts are moved closer to the front of the hopper. Production management loves to spare every expense, and the use of gravity can do that. For example, consider a box that is 24 × 12 × 6 inches lying flat on the table. The average part in that box (the only part the designer is interested in) is 12 inches back, 6 inches over, and 3 inches down the exact middle of the box. Now if you get a scrap 2 × 4-inch board out of the trash and place it under the rear end of the box and raise it up 4 to 5 inches, the parts will slide down to the front of the box as the parts are used. The operator's reach has been reduced from 12 to 3 inches from the front lip of the box—a significant cost reduction on both reaches to the part and the moves back. This is a continual saving of about $2.20 per 1,000 parts. Large boxes of parts can be moved into and out of workstations using gravity rollers and skate wheel conveyors. Parts can be moved between workstations on gravity slides made of sheet metal, plastic, and even wood.

Gravity can also be used to remove finished parts from the workstation. Dropping parts into chutes or slides that carry the parts down and away from the workstation can save time, operator fatigue, and workstation space. Slide chutes can carry punch press parts away from the die without operator assistance by using jet blasts of air, mechanical wipers, or even the next part pushing the finished part from the die.

Gravity use is everywhere. Workstation designers should try to incorporate it in their designs as much as possible. Designing the use of gravity into your workstation is a challenge, and it is fun. Opportunities are everywhere. Find them!

Principle 6: Operator Safety and Health Considerations

Keep safety hazards in mind and anticipate emergency action requirements while designing the workstation. Operator safety and health is your responsibility. You must consider the anthropometric dimensions of the workforce while designing the workstation. Design the workstation to eliminate straining of the neck to look at things, to eliminate stooping or bending, to eliminate turning sideways or turning around, and to eliminate excessive reaches and moves.

Operators become efficient and stay healthy if they are allowed to work at the right height, given the opportunity to work while both sitting or standing, given enough light to work by, and given adequate space to perform their tasks.

The *correct work height* is elbow height plus or minus 2 inches. Light work can be 2 inches above elbow height, whereas heavy work should be 2 inches below elbow height. Elbow height is measured with the forearm held parallel to the ground and the upper arm held straight down; measure the elbow height to the floor. This is the work height. A job should be designed for sitting or standing, but the elbow height must be the same. This requires the designer to calculate working height while standing, then to provide a chair that will accommodate that height while sitting. Many workstations need to be used by several people. To maintain the correct work height, have adjustable workstations, design the station for the tallest person to be operating the station and provide platforms for the shorter people, or adjust the work height on top of the workstation.

The industrial chair will need to be adjustable to maintain proper work height. Because work height is dependent on the individual, chairs and tables will have to be adjustable for efficient operation. These facilities are readily available commercially. The chair must also be comfortable. This usually means that it supports the back, and a foot ring aids comfort and reduces lower back fatigue. Comfortable chairs and the option of working while sitting or standing give the operator a chance to move around and reduce the effects of fatigue. Foot pedals, controls, or knee-operated devices can eliminate hand motions, but avoid the use of foot pedals or controls unless the operator is sitting.

Adequate lighting may not be available in the normal lighting of a manufacturing department, so additional lighting should be added—much like a desk lamp. The closer the work is, the more need there is for light. Where to place this lighting is the problem. The best place is over the work and slightly over the back, but not casting a shadow. Much lighting is placed in front of the work, but this causes glare from the reflection. Auxiliary lights could be placed to the left or right of the work as well.

Operator space should be 3×3 feet, which is normal unless the workstation is wider, but 3 feet times the width of the workstation may be needed. Three feet off the aisle is adequate for safety, and 3 feet from side to side allows parts to be placed comfortably next to the operator. If two people are working back to back, then 5 feet between stations is recommended. If machines need maintenance and cleanup, a 2-foot access should be allowed around the machine. Movable equipment can be placed in this access area if needed for efficient operation.

SPACE DETERMINATION

The *space determination procedure* for most production departments starts with the workstation design. From each workstation layout, measure the length and width to determine the square footage of each station. The following data resulted from the workstation layouts in Figure 7–5 to 7–11, and Figure 4–12 and 4–13 in Chapter 4.

Multiplying the total square feet by 150 percent allows extra space (this could be 200 percent if management wants to provide a spacious layout, or a larger contingency allowance) for the aisle, work in process, and a small amount of miscellaneous extra room. It does not include restrooms, lunchrooms, first aid, tool

	Length	×	*Width*	=	*Sq Ft*	×	*No. of Stations*	*Total Square Feet*	*Figure*
Strip shear	12	×	8.5		102		2	204	7–5
Chop shear	15	×	5		75		4	300	7–7
Punch press	11	×	8		88		3	264	7–6
Press brake	11	×	8		88		6	528	7–8
Roll former	17	×	6		102		1	102	7–9
paint system	100	×	28		2,800		1	2,800	7–11
Spot welding	34	×	28		952		1	952	4-12
assembly	38	×	16		608		1	608	4-13
								Total square feet 5,758	
								× 150 percent = 8,637 square feet are required	

rooms, maintenance, offices, stores, warehouse, shipping, or receiving. These area requirements will be discussed in Chapters 8 and 9. The extra 50 to 100 percent space added to the equipment space requirement will be used mostly for aisles. Aisles can be very space consuming; for example, let us lay out a 100 × 100-foot plant as follows:

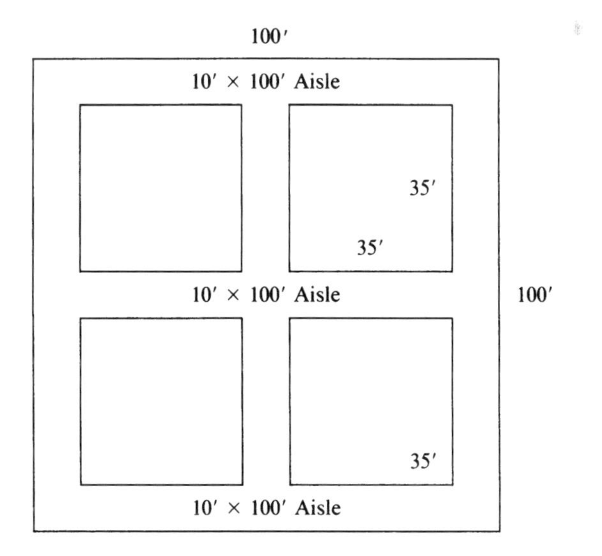

A 10-foot aisle around the outside of the production area will eliminate clutter next to the walls. But that leaves you with an 80 × 80-foot area with no aisles. Put in 10-foot cross aisles. How much room did you use?

(3) 100 feet long, 10 feet wide aisles	=	3,000 ft^2
(3) 70 feet long, 10 feet wide aisles	=	2,100 ft^2
Total Aisle Square Feet		5,100 ft
Total Square Feet (100 feet × 100 feet)		10,000 ft^2

$$\frac{5{,}100 \text{ ft}^2}{10{,}000 \text{ ft}^2} = 51 \text{ percent aisles}$$

Given that this layout requires 51% for aisle space, the 50% rule-of-thumb aisle space allocation would therefore not be sufficient. A better aisle plan might be similar to the following:

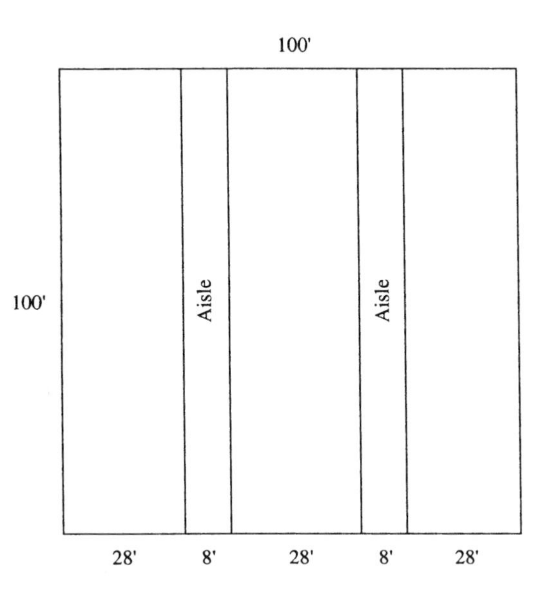

Two 100-foot aisles 8 feet wide equals 1,600 square feet.

$$\frac{1{,}600 \text{ ft}^2}{10{,}000 \text{ ft}^2} = 16 \text{ percent aisles}$$

This may be too tight, but notice the improvement from 51 percent down to 16 percent. You also have better access to areas (35-foot- vs. 28-foot-wide areas).

Small space-consuming items such as an air compressor or drinking fountain may be included in this 50 percent extra area, but large area requirements must be designed and planned. The next chapter addresses those other areas that require space designs.

An electronic copy of the machinery and equipment layout data sheet has been provided for your use. This form may be downloaded and used for your project.

QUESTIONS

1. Where do you start on workstation design? Why?
2. What is the starting point of workstation design? Why?
3. What must be included in a workstation design?
4. What are the principles of motion economy?
5. What is effectiveness?
6. What is efficiency?
7. What is the extra 50 percent space added to the workstation space requirement?
8. Design workstations for your project and develop the fabrication area requirement in square feet.
9. It is said that "A good job, like a good machine, must be designed." Explain this statement. What does it mean in terms of workstation design?
10. Define *ergonomics* and explain its importance to workstation design.
11. What is anthropometry? What are the anthropometric considerations in work design?

CHAPTER 

Auxiliary Services Requirement Space

OBJECTIVES:

Upon the completion of this chapter, the reader should:

- Understand the need for support activities in a manufacturing enterprise
- Be able to identify support activity departments such as receiving, storage, maintenance, and so on
- Be able to calculate the space requirements for such support functions

Manufacturing departments need support services, and these services need space. The purpose of this chapter is to identify these services, define the purposes of these services, determine the facilities requirements, and determine the space requirement. There are many service functions to consider in a manufacturing plant, but the activity centers that require a lion's share of space are

1. Receiving and shipping
2. Storage
3. Warehousing
4. Maintenance and tool room
5. Utilities, heating, and air conditioning

RECEIVING AND SHIPPING

Receiving and shipping are two separate departments, but they have very similar people, equipment, and space requirements. Receiving and shipping could be placed next to each other or across the plant from each other. The placement of the receiving and shipping departments has a big effect on the flow of material in the plant. The receiving department is the start of the material flow, whereas the shipping department is the end of the material flow.

Advantages and Disadvantages of Centralized Receiving and Shipping

A centralized receiving and shipping point would have the following advantages:

1. Common equipment
2. Common personnel
3. Improved space utilization
4. Reduced facility costs (fewer outside construction costs)

Loading and unloading trucks are very similar functions, so the facilities are similar. Dock doors, dock plates, fork trucks, and aisles are needed for both receiving and shipping. In some plants, it could be the same dock. Personnel requirements are also similar. Responsible people who know the value of proper counts, proper identification, and control of the company's most valuable assets are receiving and shipping clerks.

The disadvantages of centralized shipping and receiving are space congestion and material flow. Space congestion can cause injury, product damage, and lost materials. It would be a costly mistake to ship out some of the newly received parts. Material flow is more efficient if the material could flow straight through the plant: receiving on one side of the plant and shipping on the other side.

Receiving in more than one place is also a possibility. Steel plate could come in to the plant via its own area, finished parts could enter the plant close to assembly, whereas all other raw material comes in a third receiving area. The most cost-efficient method is the correct choice.

Choosing to place shipping and receiving close together or across the plant from each other is a difficult decision based on balancing the advantages and the disadvantages. The result will be an activity code of A or X. The facilities planner and management will have to choose, and that choice will dictate the flow of material through the plant.

The Trucking Industry's Effect on Receiving and Shipping

The trucking industry can affect receiving and shipping departments. The trucking industry is organized nationally to deliver raw materials and parts to industry in the morning and pick up shipments in the afternoon. This is known as *less than truck load* (LTL) quantities. Full truck loads are handled differently, but if you look at the sources of raw materials, it could come from hundreds of sources. No one would expect a truck to show up at the dock with one box of parts, and a full truck load could be years' worth of inventory, so plants use common carriers. A truck arrives in town with many orders for many plants. That truck and many more are unloaded at a local trucking company's warehouse. The materials are sorted by company to be delivered the next morning. Overnight, the local trucks are loaded for delivery. Material for several plants could be loaded on the same truck with the first stop loaded last and the last stop loaded first. The truck stops by the receiving department and drops off many raw materials and parts orders for the day. In the afternoon, the same truck could return and pick up shipments. One truck could pick up 50,000 pounds of shipment. The shipment may then be taken to a distribution center or a hub where it is sorted according to its destination. Subsequently, interstate trucks pick up the shipment as they pass through a given locality.

Functions of a Receiving Department

The functions of a receiving department include

1. Assisting in locating a trailer at the receiving dock door
2. Assisting in the unloading of material
3. Recording the receipt of the number of containers
4. Opening, separating, inspecting, and counting the material being received
5. Preparing overage, shortage, or damage reports as needed
6. Developing a receiving report
7. Sending material to raw material stores or straight to production if needed

Receiving Trailers

Trailers are backed up to the receiving dock doors, the tires are chocked, the trailer doors are opened, a dock board or dock plate is positioned between the trailer and the floor of the plant, and the driver gives the receiving clerk a manifest that tells the receiving clerk what to unload.

Unloading

The material is removed from the trailer and placed on the dock in the holding area. The receiving clerk signs the trucker's manifest (acknowledging the receipt of so many containers) and the trucker leaves. No count of material or quality check needs to take place before the driver leaves, but visible carton damage should be noted on the driver's paperwork.

Recording Receipts

When material is unloaded, it is checked in on a log. This log is often called a *Bates log* after the name of a sequencing number stamp called a *Bates stamp*. The Bates stamp has the ability to stamp the same number three times before advancing to the next number. This number is stamped on the Bates log, the packing slip, and the receiving report. The Bates log is simply the sequential record of the truck's receipt. Starting with the Julian calendar date (a three-digit number indicating the day of the year), the next three digits are the order that trucks came in that day. For example, July 3 is the 185th day of the year and this is the 21st truck arriving today. The Bates log would show the following:

Bates Number	*Trucking Company*	*No. of Containers*
185021	Arkansas Best Freight	15
185022	Allied	4

Opening, Separating, Inspecting, and Counting

During the first hours of the day, there may not have been time to open a single container to check in the merchandise officially, but before the day is complete, everything received today must be opened, separated, inspected, and counted. Opening each container to check the contents is a must. The first check is to make sure that everything in the container is the same part number. If they are not the same item, then each part must be separated and categorized by number so that it can be stored separately. After separation, a quality check must be made to see if this is what the company ordered. A visual, as well as a thorough examination of the materials may be required to ensure conformance to chemical, mechanical, or other physical standards and specifications. In this case, the quality control department may have a large facility requirement for the receiving area. The quantity must also be checked. If the vendor (supplier) said it shipped 10,000 and receiving did not count the parts, the company could pay for parts that were never received.

Preparing Overage, Shortage, and Damage Reports (OS&D)

If the count is either over or under, an OS&D report is prepared and sent to purchasing for resolution. Damage suffered in shipment and quality problems are also reported on this form. Each problem becomes a project for the purchasing department which has to work it out with the supplier, but the "eyes and ears" of the company are with the receiving department. It is said that receiving is the key to the company's bank because sloppy receiving can give away thousands of dollars.

Preparing Receiving Reports

The receiving report is the notice to the rest of the company that a product has been received. Suppliers receive purchase orders for some of their products. The suppliers in turn create a shipper, fill the order, and attach a copy of their shipping order to the box. This is called a *packing list.* At almost the same time that the product is shipped, an invoice (bill) is sent in the mail. Some companies use the customer's packing list for a receiving report, but it is better to have your own uniform report for checking in things; it also provides a record of the receipt. After checking quality and quantity, the receiving department sends the receiving report to accounting. The accounting department (accounts payable) collects copies of the purchase order, receiving report, and invoice. Only after all three documents are received is the bill paid—the company pays for only what receiving said they received. Errors can be very costly. The receiving report contains the following information: (1) the purchase order (P.O.) number, (2) the vendor's name and address, (3) the date, (4) the part number(s), (5) the part name(s), (6) the quantity, (7) the Bates log number, and (8) the packing list number.

Sending to Stores or Production

Once all receiving functions are complete, the product is set in an area between receiving and stores awaiting disposition to production stores or to production operations, depending on urgency. This is the holding area awaiting fork truck drivers to move the material off the receiving dock.

A significant portion of the problems associated with these manual operations of identification, counting, sorting, routing, and inventory management, and the resulting human errors can be alleviated through the use of automatic identification and data capture (AIDC) technologies. Customers and vendors can agree upon a common AIDC technology such as PDF417, a two-dimensional bar code. The bar code, which can be generated by the supplier and will accompany the shipment, contains all the relevant information such as the part number, quantity, price, destination, electronic data interchange (EDI), transactions, and any other data that may be required by the customer. Upon receipt of the shipment, the data contained in the bar code can be scanned and referenced to the host computers' database of associated information about the shipment. The use of this one AIDC technology (there are over 20 different AIDC technologies) can increase efficiency and throughput in the receiving and other departments, as well as significantly reduce or eliminate human keyboard errors.

Facilities Required for a Receiving Department

Dock doors, dock plates, aisles, outside parking lots, maneuvering space, roadways, and offices are a few examples of facilities needed in receiving departments. The number and size of these facilities depend on the product or products, their size, and the quantities received.

Dock Doors

The number of dock doors needed depends on the arrival rate (trucks per hour) at peak time, and the service rate (unloading time). For example, if 12 trucks arrive during a peak hour, and it takes 15 minutes to unload an average truck, three dock doors would be needed. Fifteen minutes per truck would allow you to unload four trucks per hour per door, so three doors would be needed.

Dock Plates, Dock Levelers, and Dock Boards

These are all tools used to bridge the gap between the floors of buildings and the floors of trailers so that material can be moved on and off the trailer easily. There is a big difference in the cost of these facilities. They will be discussed further in Chapter 11 on material handling equipment.

Aisles

Aisles leading from the trailers into the plant must be sized for the material handling equipment, the material being moved, and the frequency of trips. Generally, aisles into trailers are 8 feet wide because that is the width of a trailer, but sometimes a trailer is unloaded from the side or with overhead bridge cranes. Plan for such differences.

Outside Areas

The area around the outside of the loading dock should be well planned (see Figure 8–1). Space considerations should take into account the following:

1. Trailer parking alone can take up 65 feet out from the plant wall.
2. Maneuvering space is the space between the road and parking area and is usually about 45 feet.
3. Roadways are 11 feet one way or 22 feet for two-way traffic.

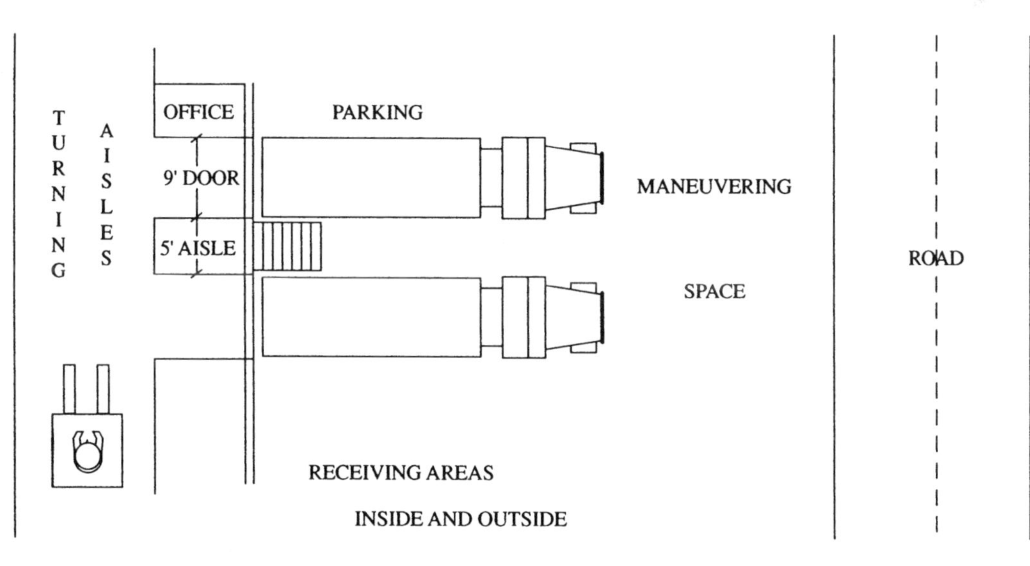

Figure 8–1 Receiving area.

Offices

Offices on the receiving dock are normally very small. Space for a desk, files for purchase orders, Bates logs, receiving reports, and over, shortage, and damage reports are all needed. Depending on the number of people assigned to the receiving area, 100 square feet per clerk is necessary.

Space Requirements for a Receiving Department

The first method of determining receiving dock space calls for visualization of the receiving job based on the number of finished products produced per day and the weight of those units. For example, if you are making 2,000 toolboxes per day and those toolboxes weigh 5 pounds each, 10,000 pounds of steel will be required every day. So, on the average, 10,000 pounds will be received and shipped every day. Some days it will be 5,000 pounds, other days 15,000 pounds, but on the average 10,000 pounds per day. The receiving dock will be sized to receive 10,000 pounds. What does 10,000 pounds of steel look like? Consider that 40,000 pounds is a truck load. You need only one-fourth of a truck load space. A semitrailer is 8 feet wide by 40 feet long and steel would be stacked only a few feet high, so 10,000 pounds would be one-fourth of 8 feet by 40 feet, or 80 square feet. Multiply this by 2 to allow for aisles, office, and so on, and the dock is 160 square feet, about 12 feet by 13 feet, a very small area that could have only one door. The outside area for parking is extra. Figure 8–2 and Figure 8–3 are examples of receiving department space requirements.

The second method of receiving department space determination is the *facility approach*. You will need the following data: (1) dock doors; (2) aisles; (3) unloading hold area; (4) working area to open, separate, count, and check quality; (5) office area; and (6) holding area for stores.

The holding area would still be 10,000 pounds' worth and could be slightly larger for a work area that would move with progress through the stack of holding area. The office area is 100 feet per person (in this case no more than one person).

Functions of a Shipping Department

The functions of the shipping department include

1. Packaging finished goods for shipping
2. Addressing cartons or containers
3. Weighing each container
4. Collecting orders for shipping (stage)
5. Spotting trailers
6. Loading trailers
7. Creating bills of lading

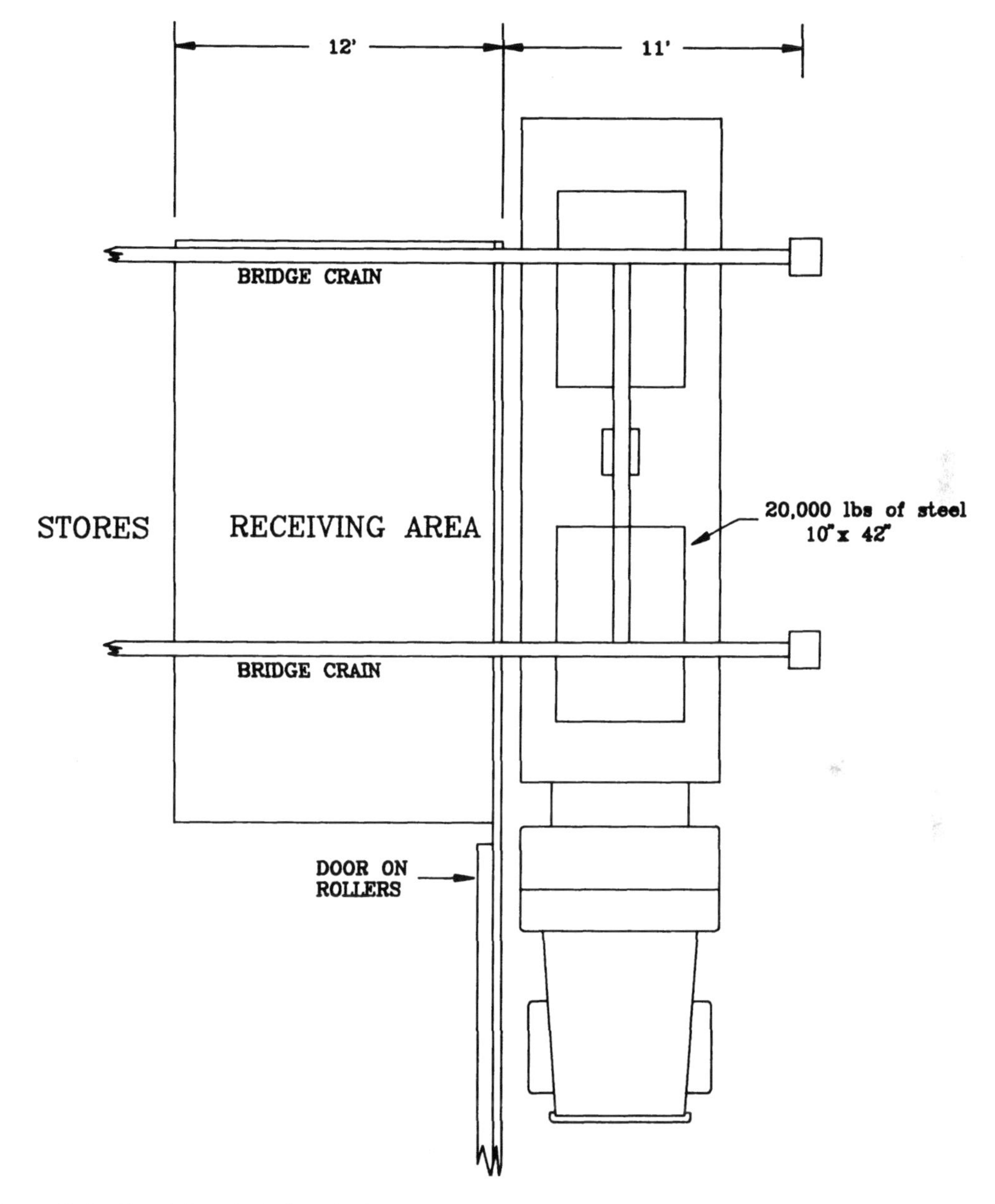

Figure 8–2 Receiving dock for steel.

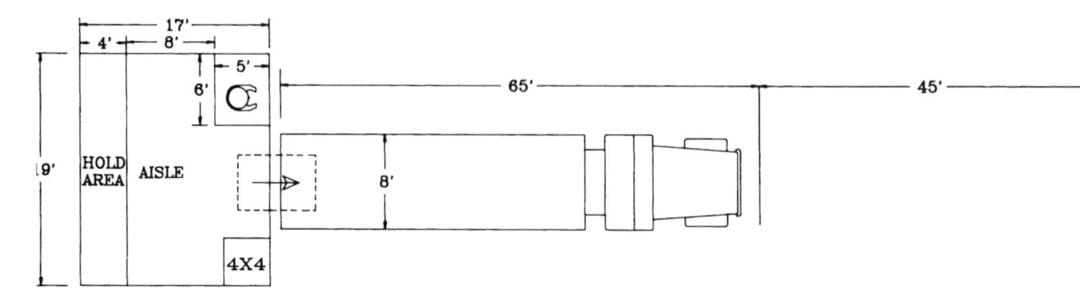

Figure 8–3 The exterior of the receiving dock.

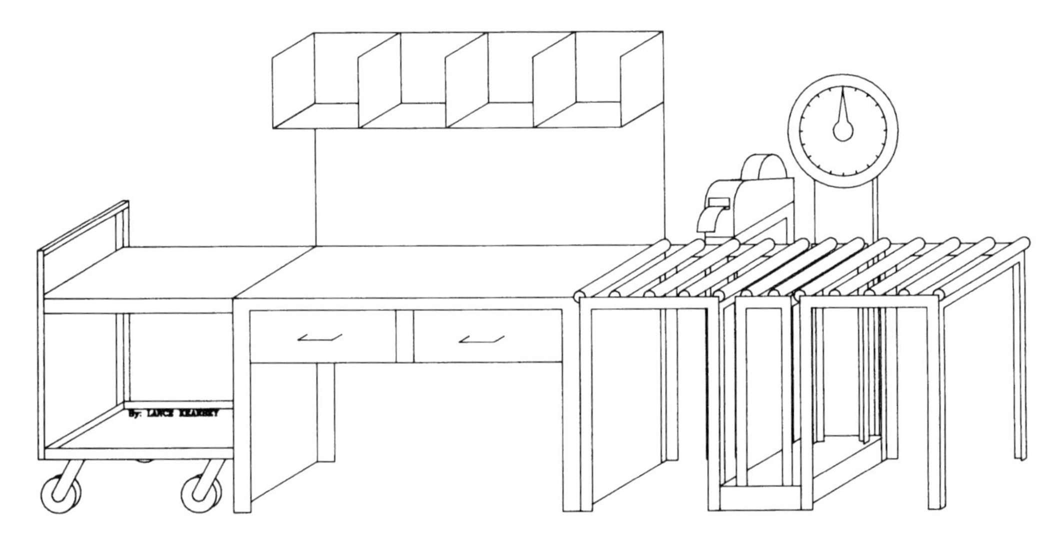

Figure 8–4 Packaging workstation.

Packaging Finished Goods for Shipping

This process varies with the product and the kind of company. One company may have thousands of products with one customer ordering a few hundred items. These items are pulled together and packaged. The package may be a box or a pallet or even a cargo container. Let us consider a hand tool company. They would pack their orders in heavy-duty cardboard boxes. Packaging must include careful placement of individual items so that they are not damaged in shipment. This may require wrapping, stuffing, nesting, and even specially designed shock absorption material. The weight of the container must be compatible with the customer's ability to unload the shipment. Packaging workstation design must also consider the principles of motion economy. Proper work height, good lighting, and all tools and materials located conveniently are only a few of the principles of motion economy that must be considered. See Figure 8–4 for a typical packaging station.

Addressing Cartons or Containers

This is required if the order goes by common carrier (e.g., LTL). Just like a letter, the order (boxes) goes into a system with many other orders, thus each box must be addressed. When many boxes are going to the same customer, a stencil may be used to mass produce the address. Some systems have computer-generated shipping labels and others use a copy of the shipper as an address label. The important point is that every container must be addressed. Efficiency (or cost reduction) will determine which addressing system to use.

Weighing Each Container

This process is required for several reasons. First, the trucking company will charge by the pound, so you need to know the weight to determine trucking costs. Second, a quality control technique is used to compare the weight of each order to the individual weight of each part shipped. If the container does not weigh enough, something must have been left out. If the container weighs too much, something extra is probably placed in the box. When customers receive the shipment and claim a shortage, you can check the weight to verify the shortage. If the weight checks out, you ask the customers what they got instead because the weight was correct. Third, trucks can haul only specific maximum weights. You must ensure that you do not overload the trucks. Last, you can use the weight as an output figure for productivity calculations. In a warehouse, the pounds shipped can be divided by the hours worked to create a performance indicator of pounds shipped per person-hour. Pounds shipped per person-hour is a good indication of performance.

Weight scales can be built into the conveyor line as shown in Figure 8–4, or the scale could be built into the floor so fork trucks can weigh whole pallets. (See Figure 8–5 and Figure 8–6.)

Collecting Orders for Shipping

This is often called *staging orders.* The company may use four trucking companies to move the freight: one for all freight going north, another for all freight going west, a third company for freight going south, and the final company for products going east. All day long as orders are filled and packed, the finished packaging is placed in the proper staging area for that truck line.

Spotting Trailers

The trucking company sends a trailer in the afternoon to pick up the freight. This is called spotting a trailer. Some big shippers may talk the trucking company into leaving a trailer at the plant all day. Then you can stage the shipments on the trailer and save plant space.

Loading Trailers

Loading the trailer can be done very quickly if pallets are used. Most trailers will hold a maximum of 18 pallets, 36 if stacked two high.

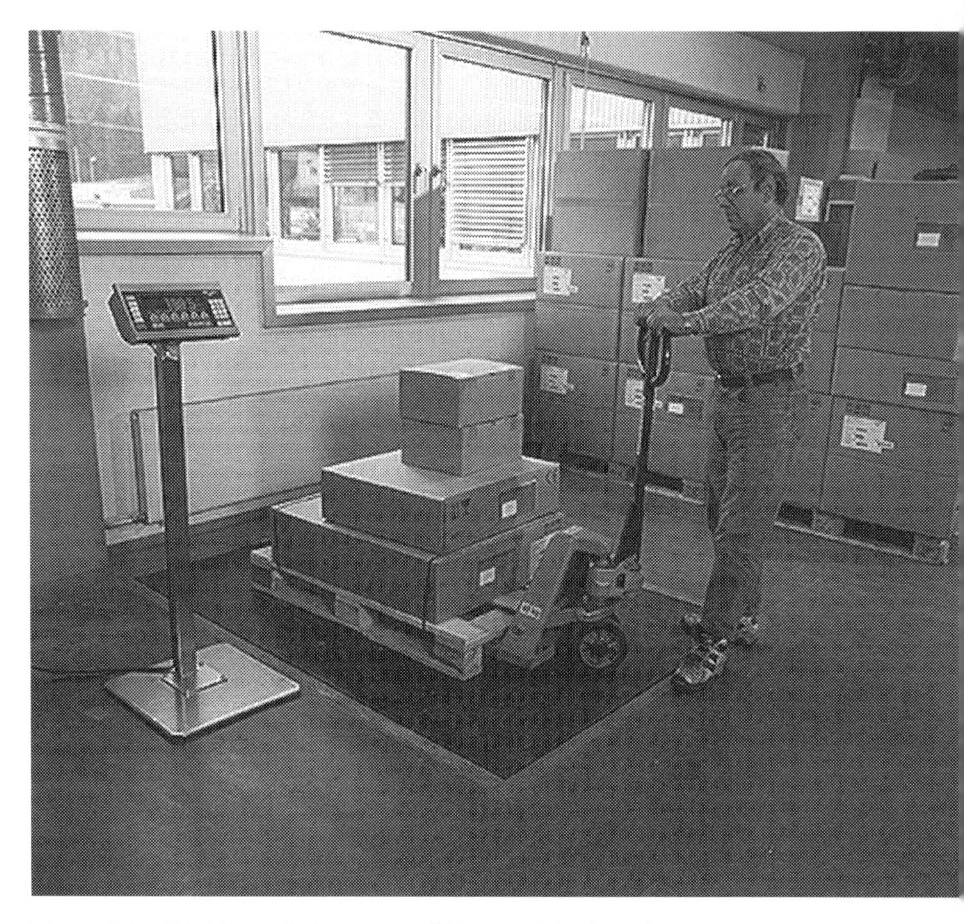

Figure 8–5 Weight scale (courtesy of Mettler-Toledo Inc).

Creating Bills of Lading

As the trailer is loaded, the bill of lading is created. The bill of lading lists every order and the weight of that product. The bill of lading is the truck driver's authorization to remove the product from the plant, and it will eventually come back as a bill for the trucking service.

Space Requirements for a Shipping Department

As was the case with the receiving department, the shipping department and subsequently the customers can significantly enhance their operations and reduce human error as the result of the application of automatic identification and data capture (AIDC) technologies. Use of a bar code can simplify the item tracking process and ensure that relevant and accurate information accompanies the shipment.

Figure 8–6 Packaging for shipping (courtesy of Hytrol Conveyor Co.)

Necessary information can be entered via a keypad or scanned and a bar code can be printed and attached to each package.

Space for shipping must include areas for packaging, staging, aisles, trailer parking, roadways, and offices. Sometimes, lounges for truckers and restrooms are included. As in the receiving department, the overall weight of the shipment will help you to visualize the size of daily shipments. Two thousand toolboxes per day times 5 pounds per toolbox equals 10,000 pounds per day. But there is a lot of air in a toolbox, so how many cubic feet do 2,000 toolboxes take up?

$$\frac{8 \times 8 \times 18 \text{ inches}}{1{,}728 \text{ cubic inches per foot}} = .66 \text{ cubic feet} \times 2{,}000 = 1{,}333 \text{ cubic feet per day}$$

A trailer is 8 feet wide times 40 feet long times 7 feet high, or 2,240 cubic feet.

$$\frac{1{,}333 \text{ cubic feet required}}{2{,}240 \text{ cubic feet per trailer}} = .6 \text{ trailer per day}$$

One dock door is required. Space to store (stage) a day's supply of shipments (1,333 cubic feet) is required. A space of 8 feet times 40 feet times 60 percent equals 192 square feet for staging. (The total trailer capacity is 8 feet times 40 feet equals 320 square feet. Only 60 percent [.6] of the total capacity is utilized; therefore, you multiply the total capacity by 60 percent in this case.)

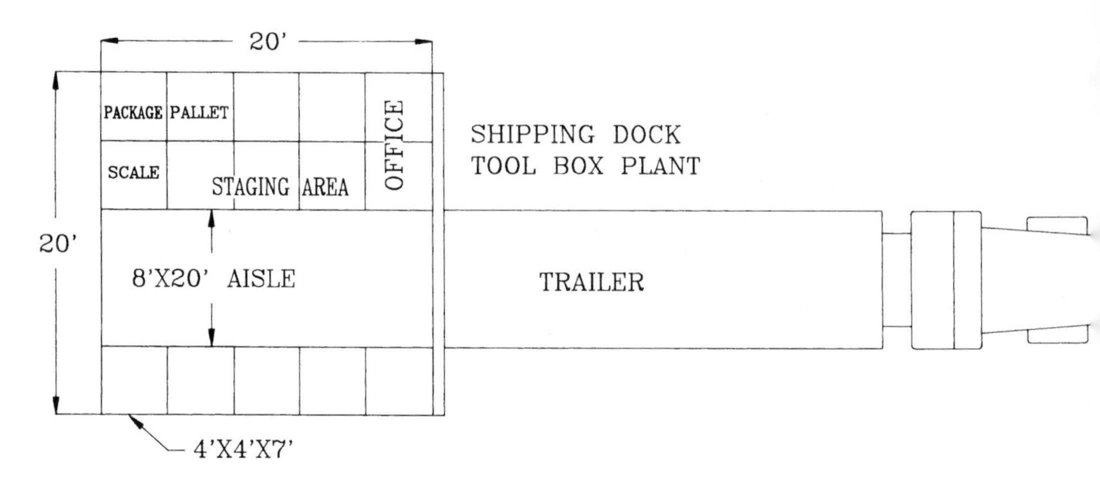

Figure 8–7 Square footage of shipping department.

Multiplying this time by 200 percent will put in the extra space needed for aisles and offices, but not for packaging. Packaging is based on the workstation layout (like production) but the toolbox example does not need much packing, just addressing and weighing. The toolbox plant's shipping department will be about 400 square feet inside the plant plus parking for one trailer. See the example in Figure 8–7 for the plant's shipping department.

STORAGE

Stores is a term used to denote an area set aside to hold raw materials, parts, and supplies. There are many different types:

- Raw material stores
- Finished parts stores
- Office supplies stores
- Maintenance supplies stores
- Janitorial supply stores

Each of these stores requires space and must be considered when calculating total space requirements, but raw material stores and finished parts stores are the biggest users of space. Your primary interest will be in raw material stores, but the same procedure can be used in calculating space for other stores.

The space requirements for stores is dependent on the stated inventory policy of the company. The policy could be as straightforward as providing space to store a one-month supply of raw material, or a more creative policy might be to provide an area to store a one-week supply of A items, 2 weeks of B items, and a one-month supply of C items. A items are those parts that account for 80 percent of the inventory value. Usually 20 percent of the part numbers makes up 80 percent of the dollar value. In an automobile assembly plant, the engine and transmission are the most expensive parts of the automobile. Assuming that each can cost $4,000 of the $24,000 total cost (or 17 percent) and that there may be over 2,000 parts to any car:

Inventory Classification	*Percentage of Parts*	*Percentage of $*	*Inventory Policy*
A	20	80	One-week supply
B	20	15	Two-weeks supply
C	60	5	One-month supply

In the toolbox example, if you make 2,000 toolboxes per day at a material cost of $5 each and inventory a 20-day supply, you would have $200,000 in inventory. A carrying cost of 25 percent per year is normal, so the cost of carrying a one-month supply of inventory is $50,000 per year. If you redesigned the system and instituted an ABC inventory system, you would reduce costs to 25 percent carrying cost times $65,000 equals $16,250.00 per year carrying cost.

A	80% for one week	$40,000*
B	15% for two weeks	15,000
C	5% for one month	10,000
Total Inventory Value		$65,000

*Eighty percent of $5 per unit times 2,000 boxes times 5 days equals $40,000.

You have saved $33,750 in inventory carrying cost. The less inventory you carry, the lower the costs if you do not run out of material. Large inventory allows production management to be very comfortable—it does not need to worry about running out of material as often, but at what cost? Carrying cost measures the cost of carrying inventory. The 25 percent includes

1. The cost of maintaining the inventory of raw material (say, about 12 percent)
2. The space for storing, heating, cooling, and lighting the material (about 8 percent)
3. The cost of taxes, insurance, damage, obsolescence, and so on (about 5 percent)

These costs are real costs that add no value to the product.

The cost of running out of a single item of inventory used on the production line could amount to shutting down the whole plant, so some inventory is needed. How much inventory is a management decision. Looking at the A item again, 20 percent of the part numbers accounts for 80 percent of the inventory cost. The philosophy is that the less you have of this most expensive class, the better it is. But you will need to reorder it four times as often as a C item. This also means four shipments, four receivings, four orders, and so on; therefore, ordering cost will increase, but only on the most important 20 percent of the part numbers.

Just-in-Time Inventories

Just-in-time (JIT) is the inventory policy that has been made famous in Japan. Primary manufacturers depend on their suppliers to deliver parts as often as every 4 hours, thereby eliminating the need for raw material inventory storage area. JIT depends on unfailing vendor performance. Vendors at far distances from the plant would have to warehouse their product in your area. This type of inventory system takes total corporate commitment and very special relationships with vendors. JIT will affect the plant layout in many ways. You will be able to

1. Adjust or eliminate receiving, receiving reports, and so on
2. Eliminate incoming quality control checks
3. Eliminate or greatly reduce stores area requirements

In this text, we will not consider JIT because designing a layout for a non-JIT system is more difficult and, unfortunately, is more common.

The goals of any stores department should be

1. To maximize the use of the cubic space
2. To provide immediate access to everything (selectivity)
3. To provide for the safekeeping of the inventory including damage and count control

Maximizing the Use of the Cubic Space

Maximizing the use of the cubic space requires the use of racks, shelves, and mezzanines, and minimizing aisle space and empty space. This brings up the number-one design criterion for a storeroom: *Leave room to store only half the required inventory.*

To explain this design criterion, we need an inventory graph (see Figure 8–8). Inventory graph terms include

1. *Units on hand.* The *y* axis (vertical axis); it measures how many units of this part number remain in the inventory.

2. *Days.* The *x* axis (horizontal axis); it measures the day of the year this day represents. In the life of a product, this axis could be long, but a year's worth of data would be very useful.

3. *Order quantity.* How many units you order at a time. If you order a week's worth of toolbox parts, you would order 10,000 sets of parts (2,000 per day). When this material comes in, you add these 10,000 units to the inventory on hand. This would create a vertical line 10,000 units high from the on-hand inventory of the day.

4. *Normal usage.* A trend line indicating balance on hand at the end of each day. The toolbox plant would be using up the parts at the rate of 2,000 sets per day.

5. *Minimum usage.* The slowest rate at which you use up parts. Normally, this would be only slightly less than normal usage; otherwise, you would not hit the

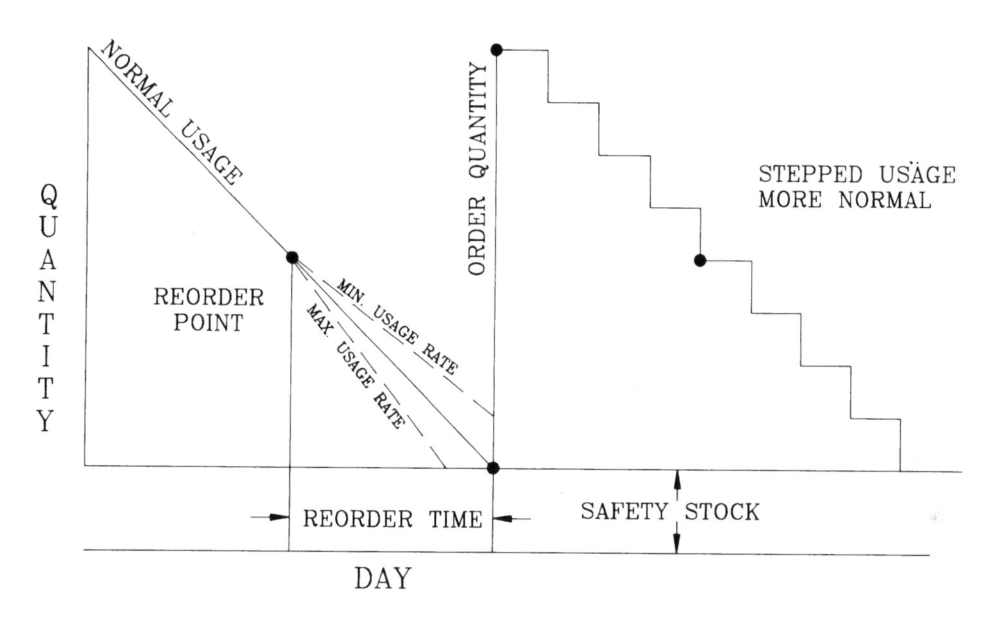

Figure 8–8 Inventory curve—One curve per part number.

2,000 units per day goal. If you fall behind schedule, you would probably work Saturday to catch up (and use up the inventory).

6. *Maximum usage.* The fastest rate you would use parts. Again, if this was much faster than your plan, you would be building up inventory of finished goods in the warehouse and an adjustment to the schedule would have to be made; otherwise, you would run out of parts. You want to carry a little extra stock so that you will not run out. (See safety stock, item 8.)

7. *Normal distribution between minimum usage rate and maximum usage rate.* The normal usage rate is like any other normal distribution curve. This would indicate that the usage rate is faster than normal about half the time and slower than normal about half the time, but not by very much. To keep from running out, consider the maximum usage rate in determining the safety stock.

8. *Safety stock.* Necessitated due to variation in usage rate and the lead time, it is the extra inventory you carry so that you will not run out of inventory, or run out only once in 100 order periods (or 1 percent outages). The distribution curve will tell you how big this safety stock must be in order to satisfy any level of service you choose.

9. *Reorder points.* That inventory level (in units on hand) where you need to reorder material to prevent a stock outage. While the order is being processed and shipped into the plant, continue using inventory (depleting the stock). The reorder point is calculated by using the usage rate and the reorder time.

10. *Reorder time.* Also referred to as *lead time,* it is that time (in days) between the ordering of new material and the receipt of that material in the stores. If it

takes 10 days to create a requisition, to type a purchase order, and to mail the order to the supplying company, that company will fill the order, ship it, you in turn receive it and put it in your storeroom. Then you need 10 days of material on hand at the time of reorder. In the toolbox example of 2,000 per day and a safety stock of 1,000 units, the reorder point would be 21,000 units (2,000 times 10 days plus 1,000 units). When the inventory drops below 21,000 units, you would reorder another quantity. The order quantity would be calculated by using a formula to minimize total cost, but that is a subject for a production inventory control class.

11. *Stepped usage.* More realistic. As production needs parts, it requests a day's supply at a time. The inventory level drops all at once by a day's supply—not by one unit at a time.

The inventory curve explains why and how you can provide room for only 50 percent of the required inventory. Look at the inventory curve (see Figure 8–8). How much inventory do you have on the day a new order arrived? How much inventory do you have on the day before the inventory came in? The answers are maximum or minimum. How much inventory do you have on the average? Answer, 50 percent. Now if you assign a spot in the storeroom for the maximum amount of inventory, how full will the storeroom be? On the average, only 50 percent full or one-half full. This is not good cube utilization. To get better use of the building cube, allow room for only approximately 50 percent. So, you cannot assign a part to any one location because there will not be enough room when the new supply arrives. Items are, therefore, stored at random locations in the storeroom depending on the availability of space when a given inventory item arrives. Special item locator files keep track of the location of each inventory in the storeroom. The locator files may be either simple paper-tracking systems, or the data may be stored in electronic media. The use of the bar code and other AIDC technologies can be extended to track item location and inventory level in the storeroom and to allow the system to issue automatically purchase orders based on the predetermined reorder points.

Providing Immediate Access to Everything (Selectivity)

The second design criterion for stores layout deals with *random locations.* Put anything anywhere, but keep track of where you put it. For clarity, pallets go in pallet racks, not on shelves, and steel storage is in another area. But within the racks, you can put anything anywhere.

A *location system* is needed to keep track of what you put where. A simple locator system would letter each aisle. Number each pallet location such as those in Figure Figure 8–9.

Rows A, B, C, D, E, and F are pallet racks (see Figure 8–10a and Figure 8–10b).

Rows G, H, I, and J are shelves (see Figure 8–11).

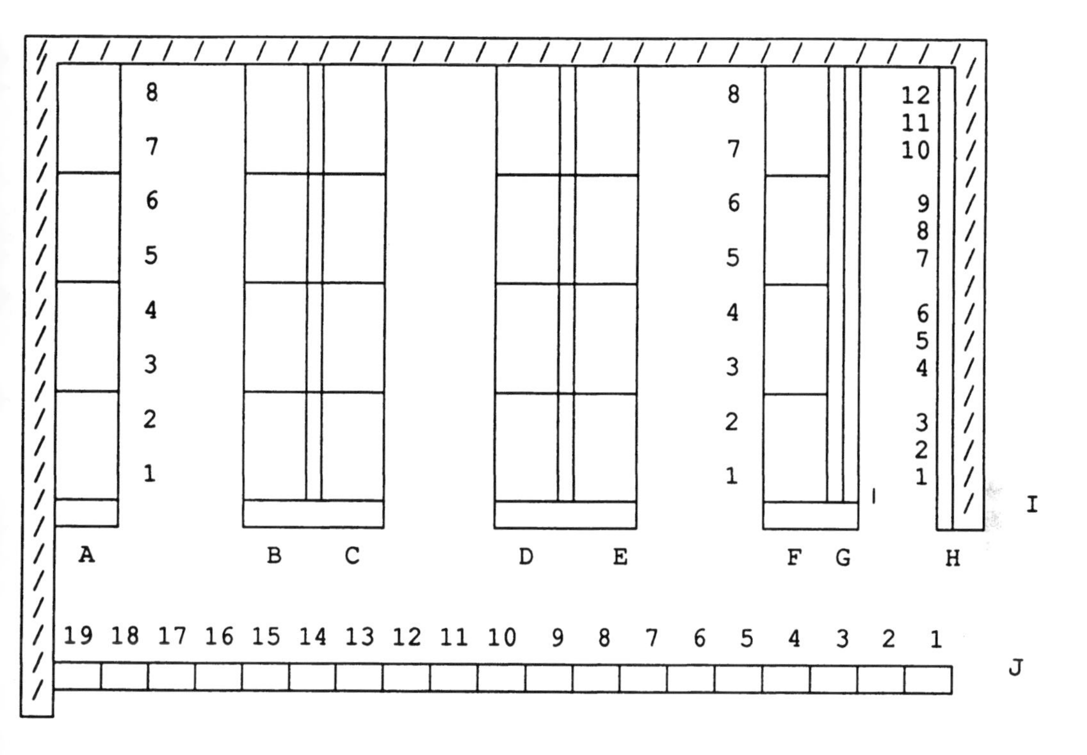

Figure 8–9 Storeroom layout—Location system.

With four racks per row, two pallets per tier, and five tiers high, each row would hold 40 pallets. Row C, pallet six, "b" level, would be six pallets down row C and the second pallet up. Vertically, "a" would be the floor, "b" the second high, and "e" the top tier. Tier "e" would always be the top, and "a" would always be the bottom.

Figure 8–11 is a six-tier shelf with each shelf measuring 3 feet wide, 1 foot deep, and 1 foot high. Rows G, H, I, and J are shelves. Row I is a row of shelves on the end of racks.

Each location in the storeroom now has a location code. The storekeeper is asked to put a pallet load of 1,500 part number 1750-1220 parts away. The driver drives to the first open spot and deposits the pallet. Then the storekeepers make out a location ticket such as that shown in Figure 8–12. Two copies are needed: one copy is attached to the pallet and one copy is kept at the stores' control desk in part number order.

Production now needs some of part number 1750-1220. The request comes to the inventory control desk. The storekeeper looks up part number 1750-1220 in the card file, finds the pallet with the closest quantity to that requested or the oldest ticket, and goes to that location to retrieve the goods. The ticket can be pulled and sent to data processing to reduce the inventory. The inventory control department had previously added this inventory from a receiving report.

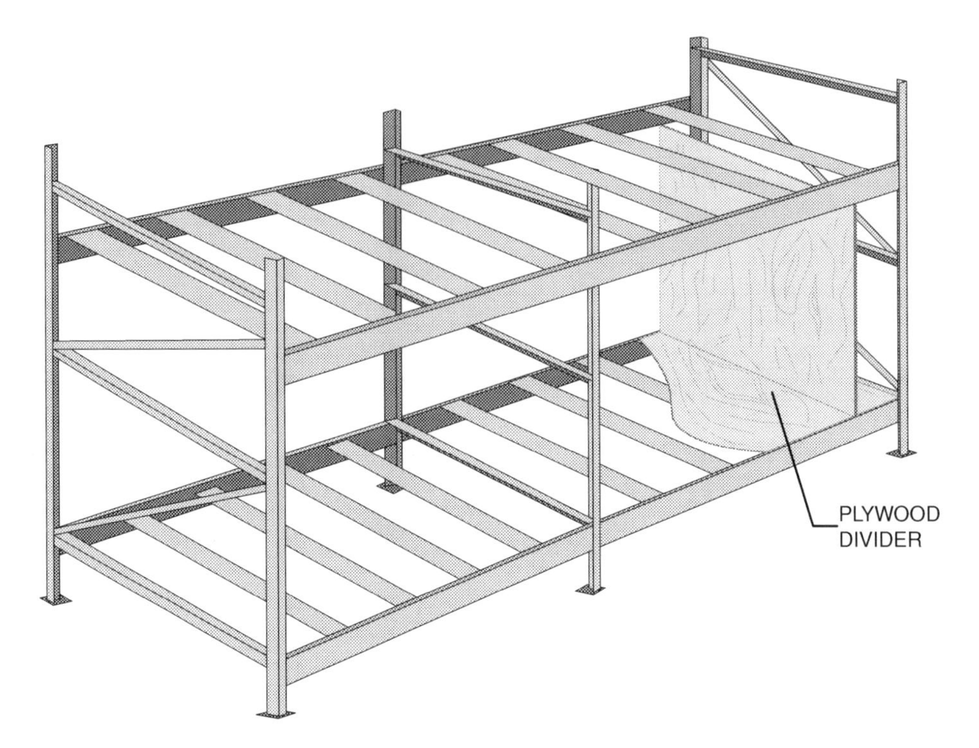

Figure 8–10a Industrial pallet rack.

Figure 8–10b Storage rack (courtesy of White Storage & Retrieval System, Inc.).

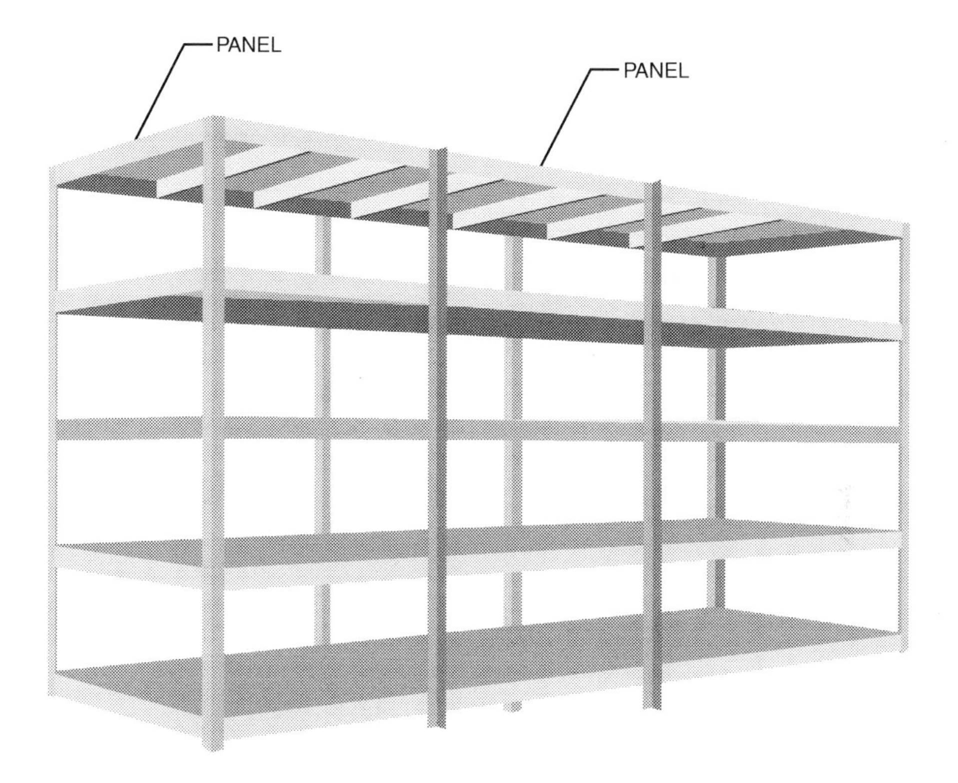

Figure 8–11 Industrial shelves.

Storage Facilities Requirements Spreadsheet

Every part must be measured for cubic size, multiplied by the number of parts to be stored, and converted to cubic feet (see Figure 8–13). The procedure for calculating storeroom size starts with an analysis of storage space needs as follows:

1. List all the raw materials and buyout parts. This will be column 1 (part number) and column 2 (part name).

2. After each part, list the length, width, height, and cubic inches of each part (columns 3, 4, 5, and 6).

3. Column 7 lists the quantity stated in the inventory policy divided by 2 (leaving room for only half the inventory).

4. Column 8 shows the cubic feet required. This is a result of multiplying column 7 by column 6 and dividing by 1,728 (cubic inches in a cubic foot).

5. Columns 9, 10, and 11 list the number of storage units required for each part. Column 9 would be shelf storage. Cubic footages under 10 cubic feet would be placed on shelves. Shelves are 3 cubic feet each (1 × 1 × 3 feet). Column 10 would be for pallets. Storage space requirements over 10 cubic feet, up to 192 cubic feet, would be placed on pallets in the pallet racks (a pallet is 4 × 4 × 4 feet high or 64

BLANK TICKET

PART #____________________

QUANTITY____________________

DATE____________________

LOCATION____________________

FILLED OUT

PART # 1750-1220

QUANTITY 1500

DATE 12/3/XX

LOCATION B1C

Figure 8–12 Location ticket.

1	*2*	*3*		*4*		*5*	*6*	*7*	*8*	*9*	*10*	*11*
Part No.	*Part Name*	*L*	×	*W*	×	*H*	= *in.*^{3}	*Q/2*	*ft*^{3}	*Shelf 3 ft*^{3}	*Pallet 64 ft*^{3}	*Floor 576 ft*^{3}
1	Bracket	18		½		½		10,000				
2	Body	12		6		2		5,000				
3	Washer	½		DA.		⅛		20,000				
4	Nut	½		DA.		¼		20,000				
5	Bolt	¼		DA		2		20,000				
6	Lid	12		6		1		5,000				
7	Hinge	6		1		1		10,000				
8	Handle	7		½		3		10,000				
9	Rivet	¼		DA.		⅜		100,000				
10	Carton	24		16		¼		10,000				
	etc. 990 other parts											
1,000	Booklet	8½		11		.020		20,000				
	Total Storage Units									1,200	1,000	20

Figure 8–13 Storage facilities requirements spreadsheet.

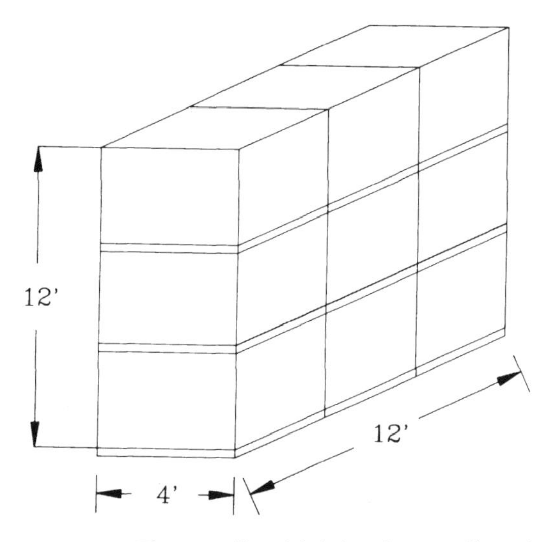

Figure 8–14a Floor storage—Three pallets high by three pallets deep (cubic feet).

Figure 8–14b Floor storage (courtesy of White Storage & Retrieval System, Inc.).

cubic feet per unit load). Some items could be placed on the floor and stacked three pallets high and three pallets deep (see Figure 8–14a and Figure 8–14b).

The results of the storage facilities requirements spreadsheet is the number of shelves, pallet racks, and bulk storage areas needed. In Figure 8–13, 1,200 shelves, 1,000 pallet spaces, and 20 bulk storage areas are necessary. The next step is to determine how many shelves to buy and how many pallet racks to set up. A shelving unit was pictured earlier in Figure 8–11. How many of these shelving units are required? (The 1,200 shelves divided by 6 shelves per unit equals 200 shelving units.) The same thinking is used for pallet racks. Figure 8–10 showed 10 pallets per pallet rack. You need to store 1,000 pallets, so 100 pallet racks are required. A storeroom layout is very close now. You know it is necessary to have 200 shelving units, 100 pallet racks, and 20 bulk storage units. How will you lay this out?

Figure 8–15 (p. 234) shows how the storage space requirements for the toolbox plant were developed. Figure 8–16 (p. 235) illustrates the resulting stores layout in the toolbox plant.

Aisle Feet

The concept of aisle feet is very useful. *Aisle feet* will help determine the space needed. Visualize one shelf (use Figure 8–11 if necessary). One shelving unit is 3 feet wide. You must place this open 3 feet on the aisle; therefore, one shelf has a need for 3 aisle feet. You need 200 shelves with 3 aisle feet each, so 600 aisle feet will be required. Another way of thinking about this is if you assemble 200 shelving units and place them side by side, they would stretch out to be 600 feet long. A 600-foot row is too long, but how about two 300-foot rows or ten 60-foot rows? There is almost unlimited flexibility of layout. Aisles for serving shelves can be much smaller than aisles serving pallet racks, so use a 4-foot-wide aisle (we will discuss aisles further in a later chapter).

How many aisle feet of pallet racks do you need? (Compute 100 pallet racks times 9 feet wide each equals 900 aisle feet.) Again, six 150-foot rows or fifteen 60-foot rows could be used. Floor storage units are 4 feet wide in the example. Twenty floor storage units are required (4 times 20 feet equals 80 aisle feet).

Fork trucks are needed to service pallet racks and floor storage areas and 8-feet-wide aisles are required for the equipment. At this time, all the information is available for you to lay out the storeroom. That information shows 600 aisle feet of shelving, 900 aisle feet of racking, 80 aisle feet of floor storage, and 4- and 8-foot aisles.

Step 1. Start with a wall, placing the floor storage against the wall (see Figure 8–17).

Step 2. Place 900 aisle feet of pallet rack with 8-foot service aisles. (Remember, racks are in multiples of 9 feet. Eighty-one feet have nine sections of 9-foot lengths. One hundred rows are not possible because 100 is not divisible by 9, 99, or 108 feet.)

Part No. Box No.	Part Name	L ×	W ×	H =	in.³	Q/2	ft³ Needed	Shelf 1 × 1 × 3 ft	Pallet 4 × 4 × 4 ft	Floor 10 × 3.5 × 3 ft High
1	Handle	6	1	1	= 6.000	22,000	76		2	
2	Handle clip	1	1	½	= .5	44,000	12.7	5		
4	Rivet	6	6	6	(10,000)	88,000	1.1	1		
2	Catch	¾	½	¼	= .094	44,000	2.3	1		
2	Strike	1	¾	¼	= .188	44,000	5	2		
8	Rivet	6	6	6	(10,000)	176,000	2.2	1		
2	Hinge	6	½	⅛	= .375	44,000	9.5	4		
4	Rivet	6	6	6	(10,000)	88,000	1.1	1		
1	Name tag	3	1	.02	= .06	22,000	.7	1		
1	Packing list	8½	5½	.005	= .234	22,000	3	1		
1	Booklet	8½	5½	.05	= 2.34	22,000	30		1	
1	Carton	36	24	¼	= 216	22,000	2,750		43	
1	Plastic bag	10	6	.03	= 1.8	22,000	23		1	
#5	Steel*	5 lbs	22 ga.			110,000	220			2
				Total				17	45	2

Our toolbox warehouse needs only:

Three shelving units = 9 aisle feet
Five pallet racks = 45 aisle feet
Two floor storage areas = 20 aisle feet

A very simple layout, but steel will be received and stored in different areas.

*Steel weighs 500 pounds per cubic feet and comes in 42 × 120 × 18 inches high.

Figure 8–15 Storage space requirements—22-day supply, 2,000 toolboxes per day.

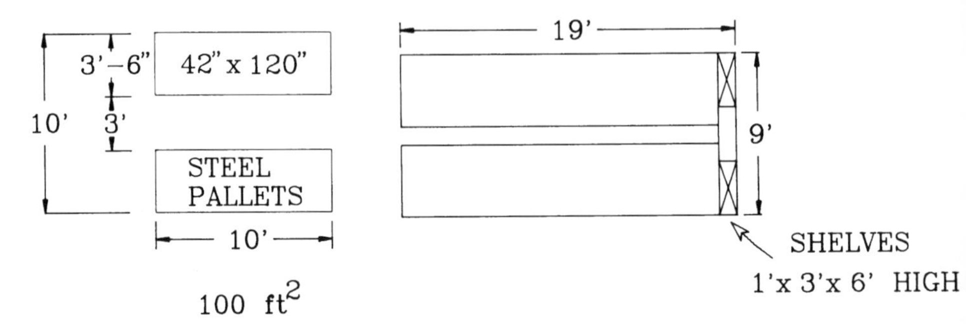

Figure 8–16 Toolbox stores layout.

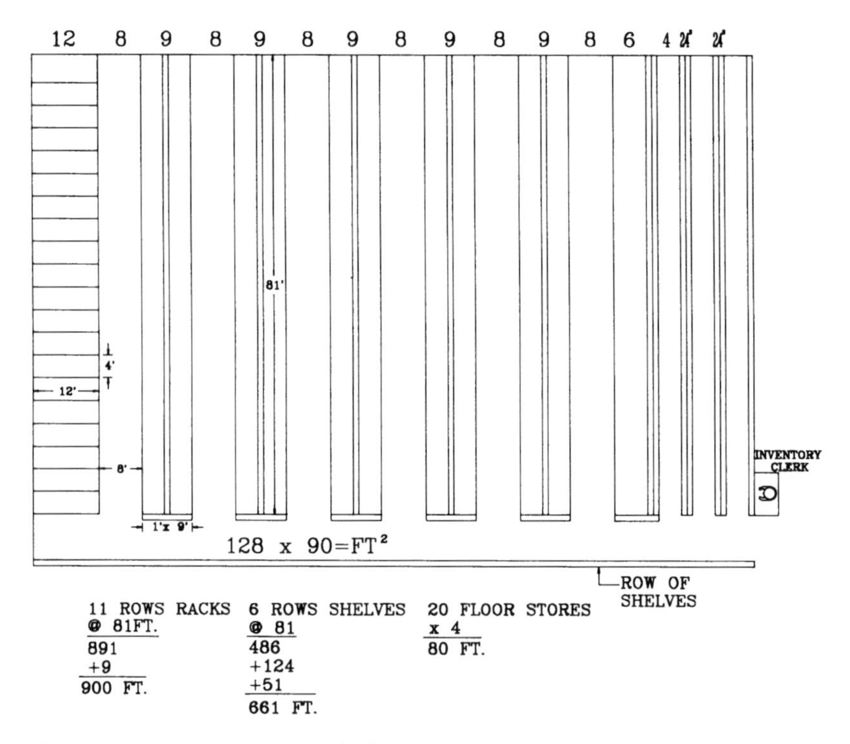

Figure 8–17 Storage space design.

Step 3. Place 600 aisle feet of shelving serviced by 4-foot aisles. Notice in Figure 8–17 that 124 feet of shelving are used to create a wall between production and stores. This was designed to create a security system in which all movement is through a controlled door. Also note that 9 feet of shelving are used as end caps to the rows of pallet racks. This is just good use of the aisles. Using both sides of an aisle is much more efficient.

Providing immediate access to everything is the second goal of a stores department. At receiving, everything is separated and checked in. The stores must keep items separated and provide a separate location for different parts. The purpose of this goal is improved efficiency. When something is needed, the storekeeper should not have to stop, sort, and then move the parts to production. This would take too much time.

Providing Safekeeping

As seen earlier, inventory is valuable. Good storage will provide safekeeping of this valuable asset. Having proper storage equipment like racks, shelves, and trucks will protect the products. Good containers can prevent dust and grime. The other part of safekeeping is preventing the unauthorized removal of inventory. Even the best-intentioned supervisors can create inventory outages if they remove inventory without adjusting the inventory records. A security checkpoint and restrictions to entry are important parts of a storeroom design.

Flat steel stock normally comes into the plant on 4×120-inch pallets. Tubing and bar stocks come in 12-foot lengths. Special racks and special floor storage areas are needed for this material. Also, special material handling equipment will be required. We will discuss material handling in Chapters 10 and 11.

WAREHOUSING

Warehousing is the storage of finished products. As in the storeroom, the area requirement will depend on management policy. Seasonality could require stockpiling finished products for months in order to meet market demands. Sometimes outside warehousing is leased to carry the overload. No one would expect manufacturing to produce all the charcoal grills one month before the spring selling season. They have to be stored somewhere. Management must tell facilities planners how many units or how many days' supply to allow space for.

A warehouse can be a department or an entire building. Our primary discussion will be about the department, but every engineer and manager must know the important differences between these two warehouses. The *warehouse building* is where the company (which could have many manufacturing plants) sends its finished product. The company may have many outside warehouses as well. Many manufacturing plants sending their product to warehouses in order to service the company's customers is a function called *distribution*. The distribution system of a company tries to minimize the cost of moving its product to customers while maintaining superior customer service. A warehouse building will have a receiving department, a stores department, a warehouse department, a shipping department, and an office. The warehouse department in a warehouse building will have the same purpose as the warehouse department in a manufacturing plant.

The *warehouse department* (called just warehouse from now on) has the primary purpose of safekeeping the company's finished product. The stores department keeps raw materials and supplies, whereas the warehouse keeps finished goods. After assembly and packout, finished products are moved to the warehouse where they are kept until ready to be shipped to the customers.

Warehouse Design Criteria

Warehousing is the storage, order filling, and preparation for shipping of products. Order filling is the most labor-intensive portion of the job and affects layout the most. Two design criteria are important to a warehouse layout:

1. Fixed locations
2. Small amount of everything

No layout will ever be a single product layout. For example, a swingset manufacturing company made two basic types of swingsets called "Big T" and "A Frame." Within each of these two groups, 50 different swingsets were sold.

The first warehouse design criterion (fixed locations) means every product must be assigned a fixed location so that the warehouse person can find that product quickly. Placing products in part number order is the simplest way, but not the most efficient. To increase productivity, the most popular items should be in the most convenient location.

The second design criterion is a direct result of the first criterion. By keeping only a small amount of everything in the fixed location, the order picker can pass all the products in fewer feet of travel. If you kept only one pallet of every tool in the warehouse, 4 feet times 8,000 items would require a trip of 32,000 feet to pick one order. That is 6 miles! Be smarter and place these tools on 3-foot-wide shelves that stand 7 feet high. Now you would have to pass only 1,000 shelves, 3 feet wide, or 3,000 feet. If you placed the shelves across the aisle from each other, only a 1,500-foot trip would be required.

To reduce further the travel distance required to pick an order, an analysis of inventory can identify the most popular and profitable items and place these items in more convenient locations. This analysis is called *ABC inventory analysis.*

Figure 8–18 shows a simple warehouse. The top drawing shows a standard layout where the average part is in the middle of the warehouse. This would require a movement of 60 feet from the middle of shipping to the middle of the warehouse in order to pick up a typical product to be shipped (120 feet round-trip). An ABC analysis (the bottom layout in Figure 8–18) would place the most important inventory (the A items) closest to shipping (20 feet away) and the least important parts in the back of the warehouse (90 feet away). The average distance to pick a product

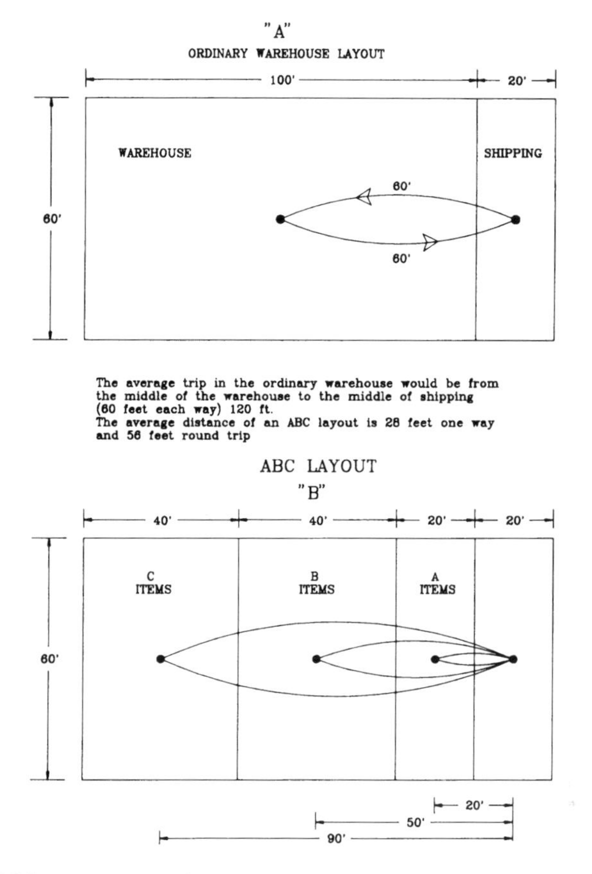

Figure 8–18 ABC layout cost savings.

now becomes 28 or 56 feet round-trip—a 50 percent savings. This was calculated as follows:

- A items account for 80 percent of the sales dollar and only 20 percent of the part number.
- B items account for 15 percent of the sales dollar and 40 percent of the part number.
- C items account for only 5 percent of the sales dollar but 40 percent of the part number.

A items = 20 feet @ 80% = 16.0 feet
B items = 50 feet @ 15% = 7.5 feet
C items = 90 feet @ 5% = 4.5 feet
Total distance for Average Part = 28.0 feet (56 feet round-trip)

FUNCTIONS OF A WAREHOUSE

The three basic functions of a warehouse are

1. To safekeep the finished product
2. To maintain some stock of every product sold by the company
3. To prepare customer orders for shipment

The safekeeping of the finished product must consider pilferage as well as damage due to material handling and storage facilities. Containers, shelves, racks, fences, gates, control desks, and inventory control systems are all part of this safekeeping requirement and the responsibility of warehousing.

Picking orders as requested by the customers is a function of warehousing that affects the layout of that warehouse the most. The efficiency of the warehouse will be determined by the layout. An example of a warehousing job may be a book publisher's warehouse. The warehouse may have 4,000 different titles. Each title is called a *stockkeeping unit* (SKU), so the book warehouse would have 4,000 SKUs. The big question is, How do you lay out these 4,000 SKUs in order to be able to pick customers' orders efficiently? A simple-minded solution would be to place the books on pallets, and place these 4,000 pallets next to each other. Four thousand 4 × 4-foot pallets lined up would be 16,000 feet long even if you used both sides of the aisle. An 8,000-foot aisle would be needed. Order pickers would have to walk 8,000 feet down the aisle and 8,000 feet back the other side to pass every book title. Over 3 miles of walking per order is not good use of people, so the first design criterion of warehouse layout is to *keep a small amount of everything in a small fixed location.*

A "small amount" may be defined as a one- to five-days' supply. This inventory could be placed on shelving or, better yet, in flow racks (see Figure 8–19).

One 6-foot-wide flow rack, six high would have 36 different SKUs (titles) in one 6-foot area, so that 112 of these racks could store 4,000 SKUs.

$$\frac{4{,}000 \text{ SKUs}}{36 \text{ SKUs per rack}} = 111.11 \text{ or } 112 \text{ racks}$$

$$112 \text{ racks} \times 6 \text{ feet per rack} = 672 \text{ feet of rack}$$

Figure 8–19 shows a layout that greatly reduces the 16,000 feet of travel required in the preview to only 678 feet of travel per order. This is still too much, but it is an improvement.

The next improvement in ABC analysis is called the *80/20 rule* or *Pareto analysis.* Basically, they mean the same thing. The 80/20 rule states that 80 percent of sales (measured in dollars) comes from 20 percent of product (book titles, for example). To maximize efficiency, you want to identify those products that account for most of the sales.

This rule divides inventory into three categories:

Class	*Percentage of $*	*Percentage of Parts*
A	80	20
B	15	40
C	5	40

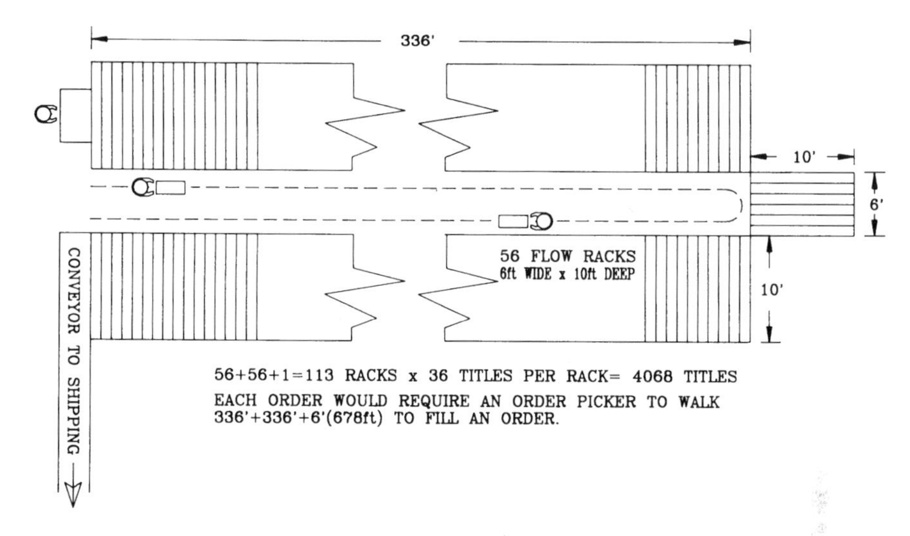

Figure 8–19 Flow racks—Layout for book publisher.

The A category of inventory is exactly like the 80/20 talked about already, but the less popular 80 percent of the products are further divided into B items and C items.

Now, the distance of travel to pick the average order is as follows:

A items = 80% × 138 feet	=	110.4 feet
B items = 15% × 100 feet	=	15.0 feet
C items = 5% × 270 feet	=	13.5 feet
Total Distance of Travel	=	138.9 feet

A comparison of methods shows that you have reduced the walking from 16,000 feet using pallets alone to 678 feet using flow racks only, to 138.9 feet using ABC analysis, flow racks, and pallets. Work smarter, not harder.

Procedure for Sales Analysis of ABC Inventory

To conduct a sales analysis using ABC inventory,

1. List all products with their unit price and average monthly demand (sales).
2. Multiply the price times the average monthly demand.
3. List the product in order of the most monthly sales dollars first and the least monthly sales dollars last.
4. Add up all the monthly sales (total sales).
5. Run a cumulative column after the total monthly sales, then add all the previous totals to each line.

Part No.	$ per Unit	Monthly Sales	Total $	Cumulative $	Percentage of Total
1650	34.50	2,000	69,000	69,000	28
1725	49.90	1,000	49,900	118,900	49
1400	45.00	1,000	45,000	163,900	67
0390	20.50	2,000	41,000	204,900	84
1450	39.00	1,000	39,000	243,900	100
			243,900		

6. The Percentage of Total column is the cumulative dollars divided by the total dollars. In a real example, you would see that only 20 percent of the part numbers accounts for 80 percent (the cumulative percent column) of the sales dollars.

No. of Book Titles	Total $	Percentage of $	Percentage of Book Titles
800	8,000,000	80	20
3,200	2,000,000	20	80

To lay this out, you would place these 800 books close to the shipping department (see Figure 8–20).

Placing the product in part number order is the simplest way of laying out a warehouse, so when a customer order comes into the warehouse printed in part number order, the picker goes to the first part number, then to the second, and so on. The products are easy to find because they are in part number order. The problem with this organization of the warehouse is that slow-moving parts are right next to fast-moving parts. To fix this problem, you can number every warehouse location and assign a product number to any location. In this system, the most popular items can be placed in the most convenient locations. As the ABC analysis is made, the A items are given convenient locations, whereas the C items are located in the back of the warehouse because they are picked only 5 percent of the time; the C area makes up 40 percent of the warehouse. When the customer's order comes into the warehouse, the product is in picking location order: The picker is told to go to location number 0529 and pick up six part number 1650-1900s, then go to location number 0533 and pick up 12 part number 1700-1550s, and so on.

ABC Inventory Layout of a Hand Tool Manufacturing Company's Warehouse

This company provides 8,000 different tools to its customers. The company markets three different brand names. The only difference in some tools is the names. Their old layout was divided into three areas (one for each brand name), and within each

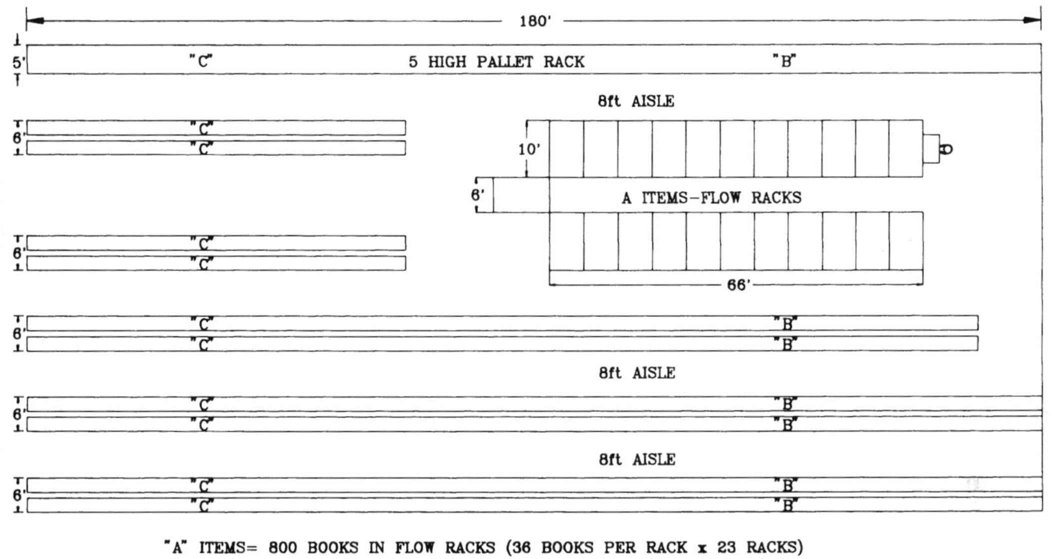

Figure 8–20 ABC book warehouse layout.

area, the tools were stocked on 3-foot-wide shelves 1-1/2 feet deep and 1 foot high. The shelves were six high and the tools were placed on these shelves in part number order. Part number 1 was the first tool on the shelf, and many aisles later, part number 9,999 was the last tool in the warehouse. Figure 8–21 shows a layout of one of three brands of tools. An order picker would pick up a customer's order from the warehouse supervisor's desk and walk 3,000 feet through the entire section of shelving to pick an order. Figure 8–22 is an improved ABC inventory layout for the same section (brand name of tools). In this layout, the A items were located on the main aisles, the B items were located on the side aisles but close to the A, and the C items were located behind the B items. Locations were numbered and the customer orders came out of data processing in location number order.

The proposed layout (see Figure 8–22) requires the picker to walk only 5.4 feet into each side aisle. Forty-two aisles times 5.4 feet equals 227 feet, plus 330 feet up the main aisle and back equals 557 feet. Compare this to 3,000 feet of travel in the present layout.

A items	=	80% walk	3 feet	=	2.4 feet
B items	=	15% walk	12 feet	=	1.8 feet
C items	=	5% walk	24 feet	=	1.5 feet
		Total Average Distance		=	5.4 feet

The pickers will need to walk only 18.6 percent of the distance they used to walk (less than one-fifth of the old distance). This will result in needing fewer pickers and

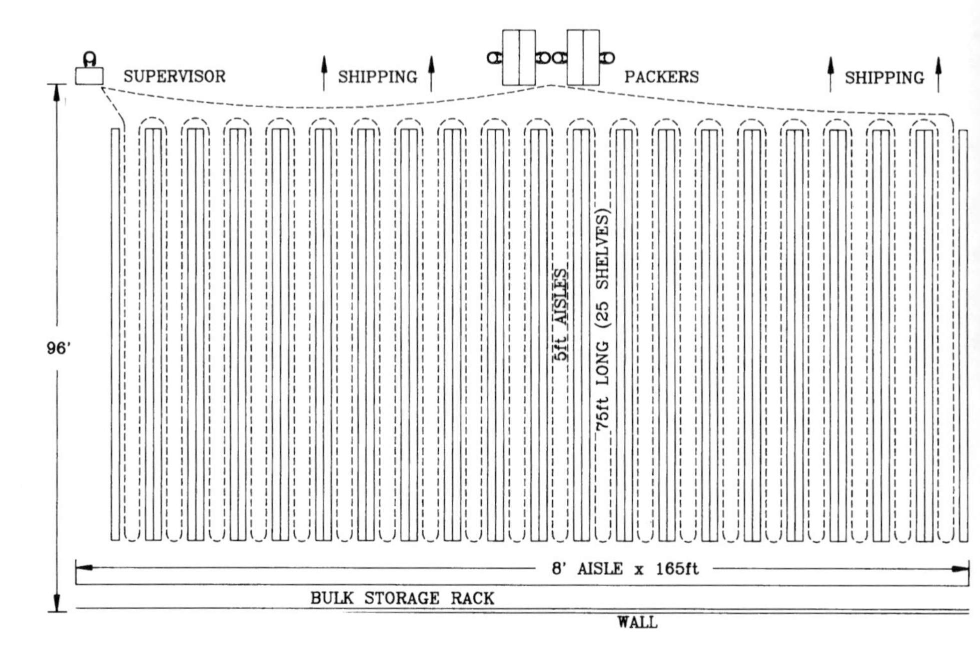

Figure 8–21 Hand tool company's warehouse—Present layout (part no. order—8,000 items on1,000 3 × 1.6 × 7-foot-high shelves; 3,000 aisle feet are needed or 40 [75-foot] rows of shelves).

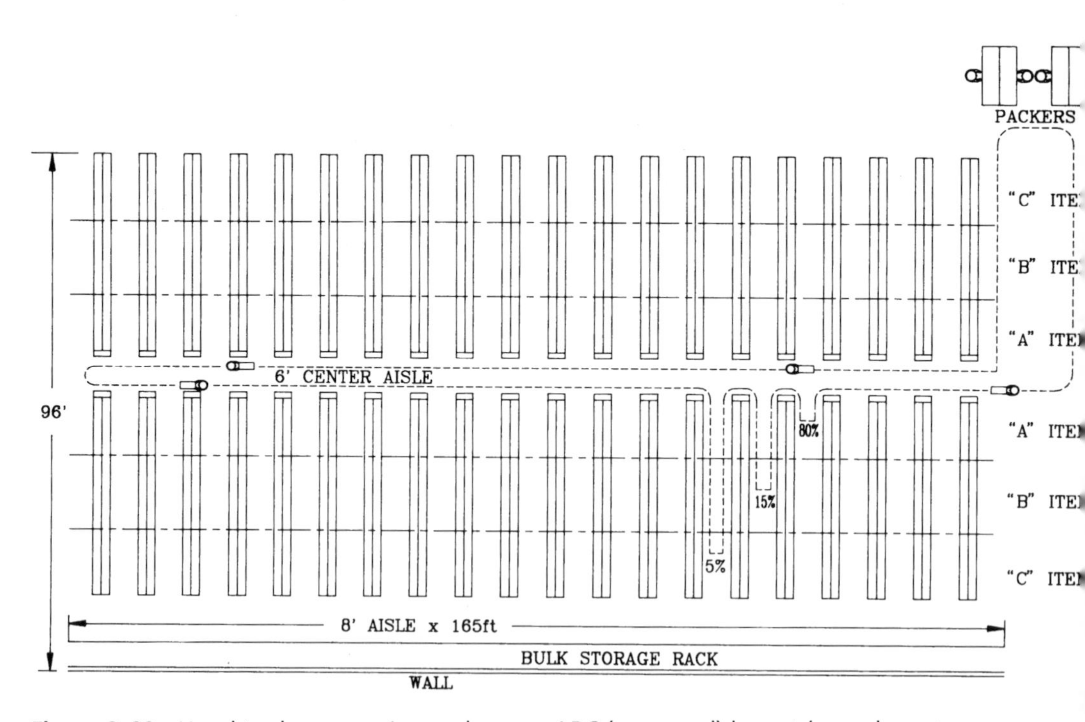

Figure 8–22 Hand tool company's warehouse—ABC (proposed) layout (same layout as Figure 8-21, with two sections of shelves removed to create new aisle and place all A items in the first 3 feet off the aisle).

less space. A world-class distribution center means that you can compete with the best warehouses in the world. You must work both smarter and harder to be the best.

A small amount of everything is the prime criterion for warehouse layout. A "small amount" could mean a one-day supply up to a one-week supply, but never everything you have of that part number. If you do not leave room for everything in the warehouse, where does the excess inventory go? You may have a 30-day supply of one part number and the warehouse is designed to hold only a one-day supply. This excess inventory would be called *bulk stock* or *backup stock* and could be kept in the raw material storeroom. Remember that stores use a random location so that you can put it anywhere. Special sections of the warehouse can be set up as bulk stock or backup stock areas. These areas would be laid out and controlled just like storerooms.

Keeping the shelves full is the job and responsibility of a group of warehouse staff other than the order pickers. These employees move material from the bulk areas to the picking area. Sometimes these warehouse stockers pull very large orders directly from the bulk areas to avoid depleting the shelf stock.

Warehousing is judged by service level and pounds shipped per labor hour. Whatever can be done to improve these figures will be good for the company.

Warehouse Space Determination

The size of the finished or packaged product multiplied by the quantity manufactured each day times the number of days' supply will equal the cubic footage of warehouse space required.

Example: Provide a warehouse to store a 30-day supply of toolboxes at the rate of 2,000 units per day.

$$\frac{18 \times 8 \times 8 \text{ inches}}{1{,}728 \text{ inches per cubic foot}} = .666 \text{ cubic foot each} \times 2{,}000 \times 30 \text{ days}$$

$$40{,}000 \text{ cubic feet} + \text{pallets}$$

See the layout in Figure 8–23 for a pallet pattern and a typical pallet.

$$\frac{42 \times 42 \times 54 \text{ inches}}{1{,}728 \text{ inches per cubic foot}} = 55.125 \text{ cubic feet per pallet}$$

$$72 \text{ toolboxes per pallet}$$

$$2{,}000 \text{ per day} \times 30 \text{ days} = 60{,}000 \text{ toolboxes}$$

$$\frac{60{,}000}{72} = 833 \text{ pallets}$$

Figure 8–24 illustrates a layout for the toolbox plant's warehouse. Note that the pallets are 42 × 48 inches (a standard width) and that eight pallets deep is only 28 feet. Pallets are stacked next to each other (with no room in between) because the cartons are all the same size and stack solidly. This is a very simple one-product layout. More complicated layouts are just more of the same procedure. If you calculate the cubic space required for each product and total them, you will have the total

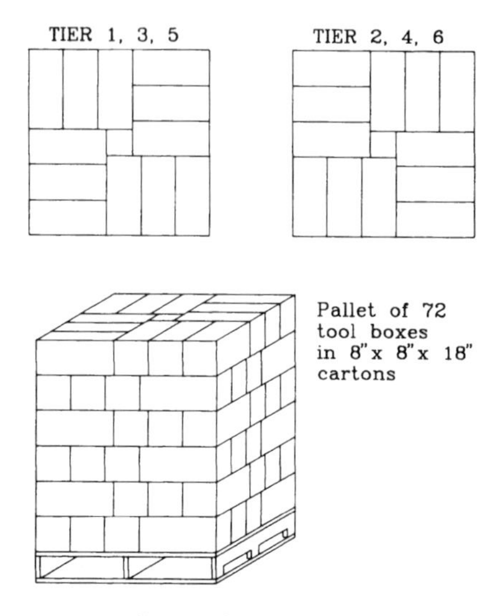

Figure 8–23 Pallet pattern—Toolboxes (12 per tier, 72 per pallet).

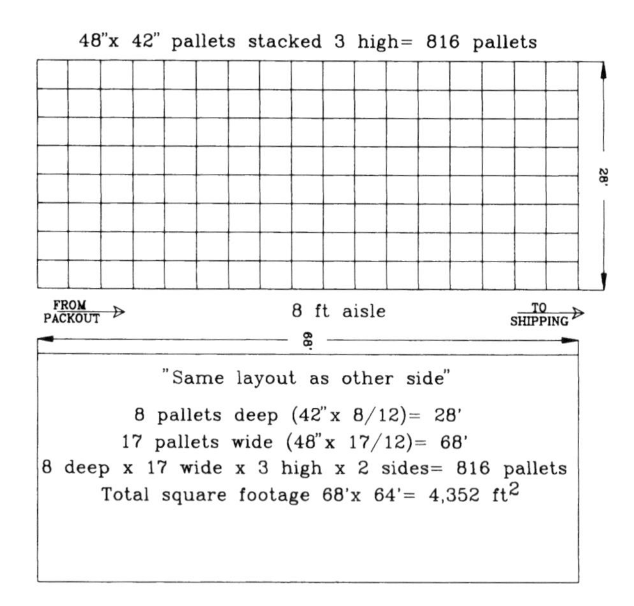

Figure 8–24 Warehouse layout—Toolbox plant.

storage space. Doubling this space will allow for aisles, and 50 percent aisles is more normal than in the example. When you have only one product, you can store deep (eight pallets from the aisle). Normally, you can store only one or two pallets deep, thereby requiring much more aisle space.

Warehouse Equipment

Shelves resembling those used in a library are most common for picking areas (see Figure 8-11). Whereas all the shelves are usually the same standard dimensions, the heights can be adjusted to allow for different sizes and storage quantities. For more room, more shelves must be used. Sometimes three or four different parts can be stored on one shelf. Tool warehouses use heavy-duty shelving, which measures 3 feet wide, feet deep, and 1 foot high, with an average of seven shelves high. Each shelf will hold cubic feet of parts. A one-week supply is warehoused on the shelves. The overstock is kept in the storeroom.

A *mezzanine,* a form of balcony, can be built over a shelving area for additional shelves. Slower moving inventory can be stored upstairs to make good use of an otherwise poorly used space.

Two-wheeled hand carts are often used to stock shelves. Boxes of material may be brought to the warehouse department by *fork truck,* but the aisles are not big enough to allow fork trucks, therefore hand carts are used. The cartons are moved to the shelves. Heavy (over 25 pounds) cartons are unpacked and placed on the shelves by hand.

Picking carts are four-wheeled shelf carts that are pushed around the shelves to pick customers' orders. The carts are unloaded as the packer fills cartons to ship to the customers.

Racks are used for larger products. The spacing between shelves can be as large or small as needed, but 2 or 3 feet is common. Two or three high is all that can be stacked because of packing height restrictions. Toolboxes are stored on racks in the tool company warehouse.

Flow racks allow for many parts to be warehoused in a small location. In a drug warehouse, 2,000 of the most popular drugs were warehoused on a 50-foot aisle. Eighty percent of the sales dollars was shipped out of this very small part of the drug warehouse (see Figure 8–25).

Conclusion

Warehousing is an area where a little planning and creative thought can save a lot of space and improve efficiency. The main design criteria are

1. Allow a small fixed location for everything.
2. Divide the inventory into ABC classifications.
3. In the following example of ABC Warehouse, locate the A items closest to shipping in the most convenient location.

Class	*Percentage of $*	*Percentage of Part*	*Warehouse Days Supply*
A	80	20	2
B	15	40	5
C	5	40	10

Figure 8–25 Flow rack picking area (courtesy of S.I. Handling Systems, Inc.).

4. Calculate the storage space required for each item in the warehouse and multiply the unit cubic foot by the number of days supply. Examples:

   1.5 cubic feet would be one-third of a shelf
   6.0 cubic feet would be one and one-third shelves

5. Calculate the total number of shelves.
6. Determine aisle size.
   a. One-way aisles should be 3 to 4 feet depending on the size of material.
   b. Two-way aisles should be one foot wider than two pieces of material handling equipment. An 18-inch picking cart would require a 4-foot aisle.
7. Lay out the shelves and the aisles and determine the warehouse width and length.
8. Maximize the warehouse cubic space. Mezzanines and racks can best use overhead space.

MAINTENANCE AND TOOL ROOM

The maintenance and tool room function is to provide and maintain production tooling. These functions vary widely from one company to another. Tool rooms may not exist in some plants because all tools are purchased from outside sources. Some maintenance is also contracted to outsiders. For example, office equipment maintenance is usually done by an outside firm.

Sizing the maintenance and tool room is dependent on management's desire to do it in-house or to contract out all or part of these jobs. A tool room is made up of machines and an assembly area similar to production. Once management determines what the plant will do, a machinery list is determined and each machine needs a workstation design. The tool room size is the sum total of all the equipment space requirements times 200 percent. The extra space is for everything except raw material and finished tool storage areas. These areas are calculated just like every other storage area.

Maintenance is service to the company's equipment. A mobile service cart could be used to maintain equipment, but more commonly a central maintenance area is designed to include equipment, machine overhead areas, maintenance supply, and spare parts storage areas. Maintenance typically accounts for 2 to 4 percent of the plant personnel. As an extreme example, maintenance accounted for 33 percent of a paper mill's employment. If you know the size of the plant (the number of employees) and, from corporate experience (or industrial averages), that the company should have 3 maintenance people for every 100 production people, you could provide them with 400 square feet of space each. This would allow for everything except maintenance stores, which is calculated like any other storeroom.

The toolbox plant used .13702 hour per unit of the rate of 100 percent. History indicates that 85 percent performance is more realistic, so

$$\frac{.13702}{.85} = .16120 \text{ hour each}$$

$$.16120 \times 2{,}000 \text{ boxes per day} = 322.4 \text{ hours of production people}$$

Each person works 8 hours per day, so 41 people are needed. Three percent of 41 people equals 1.2 maintenance people required. Therefore, you will allow space for two maintenance people.

$$2 \times 400 \text{ square feet} = 800 \text{ square feet}$$

Add a 10 × 10-foot controlled storeroom for tools and supplies to most plant layouts. This 100-square-foot storage area is just a minimum size area for controlling supplies. The plant will buy its tools, so that no tool room is needed. Total square footage for maintenance will be 900 square feet.

UTILITIES, HEATING, AND AIR CONDITIONING

Heat, air conditioning, electrical panels, air compressors, and so on must be considered when determining space. These areas also must be kept separate from the normal traffic—electrical panels should be fenced off, heaters must be kept clean, and air compressors require special construction because they are noisy. Once these facilities have been identified, they are sized and placed in an appropriate area of the plant. Many times they can be placed out of the way (on the roof or in the trusses) so that they do not interfere with material flow. But remember, whether in plain site or tucked out of the way, utilities must not be overlooked when determining plant space.

QUESTIONS

1. What are auxiliary services (support services)?
2. What do shipping and receiving have in common?
3. What are the advantages of a common receiving and shipping department?
4. What are the disadvantages?
5. Should a company have only one receiving area?
6. What effect does the trucking industry have on receiving and shipping docks?
7. What is LTL?
8. Why would you use common carriers?
9. What are the functions of a receiving department?
10. What is a Bates log?
11. What is a Julian calendar date?
12. What is an OS&D report?
13. What is a receiving report?
14. How many dock doors should you have?
15. What does arrival rate mean?
16. What outside areas are needed for receiving and shipping departments?
17. What is the visualization method of determining receiving department space requirements?
18. What are the functions of a shipping department?
19. Why do you weigh shipping containers?
20. What is a bill of lading?
21. What is a store?
22. What are the different types of stores?
23. What determines the store size?
24. What is ABC classification?
25. What is an inventory carrying cost?
26. What is JIT?
27. What are the goals of a stores department?

28. Review Figure 8–8 (the inventory curve) and identify
- **a.** the order quantity
- **b.** the usage rates (normal, maximum, and minimum)
- **c.** safety stock
- **d.** reorder point
- **e.** reorder time

29. How can you get away with leaving room for only 50 percent of the inventory?

30. How does random location work?

31. What is an aisle foot?

32. Lay out a storeroom with 18 bulk storage areas (4 feet on the aisle × 12 feet deep) + 800 aisle feet of pallet racks for 4 × 4-foot pallets + 400 feet of shelving (1 × 7 × 3 feet wide). Use 8-foot aisles for fork trucks and 4-foot aisles for shelves. Calculate the square footage.

33. What is a warehouse?

34. What are the two design criteria for a warehouse?

35. What are the two functions of a warehouse?

36. What is order picking?

37. How does ABC inventory analysis help you to lay out the warehouse?

38. What is a pallet pattern?

39. What is a mezzanine?

40. How many maintenance people should a plant have?

41. Explain how an automatic identification and data capture (AIDC) system can aid with your receiving, storage, and inventory tracking problems.

42. Explain how the Pareto analysis works and how it can be applied to the organization of your warehouse.

43. Explain the concept of random location for various items in the storeroom. How does it help with space utilization?

44. What is an item locator file and what purpose does it serve?

45. What necessitates the use of safety stock?

46. Given the following inventory items, (a) calculate the required number of pallets for each item, and (b) determine the total required aisle feet for an average inventory level. Pallet capacity is 4 × 4 × 4 feet. Racks are 14 feet wide and each one can store three pallets side by side and four pallets high.

*Length**	*Width**	*Height**	*Maximum Quantity*
25	24	4	5,000
12	10	3	7,000
36	12	8	9,000
24	8	8	8,000

*In inches.

CHAPTER 

Employee Services—Space Requirements

OBJECTIVES:

Upon the completion of this chapter, the reader should:

- Be able to recognize and identify employee needs and requirements
- Be able to identify facilities such as parking lot, cafeteria, and so on, in support of such needs
- Be able to calculate space requirements in fulfillment of such requirements

Employees have needs, and employee services describe the various needs. This chapter will discuss the following:

1. Parking lots
2. Employee entrance
3. Locker rooms
4. Toilets and restrooms
5. Cafeteria or lunchroom
6. Recreational facilities
7. Drinking fountains
8. Aisles
9. Medical facilities
10. Break areas and lounges
11. Miscellaneous employee services

These services require quite a bit of space. Their locations will affect the efficiency and productivity of the employees and the quality of these services will affect the quality of work life and the employees' relationship with the company's management. It is said that if you want to "see" management's attitude toward its employees, look in the restroom. If it is untidy or in disrepair, a poor attitude exists. A neat, clean restroom indicates a positive attitude.

PARKING LOTS

The interface between the outside world and the plant is the driveways and parking lots. The goal is to provide adequate space with a convenient location. Three parking lots may be needed. They could be broken down by usage as follows:

1. Manufacturing employee parking
2. Office employee parking
3. Visitor parking

Convenience and the efficient use of space are very important considerations when determining parking lot design. The entrances to the plant will determine where the parking lots are located. Parking as close to the entrance as possible should be your goal, but remember not everyone can park in the same space. One thousand feet takes an average of 4 minutes to walk. This distance should be the farthest point for either employee or visitor to walk. Large plants could have trouble with this requirement due to space restrictions, but it still should remain a goal.

The size of the parking lot is directly proportional to the number of employees. If the company were located out in the country and the employees drove to work, you might allow one parking place for every one and one-half employees. If it were closer to town and property was costly, you might allow one parking space for every two employees. You must consider plant location, the number of employees, and management's attitude toward carpools and the like and then decide on the parking space-to-employee ratio. In conjunction with plant location, special consideration must also be given to the availability of public transportation and municipal parking facilities in the vicinity of the plant.

No. of Employees		*Parking Space*	*Spaces per 100 Employees*
1.25	to	1	80
1.5	to	1	67
1.75	to	1	57
2.0	to	1	50

Office parking may be different from factory parking because you can incorporate the visitor parking spaces in this area. A ratio of 1:1 may be appropriate, and we could assign the closest parking spaces to visitors' parking.

Assigned parking is normally a bad idea, especially if employees who arrive early have to park farther away from the employee entrance and pass the parking places of these "big shots." Assigned parking indicates an air of superiority and could lead to poor employee relations and morale. The best assignment is first come, first served for the best spot. Another way of improving parking and employee relations is to have more than one entrance (i.e., a factory entrance and an office entrance). This way, the office people who arrive late still get a chance at a good spot. The only two reasons for assigned parking spaces are for company cars and carpool vehicles. The car that is used to run errands should be conveniently located close to the entrance to promote productivity. Carpooling promotes cost efficiency.

The facilities planner must incorporate the requirements of the Americans with Disabilities Act (ADA) of 1989 in all aspects of planning and design. Providing special and properly designed parking spaces and creating a barrier-free environment in all aspects of the facility should be of prime concern to the planner. Complying with the ADA requirements is a matter of the law. ADA guidelines must be consulted and the planner must understand the intent of the law and the "barrier-free" concept while planning parking facilities, entrances, restrooms, offices, and most areas of personnel services. Other local zoning ordinances and laws may also have an effect on the location and planning of the parking facilities.

Once the number of parking lots and parking spaces has been determined, there is almost an unlimited number of ways to arrange parking. The size and shape of the available space may be the deciding factor, but some general statistics follcw:

Parking Stalls	*Width**	*Length**
Small cars	8	15.0
Medium cars	9	17.5
Large cars	10	20.0

* In feet.

Width of Driveways
Single lane—11 feet
Double lane—22 feet

1. As the angle, A, of the parking spaces increases, the required width of the aisle increases (see Figure 9–2). For example, a narrower aisle is needed when $A = 45°$. However, as A approaches 90°, a wider aisle is required.
2. As the width of the parking spaces increases, the required width of the aisle decreases.
3. The wider the parking spaces (to 10 feet) are, the less door damage there is in the parking lot.
4. Local building codes often dictate parking space size.
5. Local building codes often dictate the number and location of handicapped spaces.
6. As a general rule of thumb, a parking lot will be 250 square feet per number of parking spaces needed (see Figure 9–1 and Figure 9–2).

It is also imperative to realize that regardless of the size and the configuration of the parking lot, one must consider accessibility to the roadways for easy and safe entrance into and exit from the parking lot.

It may be interesting to analyze and compare the driveway and entrance/exit requirements for the two alternatives presented in Figures 9-1 and 9-2. While in Figure 9-1 only one entrance/exit is required at each end, the driveway must be wide enough (approximately 22 feet) to accommodate two cars simultaneously for a two-way traffic flow (a total of 44 feet). Whereas in Figure 9-2, a total of four entrances and exits are required, each aisle requiring one entrance and one exit at each end. However, since in Figure 9-2 the flow in each aisle is unidirectional, the required width is only approximately 11 feet (a total of 44 feet).

EMPLOYEE ENTRANCE

Where the employees enter the plant will have an effect on the placement of parking, the locker room, time card racks, restrooms, and cafeterias. The flow of people into a factory is from their cars into the plant via the employee entrance to their lockers and to the cafeteria to wait for the start of their shifts.

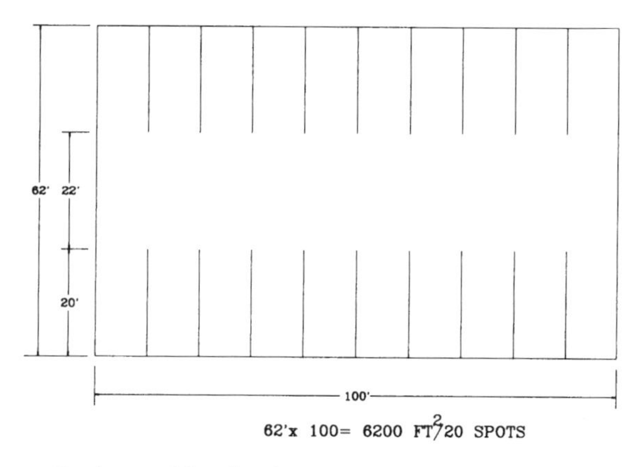

Figure 9–1 Perpendicular parking lot (most space consumed).

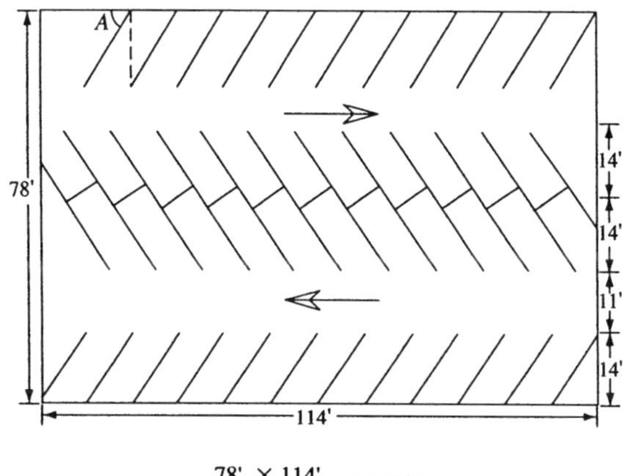

Figure 9–2 Angular parking lot.

The employee entrance is where security, time cards, bulletin boards, and sometimes the personnel departments are located. Depending on management's attitude and corporate requirements, the employee entrance can vary from a simple doorway with a time card rack and time clock to a series of offices and gates through which to pass.

The size of the employee entrance must consider individual requirements. How many persons will be using this door at any given time? The door could measure from 3 feet to 6 feet with an aisle or walkway leading into the plant. Allow for traffic flow.

Personnel offices and security offices will be sized at 200 square feet per office person. About one personnel person per 100 employees and one security person per 300 employees are normal.

In the toolbox plant example, the employee entrance consists of a 4-foot door and a 6-foot aisle such as shown in the layout in Figure 9–3. The time clock and time card racks are mounted on a wall, and on the other side of the aisle, the bulletin board is mounted on a wall. Figure 9–4 shows a larger plant's employee entrance feet or 90 square feet with security.

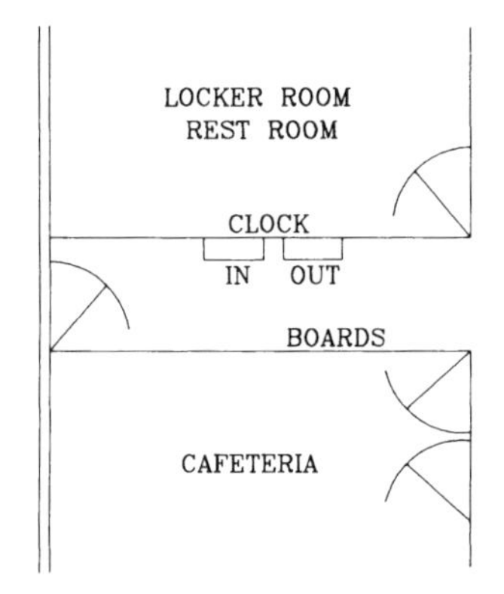

Figure 9–3 Simple employee entrance.

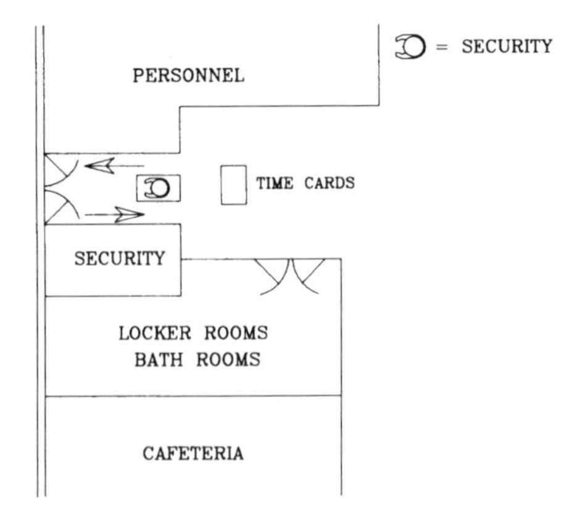

Figure 9–4 Employee entrance with security.

LOCKER ROOMS

Locker rooms give the employees space to change from their street clothes to their work clothes and a place to keep their personal effects while working. Their coats, lunches, street shoes, and so forth will be kept in their lockers. Locker rooms are very much like gym locker rooms. Showers, toilets, washbasins, lockers, and benches are all part of a well-equipped locker room, but we will consider toilets and washbasins in the next section.

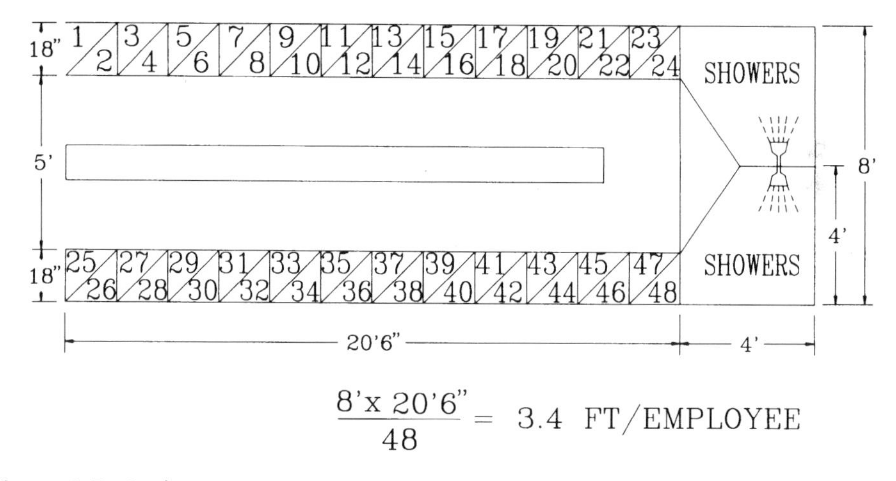

$$\frac{8' \text{x } 20'6"}{48} = 3.4 \text{ FT/EMPLOYEE}$$

Figure 9–5 Locker room.

The size of a locker room is directly proportional to the number of employees. Figure 9–5 shows a typical small locker room. The top tier could be for the day shift and the bottom tier for the night shift. Staggered shifts could reduce congestion in the locker room. For example, consider the following schedules:

7:00–11:00	Lunch	11:30–3:30
7:30–11:30	Lunch	12:00–4:00
8:00–12:00	Lunch	12:30–4:30
3:30–7:30	Lunch	8:00–12:00
4:00–8:00	Lunch	8:30–12:30
4:30–8:30	Lunch	9:00–1:00

In a plant of 48 people, only 8 of them would be in the locker room at one time. If this were a plant of 96 people (48 men and 48 women), two identical locker rooms would be provided.

The size of a locker room can be initially sized by multiplying the number of employees by 4 square feet per employee.

RESTROOMS AND TOILETS

Personal hygiene facilities are usually called restrooms. How many restrooms are required per employee may be the first question. As a rule of thumb, one toilet is required per every 20 employees, and restrooms should be no farther than 200 feet away from any employee. Furthermore, one sink per toilet must be installed in every restroom. Special accommodations and provisions must also be made for employees with disabilities as required by the ADA. At a minimum, there should be a men's restroom and a women's restroom in the office and in the factory.

The number of toilets required depends on how many employees work on the major shift. The local building code may dictate how many toilets are necessary. The number of washbasins is equal to the number of toilets. See Figure 9–6 for a partial example of a local building code.

The size of a restroom is 15 square feet per toilet, washbasin, and entryway, and 9 square feet for urinals.

If the plant has 50 male employees and 50 female employees, then the two restrooms would look like this:

	50 Men	*50 Women*
Toilets	2 × 15 = 30	3 × 15 = 45
Washbasins	3 × 15 = 45	3 × 15 = 45
Urinals	1 × 9 = 9	
Reclining area		1 × 15 = 15
Door	1 × 15 = 15	1 × 15 = 15
Total	99	120
× 150 percent	149	180

See Figure 9–7 for a diagram of this men's restroom.

CAFETERIAS OR LUNCHROOMS

A typical plant could have any of these five types of eating facilities:

1. Cafeterias with serving lines
2. Vending machines
3. Mobile vendors
4. Dining rooms (executive)
5. Off-site diners (lunch counters)

A cafeteria feeds a lot of people in a short time. Schools, military installations, and family picnics use this type of service. Many people line up at a serving line and are given food as they pass different stations. Assume, for instance, that one serving line can service nine employees per minute (about 7 seconds each). One line is 30

Sec. 31-37 Toilet Accommodations in Manufacturing, etc.:

"... Provide adequate toilet accommodations, so arranged as to secure reasonable privacy for both sexes ... inside such establishment when ... Practicable ... Adequate fixtures ... good repair ... clean and sanitary conditions, adequately ventilated with windows or suitable ventilators opening to the outside ... provide convenient means for artificial lighting ... clearly marked ... to indicate the sex for which ... intended for use ... and ... partition ... to be solidly constructed from the floor to the ceiling."

Sec. 177-4-5 Toilet Facilities:

a. Water-closets required and sex designation:
   "Separate water-closet ... for each gender ... clearly marked. ..."
b. Number:
   "Water or toilet closets shall be provided for each sex at the rate of one closet to twenty persons or fraction thereof, up to 100, and thereafter at the rate of one closet for every twenty-five persons."
c. Location:
   "Such closets and urinals must be readily accessible to the persons for whose use they are designed. In no case may a closet be located more than three hundred feet distance from the regular place of work of the persons for whose use it was designed. ..."

Sec. 177-6-6 Privacy:

b. New installations:
   1. "Every water-closet compartment ... shall be located in a toilet room, or shall be built with a vestibule and door to screen the interior from view."
   2. "The door of every toilet-room and of every water-closet compartment, which is not located in a toilet-room, shall be fitted with an effective self-closing device to keep it closed."
   3. "... male and female ... adjoining compartments (must have) solid plaster or metal covered partitions ... extending from floor to ceiling."

Sec. 177-4-7 Construction:

b. New installations:
   1. "The floor ... and the side walls to a height of not less than six inches ... shall be marble, portland cement, tile, glazed brick, or other approved waterproof material."

Sec. 177-4-12 Urinals:

a. "In establishments or departments employing ten or more male employees, one urinal shall be installed for every forty males or fractional part thereof up to two hundred and thereafter, an additional urinal for every sixty males or fractional part thereof. Two feet of an approved trough urinal shall be equivalent to one individual urinal. ... Urinals may substitute for up to 50% of men's toilets."

Sec. 177-4-4 Washing Facilities:

a. "All water supplied by any establishment for washing purposes shall be potable (pure, clean water suitable for drinking or for washing purposes)."
b. "Every establishment should furnish for each sex at least one standard wash basin or its equivalent for every 20 such employees, or, fractional part thereof, up to 100. Beyond 100 the ratio may be one basin or equivalent to each 25 employees of either sex, or fractional part thereof."
c. "If washing sinks or troughs are furnished, each 2-1/2 feet of trough or sink equipped with a hot water and a cold water faucet or a single faucet carrying tempered water may be counted equal to one basin. Where washing fountains are furnished, 2-1/2 feet of the circumference of such fountains shall be equivalent to one wash basin.

Sec. 31-41 Order to Remove Excessive Dust:

"Each employer whose business requires the operation or use of any emery, tripoli, rouge, corundum, stone, carborundum or other abrasive, polishing or buffing wheel in the manufacture of articles of metal or iridium or whose business includes any process which generates an excessive amount of dust, shall install and maintain ... such devices ... necessary ... to remove from the atmosphere any dust created by such process. ..."

Figure 9–6 Sample building codes for toilet accommodations.

Sec. 31-45 Emergency Kits Required by Factories:
"Each . . . firm . . . employing persons to work . . . with dangerous machines . . . , except those maintaining equipped first-aid-to-the-injured rooms, shall . . . (place) where such machinery is operated . . . an emergency kit for use in case of accidents . . . such (kits) shall be kept in a dustproof case or cabinet within easy access. . . ."

Par. 716.3 Locks and fastenings on required exit doors shall be readily opened from the inner side without the use of keys. Draw bolts, hooks, and other similar devices shall be prohibited on all required exit doors.

728.2 All exit signs shall be generally located at doors or exit ways so as to be readily visible and not subject to obliterations by smoke. They shall be illuminated at all times when the building is occupied from an independently controlled electric circuit or other source of power.

Figure 9–6 (continued) Sample building codes for toilet accommodations.

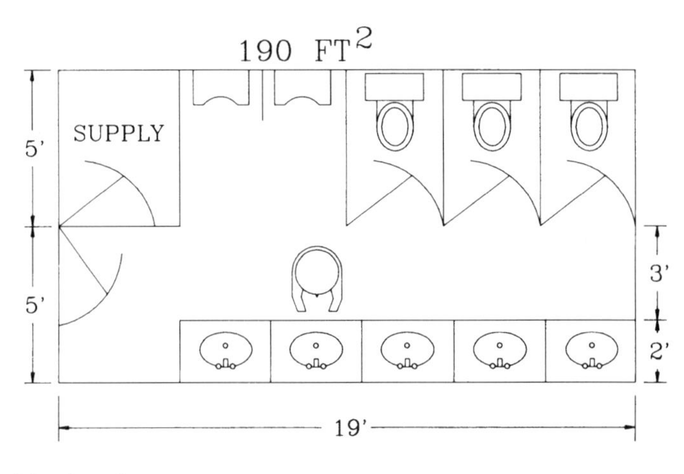

Figure 9–7 Men's restroom.

feet long and 10 feet wide. Employees would not spend more than 10 minutes in line, so that 90 people could be served every 10 minutes. If lunch periods were staggered, 540 people per hour could be served. Cafeterias are generally used in big plants.

Vending machines can serve very complete meals. A vending machine with a microwave oven for special foods can provide employees with many meal choices. Vending machines are generally used for small plant lunchrooms. The machines generally are lined up against a wall. Room for customer lines must be allowed in front of the machines. Vending machines earn money. Most companies use these profits for employee benefit programs, but these profits could be used to buy services from the vending company (such as a custodian who would be available to service the machines in high-use periods). These people can also be used to keep the lunchroom clean and in good order.

Mobile vendors are outside vendors who drive their specially built pickup trucks to the back door and honk their horns signaling the start of lunch periods or breaks. Only very small plants could use this service. Office buildings also use smaller mobile carts for serving coffee and donuts. They can go from office to office and even from floor to floor. The mobile vendors are affectionately called "Roach Coaches." Whatever they are called, they provide a useful service.

Executive dining rooms are used to entertain special customers, vendors, and stockholders. They usually provide a limited selection from a menu with the meal being cooked on site. Dining rooms cannot serve many people at a time and usually eating takes more time.

Off-site dining at local diners is attractive to many employees. It allows them to get away from the job. Private businesspeople develop local clientele for lunch and create a comfortable environment in which to eat. But most factory workers are not given enough time to leave the plant for lunch. Companies discourage employees from leaving the plant at lunchtime because the companies lose control (e.g., lunch may include an alcoholic beverage, etc.).

Lunchrooms should provide a comfortable, pleasing environment in which to recuperate from work and to eat lunch. Nice facilities show respect for employees and improve the productivity of the workforce by allowing workers to regenerate their energy for the next work period. Comfort, attractiveness, speed of service, and convenient location are all important in the design of a lunchroom. The location is analyzed in the activity relationship diagram studied in Chapter 6, but two more factors should be considered:

1. An outside eating facility or area would allow employees more flexibility. During nice weather, this facility would increase morale. An outside wall gives easy access to food delivery and trash removal.
2. The employee flow typically calls for the employees to wash up before eating and to get their lunches from their lockers, so restrooms and locker rooms should be close to the lunchroom.

The size of the lunchroom will depend on (1) the number of employees, (2) the type of service provided, and (3) the facilities included.

Per Person or Unit	*3 Periods 100 people*	*5 Periods 500 people*	*7 Periods 1,000 people*
Cafeteria serving line (300 ft.)	—	300	600
Waiting line (4 ft^2)	120	180	320
Vending machines (20 ft^2)	100	—	—
Eating area	495	1,995	3,000
Waste (½ ft^2)	50	250	500
Food storage (½ to 1 ft^2)	50	500	1,000
Food prep. (2 ft/meal)	—	Catered	2,000
Dish washing (½ to 1 ft^2)	—	500	750
Total	815	3,725	8,170
Aisles and misc. +25 percent	204	931	2,043
Grand total	1,019 ft^2	4,656 ft^2	10,213 ft^2
Square ft per person	10.2 ft^2	9.3 ft^2	10.2 ft^2
Size	22 × 46	48 × 96	71 × 143

Figure 9–8 Lunchroom space determination.

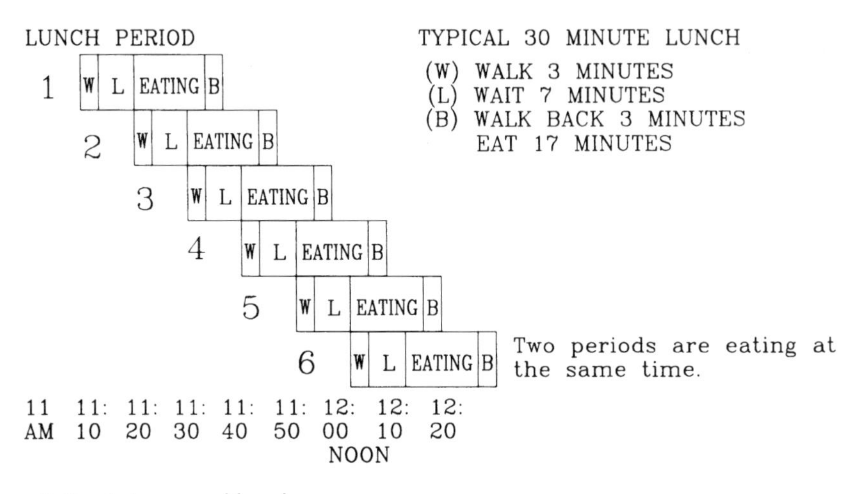

Figure 9–9 A staggered lunchroom.

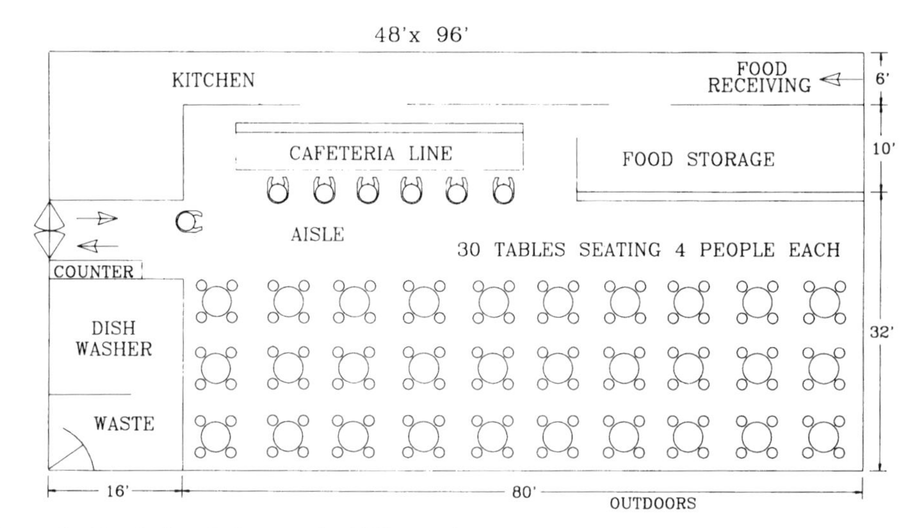

Figure 9–10 Cafeteria layout for 500 people.

Figure 9–8 shows the backup data for three different-sized lunchrooms. As can be seen, they are very similar in size per person (10 square feet). Cafeteria space can be saved if the food is cooked off-site and carried into the plant at lunchtime. Space can be saved by overlapping lunch periods. A lunch period starting every 10 minutes will reduce waiting time or the number of facilities required in a 30-minute lunch period (15 to 20 minutes of seating is all that is used). Keep in mind that 10 to 15 minutes of employee time can be used in walking to and from the workstation and waiting in line (see Figure 9–9). A good estimate for a lunchroom would be 10 square feet per employee. Figure 9–10 shows an example layout for a 500-person lunchroom.

RECREATIONAL FACILITIES

Recreational facilities are becoming more important every year. Health-conscious employees are better employees and companies are recognizing this fact. Health clubs, tracks, locker rooms, as well as Ping-Pong tables, card games, and social clubs are becoming part of today's plants. These facilities take space, and the plant layout designer must talk with management to understand what facilities need to be included. A workstation layout drawing must be made for each facility, and the individual space must be determined and included in the plan.

DRINKING FOUNTAINS

Drinking fountains should be located within 200 feet of every employee and on an aisle for easy access. Each drinking fountain will include space for the drinking fountain and a person getting a drink. Fifteen square feet (3 × 5 feet) should be allowed for each drinking fountain.

AISLES

Aisles can be the greatest consumers of plant space if you are not careful. Aisles are for movement of people, equipment, and material and must be sized for that use. For example, two-way fork truck traffic means that the aisle must be able to handle two trucks passing each other plus a safety cushion (4 + 4 + 2 feet). In this case, 10-foot aisles would be needed. Two-way people aisles must be at least 5 feet wide. Every workstation shelf, or rack must have aisle access.

Aisles should be long and straight. The major aisle of the plant may run from the receiving dock straight through the plant to the shipping dock. Side aisles may be smaller but perpendicular to the main aisle (see Figure 9–11).

The percentage of the plant's total square footage used as aisles (square footage of aisles divided by total plant square footage) is a valuable measurement. This percentage should be plotted on a graph at least yearly. The objective is to reduce this percentage. Here are a couple of ideas to reduce aisles:

1. Use stand-up reach trucks instead of fork trucks because of their shorter turning radius.
2. Use double-deep pallet racks or drive-in pallet racks, thereby reducing the number of aisles at least in half.

Space allocation for the production aisles is accomplished by increasing the total production equipment space area by a factor of 50 percent.

Warehouse aisle space is calculated with the number of aisle-feet of storage units (see Chapter 8). Access aisles around the equipment are included in the workstation layouts.

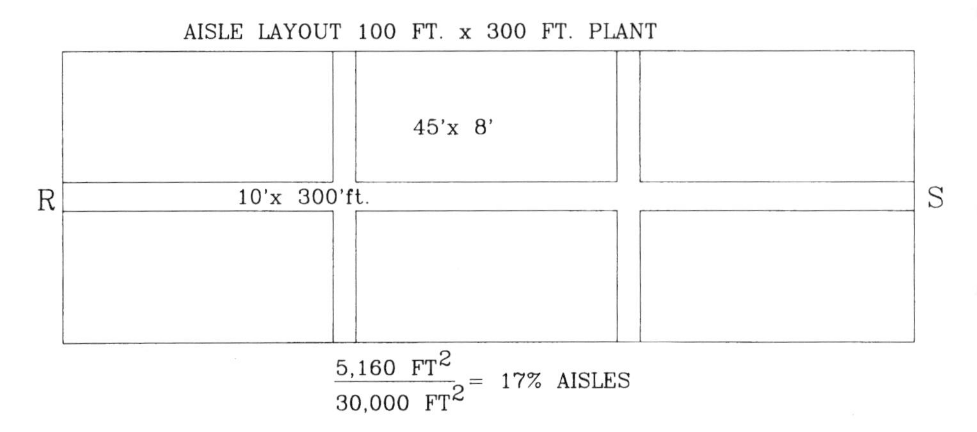

Figure 9–11 Aisle layouts.

MEDICAL FACILITIES

Medical facilities vary from 6 × 6-foot first-aid rooms to full-fledged hospitals. In smaller plants, first aid is handled by trained employees in the plant. Medical emergencies are handled by the emergency room at the local hospital or clinic. When a plant approaches 500 people, a registered nurse is usually justified. Nurses require facilities such as waiting rooms, examining room, medical supplies, and record and reclining areas. One nurse would require a 400-square-foot area. If the plant has 3,000 employees, 6 nurses would be justified and a doctor could be hired to supervise the medical staff. Depending on the number of shifts and the type of manufacturing being done, various medical facilities will be required. Some space requirements are as follows:

Office	100 ft^2/nurse
	200 ft^2/doctor
Examining room	200 ft^2/room
Waiting area	25 ft^2/nurse and doctor
Supply room	25 ft^2/nurse or doctor
Basic first-aid room	36 ft^2

Figure 9–12 shows a minimum medical facility. Figure 9–13 illustrates a layout of a larger facility.

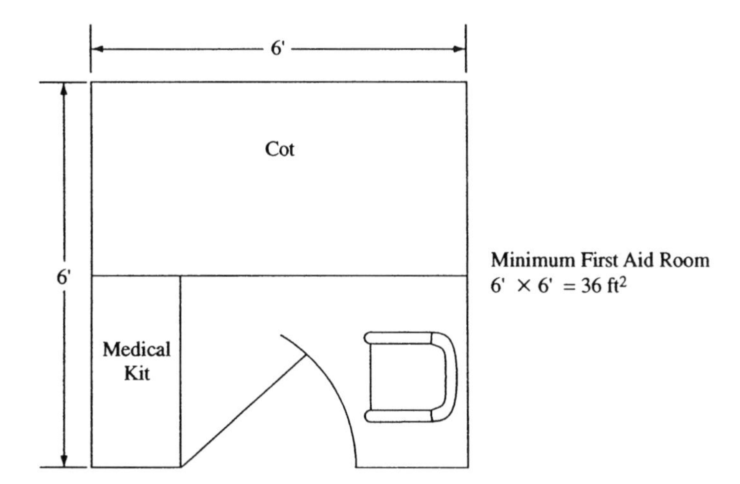

Figure 9–12 Minimum medical facility.

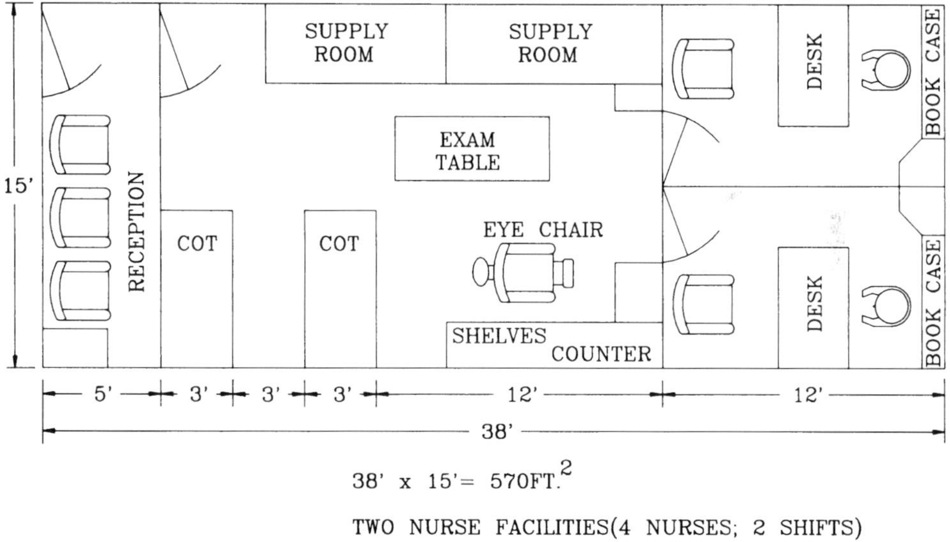

Figure 9–13 Larger medical facility.

BREAK AREAS AND LOUNGES

If the lunchroom is too far away from groups of employees, a break area should be provided. "Too far" is defined as over 500 feet. Five hundred feet takes 2 minutes to walk. On a 10-minute break, 2 minutes walking to the lunchroom and 2 minutes back leaves only 6 minutes of rest. A break area in a remote area might include a picnic table, a drinking fountain, maybe a vending machine or two, and sometimes a Ping-Pong table that folds up and rolls away. There should be enough seats for everyone on a break. Staggered breaks will reduce the need for excessive space.

Lounges are usually found in shipping and receiving areas for visiting truck drivers to wait for their loads. The size of lounges will be proportional to the number of trucks arriving and leaving the plant each day. The lounges also keep the non-employees (truck drivers) out of the plant. Restrooms should be conveniently close to lounges to eliminate the need for drivers walking through the plant. The lounge should be sized by multiplying the number of drivers that could be waiting at one time by 25 square feet. If four drivers would be the most drivers expected at one time, then a 100-square-foot lounge would suffice.

MISCELLANEOUS EMPLOYEE SERVICES

The above employee services and facilities are those most commonly found in manufacturing plants. Some others include (1) training and educational facilities, (2) child-care services, (3) hairstyling services, (4) libraries, and (5) exercise and workout facilities.

See Figure 9–14 for pictures of sample employee service equipment.

Figure 9–14 Employee service equipment (courtesy of Globalindustrial.com).

QUESTIONS

1. What are the employee services that require space?
2. How many parking spaces do you need?
3. How many square feet per parking space (including aisles) are required?
4. What is included in employee entrance space?
5. What should the employee entrance be close to?
6. What is a locker room for?
7. How big is a locker room?
8. How many restrooms do you need?
9. How do you know how many toilets, urinals, and sinks are required?
10. How big is the restroom?
11. What are the five types of eating facilities?
12. Where should the eating facility be located?
13. How big should a lunchroom be?
14. How many drinking fountains should you have?
15. How big are drinking fountain areas?
16. How much of the plant's space should be taken up by aisles?
17. How many employees justify a nurse?
18. How large should a medical facility be?
19. What is ADA and what are its implications on the design of facilities?
20. In addition to the personnel requirements and the areas suggested in the text, what other areas can you suggest that may benefit your employees?

CHAPTER 

Material Handling

OBJECTIVES:
Upon the completion of this chapter, the reader should:

- Be able to justify the needs for material handling
- Understand the goals of material handling
- Understand the principles of material handling

Material handling is the function of moving the right material to the right place, at the right time, in the right amount, in sequence, and in the right position or condition to minimize production costs. The equipment to perform this function will be discussed in the next chapter. First, the principles of material handling and the control systems must be understood.

Material control systems are an integral part of modern material handling systems. Part numbering systems, location systems, inventory control systems, standardization, lot size, order quantities, safety stocks, labeling, and automatic identification techniques (bar coding) are only some of the systems required to keep industrial plants' material moving.

Material handling can be broadly defined as all movement of materials in a manufacturing environment. The American Society of Mechanical Engineers (ASME) defines "material handling" as the art and science involving the moving, packaging, and storing of substances in any form. Material handling may be thought of as having five distinct dimensions: movement, quantity, time, space, and control.

Movement involves the actual transportation or transfer of material from one point to the next. Efficiency of the move as well as the safety factor in this dimension are of prime concerns. The quantity per move dictates the type and nature of the material handling equipment and also the cost per unit for the conveyance of the goods. The time dimension determines how quickly the material can move through the facility. The amount of the work in process, excessive inventories, repeated handling of the material, and order delivery lead times are affected by this aspect of the material handling systems. The space aspect of the material handling is concerned with the required space for the storage of the material handling equipment and its movement, as well as the queuing or staging space for the material itself. The tracking of the material, positive identification, and inventory management are some aspects of the control dimension. Material handling is also an integral part of plant layout; they cannot be separated. A change in the material handling system will change the layout, and a layout change will change the material handling system.

Material can be moved by hand or by automatic methods, it can be moved one at a time or by the thousands, it can be placed in a fixed location or at random, and it can be stored on the floor or high in the sky. The variations are limitless and only by cost comparison of the many alternatives will the correct answer emerge.

The proper material handling equipment choice is the answer to all the questions in this section. A material handling equipment list will include over 500 different types (classifications) of equipment, and if you multiply this number by the different models, sizes, and brand names, several thousand pieces of equipment are available for use.

Material handling equipment has reduced the drudgery of work. It has reduced the cost of production and has improved the quality of work life for nearly every person in industry today.

But the handling of material is attributed to more than one-half of all industrial accidents. Material handling equipment can eliminate manual lifting. But, like all equipment, it also can cause injury, so material handling project engineers can never forget about the safety aspects.

On the average, material handling accounts for 50 percent of the total operations cost. In some industries, such as mining, this cost increases to 90 percent of operations cost. This fact alone justifies great effort on the part of industrial managers and facilities designers.

COST JUSTIFICATION

Material handling equipment can be expensive, so all investments should be cost-justified. The lowest overall cost per unit gives you the best answer. If a very expensive piece of equipment reduces unit cost, it is a good purchase. If it does not reduce unit cost, it is a bad purchase.

Nonpowered equipment can be very cost efficient and should always be considered. Gravity chutes, rollers, hand carts, and hand jacks are only a few of the many popular methods of moving material economically.

Safety, quality, labor, power, and equipment costs must all be included in the unit costs. If someone is expected to lift a 100-pound load while performing a task, the long-term effect of the activity, or the cumulative trauma disorder (CTD), associated with this job must be considered. Ergonomic consideration of job design dictates that some types of material handling system such as a hydraulic or pneumatic lifting device should be studied. If taken in isolation, the dollar cost may not be justifiable;

however, the long-term safety considerations will certainly prove the investment to be a wise one. An automobile manufacturer discovered that a simple manipulator device, which assisted in lifting and turning the car seats while installing them, prevented serious and chronic lower back pain and injury to the assembly line workers.

Sample Material Handling Cost Problem

An oil remanufacturing company uses clay in its manufacturing process. This clay comes into the plant in 80-pound bags stacked 40 per pallet and 50 pallets per boxcar. The railroad spur comes into the plant property, but your plant does not have a rail car siding. Two car loads per year are used. The union and the company agreed that two part-time workers would be hired for one week, twice a year at the rate of $7.50 per hour to unload these boxcars. You feel that this is a bad job and no one should have to work this hard. You look into this project.

Why is this done? You need the clay, and the railroad is by far the cheapest way to transport it. Look at it like this:

What? = 2,000 80-pound bags of clay equals a 160,000 pound boxcar load; no other size bags are available.

Where? = From the boxcar in the yard to the storeroom, which is 300 feet away.

Who? = Two temporary workers.

When? = One week, twice a year.

How? = Present method. Manually unload the pallets off the boxcar, then move these pallets into the storeroom with the fork truck you already own.

This is backbreaking work, but how much could you spend improving this job? You spend one week, twice a year with two temporary employees being paid $7.50 per hour.

$$4 \text{ weeks} \times 40 \text{ hours per week} \times \$7.50 \text{ per hour} = \$1,200$$

Currently you are spending $1,200 per year on this job. Should the current method stay the same or are there other alternatives that can be employed? Is the current method the cheapest method in the long run? Chapter 11 will provide some answers to these questions. How would you justify an expenditure of, for instance, over $2,400 (labor cost for 2 years for the current manual method) to improve a task that is performed so infrequently? Before attempting to answer these questions and concentrating on the simple direct labor cost of this operation, which is indeed a common fallacy, consider the following facts.

Cumulative trauma disorders and work-related injuries are costing business and industry real dollars and productivity. According to the U.S. Bureau of Labor Statistics (BLS), the rate of incidents for disorders associated with repeated trauma has increased steadily since 1986. According to the BLS, the rate was reported as 6.4 per 10,000 FTE (full-time equivalent) in 1986. This rate increased to 41.1 by 1994. In 1996,

25 million workers reported lower back pain, and in the same year, 25 percent of all workers missed an average of one day of work per year due to the same problem. Furthermore, 2 percent of the workforce experienced compensable back injuries in 1996. These compensations cost U.S. businesses approximately $20 billion and 12 million lost workdays annually. Given the example of the 80-pound bags, it may be of particular interest to learn that the cost of an "average" lower back case is over $5,500 in direct costs. Indirect costs may raise the total U.S. loss to well over $30 billion annually.

GOALS OF MATERIAL HANDLING

The primary goal of material handling is to reduce unit costs of production. All other goals are subordinate to this goal. But the following subgoals are a good checklist for cost reduction:

1. Maintain or improve product quality, reduce damage, and provide for protection of materials.
2. Promote safety and improve working conditions.
3. Promote productivity through the following:
   a. Material should flow in a straight line.
   b. Material should move as short a distance as possible.
   c. Use gravity! It is free power.
   d. Move more material at one time.
   e. Mechanize material handling.
   f. Automate material handling.
   g. Maintain or improve material handling/production ratios.
   h. Increase throughput by using automatic material handling equipment.
4. Promote increased use of facilities as follows:
   a. Promote the use of the building cube.
   b. Purchase versatile equipment.
   c. Standardize material handling equipment.
   d. Maximize production equipment utilization using material handling feeders.
   e. Maintain, and replace as needed, all equipment and develop a preventive maintenance program.
   f. Integrate all material handling equipment into a system.
5. Reduce tare (dead) weight.
6. Control Inventory.

TEN PRINCIPLES OF MATERIAL HANDLING

The College Industry Council on Material Handling Education (CICMHE) of the Material Handling Institute, Inc., has adapted the 10 principles of material handling found in Figure 10–1.

1 Planning Principle

All material handling should be the result of a deliberate plan where the needs, performance objectives and functional specification of the proposed methods are completely defined at the outset.

Definition: *A plan is a prescribed course of action that is defined in advance of implementation. In its simplest form a material handing plan defines the material (what) and the moves (when and where); together they define the method (how and who).*

Key Points:

- The plan should be developed in consultation between theplanner(s) and all who will use and benefit from the equipment to be employed.
- Success in planning large scale material handling projects generally requires a team approach involving suppliers, consultants when appropriate, and end user specialists from management, engineering, computer and information systems, finance and operations.
- The material handling plan should reflect the strategic objectives of the organization as well as the more immediate needs.
- The plan should document existing methods and problems, physical and economic constraints, and future requirements and goals.
- The plan should promote concurrent engineering of product, process design, process layout, and material handling methods, as opposed to independent and sequential design practices.

2 Standardization Principle

Material handling methods, equipment, controls and software should be standardized within the limits of achieving overall performance objectives and without sacrificing needed flexibility , modularity and throughput.anticipation of changing future requirements

Definition: *Standardization means less variety and customization in the methods and equipment employed.*

Key Points:

- The planner should select methods and equipment that can perform a variety of tasks under a variety of operating conditions and in
- Standardization applies to sizes of containers and other load forming components as well as operating procedures and equipment.
- Standardization, flexibility and modularity must not be incompatible.

3 Work Principle

Material handling work should be minimized without sacrificing productivity or the level of service required of the operation.

Definition: *The measure of work is material handling flow (volume, weight or count per unit of time) multiplied by the distance moved.*

Key Points:

- Simplifying processes by reducing, combining, shortening or eliminating unnecessary moves will reduce work.
- Consider each pickup and set-down, or placing material in and out of storage, as distinct moves and components of the distance moved.
- Process methods, operation sequences and process/equipment layouts should be prepared that support the work minimization objective.
- Where possible, gravity should be used to move materials or to assist in their movement while respecting consideration of safety and the potential for product damage.
- The shortest distance between two points is a straight line

Figure 10–1 The principles of material handling (Courtesy of the College Industry Council on Material Handling Education [CICMHE] of the Material Handling Industry of America [MHIA]).

4 Ergonomic Principle

Human capabilities and limitations must be recognized and respected in the design of material handling tasks and equipment to ensure safe and effective operations.

Definition: *Ergonomics is the science that seeks to adapt work or working conditions to suit the abilities of the worker.*

Key Points:

- Equipment should be selected that eliminates repetitive and strenuous manual labor and which effectively interacts with human operators and users.
- The ergonomic principle embraces both physical and mental tasks.
- The material handling workplace and the equipment employed to assist in that work must be designed so they are safe for people

5 Unit Load Principle

Unit loads shall be appropriately sized and configured in a way which achieves the material flow and inventory objectives at each stage in the supply chain.

Definition: *A unit load is one that can be stored or moved as a single entity at one time, such as a pallet, container or tote, regardless of the number of individual items that make up the load.*

Key Points:

- Less effort and work is required to collect and move many individual items as a single load than to move many items one at a time.
- Load size and composition may change as material and product moves through stages of manufacturing and the resulting distribution channels.
- Large unit loads are common both pre and post manufacturing in the form of raw materials and finished goods.
- During manufacturing, smaller unit loads, including as few as one item, yield less in-process inventory and shorter item throughput times.
- Smaller unit loads are consistent with manufacturing strategies that embrace operating objectives such as flexibility, continuous flow and just-in-time delivery.
- Unit loads composed of a mix of different items are consistent with just-in-time and/or customized supply strategies so long as item selectivity is not compromised.

6 Space Utilization Principle

Effective and efficient use must be made of all available space.

Definition: *Space in material handling is three dimensional and therefore is counted as cubic space.*

Key Points:

- In work areas, cluttered and unorganized spaces and blocked aisles should be eliminated.
- In storage areas, the objective of maximizing storage density must be balanced against accessibility and selectivity.
- When transporting loads within a facility the use of overhead space should be considered as an option.

Figure 10–1 (continued)

7 System Principle

Material movement and storage activities should be fully integrated to form a coordinated, operational system which spans receiving, inspection, storage, production, assembly, packaging, unitizing, order selection, shipping, transportation and the handling of returns.

Definition: *A system is a collection of interacting and/or interdependent entities that form a unified whole.*

Key Points:

- Systems integration should encompass the entire supply chain including reverse logistics. It should include suppliers, manufacturers, distributors and customers.
- Inventory levels should be minimized at all stages of production and distribution while respecting considerations of process variability and customer service.
- Information flow and physical material flow should be integrated and treated as concurrent activities
- Methods should be provided for easily identifying materials and products, for determining their location and status within facilities and within the supply chain and for controlling their movement.
- Customer requirements andregarding regarding quantity, quality, and on-time delivery should be met without exception. consitency and predictability, regarding quantity, quality, and on-time delivery should be met without exception.

8 Automation Principle

Material handling operations should be mechanized and/or automated where feasible to improve operational efficiency, increase responsiveness, improve consistency and predictability,

Key Points:

- Pre-existing processes and methods should be simplified and/or re-engineered before any efforts at installing mechanized or automated systems.
- Computerized material handling systems should be considered where appropriate for effective integration of material flow and information management.
- Treat all interface issues as critical to successful automation, including equipment to equipment, equipment to load, equipment to operator, and control communications.
- All items expected to be handled automatically must have features that accommodate mechanized and automated handling.

9 Environmental Principle

Environmental impact and energy consumption should be considered as criteria when designing or selecting alternative equipment and material handling systems.

Definition: *Environmental consciousness stems from a desire not to waste natural resources and to predict and eliminate the possible negative effects of our daily actions on the environment.*

Key Points:

- Containers, pallets and other products used to form and protect unit loads should be designed for reusability when possible and/or biodegradability as appropriate.
- Systems design should accommodate the handling of spent dunnage, empty containers and other by-products of material handling.
- Materials specified as hazardous have special needs with regard to spill protection, combustibility and other risks.

Figure 10–1 (continued)

10 Life Cycle Cost Principle

A thorough economic analysis should account for the entire life cycle of all material handling equipment and resulting systems.

Definition: *Life cycle costs include all cash flows that will occur between the time the first dollar is spent to plan or procure a new piece of equipment, or to put in place a new method, until that method and/or equipment is totally replaced.*

Key Points:

- Life cycle costs include capital investment, installation, setup and equipment programming, training, system testing and acceptance, operating (labor, utilities, etc.), maintenance and repair, reuse value, and ultimate disposal.
- A plan for preventive and predictive maintenance should be prepared for the equipment, and the estimated cost of maintenance and spare parts should be included in the economic analysis.
- A long-range plan for replacement of the equipment when it becomes obsolete should be prepared.
- Although measurable cost is a primary factor, it is certainly not the only factor in selecting among alternatives. Other factors of a strategic nature to the organization and which form the basis for competition in the market place should be considered and quantified whenever possible.

The experience of generations of material handling engineers has been summarized here for new practitioners. These principles are guidelines for the application of sound judgment. Some principles are in conflict with others, so only the situation being designed will determine what is correct. The principles will be a good checklist for improvement opportunities. Each of these principles will be discussed in the following section.

The following paragraphs are some additional thoughts and comments on a few material handling principles.

Planning Principle

General Dwight D. Eisenhower stated that the plan was nothing, but that planning was everything. What General Eisenhower was telling us is that the planning process (all the time and effort that go into the plan) is what is important. The plan is only our way of communicating the tremendous work (planning) that went into it. Material handling planning considers every move, every storage need, and any delay in order to minimize production costs.

Which piece of material handling equipment should you use? Which problems should be studied first? Should you do an overview before studying the individual material handling problems? These are typical questions asked by a new project engineer. Where to start is easy—just start collecting information about the product (material) and the move (job). As discussed in an earlier chapter, a series of questions that have been used for generations by reporters will serve the material handling project engineer well: Why? Who? What? Where? When? How? If you answer these questions about each move, the solution will become evident.

The material handling equation is a plan for a systematic approach to equipment solution (see Figure 10–2).

Understand the material plus the move and the proper piece of equipment will develop.

Here is a list of the specific questions to ask:

1. *Why are we making this move?* (Why?) This question is asked first because if a good answer is not forthcoming, you can eliminate the move. By combining operations, the move between operations can be eliminated. One can combine machines (called *work cells*) and eliminate moves.

2. *What are we moving?* (What?) Understanding what is being moved requires a knowledge of the size, shape, weight, and the number being moved and the kind of material. With the knowledge of what needs to be moved, you have half the information required to make an equipment selection.

3. *Where are we moving the material from and to?* (Where?) If the move is the same every time, a fixed path technique is warranted (conveyor). If the move changes from part to part, a variable path technique is used (industrial truck). If the path is short, maybe gravity can be used (e.g., slides, rollers, skate wheel).

4. *When is the move needed?* (When?) Is this move once or twice a day? If so, an industrial truck is warranted. If this is several times a minute, a conveyor is

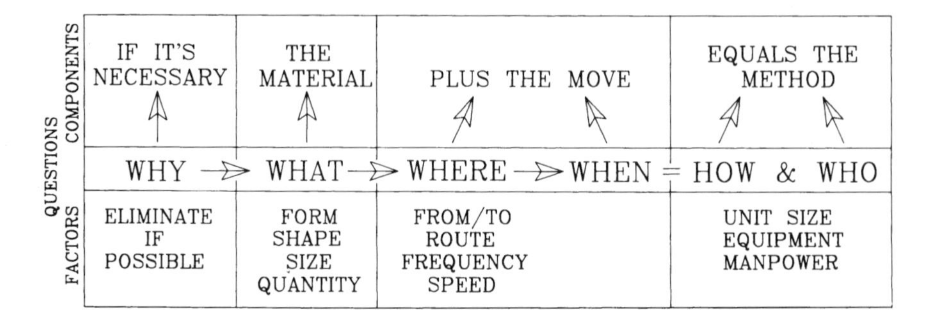

Figure 10–2 The material handling equation.

used. A few examples of analyzing the "material + move = method" follow item 5 below.

5. *How will the move be made?* (How?) Will it be by hand, or by conveyor, or by fork truck? Many options are available and the most cost-efficient method is your goal.

Here are a few examples of ways to put these questions into action.

Example 1: Move: oil from a tanker truck to a tank farm for a bottling company.

Why	=	you need oil to put into quart cans
What	=	oil
Where	=	from tank truck to tank farm
Who	=	receiving clerk
When	=	four times a day as they arrive
How	=	pump—meter and hose

Example 2: Unload 20,000 pounds of flat steel stock from a flatbed trailer into the plant.

Why	=	you need steel in the plant (maybe coil would be better)
What	=	20,000-pound loads of steel (3-1/2 feet × 10 feet × 20 inches)
Where	=	from flatbed trailer into storage area
Who	=	receiving clerk
When	=	40,000 pounds per day (one truck)
How	=	bridge crane

Example 3: Move parts one at a time from spot weld through paint to assembly.

Why	=	to move a major amount of product automatically
What	=	toolboxes and tote trays
Where	=	from spot weld to paint to assembly
Who	=	automatically
When	=	11 parts per minute, 432 minutes per shift
How	=	overhead conveyor belt

Example 4: Move all the product (except sheet metal) from the receiving dock to stores and/or to manufacturing.

Why	=	to keep a bank of material to prevent running out of raw material and parts
What	=	all raw material and purchased parts
Where	=	from receiving to stores to manufacturing
Who	=	store clerk
When	=	as material arrives and as production requests material
How	=	narrow aisle reach truck

As information is collected, the picture becomes clearer and the plan takes shape. The more you know about the material and the move, the better the job of equipment selection you will do.

Systems Principle

All material handling equipment should work together so that everything fits. This is the *systems concept.* The boxes fit the pallets, the pallets fit the rack, and the pallets fit the workstation. A toy company purchased parts manufactured on the outside, but these outside suppliers sent the parts into the toy company in the toy company's cartons. This company used only four different-sized boxes, and these boxes fit the pallets perfectly. When the parts were moved to the assembly line, the box fit into a holding device that held the box in perfect position for use.

Another example of the systems approach involves a TV manufacturer. The TV manufacturer did not make the wooden cabinet but purchased it from a supplier. The supplier built the wooden cabinet and then packaged it in a cardboard box that the TV manufacturer provided. The cabinet came into the TV plant, was removed from the carton, and was placed on a conveyor for assembly (the TV was set into the cabinet). The carton was then placed on an overhead belt conveyor that carried it to the packout department. When the TV was completed, it was placed back in the same carton in which it was received. That carton was then moved to the warehouse and shipped to the customer.

In another example, a major oil company purchased plastic bottles from an outside manufacturer. The quart bottles were packaged in a carton of 12 with separators between every bottle. These cartons were placed on a pallet and shipped to the oil company's bottling plant. In the plant, the bottles were dumped onto a filling line and filled with oil. The empty carton was conveyed to the packout end of the filling line and repacked with 12 bottles, closed, stacked on a pallet, and shipped to the customer.

The systems principle integrates as many steps in the process as possible into a single system from the vendor through your plant and out to your customers. An integrated system is where everything seems to fit together.

Work Principle

Material handling, like every other area of work, should be scrutinized for cost reduction. The work simplification formula tells us to ask four questions:

1. Can this job be eliminated? This is the first question asked because a positive answer will save the maximum amount of cost, namely, everything. Material handling tasks can often be eliminated by combining production operations.
2. *If you cannot eliminate,* can you combine this movement with other movements to reduce that cost? The *unit load concept* (a special section of this chapter) is based on this work simplification principle. If you can move two for the same cost as one, the unit cost of the move will be half. Just think, what if you could move 1,000 instead of one? Many times, moves can be effectively eliminated when combined with an automatic material handling system that moves material automatically between workstations. Conveyors are a good example of this.
3. *If you cannot eliminate or combine,* can you rearrange the operations to reduce the travel distances? Rearranging the equipment to make the travel distances less will reduce the material handling costs.
4. *If you cannot eliminate, combine, or reroute,* can you simplify? Simplification is making the job easier. More than any other type of equipment, transportation or material handling equipment has taken the drudgery out of work. Some simplification ideas for material handling include:

   a. carts instead of carrying
   b. roller conveyors to move boxes from trucks to the plant floor
   c. two-wheel hand trucks
   d. manipulators, which can make superpeople out of everyone
   e. slides or chutes
   f. rolltop tables (ball bearings)
   g. mechanization
   h. automation

Cost reduction is a part of every engineer's and manager's job. Material handling equipment makes cost reduction easier.

Space Utilization Principle

A goal of material handling is to maximize the building cube. A *building cube* is the cubic feet of the building volume resulting from multiplying the building's length times height. Racks, mezzanines, and overhead conveyors are a few of the material handling devices that promote this goal. Purchasing or leasing land and building the plant are significant costs that are always on the increase. The better you use the building cube, the less space you need to buy or rent.

Unit Load Principle

A *unit load* is a load of many parts that move as one. The advantages of a unit load are that it is faster and cheaper than moving parts one at a time. The disadvantages are

1. The cost of making the unit loads and deunitizing
2. Tare weight (the weight of boxes, pallets, and the like)
3. The problem of what to do with the empties
4. The need for heavy equipment and its space requirements

Of course, the advantages must outweigh the disadvantages before you consider a unit load system.

The most common unit load is the pallet. Almost anything can be stacked on a pallet tied with bonding or plastic wrap and moved around the plant or world as one unit. Pallets are made of a variety of materials with greatly differing costs.

Cardboard pallets @ $1.00 each will make one trip.

Plastic pallets @ $4.00 each will make 20 trips.

Wooden pallets @ $20.00 each will make 100 trips.

Steel skids @ $150.00 each will make 2,000 trips.

If you had no chance of getting your pallet or the cost of the pallet back, you would use a cardboard pallet. If you used only pallets within the plant, you would choose the steel pallet because its cost per move would be only one-third a wooden or plastic pallet. Strength, durability, versatility, weight, size, cost, and ease of use must all be considered when choosing a unit load technique. Wooden pallets are the most popular because the trucking industry trades pallets. When truckers drop 18 full pallet loads of material, they pick up 18 empties and return them to the supplier. Tens of thousands of dollars per year can be lost without a pallet control system.

The pallet is only one of the *under mass techniques* of unit loading where the pallet supports the load by being placed under the mass (load). Others exclude boxes, tubs, and slip sheets. Still others are *squeezing* and *suspending* methods of handling the unit load.

Squeezing the load is performed by a clamp truck. The product is stacked on the floor into *pallet patterns* just like on pallets (see Figure 10–3). When the stack is complete a fork truck with two vertical plates (about 4 × 4 feet) drives up to the stack with one plate on the right side and the other plate on the left side. The two plates are pulled toward each other squeezing the material between the plates. The load can now be moved. This load can be placed on top of another stack of similar products right up to the rafters. The advantage is no pallet cost or space. Trailers can be loaded and unloaded with no pallets needed.

In the process of suspending unit loads from bridge cranes or jib cranes, a hook suspends from a lifting motor and attaches to chains or cables around the load. Lumber, steel coils, and steel plates are often moved this way. A monorail conveyor can also move many parts at a time.

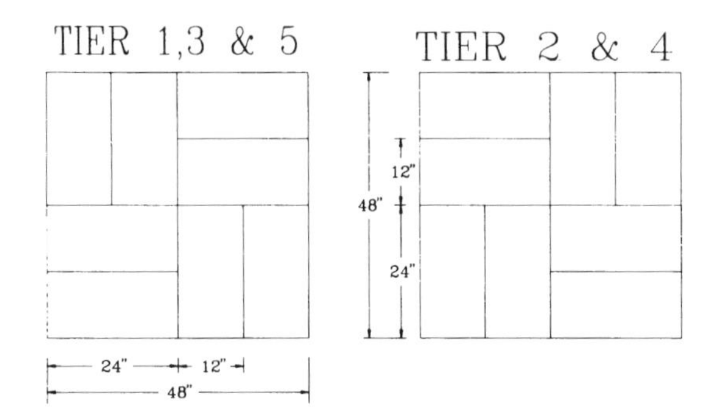

Figure 10–3 Pallet pattern.

Automation Principle

The automation principle makes moves automatic. Many new systems are completely automatic. Automatic storage and retrieval systems place material into storage racks automatically (no people assistance) and remove it when needed. Many machines are automatic because material handling equipment loads and unloads the machine. Automation is the way of the future, thus even users of the manual system must consider when it can be justified.

An engine block is automatically moved from machine to machine for processing. Machine centers are arranged around an indexing table. When all machines finish their function, the table advances one station and the machines go back to work. The finished parts can be removed by gravity, or a robot can pick up the finished part and place it in a container. This principle is fun to work with because your creative efforts will be well rewarded and personally gratifying.

Standardization Principle

There are many types of material handling equipment—shop boxes, bins, pallets, shelves, racks, conveyors, trucks, and the like—and in every area, you want to standardize on one (or as few as possible) size, type, and even brand name. The reasons are many, and they change with the type of equipment, but if you have a special piece of equipment for every move or storage, you have too many different types and sizes to inventory and control. Material handling moving equipment (like fork trucks) is manufactured by many companies. You need to choose just one and then stay with that brand, type, and size because spare parts inventory, maintenance, and operation of this equipment will be most cost efficient. Equipment selection and standardization should not be based on the initial purchase cost alone. Costs of material handling systems can be grouped into two categories—the cost of the ownership of the system, which includes the initial purchase price and the subsequent maintenance costs, and the cost of the operation of the system. This latter cost includes the cost of training personnel to use the system safely, energy cost, and other direct or indirect costs associated with the use of the system.

Having only a few sizes of cartons will simplify the storage area. You may put these few sizes of cartons onto a single-sized pallet and into a uniform-sized rack, which is serviced by one type of lift truck.

THE MATERIAL HANDLING PROBLEM-SOLVING PROCEDURE

Use the following steps and checklist to problem solve, improve efficiency, and reduce costs in material handling.

Step 1. Analyze the requirements to define the problem. Be sure the move is required.
Step 2. Determine the magnitude of the problem. Cost analysis is best.
Step 3. Collect as much information as possible—why, who, what, where, when, and how.
Step 4. Search for vendors. Suppliers often provide outstanding engineering and cost justification assistance.
Step 5. Develop viable alternatives.
Step 6. Collect costs and savings data on all alternatives.
Step 7. Select the best method.
Step 8. Select a supplier.
Step 9. Prepare the cost justification.
Step 10. Prepare a formal report.
Step 11. Make a presentation to management.
Step 12. Obtain approvals (adjust as needed).
Step 13. Place an order.
Step 14. Receive and install equipment.
Step 15. Train employees.
Step 16. Debug (make it work) and revise as necessary.
Step 17. Place into production.
Step 18. Follow up to see that it is working as planned.
Step 19. Audit performance to see that payback was realized.

MATERIAL HANDLING CHECKLIST

100 Areas of Cost Reduction

Yes	*No*	
_____	_____	1. Are the receiving and shipping docks protected from the weather?
_____	_____	2. Are dock boards adequate?
_____	_____	3. Are trailers loaded and unloaded by hand?
_____	_____	4. Is incoming material packaged for economical use?
_____	_____	5. Do you replace obsolete equipment?

100 Areas of Cost Reduction (continued)

Yes	*No*	
_____	_____	6. Do you standardize on material handling equipment to reduce spare parts needs?
_____	_____	7. Do you have a preventive maintenance program for every piece of material handling equipment?
_____	_____	8. Do you have a pallet repair area?
_____	_____	9. Do you have a pallet control program?
_____	_____	10. Do you measure and track the ratio of material handlers to direct labor?
_____	_____	11. Do you have a material handling training program?
_____	_____	12. Do you maintain safety records for material handling?
_____	_____	13. Do any of your employees lift 50 pounds or more manually?
_____	_____	14. Are there any material handling jobs that require more than one person to lift?
_____	_____	15. Are you using the overhead space in
_____	_____	a. stores?
_____	_____	b. fabrication?
_____	_____	c. paint?
_____	_____	d. assembly and packout?
_____	_____	e. warehouse?
_____	_____	f. offices?
_____	_____	16. Is weight control measured and recorded automatically?
_____	_____	17. Do you still receive raw material (like plastics) in 50- to 100-pound bags when usage would justify bulk handling equipment?
_____	_____	18. Are you storing material that is available locally?
_____	_____	19. Are you using the building cube?
_____	_____	a. Are you storing only 8 feet high?
_____	_____	b. Are you picking orders only 6 inches high?
_____	_____	c. Are you using overhead conveyors?
_____	_____	d. Are your ovens off the floor?
_____	_____	e. Are you using overaisle storage?
_____	_____	f. Are you stacking two or more deep?
_____	_____	g. Is more than 30 percent of your plant taken up by aisles?
_____	_____	20. Are you using powered equipment when gravity would do the job?
_____	_____	21. Do you use the material handling equipment to do secondary operations automatically, such as
_____	_____	a. counting?
_____	_____	b. weighing?
_____	_____	c. branding or numbering?
_____	_____	d. segregating?

100 Areas of Cost Reduction (continued)

Yes	*No*	
_____	_____	e. slitting bags?
_____	_____	f. opening and closing lids?
_____	_____	g. gluing boxes closed?
_____	_____	h. banding boxes?
_____	_____	22. Do you automatically move material to the point of use, then hand feed?
_____	_____	23. Is the maintenance and service area for mobile equipment conveniently located?
_____	_____	24. Are skilled employees spending their time handling material?
_____	_____	25. Does the assembly line stop when delivering and removing material?
_____	_____	26. Do operators have to load their own hoppers?
_____	_____	27. Do operators need to stop work when material is being loaded into their workstation?
_____	_____	28. Are material storage areas congested?
_____	_____	29. Do you measure the utilization of material handling equipment?
_____	_____	30. Do you encourage backhauling?
_____	_____	31. Does your equipment move empty more than 20 percent of the time?
_____	_____	32. Do your shipping clerks load carriers' trucks?
_____	_____	33. Do you load material handlers with work
_____	_____	a. by past practices?
_____	_____	b. by time standards?
_____	_____	c. by guess?
_____	_____	34. Do you pay damage charges?
_____	_____	35. Do you know your floor loadings?
_____	_____	36. Do your products get damaged during material handling?
_____	_____	a. Do you know which equipment is responsible?
_____	_____	b. Do you know how much money is lost?
_____	_____	c. Do you know which people are responsible?
_____	_____	37. Do your truck drivers use two-way radios?
_____	_____	38. Do workers handle material too many times?
_____	_____	39. Are single pieces being moved where two or more could be moved?
_____	_____	40. Are the floors smooth and clean?
_____	_____	41. Do you know the capacity (pounds) of your equipment?
_____	_____	a. Do your material handlers know?
_____	_____	b. Is your equipment marked for capacity?
_____	_____	42. Do you ever change material from one container to another?

100 Areas of Cost Reduction (continued)

Yes	*No*	
_____	_____	43. Are the aisles over 8 feet wide?
_____	_____	44. Have you flow process charted your products?
_____	_____	45. Do you use certified truck load weight?
_____	_____	46. Do you use point of use receiving?
_____	_____	47. Do you ever store material in aisles?
_____	_____	48. Do you use outdoor storage when practical?
_____	_____	49. Is your safety stock too large?
_____	_____	50. Are the doors too small?
_____	_____	51. Do you have too many doors?
_____	_____	52. Do you control the movement of material?
_____	_____	53. Do you have a locator system?
_____	_____	54. Do you use drums instead of tanks?
_____	_____	55. Do you use trailers for long hauls in the plant?
_____	_____	56. Are you ground loading or unloading trailers or boxcars?
_____	_____	57. Can your customers unload your trailers?
_____	_____	58. Can you eliminate pallets?
_____	_____	59. Can you use expendable pallets (one way)?
_____	_____	60. Can you build material handling devices into your finished package?
_____	_____	61. Are you using fork trucks where narrow aisle trucks would be better?
_____	_____	62. Do you use the receiving container to ship?
_____	_____	63. Do you have material handlers waiting on assembly and packout liner?
_____	_____	64. Have you eliminated backtracking?
_____	_____	65. Have you eliminated cross traffic?
_____	_____	66. Have you reduced the distance of travel to a minimum?
_____	_____	67. Are materials too far from their point of use?
_____	_____	68. Are the containers at a workstation large enough to keep the station working for 2 hours or more?
_____	_____	69. Are inexpensive parts stored close to the workstation?
_____	_____	70. Is scrap disposal
_____	_____	a. separated at origin?
_____	_____	b. stored?
_____	_____	c. sold (not given away)?
_____	_____	71. Can the shipping container start the production line?
_____	_____	72. Are you bulk-storing finished goods to maximize the building cube?
_____	_____	73. Can turntables eliminate steps?
_____	_____	74. Do you use light signals to notify material handlers of stock needs?

100 Areas of Cost Reduction (continued)

Yes	*No*	
_____	_____	75. Do you have the high-use items conveniently located?
_____	_____	76. Have you done an ABC analysis of inventory?
_____	_____	77. Are you breaking carton quantities for shipments?
_____	_____	78. Are you using hydraulic hand carts for short trips?
_____	_____	79. Are your maintenance people mobile?
_____	_____	80. Are you continually looking for ways to mechanize?
_____	_____	81. Do you ask operators' opinions on what would make their jobs easier?
_____	_____	82. When an employee's suggestion is rejected, do you formally tell the employee why?
_____	_____	83. Do you listen to material handling vendors (salespeople)?
_____	_____	84. Have you asked vendors to package raw materials more conveniently?
_____	_____	85. Do you prequalify vendors so that their material will not need inspection upon receipt?
_____	_____	86. Do you have the best systems for mailing, strapping, taping, stapling, labeling, and marking your products?
_____	_____	87. Are you using conveyors as much as possible
_____	_____	a. from receiving to stores?
_____	_____	b. from stores to production?
_____	_____	c. from fabrication through clean, paint, and bake to assembly?
_____	_____	d. between operations?
_____	_____	e. through heat treatment?
_____	_____	88. Are you loading machines with walking beams?
_____	_____	89. Do you use indexing tables?
_____	_____	90. Are you using computer-generated shipping labels?
_____	_____	91. Are you generating computer weights for shipments?
_____	_____	92. Are you using computer location/numbered layouts for warehousing?
_____	_____	93. Are all doors equipped with automatic approach openers?
_____	_____	94. Are corners protected with signals and mirrors?
_____	_____	95. Are you using remote controllers and indicators to eliminate climbing and walking to remote areas?
_____	_____	96. Is equipment protected with panic buttons?
_____	_____	97. Are switches foot-, knee-, or leg-controlled so hands are free to work?
_____	_____	98. Are you using air, electric, or hydraulic clamps to eliminate the need for handholding?
_____	_____	99. Are jigs being used to hold parts in position for welding, cementing, or assembly?
_____	_____	100. Are handheld power tools being used instead of hand tools?

QUESTIONS

1. What is material handling?
2. What are some of the material control systems?
3. Which is the best piece of material handling equipment for a specific job?
4. What are the goals of material handling?
5. From where did the 10 principles of material handling come?
6. What are the 10 principles of material handling?
7. What is the material handling equation?
8. What is the material handling problem-solving procedure?
9. How can material handling be combined with other production activities?
10. How can automatic identification and data capture (AIDC) be incorporated into material handling systems?
11. State some industrial situations in which AIDC can improve efficiency when incorporated into a material handling system.
12. What factors, other than cost, could be considered important when selecting material handling equipment?
13. What are the two categories of costs associated with the selection and acquisition of material handling equipment?
14. What would be an example of ergonomic consideration when selecting material handling equipment?
15. How can a material handling system increase plant space use? How can it reduce work-in-process (WIP) inventory?
16. What are the five dimensions of a material handling system?
17. What does cumulative trauma disorder (CTD) mean?
18. How can CTD risks be reduced through the use of material handling systems?

CHAPTER 11

Material Handling Equipment

OBJECTIVES:
Upon the completion of this chapter, the reader should:

- Understand various classifications of material handling equipment
- Be able to identify various classifications of material handling equipment
- Be able to select the appropriate material handling equipment for a given task

There are literally thousands of pieces of material handling devices. These pieces of equipment vary from the most basic manual tools to the most sophisticated computer-controlled material handling systems that can incorporate a vast array of other manufacturing and control functions. Almost as varied and numerous are the classification strategies and methods of material handling equipment.

Traditionally, material handling equipment has been grouped into four general categories. The first category is the *fixed-path* or *point-to-point* equipment. This class of equipment serves the material handling need along a predetermined, or a fixed path. The most common and familiar example of a fixed-path system is the train and the railroad track. The train can travel from any point to any other point and serve any point along the track system. Conveyor systems, powered, gravity-fed, or otherwise operated, fall into this classification. Fixed-path material handling systems are also referred to as *continuous-flow systems.* Automated guided vehicles (AGVs) fall into this group.

The *fixed-area* material handling system can serve any point within a three-dimensional area or cube. Jib cranes or bridge cranes are examples of fixed-area systems. A jib crane installed on a floor pedestal can move parts and other material from any point to any other point in the *x, y,* and *z* direction; however, this ability is limited within the confines of the equipment. Automated storage and retrieval systems (ASRSs) also fall into this category.

The material handling equipment that can move to any area of the facility is referred to as *variable-path variable-area* equipment. All manual carts, motorized vehicles, and fork trucks can be pushed, dragged, or driven throughout the plant. What, then, would a jib crane that is installed on a mobile pedestal be called? Obviously, this is a compound material handling system. The crane is a fixed-area system and the pedestal is a variable-path vehicle. When the base is stationary the crane is confined within its reach.

The fourth category consists of all *auxiliary tools and equipment* such as pallets, skids, automatic data collection systems, and containers.

In the following sections of this chapter, we will examine each category of the material handling systems in various applications and areas of the facility. You will notice that any material handling device may have several applications in various departments in the manufacturing facility, and that the device can be easily placed in one of the four classifications.

How do you choose the proper piece of equipment from the thousands of material handling devices available? For the experienced project engineer or manager, this problem is not as great as it is for the novice. The new facilities planner should use an organized approach in determining equipment needs, following the flow of material from the receipt of material to the warehousing of that material.

1. Receiving and shipping
2. Stores
3. Fabrication
4. Assembly and paint
5. Packout
6. Warehousing

Two additional areas of material handling need to be discussed because of their importance: bulk material handling and automated storage and retrieval systems.

The systems principle of material handling states that material handling devices should be used in as many areas as possible and that everything fits (works) together. In the following discussion of material handling equipment, each piece of equipment will be discussed in the primary area of use.

RECEIVING AND SHIPPING

Material handling equipment for receiving and shipping is often the same. Sometimes receiving and shipping are accomplished through the same dock door. To save time, we will discuss both of these important departments at the same time.

Receiving and Shipping Docks

Receiving and shipping docks come in a variety of sizes and shapes. The term "dock" comes from the shipping industry where ships pulled into port, landed, were tied up, and unloaded. Industrial plants' docks are for the same purpose. Trucks, trains, and ships all can pull into plants to deliver or remove materials.

The most common type of dock is known as a *flush dock*. Flush docks are doors in an outside wall. The truck or train pulls up or backs into the doorway. Trucks are serviced from the rear door mostly, but some trucks and most boxcars are serviced from the sides. Flush docks can be designed for both rear and side service. The height of the plant floor off the driveway or railbed should be 46 inches for trucks and 54 inches for boxcars. The driveways should slope away from the plant to prevent water damage to the building foundation. Docks that slope toward the plant are commonly after-thought docks or short-term cheap construction. The negative slope docks will be continuing problems even with drains. Everything seems to blow into and collect in the wells. Figure 11–1a shows a side view of a flush rear service dock. Notice the dock height, slope away from the plant, bumpers, chalks, and the dock plate. Notice also the front wheels of the trailer. These small wheels will place a lot of weight on the driveway, so be sure the driveway can support the heavy weights of trailers. Figure 11–1b is a top view of the same dock, whereas Figure 11–1c shows a top view of a side service dock. These figures could just as easily have shown illustrations of a railroad and boxcar.

Drive-in docks are similar to flush docks except that when the door opens, the truck and trailer can back into the plant and the door can be closed. This type of dock is very expensive and quite desirable in bad weather. Drive-in docks can service both the rear end and the side. Some large plants have railroad siding where trains back cars into the plant. The Caterpillar Tractor Company actually drives its new products from the plant floor straight across to the deck of a flatcar.

Drive-through docks are a pair of doors across the plant from each other. The truck drives into the plant, the door is closed, the trailer (usually a flatbed) is unloaded or loaded, the other door is opened, and the truck drives out. The driveway and the plant floor are usually at the same level, so personnel must climb on the

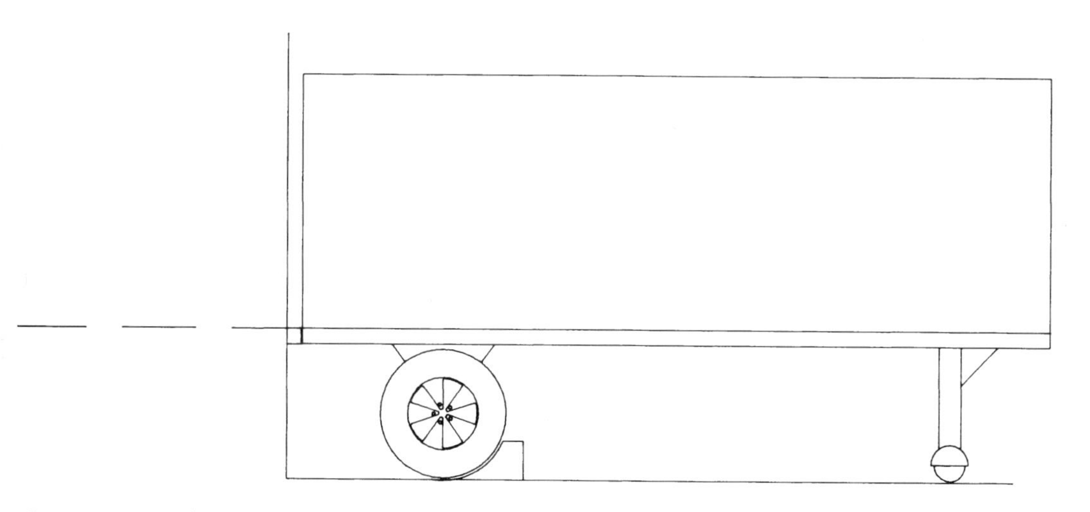

Figure 11–1a Flush dock—Side view.

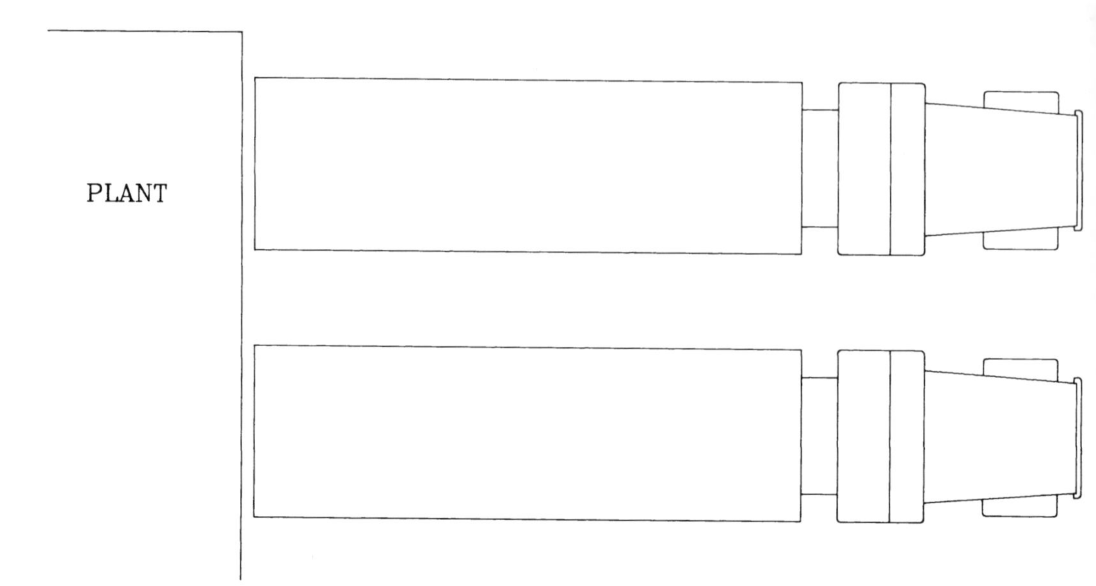

Figure 11–1b Flush dock—Top view.

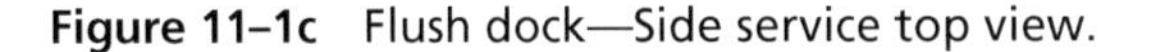

Figure 11–1c Flush dock—Side service top view.

trailer to unload. Overhead bridge cranes are often used to unload very heavy pallets of steel. This type of dock is a bigger safety hazard than the other docks, but good step ladders and operator training will minimize the danger.

Finger docks are extensions to the plant and can handle many trailers at one time. A concrete finger from 10 to 15 feet wide can be extended 100 feet out from a side of the plant. Ten trailers can then be backed up to both sides of the dock, and only one or two doors go into the plant. Freight companies use this type of dock to

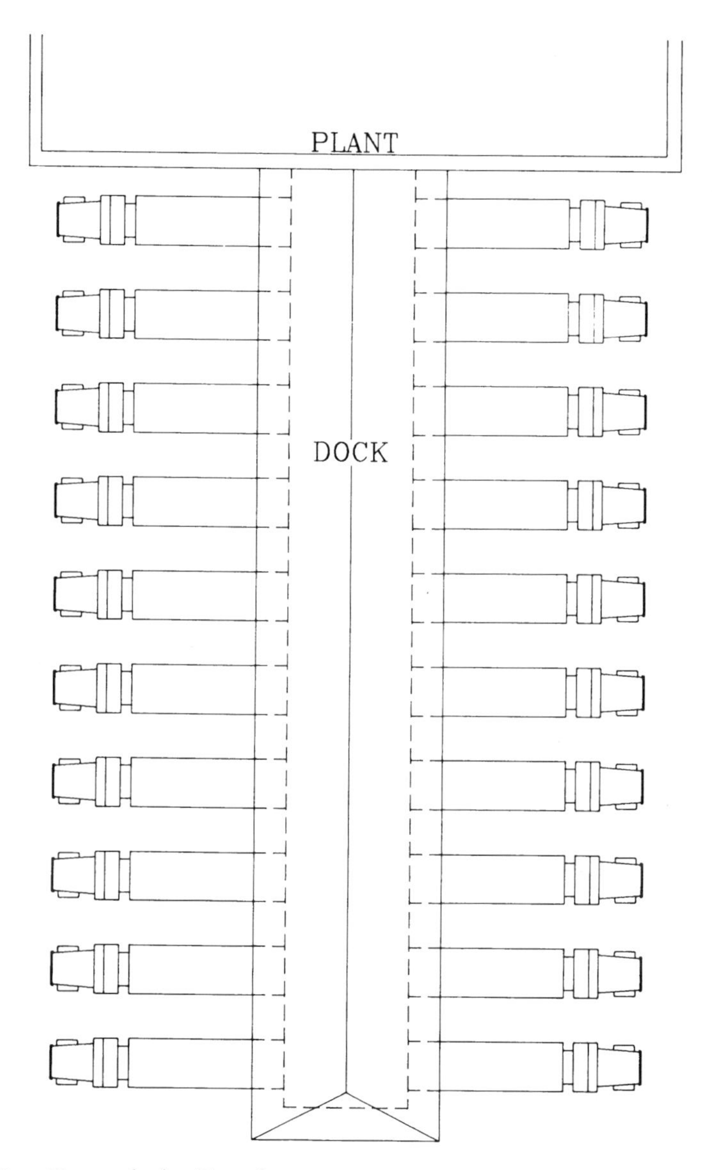

Figure 11–2a Finger dock—Top view.

unload and load many trailers at the same time. Figure 11–2a is a drawing of a finger dock for 20 trucks. A carport-like cover over the finger dock is important for weather protection (see Figure 11–2b).

There are other types of docks, some good, some bad, but these four types represent over 90 percent of all docks used for trucks or railcars.

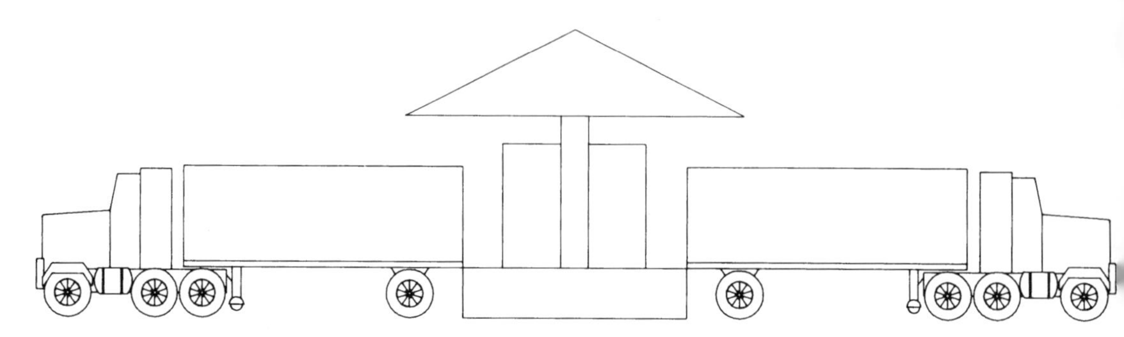

Figure 11–2b Finger dock—Side view.

Dock Equipment

A door is part of the dock. A trailer (end service) dock door is typically a 9 × 9-foot rollup. The door is electrically or manually lifted into a container mounted above the door opening. Around the door, a seal is built of compressible material that will seal off the outside weather from the plant environment. Bumpers are placed outside the dock door below the floor level to stop the trailer and to protect both the building and the trailer from collision damage. Once the trailer is in position and the plant and the trailer doors are open, a dock plate is placed between the plant floor and the trailer floor to enable driving on and off the trailer. An automatic dock leveler can be built into the dock door floor to make placing the dock plate automatic.

An awning or porch over the rear of the trailer extending from the plant wall will help keep rain and snow out of the plant. Sometimes air curtains and plastic curtains are placed in doorways to minimize air loss from the plant.

Extra lighting is often needed inside trailers, so portable lighting is often useful. Figure 11–3 shows a collection of dock equipment.

Moving Equipment

Hand Carts

Literally hundreds of different hand carts are available today. A few of the most versatile and popular pieces of equipment follow:

1. *Two-wheel hand truck* (see Figure 11–4). Weights up to 500 pounds can be moved by a single person and a good two-wheel hand truck. Hand trucks are used in about every area of business, even in the office.

2. *Pallet hand jack or pallet truck-hydraulic lift* (also called just hand jacks, see Figure 11–5). Hand jacks are rolled under a pallet, the handle is pumped (hydraulic pump handle), the pallet is lifted off the floor a few inches, and now the pallet and up to 2,000 pounds of material can be moved easily by hand.

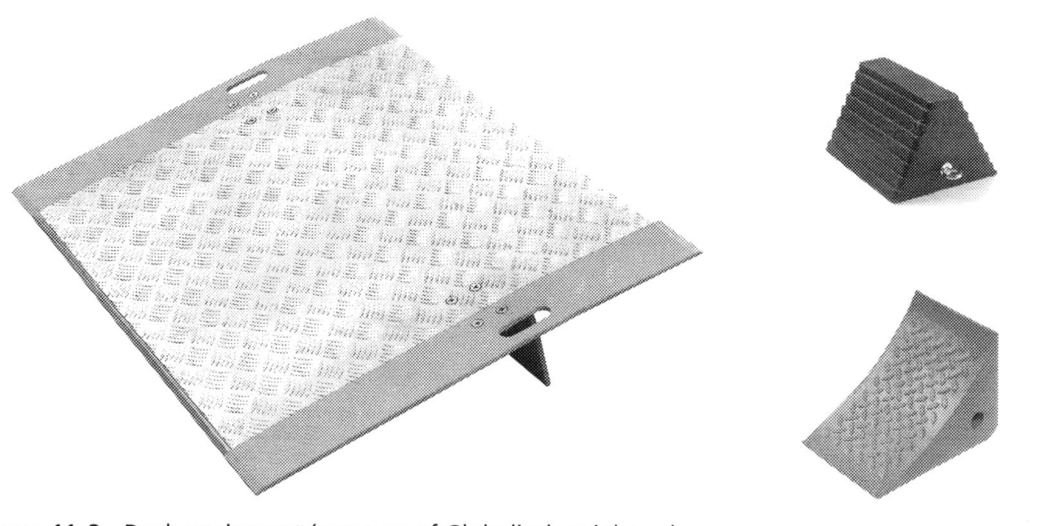

Figure 11–3 Dock equipment (courtesy of Globalindustrial.com).

Figure 11–4 Two-wheel hand trucks (courtesy of Globalindustrial.com).

3. *Four-wheel hand carts* (see Figure 11–6). There are hundreds of shapes, sizes, and uses of hand carts. You can build any shape on the platforms and move very special material. The examples shown in Figure 11–6 are very versatile. Many things can be loaded and moved nearly anywhere.

4. *Pallets* (see Figures 11–7 and 11–8). As discussed in Chapter 10, the pallet is an important piece of material handling equipment.

Fork Trucks

Fork trucks are by far the most popular piece of material handling equipment for unloading and loading trucks and railcars (see Figures 11–9a and 11–9b). Every department in the plant can use them, but they are probably the most misused of all plant equipment. In most cases, it is too much equipment moving too lit-

Figure 11–5 Hand pallet truck (courtesy of Crown Equipment Corp.).

Figure 11–6 Four-wheel hand cart (courtesy of Globalindustrial.com).

Figure 11–7 Pallets in pallet rack (courtesy of Globalindustrial.com).

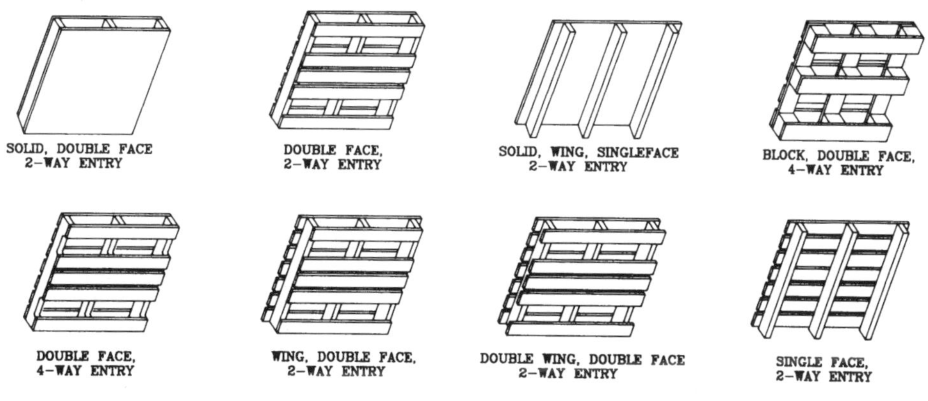

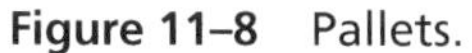

Figure 11–8 Pallets.

tle weight. There is almost always a better choice than a fork truck. Fork trucks have only one redeeming quality—their versatility. They can go about *anywhere* and can move *anything*. We will talk about narrow aisle trucks that can turn in a much shorter radius in the stores section of this chapter.

Figure 11–9a Fork truck (courtesy of Crown Equipment Corp.).

Figure 11–9b Fork truck (courtesy of Yale Materials Handling Corp.).

Fork truck attachments are available for more specific jobs. Standard forks are not appropriate for moving paper rolls, carpet rolls, drums, trash, or many other parts and containers, but fork truck attachments are available to create a unique lifting and moving device. A boat marina uses extended forks to put motor boats into a storage rack. Oil, paint, scrap, and parts can be dumped by using a special dumping attachment.

Multipurpose Equipment

In an attempt to standardize material handling equipment, diversified multipurpose equipment must be given special consideration. Figure 11–10 shows a universal lift system. In the receiving and shipping areas, as well as in storage and warehousing, this material handling system can assist in loading and unloading trucks, and in lifting pallets, boxes, and other containers. It is capable of reaching high and hard-to-reach places, lifting or lowering loads well below the ground level, and performing a variety of other activities. The same equipment can also be quite useful on the factory floor. It can simplify the handling of dies and molds, handle large coils, and uncoil sheet metals.

Bridge cranes are so named because they bridge a bay (wall to wall). Columns are placed at intervals, say 40 × 60 feet. The bay will be 60 feet wide. Two rails (like a big railroad track) are mounted to the columns and can run the full length of the bay and even outside at 60 feet apart. The bridge then runs on wheels on these two tracks. On the bridge, a lifting motor is attached. This lifting motor travels back and forth under the bridge. The crane can be operated from the floor or, on bigger units, an operator rides in a cab mounted to the bridge.

Bridge cranes can lift and move very heavy loads—up to 100,000 pounds or more. Steel, bar stock, major subassemblies, and the like are moved with bridge cranes.

Bridge cranes mounted on the wall on one side and a leg on the other side are called *single gantry cranes.* For outside use, both ends are attached to legs; this is a double gantry crane. Cargo ships are loaded using very large double gantry cranes (see Figure 11–11).

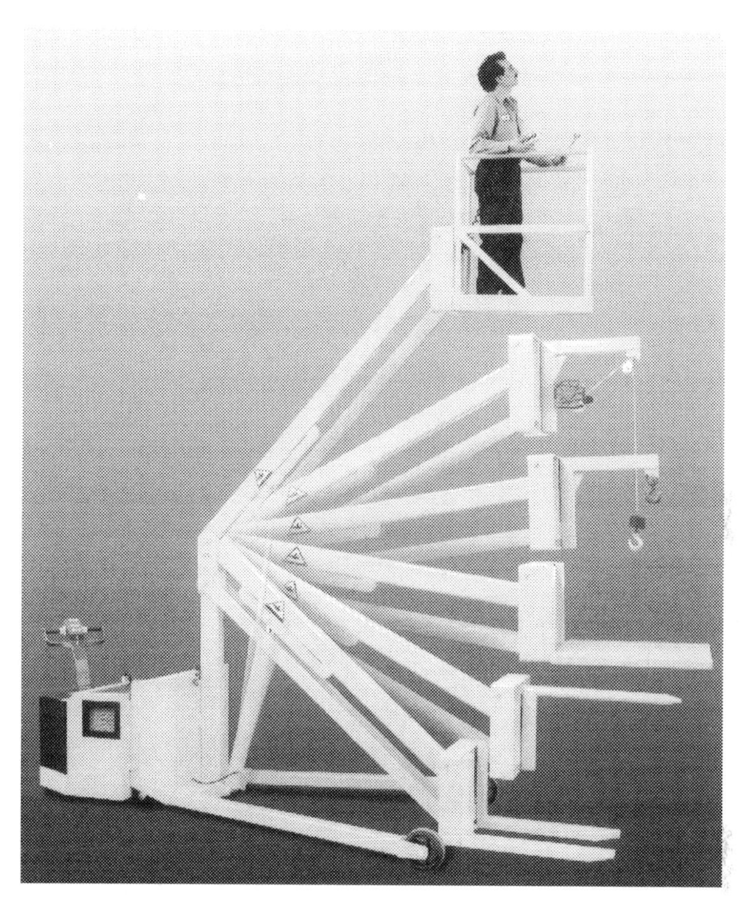

Figure 11–10 Universal lift system (courtesy of Air Technical Industries).

Telescopic Conveyor

Telescopic conveyors have several sections of conveyor that extend as needed (see Figure 11–12). When unloading a truck of small cartons, the first cartons are next to the door, but as the truck is unloaded, the cartons get farther and farther away from the door. A telescopic conveyor can move into the trailer as the work dictates. Telescopic conveyors can save many feet of travel. The Sears warehouses receive much of their incoming goods by way of a telescopic conveyor. One conveyor covers two dock doors. The telescopic conveyors are connected to additional flat belt conveyors, but flat belt conveyors will be discussed in the assembly department section of this chapter because the belt conveyor almost describes assembly material handling.

Figure 11–11 Double gantry crane (A-frame) (courtesy of Air Technical Industries).

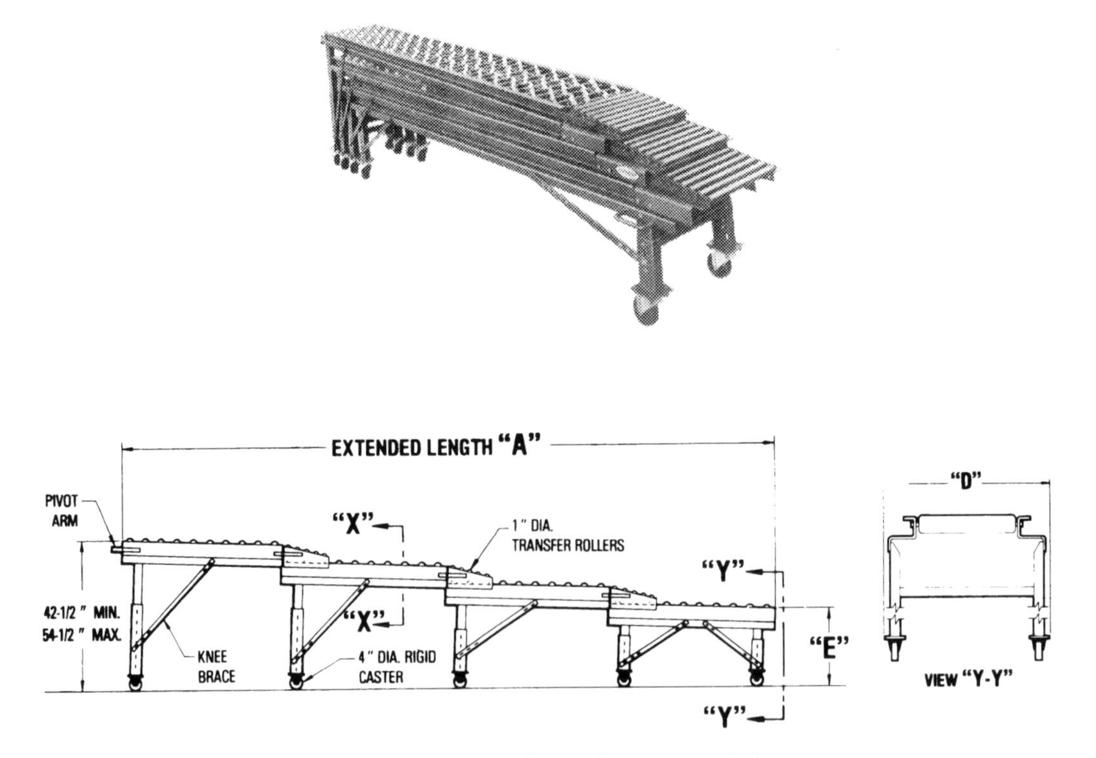

Figure 11–12 Telescopic conveyor (courtesy of Hytrol Conveyor Co.).

Weight Scale

Weight scales are valuable receiving and shipping tools that are built into the material handling system (see Figure 11–13). On receiving docks, scales are used for counting incoming material. Drive-on scales are used when fork trucks place a pallet of material onto the scale to weigh it before shipping. Pedestal scales can be built into a conveyor line to weigh material automatically. Weight scales assist in the quality control of receiving and shipping counts.

Systems Required on Receiving and Shipping Docks

Systems on the receiving and shipping docks should include

1. Part numbering systems that allow for identification of inventory
2. Purchase order system authorizing the receiving of material (see Figure 12–12 in the next chapter.)
3. Customer order system authorizing the shipment of material
4. Bill of lading authorizing a trucking company to move material and to bill for their services

Figure 11–13 Low-profile scale (courtesy of Globalindustrial.com).

STORES

"Stores" is the term used to describe the room where raw materials and supplies are held until they are needed by the operations department. The raw material stores is usually the largest, but maintenance and office supplies stores can be large as well. The material handling equipment in the stores area tends to be very expensive.

Storage Units

Storage units can include the following:

1. *Shelves* store small parts. A typical shelving unit resembles a bookshelf with six 1 × 1 × 3-foot shelves one over the other (see Figure 11–14a).

Figure 11–14a Industrial shelves (courtesy of Yale Industrial Trucks).

2. *Racks* are generally used to store palletized material on pallet racks. A typical pallet rack is 9 feet wide with five tiers for a height of 22 feet. With two pallets per tier, five tiers high equals 10 pallets per pallet rack (see Figure 11–14b).

3. *Double deep pallet racks* allow for stacking 20 pallets on both sides of the aisle instead of 10 pallets. The density of storage is much better, and utilization of the building cube is improved (see Figure 11–14c).

4. *Portable racks* are racks that fit over a pallet load of soft material. Another pallet is then set on top of this portable rack. Heights can be much higher without the danger of a stack falling over.

5. *Mezzanines* can be built over shelving areas to use the space over the shelves. Additional shelves can be placed on the mezzanine, doubling the number of shelves in the stores area. Slow-moving stock can be placed on the mezzanines (see Figure 11–14d).

6. *Rolling shelves* allow for only one aisle in maybe 10 rows of shelves. This would save 9 aisles. The shelves are on wheels and tracks and can be moved to open up an aisle where no aisle exists now. Rolling shelves are popular in maintenance and office supply stores.

7. *Drawer storage units* are another popular maintenance storage unit because many small parts can be stored in a small area. One drawer may have 32 to 64 storage locations, and a 6-foot drawer unit could hold nearly 1,000 different parts (see Figure 11–14e).

Figure 11–14b Pallet racks (courtesy of Globalindustrial.com).

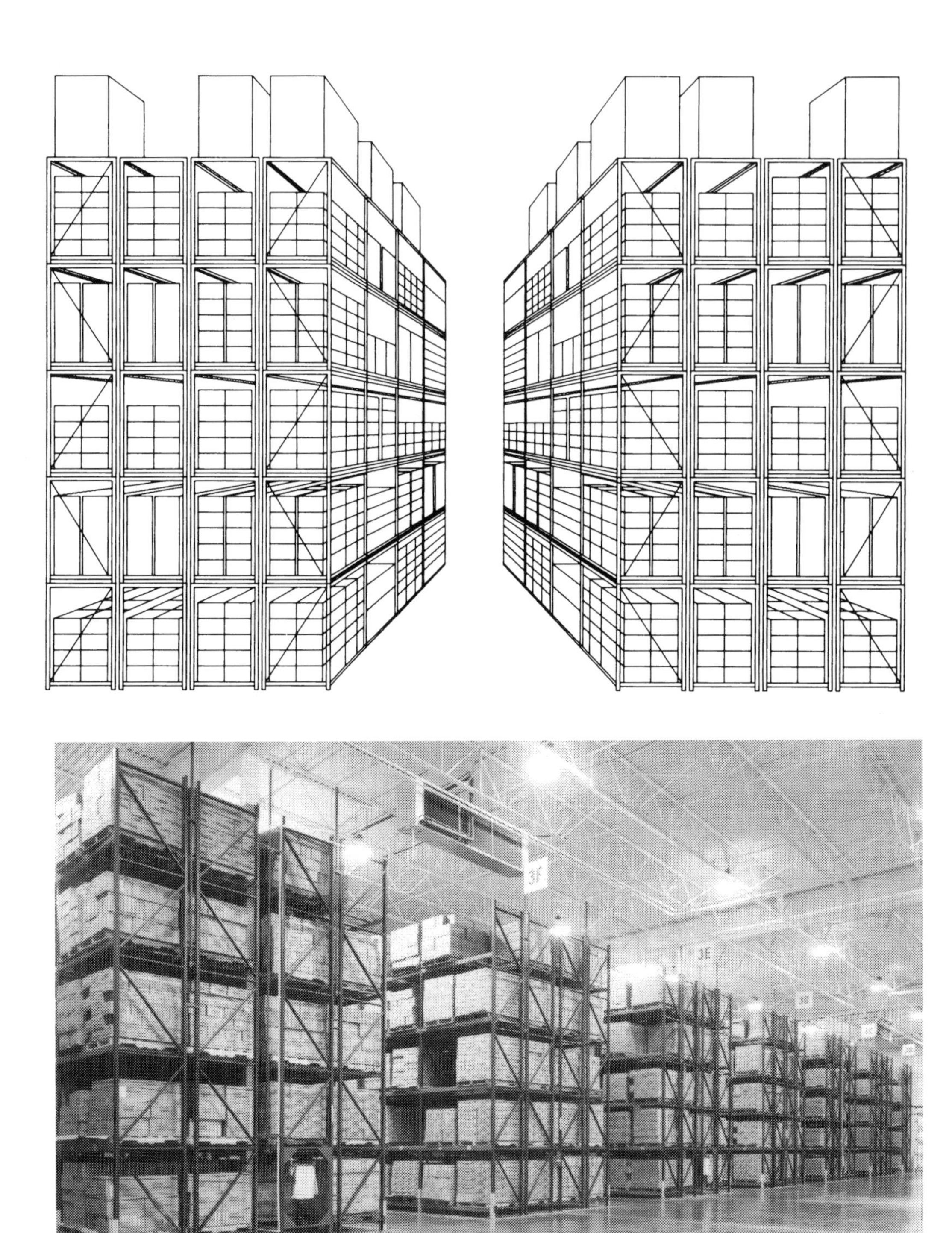

Figure 11–14c Double deep pallet racks (courtesy of Ridgu Rak).

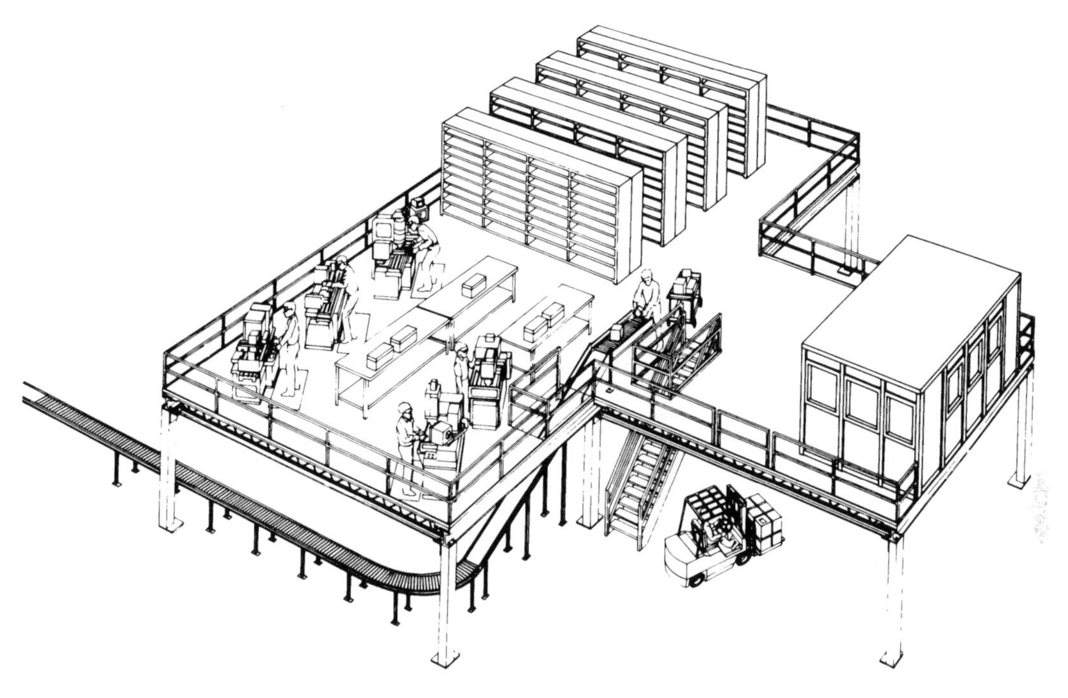

Figure 11–14d Mezzanines (courtesy of W.A. Schmidt, Inc.).

Stores Mobile Equipment

Narrow aisle reach trucks are one of the better choices for maneuvering in storage areas. Narrow aisle trucks can turn in narrow aisles, and the operator stands up. Both of these features increase productivity of the company's resources.

1. Narrow aisles save space.
2. Standing up saves operator time and makes getting off and on very easy. Fork truck drivers sit up in the air about 3 to 4 feet. Once the operators get seated up there, they do not come down except for lunch and breaks.

There are several types of narrow aisle trucks:

1. *Reach truck.* A reach truck (narrow aisle reach truck) has a scissor attachment on the forks allowing them to be extended over 4 feet (see Figure 11–15). This allows for the stacking of two pallets deep in an 8-foot-deep rack. Two pallets deep will save about 50 percent of the aisle space.

2. *Straddle truck.* The name comes from the truck's ability to straddle a pallet on the floor—one front leg on both sides of the pallet on the floor (see Figure 11–16). This allows for more stability and the ability to lift heavier loads with lighter-weight vehicles.

Figure 11–14e Drawer storage (courtesy of Globalindustrial.com).

3. *Side shifting lift trucks.* Narrow aisle side shifting lift trucks are the most space-conserving mobile equipment for storerooms (see Figure 11–17). Many different sizes and shapes of this class exist, but one of the most useful and unique is used for bar stock of 10 to 20 feet in length. How else could very long pieces be moved? This bar stock would be stored in a cantilever rack.

4. *Maintenance carts.* Maintenance carts are almost unique to the maintenance person (see Figure 11–18). Portable oil and grease carts, portable welders, portable toolboxes, and portable benches are all possible. The purpose and goal of maintenance carts is to eliminate the need to run back and forth from the maintenance department to the problem because something was forgotten. The cart is a small maintenance storeroom.

Figure 11–15 Reach truck (courtesy of Yale Materials Handling Corp.).

Figure 11–16 Straddle truck (courtesy of Crown Equipment Corp.).

5. *Dollies and casters.* Moving equipment is a common maintenance job. Dollies placed under the equipment can expedite things (see Figure 11–19). For moving a desk, for example, one would use a dolly that looks like a pallet with wheels.

6. *Maintenance tool crib.* This is used for the safekeeping of maintenance tools and supplies (see Figure 11–20).

7. *Carousel storage and retrieval systems.* Visualize the conveyor system at your dry cleaners (see Figure 11–21). When you come in to pick up your cleaning, the clerk pushes a button and a conveyor moves all the clothes until yours arrives. In a parts storeroom, the same efficient system can be achieved by using a carousel conveyor. Each bin is numbered, and when you put something away, you make a record of that item and the bin number where that part is stored. When needed, you look up the part number and find its location number. You push the button and the part comes to you. The picture in Figure 11–21 shows a horizontal carousel conveyor, whereas Figure 11–22 shows a vertical ferris wheel system.

Systems Required for the Stores Department

Locator System

As was discussed in Chapter 8, every location has an address and the warehouse person must know how to reach any address without taking time to think.

Figure 11–17 Side shifting truck (courtesy of Crown Equipment Co.).

Figure 11–18 Maintenance cart (courtesy of Globalindustrial.com).

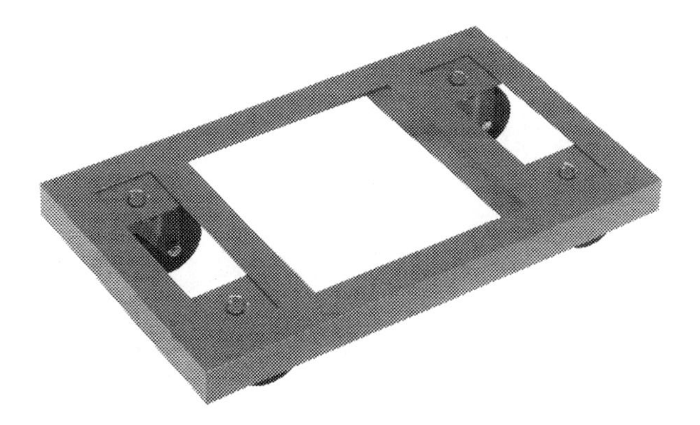

Figure 11–19 Dollies and casters (courtesy of Globalindustrial.com).

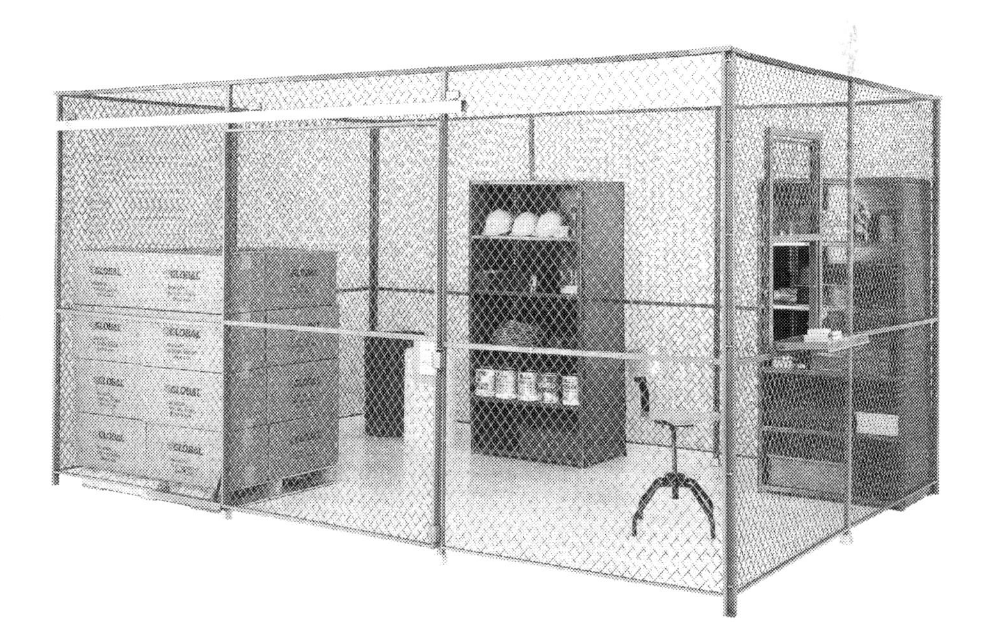

Figure 11–20 Maintenance tool crib (courtesy of Globalindustrial.com).

Figure 11–21 Carousel conveyor (courtesy of S.I. Handling Systems, Inc.).

Figure 11–22 Vertical carousel storage and retrieval system (courtesy of S.I. Handling Systems, Inc.).

Kitting System

Kitting is the process of pulling together 1,000 sets of parts (one day's supply) for tomorrow's production (see Figure 11–23). This inventory is pulled from the storeroom stock and placed on pallets or carts to be moved to the assembly line for tomorrow's work. Kitting needs space for holding the material and material handling equipment to move it out of the storeroom to production. The kitting system is important because having a full day's inventory on the assembly line means that

Figure 11–23 Carts for kitting (courtesy of Globalindustrial.com).

you will not run out. If something was missing in the warehouse, you could have 16 to 24 hours to resolve the problem.

Inventory Control System

The inventory control system controls the storeroom. Maintaining a proper level of inventory is the function of inventory control. The storeroom is sized to maintain this inventory. The movement of material into and out of the storeroom must be reported and entered into the inventory control system.

FABRICATION

The fabrication department is the department that produces parts for the assembly and/or packout lines. This fabrication starts with raw materials and ends with finished parts. The material handling facilities include containers, workstation handling devices, and mobile equipment.

Shop Containers

Shop containers are used to move parts in unit loads (see Figure 11–24). Large sheets of steel or coils of steel are chopped into smaller parts. These parts are collected in bins or boxes made of cardboard, plastic, or steel and moved to the second operation (see Figure 11–25). Shop containers are often stacked on pallets moved to the next machine and placed on the next machine. Machines can be filled by using a device that holds shop containers at the correct angle and position. Because shop containers are used over and over again, they must be durable, stackable, and portable.

Figure 11–24 Shop boxes in fabrication area (Courtesy of Streator Dependable Mfg.).

Figure 11–25 Shop containers (courtesy of Globalindustrial.com).

Tubs and Baskets

Tubs and baskets are larger shop containers (see Figure 11–26). Regular-sized tubs and baskets are 4 feet × 4 feet × 42 inches high. Parts on the bottom are often difficult (and time consuming) to retrieve. For this reason, several special tubs and baskets generally are available.

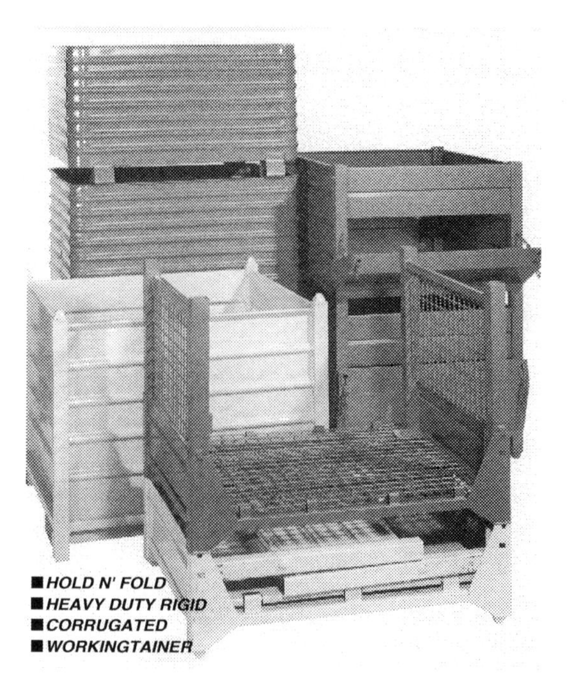

Figure 11–26 Tubs and baskets (courtesy of Steel King Industrial Containers).

Drop Bottom Tubs

A special rack holds the tub over an inclined slide (see Figure 11–27). Once on the rack, the bottom front is dropped and parts flow out the bottom of the tub, which has been elevated to a good work height for the operator.

Drop Side Tubs or Baskets

These are not as good as drop bottom tubs, but they are less costly (see Figure 11–28).

Tilt Stands

Tilt stands hold regular tubs and baskets at an angle to make retrieving parts easier (see Figures 11–29 and 11–30).

V Stands

V stands are used to hold smaller cartons at work level at a 45° angle to allow for easy access to parts (see Figure 11–31).

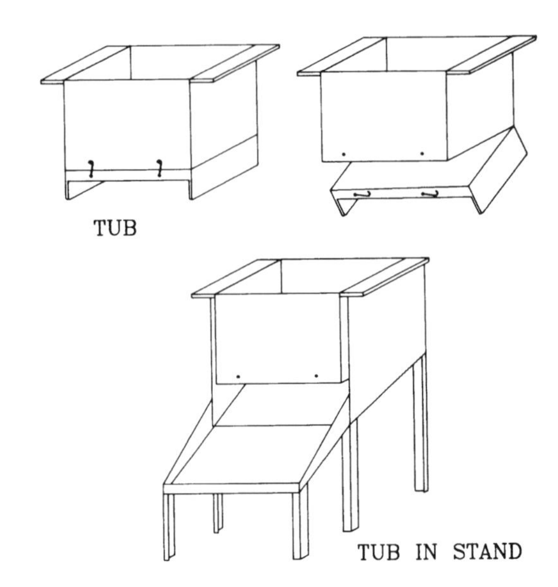

Figure 11–27 Drop bottom tub.

Figure 11–28 Drop side basket (courtesy of Steel King Industrial Containers).

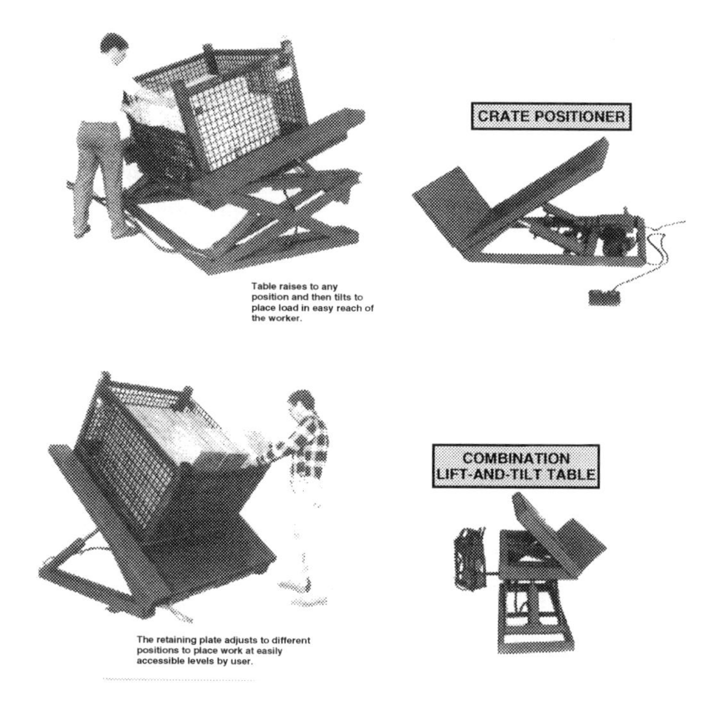

Figure 11–29 Tilt tables and stands (courtesy of Air Technical Industries).

Figure 11–30 Tilt stand (courtesy of Streator Dependable Mfg.).

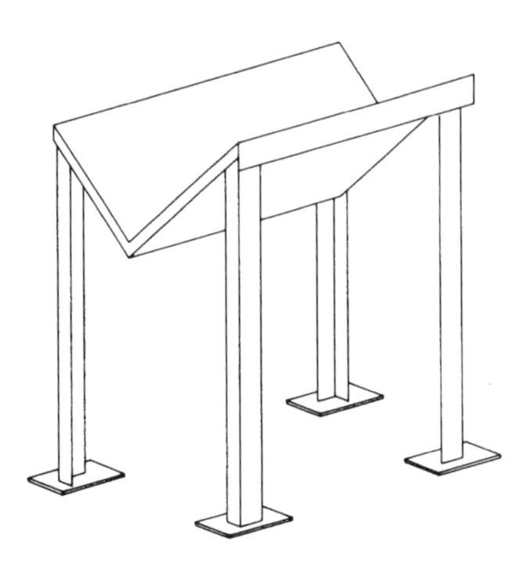

Figure 11–31 V stand.

Scissor Lifts or Hydraulic Lifts

A scissor lift will lift up a pallet of material to keep the material at a comfortable height.

Dump Hoppers

Dump hoppers can make the handling of material at a workstation almost effortless. Dump hoppers will clamp a tub or basket of parts into position, lift the tub and tilt it to a 120° angle, spilling the parts onto a slide that can bring the parts up to the point needed. They are very efficient.

Workstation Material Handling Devices

Counterbalances

Counterbalances hold tools above where they are needed and nearly eliminate the weight of the tool (see Figures 11–32a and 11–32b). They are one material handling device that takes the physical labor out of work.

Manipulators and Lifting Devices

Manipulators are positioning devices with almost human flexibility and dexterity, but with superhuman strengths. Manipulators are specially designed to perform lifting, rotating, tilting, turning, and positioning tasks that far exceed human capabilities. Whether manual, hydraulic, or pneumatic, manipulators can be installed on either a stationary pad or a portable base and can be utilized to perform a variety of tasks to enhance worker productivity and safety (see Figure 11–32c). Lifting devices assist the operator in lifting heavy parts and components, as well as serve as manipulators to position the part.

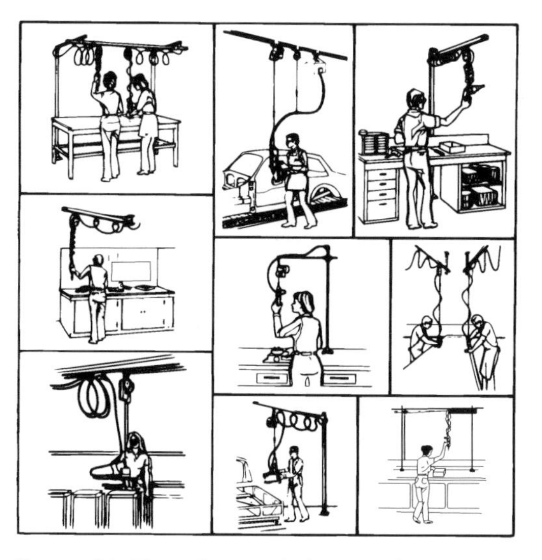

Figure 11–32a Counterbalancers (courtesy of Aero-Motive Mfg. Co.).

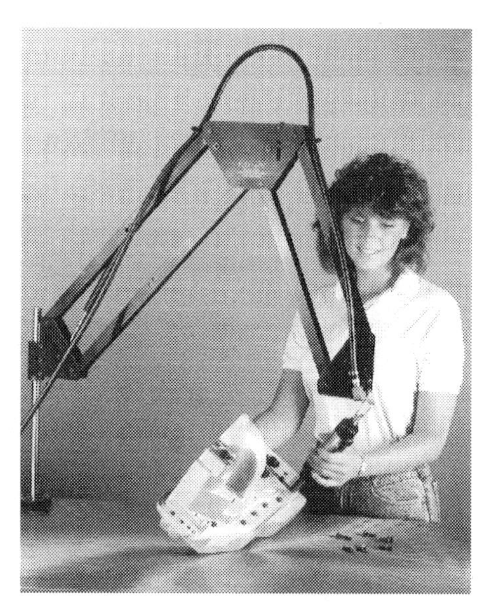

Figure 11–32b Counterbalancer (courtesy of Flex Arm).

Vibratory Feeders

Vibratory feeders orient, feed, count, and present a part to the next operator. Many machines have part feeders loading parts automatically. A toy maker needed 4 million little tires put on hubs. It set up two vibratory feeders (one for wheels and one for hubs), and two round wheels picked up a part each from the feeders and pressed the tire onto the hub automatically. A swingset manufacturer set up 20 vibratory feeders to assemble parts bags. Each feeder was connected to a control panel and the number of required parts was entered into the controller. All 20 feeders counted out their parts and stopped. When all 20 feeders had stopped, the hoppers were opened and the parts fell into a bucket on the conveyor. The conveyor advanced one feeder and the feeders started again. The bags were formed, packed, and weight checked automatically.

Rivets, eyelets, screws, and bolts are fed into machines that use these fasteners by means of vibratory feeders.

Waste Disposal

Removing waste from workstations requires special material handling equipment (see Figure 11–33a). Dump hoppers like those pictured in Figure 11–33 can be used for waste disposal. Chip removal from cutting machines removes the cutting oils and places the chips in a dump hopper. Trash compactors reduce waste removal costs, and paper bailers will turn trash costs into profits. Waste disposal is an area where material handling equipment can greatly improve performance and reduce costs.

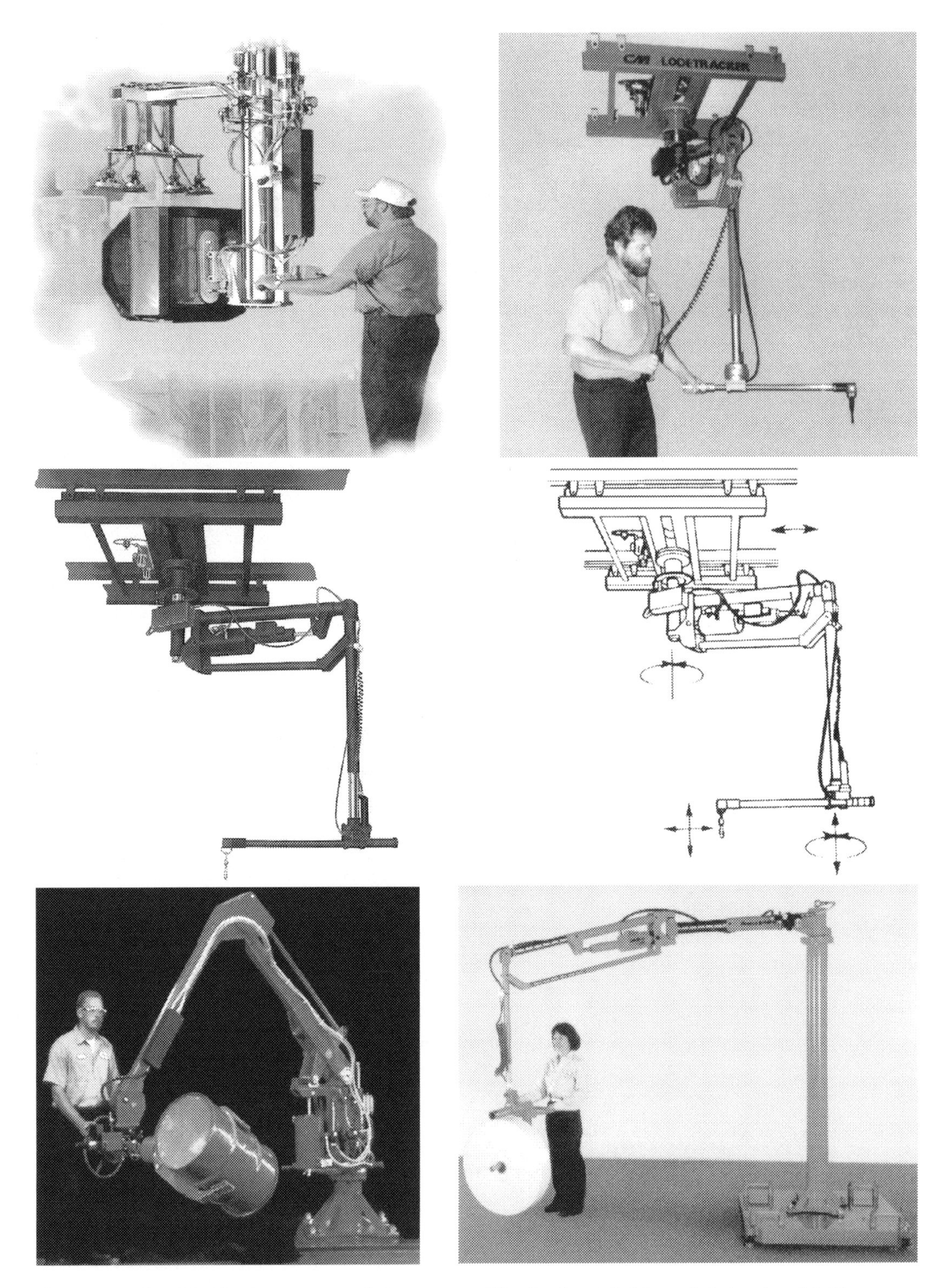

Figure 11–32c Manipulators at work (courtesy of Positech, a Division of Columbus McKinnon Corporation).

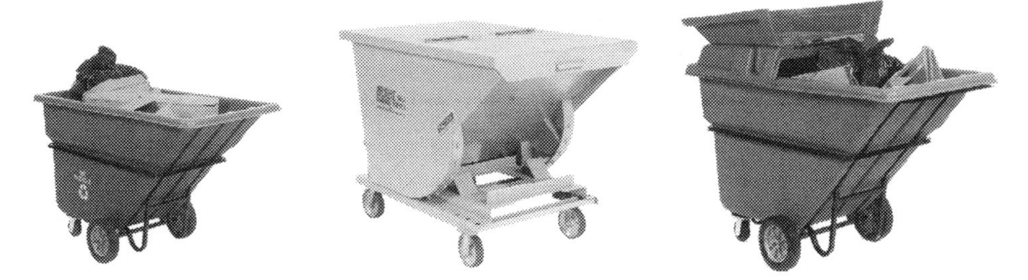

Figure 11–33a Waste disposal (courtesy of Global Equipment Co.).

Among the most difficult to handle waste by-products in machining operations are chips. The difficult, dirty, and dangerous manual work of removing machining chips can be avoided by installing special material handling systems. Figure 11–33b shows an enclosed chip removal system that uses an endless chain with attached

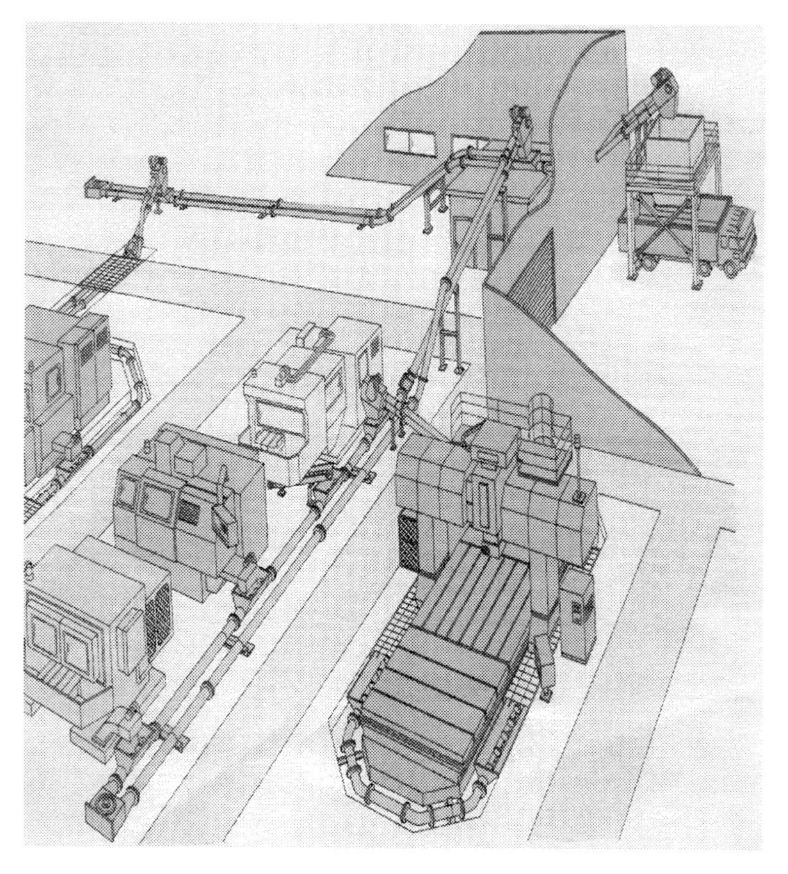

Figure 11–33b Automatic chip removal system (courtesy of LNS).

Figure 11–33c Chip discharge (courtesy of LNS).

blades to pull material through a sealed pipe to a storage hopper. Cutting fluid or other liquids can be recovered at points where the piping rises to a higher level. The system offers great labor and power saving along with safety, space, flexibility, and pollution-free transportation.

The system can be arranged to fit almost any plant layout and can be easily rearranged when machines are added, deleted, or relocated. Figure 11–33c shows chip discharge into the hopper.

Walking Beams

Walking beams continually load and unload machines, eliminating the need for an operator doing any material handling (see Figure 11–34). Walking beams pick up a part, move the part into the machine, lower the part, and return to the starting point. Two walking beams work on the same machine—one loading and one unloading.

Ball Tables

Ball tables have ball bearings mounted on a table top to allow the easy movement of heavy material (see Figure 11–35). A 200-pound sheet can be moved with only 10 pounds of force.

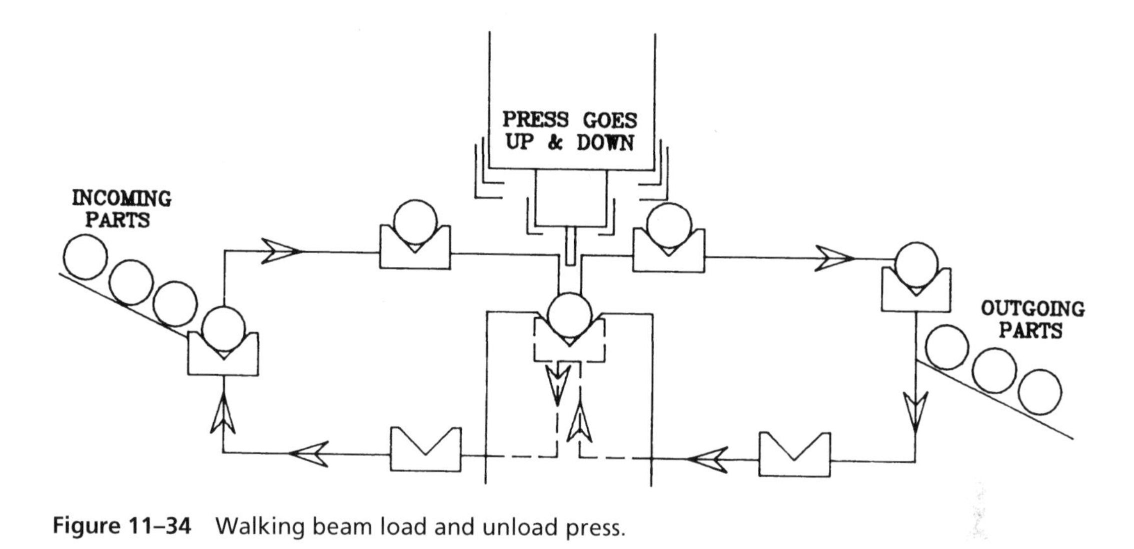

Figure 11–34 Walking beam load and unload press.

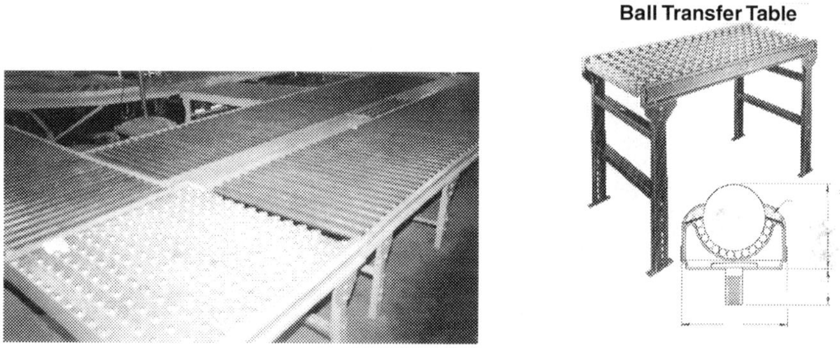

Ball Transfer Table

Figure 11–35 Ball table (courtesy of Hytrol Conveyor Co.).

Powered Round Tables

A workstation can be built on a round table and indexed automatically (rotated) (see Figure 11–36). When manufacturing golf club irons, a 2-inch hole is drilled in the hozzle (shaft hole). The hole is tapered, tapped, and the top is spot faced. All these operations are done at the same time on a round table.

Jib Cranes

Jib cranes are lifting devices attached to a boom (see Figure 11–37). The boom is mounted to the top of a mast (upright beams), which rotates 360° around the boom. The crane loads heavy parts or tools into machines. A 20-foot boom mounted between four machines can service all four machines.

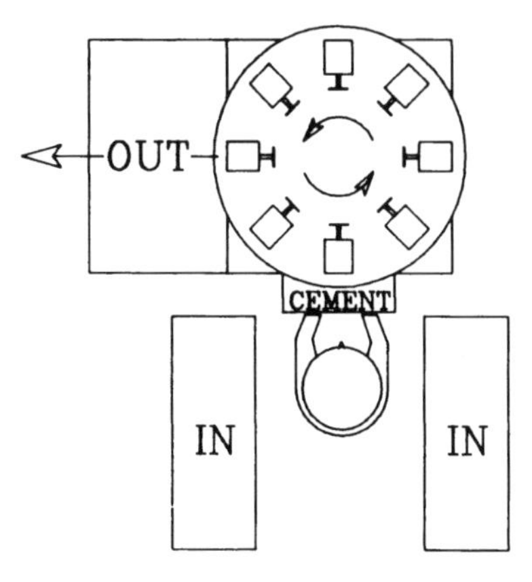

Figure 11–36 Powered round table.

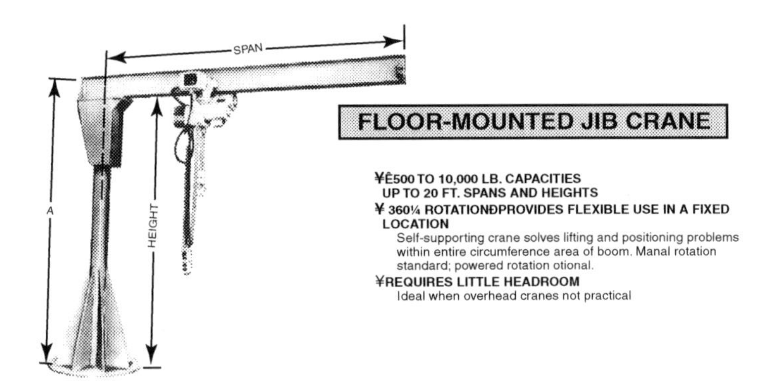

Figure 11–37 Jib crane (courtesy of Air Technical Industries).

Vacuum or Magnetic Lifts

Mounted to a jib crane or, in large plants, bridge cranes, a vacuum or magnetic lift hoists long, heavy sheets of material (see Figure 11–38). Skins for aircraft are moved with vacuum lifts or manipulators.

Robots

Robots can be used to perform a variety of tasks, including loading and unloading, painting, welding, and a vast array of material handling tasks. In addition to performing repetitive tasks with a great deal of accuracy, they are also most useful in performing dangerous and hazardous tasks in environments that would be hostile to human workers. Figure 11–39a displays a state-of-the-art robot performing a variety of manufacturing tasks. Figure 11–39b displays the robot's operations center that acts as a production supervisor designed to increase productivity by providing

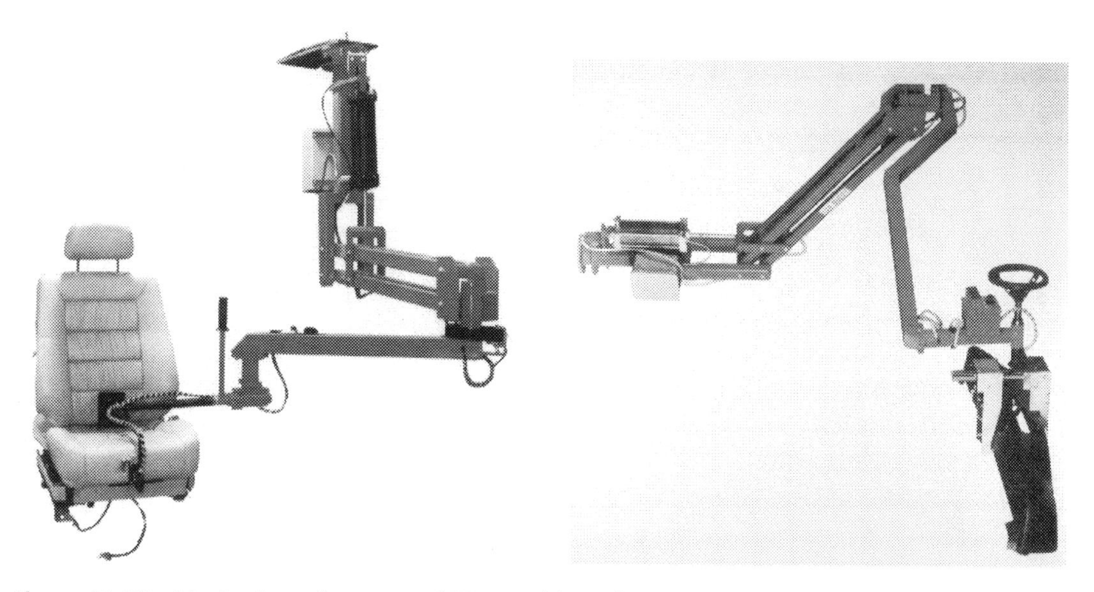

Figure 11–38 Manipulator (courtesy of TDA Buddy Inc.).

cell-level monitoring, diagnostics, and reporting. A robot's dimensions are shown in Figure 11–40.

Mobile Fabrication Equipment

Moving material around the fabrication area requires equipment that can follow various paths. These paths can change with each different manufactured part's requirements. The equipment covered in this section is not in any preset order and can be easily rearranged.

Slides and Chutes

Slides and chutes are as simple as a child's playground slide (see Figure 11–41). Material is placed on the top of the slide by the operator who has just finished the operation. The part slides down to the next operator by use of gravity. Slides and chutes can be made of wood, plastic, or steel and can be easily moved.

Skate Wheel and Roller Conveyors (Nonpowered)

Skate wheel and roller conveyors come in 10-foot sections and can be combined to make any length (see Figure 11–42). They can be easily moved for a change of direction, and the slope can be adjusted to make the parts roll. If some parts will not roll, they can be put in shop containers or on wooden boards in order to make them roll.

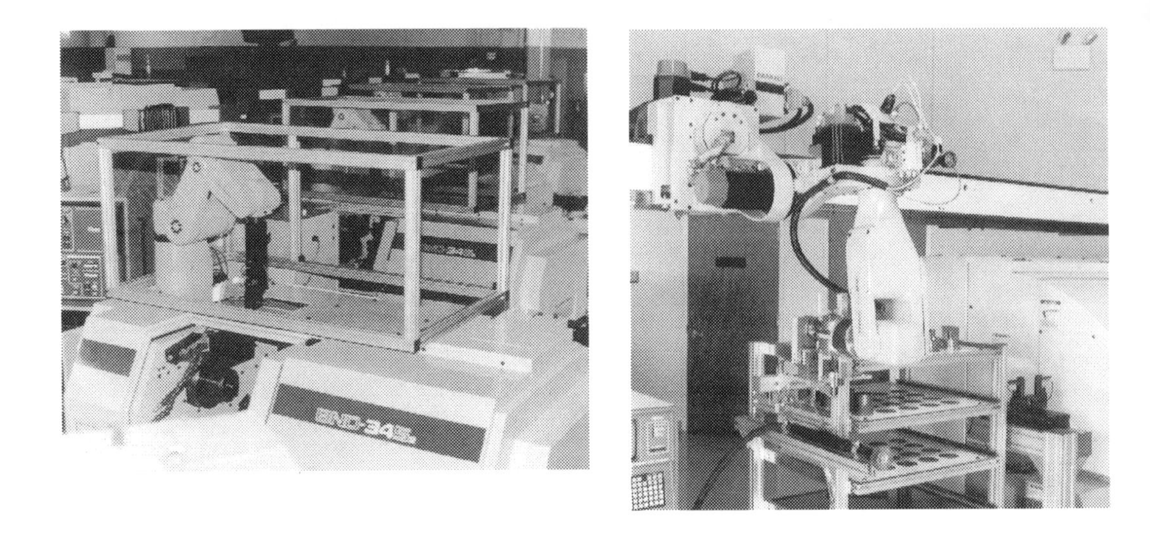

Figure 11–39a State-of-the-art robot performing manufacturing tasks (courtesy of GM Fanuc Robotics Corporation).

Skate wheel and roller conveyors are very flexible in that they can be made to follow any path. V stands (refer back to Figure 11–31) have been combined with skate wheel rollers to create a material handling system that connects two workstations and automatically feeds parts to a convenient position for the operators to place and pick up parts without walking or bending.

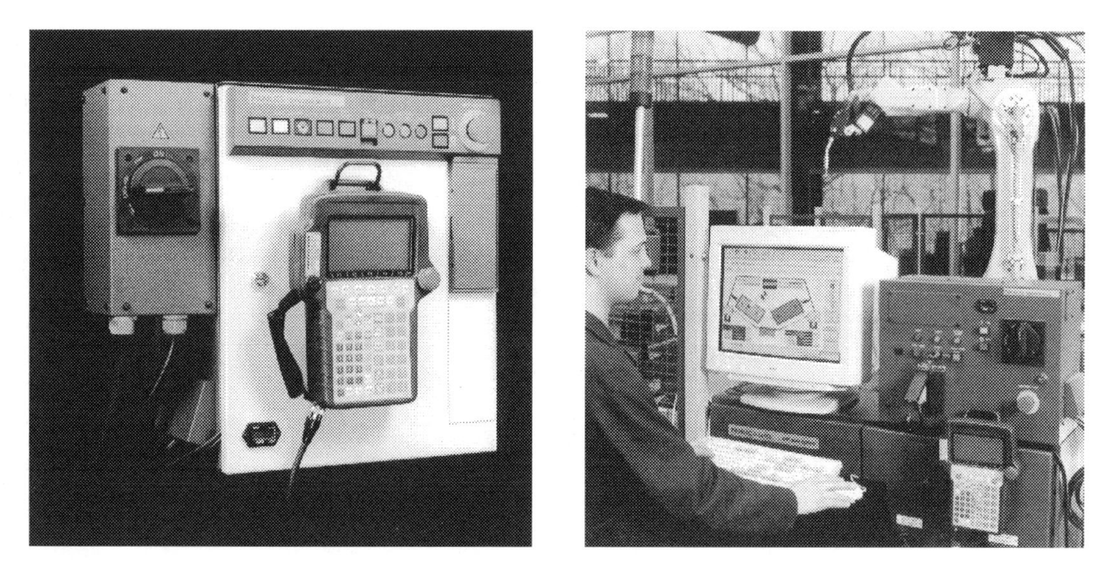

Figure 11–39b Robot's operations center (courtesy of GM Fanuc Robotics Corporation).

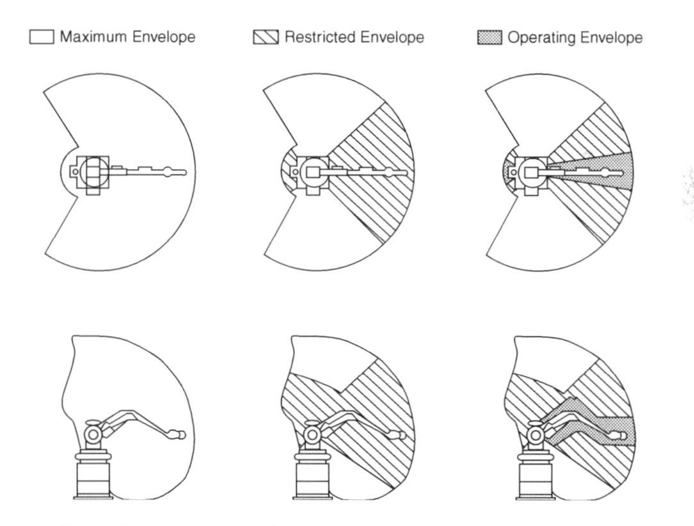

Figure 11–40 Robot dimensions and work envelope.

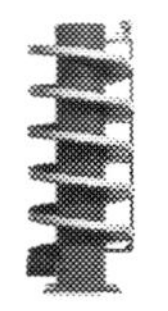

Figure 11–41 Spiral slide (courtesy of Ambaflex).

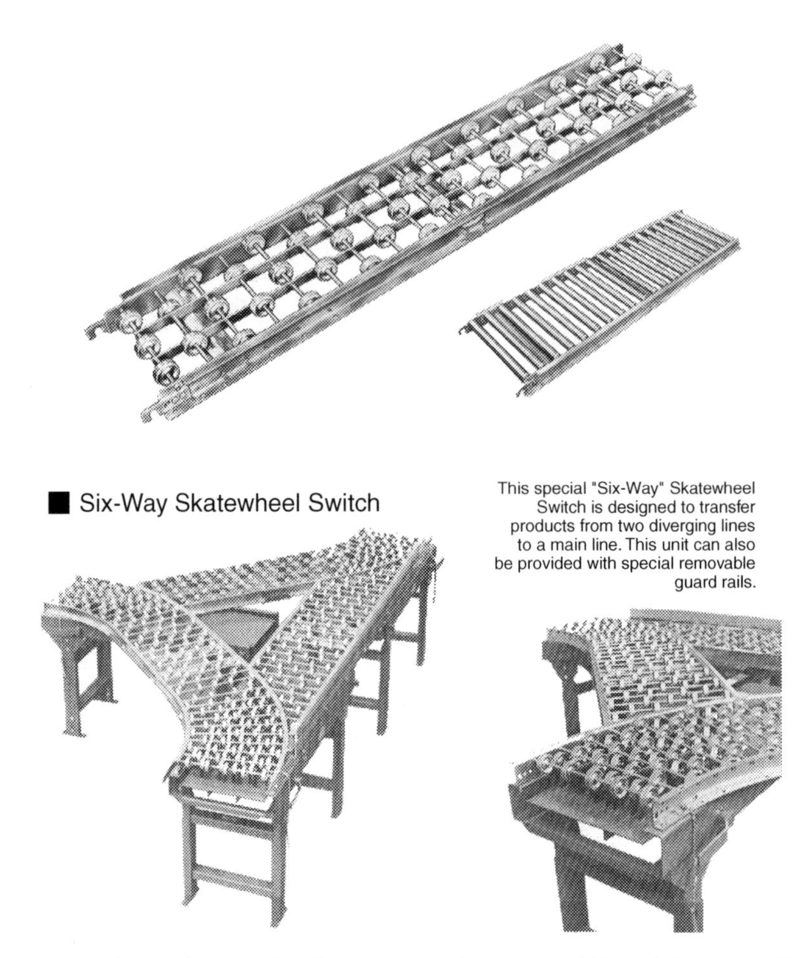

Figure 11–42 Skate wheel and roller conveyor (courtesy of Hytrol Conveyor Co.).

Lift Conveyors

Lift conveyors can move parts from near the floor to any elevated position (see Figure 11–43a). They are used on automatic machines where the part drops out of the machine on a chute to the bottom of the lift conveyor and lifted to the top of a tub where hundreds or thousands are collected. Lift conveyors (sometimes called *bucket conveyors*) can move water, grain, coal, or just about anything where a lot of volume is needed.

Adjustable Angle Conveyors

Adjustable angle conveyors (shown in Figure 11–43b) can also be included in this class of conveyors. These conveyors can be set up and easily reconfigured to meet a variety of manufacturing, assembly, and bulk handling needs. The cleated belts, with or without cleated sidewalls, are ideal for bulk or small product handling.

Figure 11–43a Lift conveyor (courtesy of Hytrol Conveyor Co.).

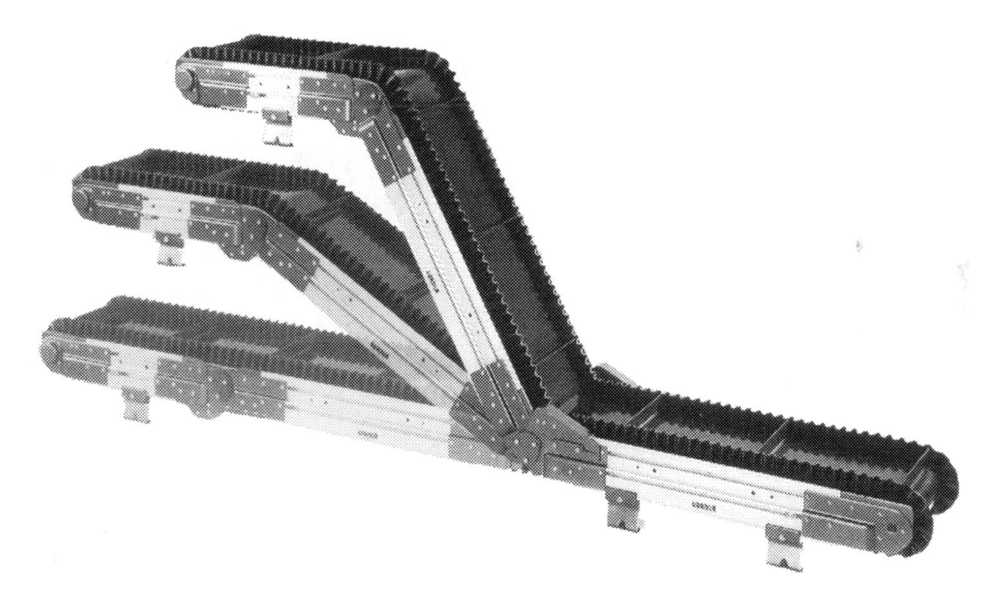

Figure 11–43b Adjustable angle conveyor (courtesy of Dorner Mfg. Corp.).

Magnetic Conveyors

Magnetic conveyors can also be used for lifting ferrous parts and components on an inclined angle cheaply and efficiently. Magnetic conveyors are also quite useful for removing parts and scrap from under the dies and in other turning or material removing operations. Vacuum conveyors are useful in the case of nonferrous materials such as film, paper, and plastic. Magnetic as well as vacuum conveyors (shown in Figure 11–43c)

Figure 11–43c Magnetic and vacuum conveyors (courtesy of Dorner Mfg. Corp.).

can assist in controlling drop and orientation and spacing of parts. The adjustable angle conveyor shown in Figure 11–43b offers flexibility and efficient part removal.

Auger or Screw Conveyors

Auger or screw conveyors are tubes with a screw inside (see Figure 11–44). The turning of the screw pulls and pushes the material in the direction of the screw rotation. Grain and wood chips are moved this way.

Vibratory Conveyors

Vibratory conveyors move parts down a chute or slide by vibration. Inclined vibratory conveyors are used in the separation of parts such as sand in casting or parts from tumbling media such as plastic pallets, corn cobs, and rocks (see Figure 11–45).

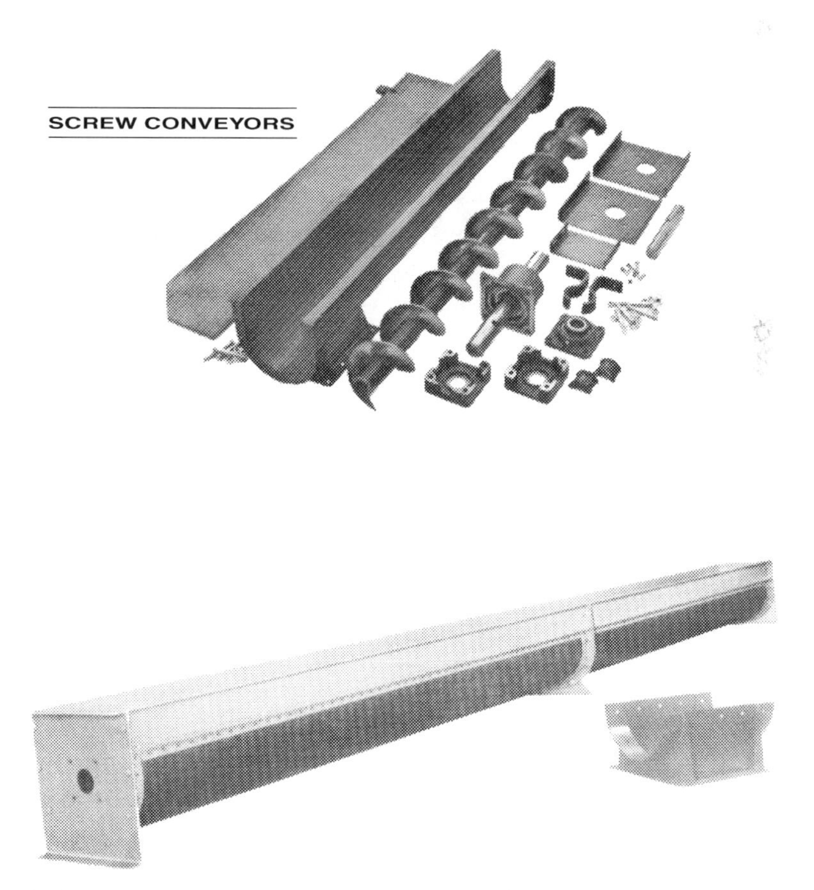

Figure 11–44 Auger or screw conveyor (Courtesy of Conveyor and Drive Equipment Co., Inc.).

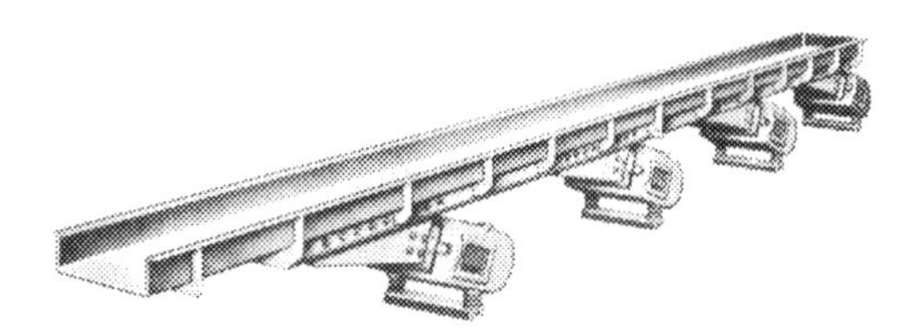

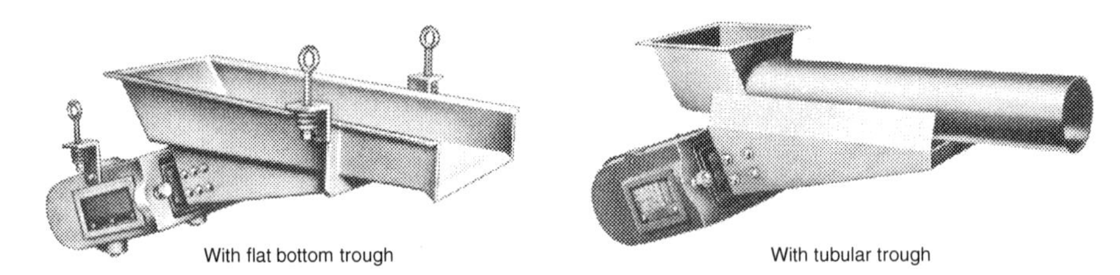

Figure 11–45 Vibrating equipment (courtesy of Conveyor and Drive Equipment Co., Inc.).

Monorail Trolley Conveyors

A monorail trolley conveyor is a single rail over the workstation or between two workstations that can move parts of tools along a fixed path (see Figure 11–46). If a heavy tool is needed anywhere along a 20-foot path, place a monorail above this path and hang the tool from the rail.

Powered Hand Trucks

Powered hand trucks resemble hand trucks except that they have a battery attached (see Figure 11–47). They can lift and move greater weights and are easier to control. For short distances (like within a department), they are more cost efficient than fork trucks.

ASSEMBLY AND PAINT

Many assembly operations, especially small assemblies and subassemblies, are just like fabrication workstations and will use the equipment discussed in the previous section (fabrication). Counterbalances (Figure 11–32), vibratory feeders, tilt stands (Figures 11–29 and 11–30), dump hoppers (Figure 11–33), shop containers (Figure 11–25), and tubs and baskets (Figure 11–26) all are used in assembly, but when speaking of assembly material handling equipment, almost everyone thinks of conveyors. There are many different conveyors. This section will cover the most popular ones.

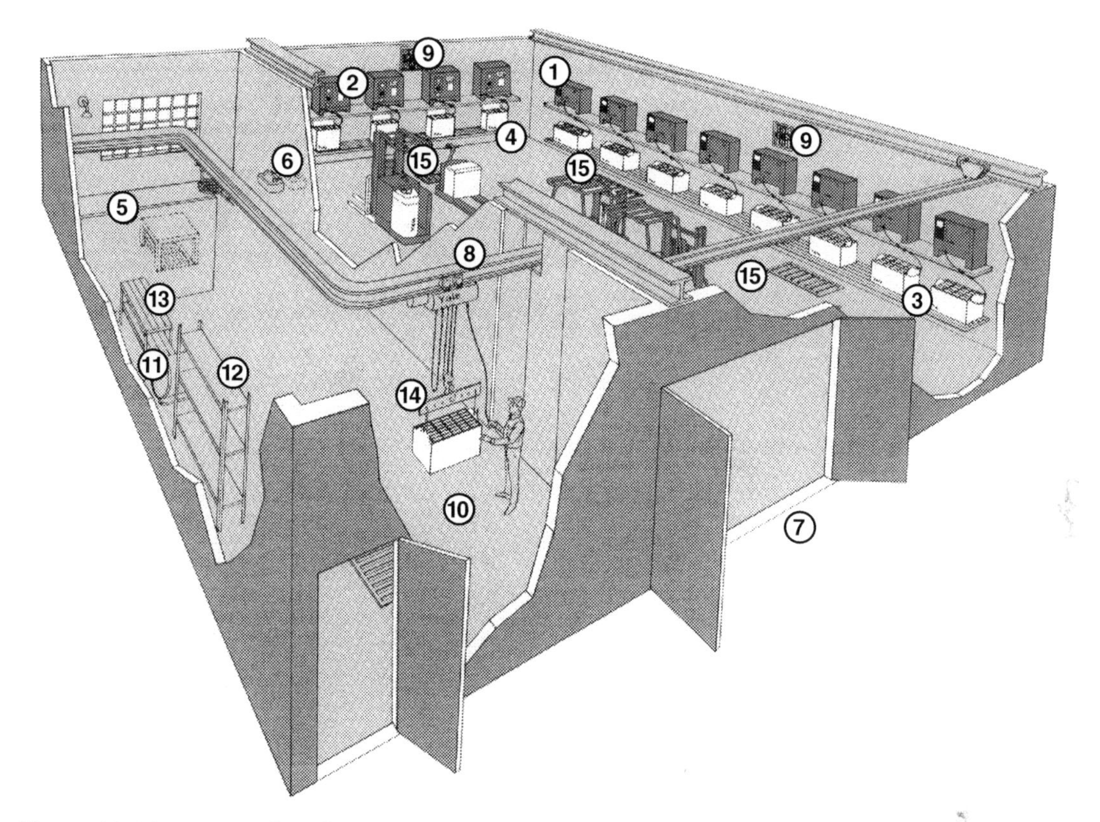

Figure 11–46 Monorail trolley conveyor (courtesy of Yale Materials Handling Corp.).

Belt Conveyors

Belt conveyors are endless loops of fabric that can be any width by any length (see Figure 11–48). Belt conveyors eliminate the need for assemblers to move assemblies into and out of their workstation. They also eliminate the need to hold the base unit. Conveyor belt speed and work height should be adjustable. Stops can be built over the belt to deliver assemblies to a workstation and to hold them until the task is complete. Belt material can be cloth or rubber, and can run over sheet metal or rollers.

Powered Roller Conveyors

Powered roller conveyors (see Figure 11–49) perform similarly to belt conveyors and resemble a nonpowered roller conveyor such as that shown in Figure 11–44. Moving boxes over a fixed path for long distances is a good use of the powered roller conveyor.

Figure 11–47 Powered hand truck (courtesy of Yale Materials Handling Corp.).

Car-Type Conveyors

A car-type conveyor can be made by attaching fixtures onto a cable and pulling the cable around a fixed path (see Figure 11–50). The car-type conveyor looks like a small railroad train with the flatcars totally filling the looped track. Think of a child's train set and visualize the track being full of flatcars. Instead of an engine pulling the cars, a cable system pulls the cars along at a uniform speed. On top of the flatcars, you can build holding fixtures to hold any shape.

Slat Conveyors

Slat conveyors are narrow slats of wood or metal attached to chains (see Figure 11–51). The slats travel down a parallel pair of chains to the end of the line and run back to the beginning under the line, just like belt conveyors. The lumber industry cuts lumber and lets it drop on a slat conveyor made out of 2 inches × 6 inches × 20 feet long slats. This belt runs 200 feet away from the saw, and along both sides, laborers separate the sizes and grades of lumber and place lumber on carts.

Figure 11–48 Belt conveyor (courtesy of Hytrol Conveyor Co.).

In drink bottling plants, bottles or cans are carried through the filling, capping, and labeling machines by slat conveyors made of thin 6 × 4-inch metal slats. TV sets are assembled on a slat conveyor 2 inches × 4 inches × 4 feet long with electrical plug-ins every 2 feet. The Caterpillar Tractor Company assembles its largest tractors on a slat conveyor built at floor level. The conveyor moves only a few feet per hour, but material can be moved on and off the slat conveyor with fork trucks. People walk on and off the conveyor without even knowing that they were on a conveyor. The Caterpillar slat conveyor is made of steel slats about 1/2 inch thick, 12 inches wide, and 20 feet long.

Tow Conveyors

Tow conveyors pull carts around a fixed path (see Figures 11–52a through 11–52c). The power can be overhead or under the floor, but both do the same job. One advantage of a tow assembly line is that one unit can be removed from the line

LIVE ROLLER CONVEYOR

The Model "190-SP" live roller "spool" conveyor is a general transport conveyor with the capabilities of accumulating products with minimum back pressure. Quiet operation, versatile design, and easy installation are standard features that make the "190-SP" conveyor a valuable component in operations requiring high performance with minimal downtime.

★ *12 Bed Widths*

★ *Minimum Back Pressure*

★ *Single Drive Powers Curves—Spurs—Straights*

★ *High Speed Capabilities*

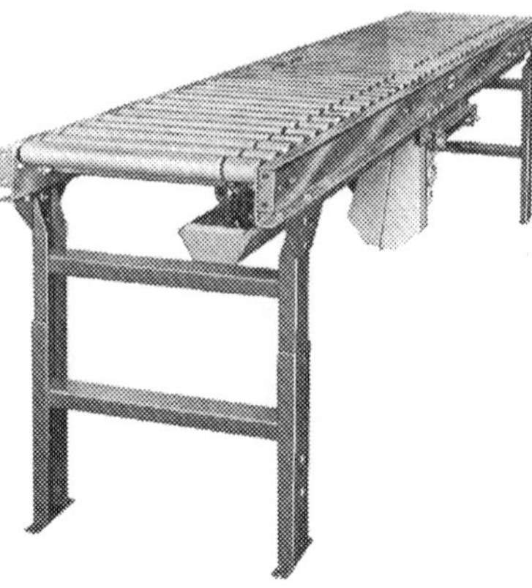

CHAIN DRIVEN LIVE ROLLER CONVEYOR (ROLL-TO-ROLL)

The heavy, rugged design of the "25-CRR" & "26-CRR" allows it to be used for conveying higher load capabilities such as loaded pallets and drums. Chain driven rollers make it ideal for wash-down operations and conveying oily parts in bottling and steel industries, foundries, etc.

★ *Center Drive*

★ *12 Bed Widths—22-1/4 in. to 54-1/4 in.*

★ *Removable Sealed Bearings*

★ *Reversible*

★ *Adjustable Floor Supports*

Figure 11–49 Roller-type conveyors (courtesy of Hytrol Conveyor Co.).

without stopping the line. The tow conveyor carts can carry a wide variety of products. The fixtures mounted to the carts will be for a specific product.

Overhead Trolley Conveyors

Overhead trolley conveyors can go anywhere (see Figures 11–53a and 11–53b). One conveyor manufacturer uses the slogan, "Anywhere and Everywhere with Unibilt." Overhead trolley systems can carry parts through heat treating, washing, painting, and drying to the assembly department. They can be loaded and unloaded at floor level, and then raised to the ceiling for traveling over the plant's equipment and employees. Trolley conveyors are cable or chain pulled through

Figure 11–50 Car-type conveyors (courtesy of Webb-Stiles Company).

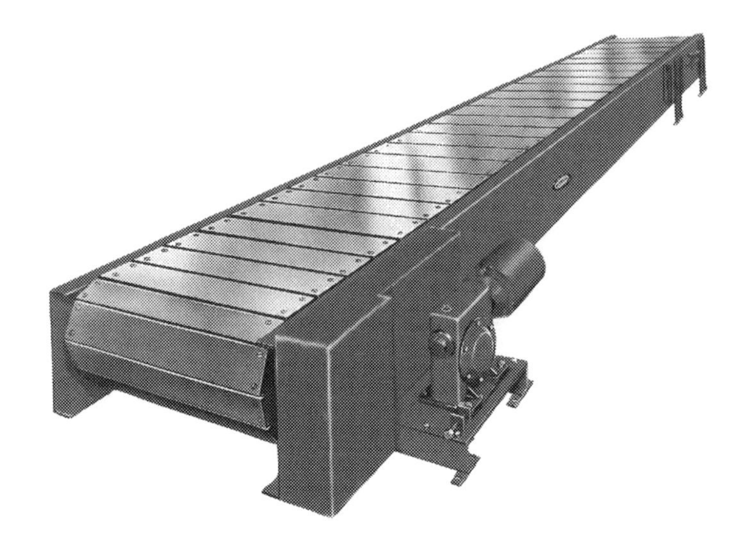

Figure 11–51 Slat Conveyor (Courtesy of Hytrol Conveyor Co.).

channels of I beams with a single drive unit. Below the trolleys there are mounted hooks and racks for carrying the parts. The study of overhead trolley conveyors alone could be a career.

A simple S hook is the most common method of hanging parts from the trolley, but the hooking systems can get quite complicated. Parts can be spun on the overhead trolley conveyor by placing a wheel on the hook and a stationary rub bar mounted along the area where you want the part to spin. Parts can be turned by placing an *X* on top of the hook and placing a stationary pin for every 90° of turn desired.

Power and Free Conveyors

Power and free conveyors are dual-track trolley conveyors with one track for the power line and the other track for carrying the trollies (see Figure 11–54). The advantage is that a single part can be stopped without stopping the line. If you are pouring molten iron into a mold, you would not want the mold moving, so you stop it long enough to pour, then reconnect it to the power line. Power and free conveyors can divert the product to different lines or place it in hold areas.

PACKOUT

Packout is usually the end of assembly and many of the same material handling devices are used here. Although packout typically involves packaging one unit for shipment, sometimes it includes putting many products into one package. Material

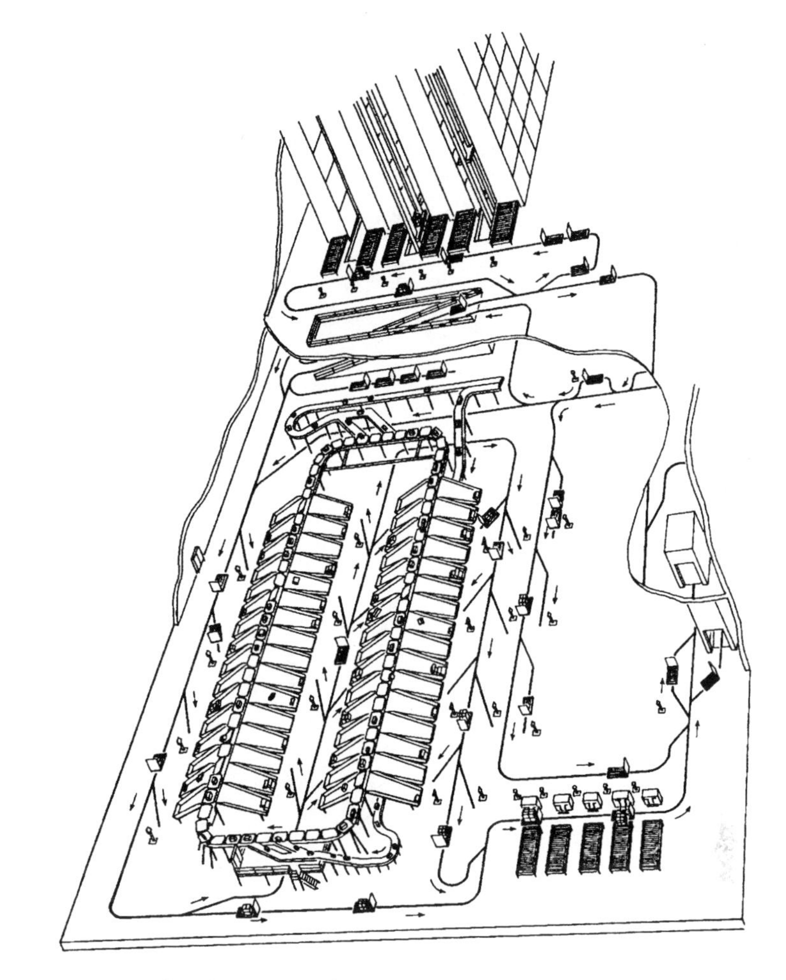

Figure 11–52a Towline (courtesy of S.I. Handling Systems, Inc.).

handling equipment has improved the quality and efficiency of packaging. The following equipment is used in the packout department.

Box Formers

Box forming can be accomplished automatically and wrapped around the product being packaged (see Figure 11–55). Soft drink bottling plants use box formers.

Automatic Taping, Gluing, and Stapling

Closing boxes and sealing them can be accomplished automatically on the packout conveyor (see Figure 11–56).

Figure 11–52b Towline conveyor (courtesy of S.I. Handling Systems, Inc.).

Figure 11–52c Towline conveyor (courtesy of S.I. Handling Systems, Inc.).

Figure 11–53a Overhead trolley conveyor (courtesy of Richards-Wilcox).

Palletizers

After the boxes are filled and closed, they are automatically stacked on pallets per a prearranged program and the full pallets are moved out to a pickoff area where a truck moves them to the warehouse (see Figure 11–57).

Pick and Place Robots

Pick and place robots also make up pallets of finished products (see Figure 11–58). The robot can handle a number of different packages at the same time, but the function is the same as a palletizer.

Banding

Banding boxes closed can be accomplished automatically by placing a bander around the packout conveyor (see Figure 11–59). Banding will also hold many packages on a pallet. Banding is used when packages cannot hold themselves on pallets. When a carton is nearly square, they will not tie (hold themselves on the pallet), so banding is needed.

Figure 11–53b Trolley conveyors (courtesy of Cirnis).

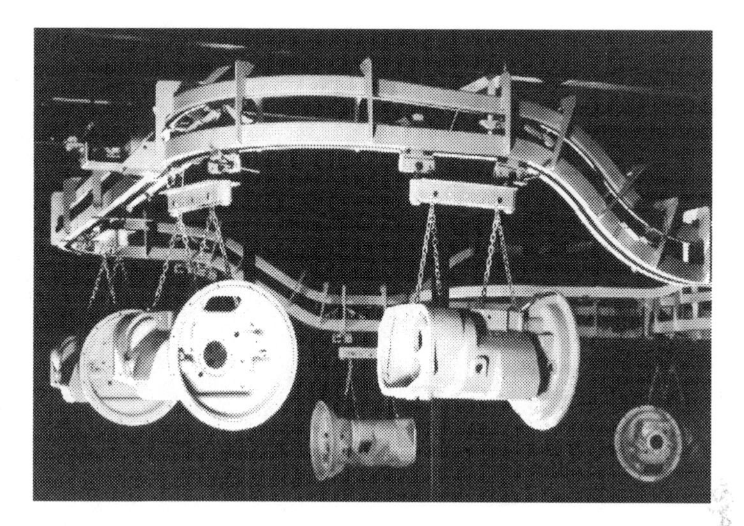

Figure 11–54 Power and free conveyor (courtesy of Richards-Wilcox).

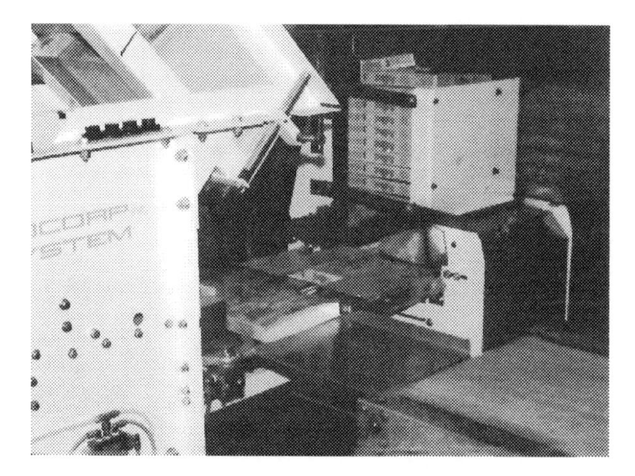

Figure 11–55 Box former.

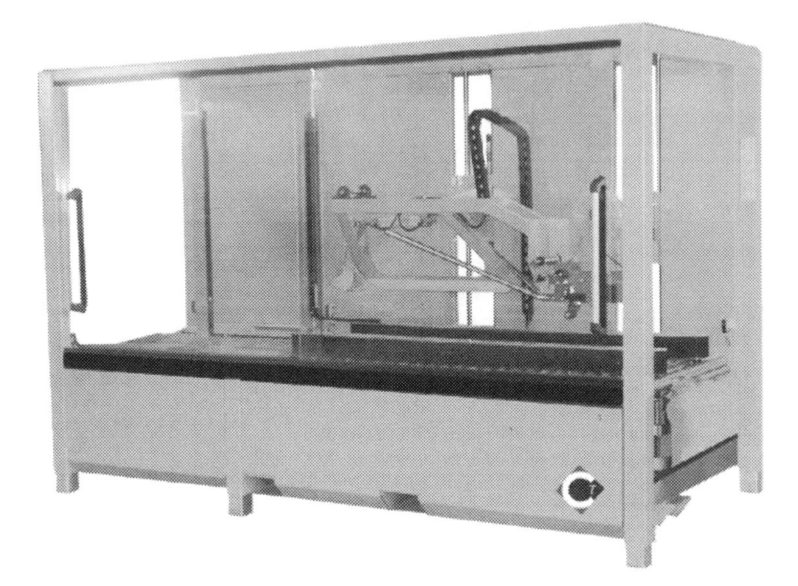

Figure 11–56 Taping (courtesy of Durable Packaging Corp.)

Figure 11–57 Robotic palletizer (courtesy of GM Fanuc Robotics Corporation).

Figure 11–58 Pick and place robot (courtesy of GM Fanuc Robotics Corporation).

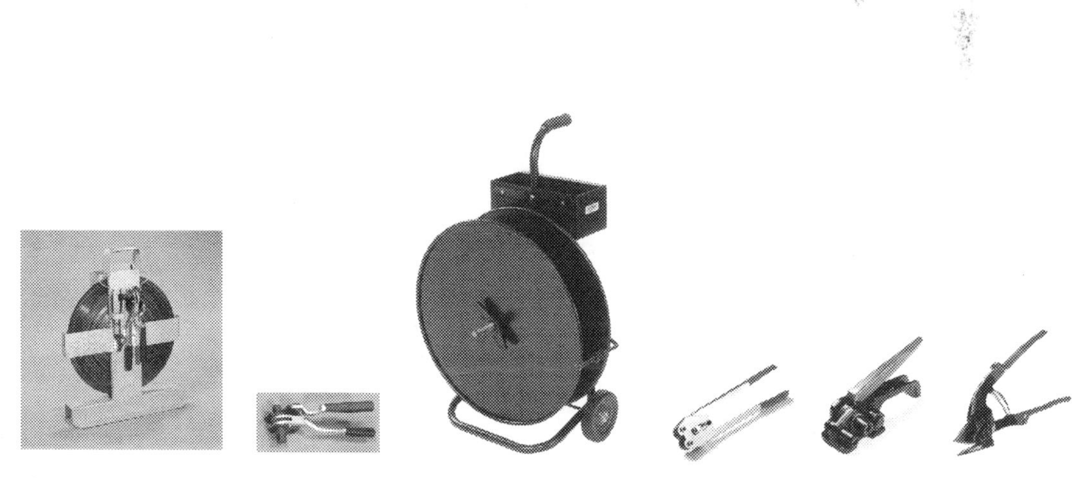

Figure 11–59 Banding (courtesy of Globalindustrial.com).

Stretch Wrap

Stretch wrap is like banding in that it holds packages together on a pallet.

WAREHOUSING

Warehousing resembles stores in that the shelves, racks, pallets, and some trucks typically are similar for both areas. Therefore, we will not cover these pieces of equipment again here. Some unique equipment used for warehousing will be the focus of this section. The functions of a warehouse are to pick customer orders and prepare them for shipping. The first group of equipment to be discussed relates to picking customer orders.

Picking Carts

Customers' orders can be picked from shelves and placed on picking carts. Tool, drug, audiotape, and book warehouses would use this kind of equipment (see Figures 11–60 and 11–61).

Gravity Flow Bins

When the product is small, and high volume is sold, many parts can be stored in a small area, reducing the pickers' need to travel great distances (see Figure 11–62). Drug warehouses use this system for their A items.

Tractor-Trailer Picking Carts

When picking larger orders, such as grocery store warehouses, an order picker would drive a tractor pulling many trailers (see Figure 11–63). Remote-control tractors are often used so that the order picker can retrieve groceries for a rear trailer and still move the tractor.

Trailers that follow one another (tracking) are very important. There are two techniques used to get trailers to track well:

1. The front wheels turn one way while the back wheels turn the opposite way.
2. Two load-bearing wheels are placed in the middle of the trailer, and pilot wheels are mounted in the middle of the front and rear. The pilot wheels just keep the trailer level.

Once the trailers have been filled, the driver takes them to shipping to be loaded on an over-the-road trailer.

Clamp Trucks

Clamp trucks are a special fork truck that eliminate the use of pallets (see Figure 11–64). Two 4 × 4-foot plates squeeze the stacked boxes together, then

Figure 11–60 Picking cart (courtesy of Globalindustrial.com).

Figure 11–61 Picking truck (courtesy of Crown Equipment Corp.).

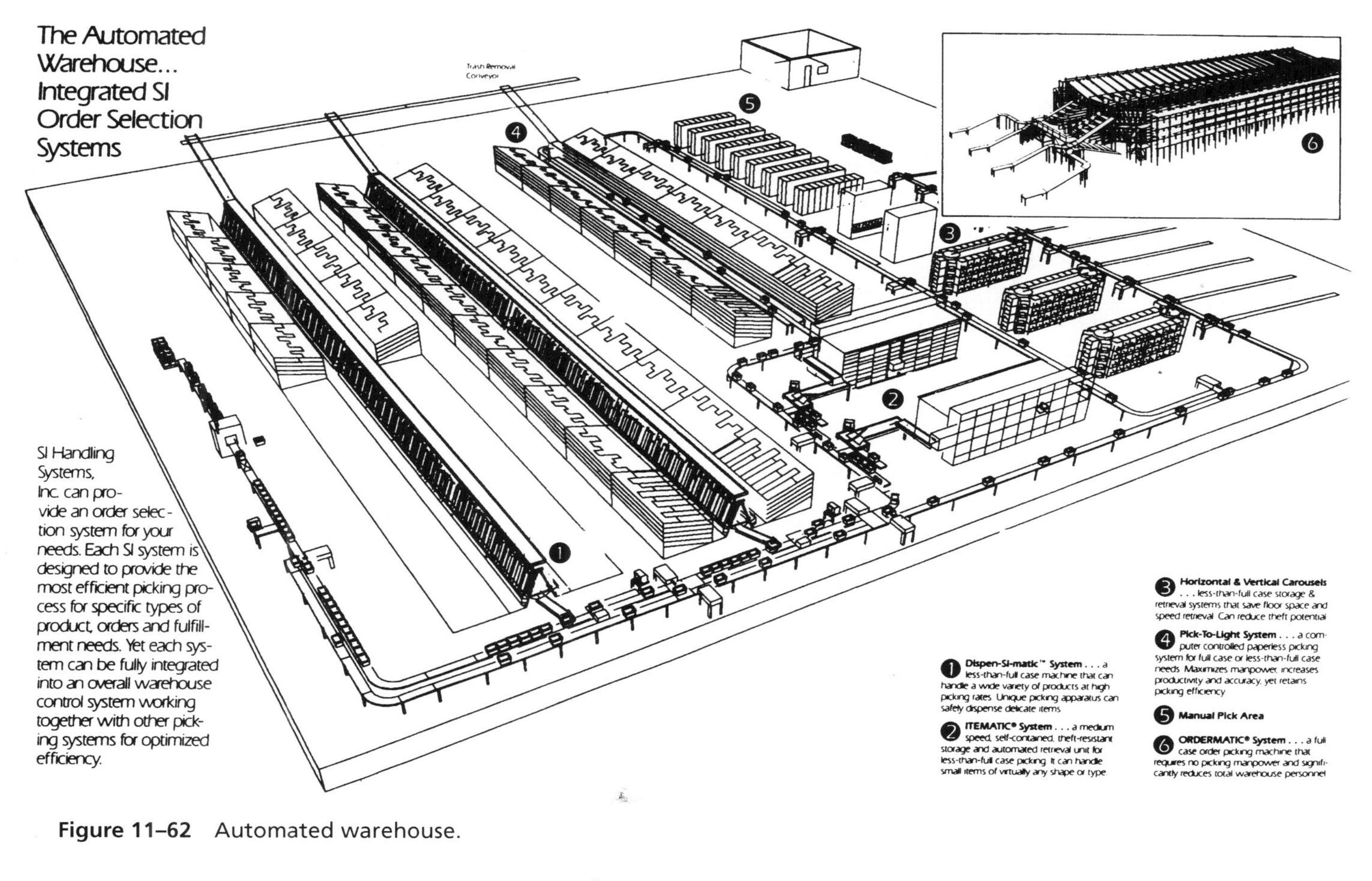

Figure 11–62 Automated warehouse.

Figure 11–63 Tractor-trailer picking (courtesy of Crown Equipment Corp.).

the truck lifts and moves the material into and out of the warehouse. Manufacturers of toys, gas grills, appliances, tackle boxes, and many more industries that produce large volumes of large items use this tool for space utilization and efficiency reasons.

Rotary Conveyor Bins

Rotary conveyor bins bring the (small) product to the picker, saving all the walking. Spare parts warehouses use this technique (see Figure 11–65).

Vertical Warehouse and Picking Cars

The vertical warehouse could have 40 shelves high that are 300 feet long on both sides of a 4-foot aisle (see Figure 11–69). Eight thousand shelves would be available for 8,000 different items. The order picker picks up all the orders in the warehouse for this group of products (they were placed in picking order first) and proceeds to the first location. The order picker rides on a pallet that goes up and down as he loads the stock. After one pass through the aisle, the picker unloads at shipping and returns for new orders. The cart does not need to be

Figure 11–64 Clamp truck (courtesy of Crown Equipment Corp.).

Figure 11–65 Rotary conveyor bins or horizontal carousel conveyor (courtesy of White Storage & Retrieval Systems, Inc.).

steered; it is on rails or has wheel guides. Large catalog distribution centers use this technique (see Figure 11–66).

Packing Station

Once all the orders have been picked, they must be packed for shipping (see Figure 11–67). Small parts must be packaged properly into larger cartons and wrapped so that they are not damaged in shipment. Although larger product packages may just need addressing, some preparation is always needed; therefore, a packaging station is required (see Figure 11–68). A weight scale built into the packaging station (refer back to Figure 11–13) is many times very desirable.

Shipping Containers

Most shipping containers are pallets or the packout carton, but sometimes they may be the size of a tractor trailer (see Figure 11–69). These are called *cargo containers* and can be shipped over the road, on railroad flatcars, and on ocean freighters. These containers can be sealed by the shipper and not opened until received by the customer.

Figure 11–66 Vertical warehouse picking (courtesy of Crown Equipment Corp.).

Figure 11–67 Vertical warehousing (courtesy of Crown Equipment Corp.).

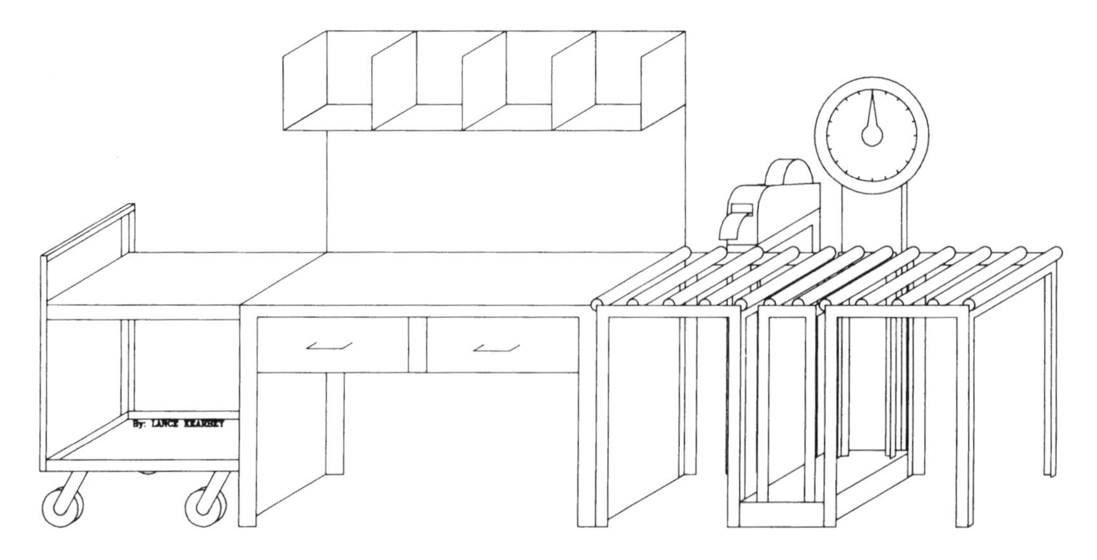

Figure 11–68 Packaging workstation.

Figure 11–69 Shipping containers (courtesy of Steel King Industries, Inc.).

BULK MATERIAL HANDLING

Bulk material handling is a very special subject. It deserves much more attention than will be given in this book due to limited space. Bulk material means a large amount of material (e.g., coal in a coal mine through the power plant, lumber and paper products from the forest through the mills and plants, ore from the ground through the mills, oil from the ground to the service station, and grain from farms through mills and plants). The one advantage these bulk mills and plants have is that one or a few materials make up their material list, and you can concentrate on this one item. Bulk material handling equipment varies in size from a pump for an oil plant to a conveyor system several miles long. The following list of bulk material handling equipment is grossly understated, but if you join one of these industries, the list of equipment will be specific to that industry, and you soon will become familiar with that group of equipment.

Bulk Material Conveyors

Troughed Belt Conveyors

Troughed belt conveyors are concave and resemble a long feed-trough (see Figure 11–70). The coal industry uses these conveyors to move coal from the face of the mine to the elevators and from the top of elevators to the coal pile. Logging industries use troughed conveyors for moving logs from rivers or lakes to the mill ponds. A sidelight is that mill ponds, which float logs to the mills, are material handling devices.

Screw Conveyors

This type of equipment (also called augers) was discussed in the fabrication area, but much more use is made of screw conveyors in processing plants such as paper mills, bakeries, and feed mills (see Figure 11–44).

Vacuum Delivery Systems

A vacuum system is a system of tubes moving pellets or powders from tank cars to storage towers to equipment (see Figure 11–71). A plastics manufacturer typically would use this system. A vacuum system makes the material handling labor free. The storage towers (or silos) are also material handling devices.

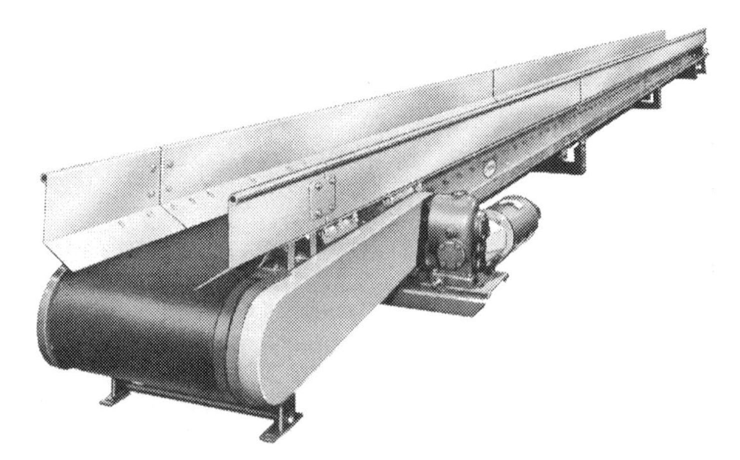

Figure 11–70 Troughed bed belt conveyor (Roller bed) (courtesy of Hytrol Conveyor Co.).

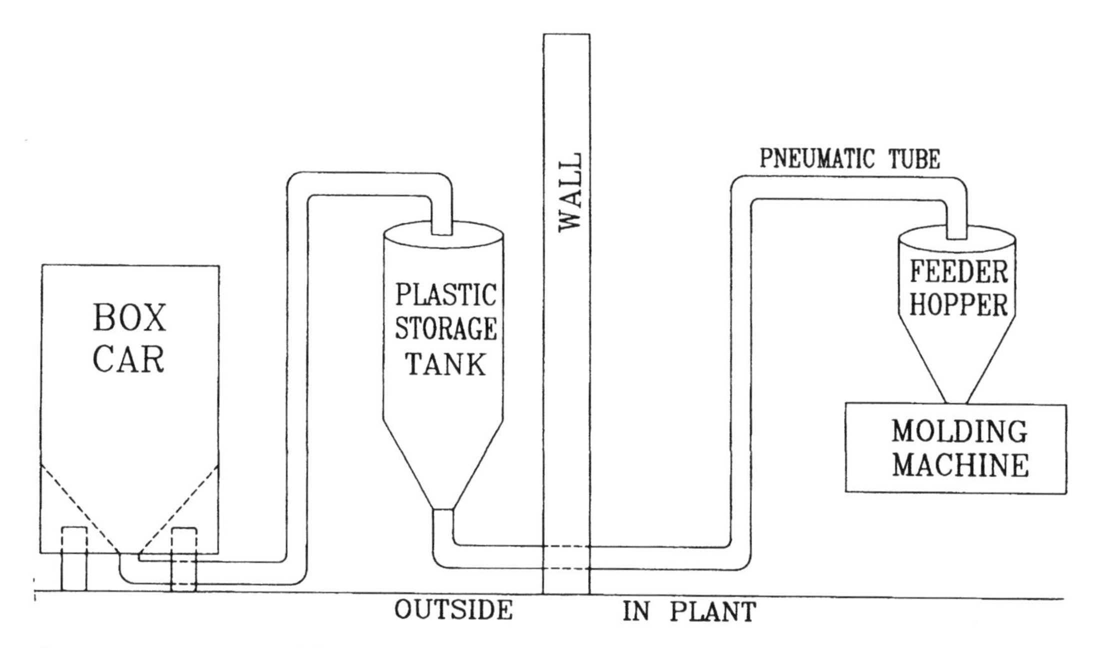

Figure 11–71 Vacuum delivery system.

Figure 11–72 Pumps and tanks.

Pumps and Tanks

Oil, drinks, most liquids, and semiliquids are moved from tankers to tanks to filling stations by pumps (see Figure 11–72). Pumps have hoses leading into and out of them, flow meters to measure volume, gauges to measure tank fullness (tanks are a part of material handling), and petcocks to tap off samples of the product for quality control.

Conveyor Systems

If your bulk products are cartons, a system of diverted conveyors may be used (Figure 11–73). UPS, Sears Distribution, and JC Penney's distribution use major conveyor systems for the distribution of many packages.

COMPUTER-INTEGRATED MATERIAL HANDLING SYSTEMS

State-of-the-art material handling systems, smart systems, and world-class material handling systems all communicate the need to keep improving cost performance every day. Manufacturers today are in worldwide competition, and material handling costs are a major component of product cost, so they need to keep improving. The technology exists to eliminate a large portion of product costs. Material handling equipment is as important as any machine making parts, and the modern technology of material handling equipment has been keeping pace with all other equipment. One piece of equipment (actually a whole system) is the *automated storage and retrieval system* (ASRS). ASRS will automatically put away the product or parts, or take out the product, move it to where required, and adjust the inventory level at both ends of the move (see Figures 11–74 and 11–75). ASRS systems are typically in very tall (60 feet and over) and very large areas (Figure 11–76). The ASRS is made up of (1) racks, (2) shuttle cars, (3) bridge cranes, (4) computer control center, and (5) conveyor systems.

Cross-Docking and Flow-Through

Integration of computers with material handling systems has also facilitated the concept of cross-docking or flow-through. Whereas cross-docking may be more applicable to a distribution center, its mechanics can be applied to any environment that deals with the incoming and outgoing flow of material—a manufacturing facility is no exception. Cross-docking differs from the traditional "move-store, move-store" method of inventory management and control in the facility. If the ultimate destination of a part or product that enters the facility can be specified, then the flow of the part or the product can be virtually uninterrupted through the facility by intervals of storage. Cross-docking facilitates product mixing and sorting operations. In such operations, products or parts from different suppliers

Figure 11–73 Bulk carton handling (courtesy of Hytrol Conveyor Co.).

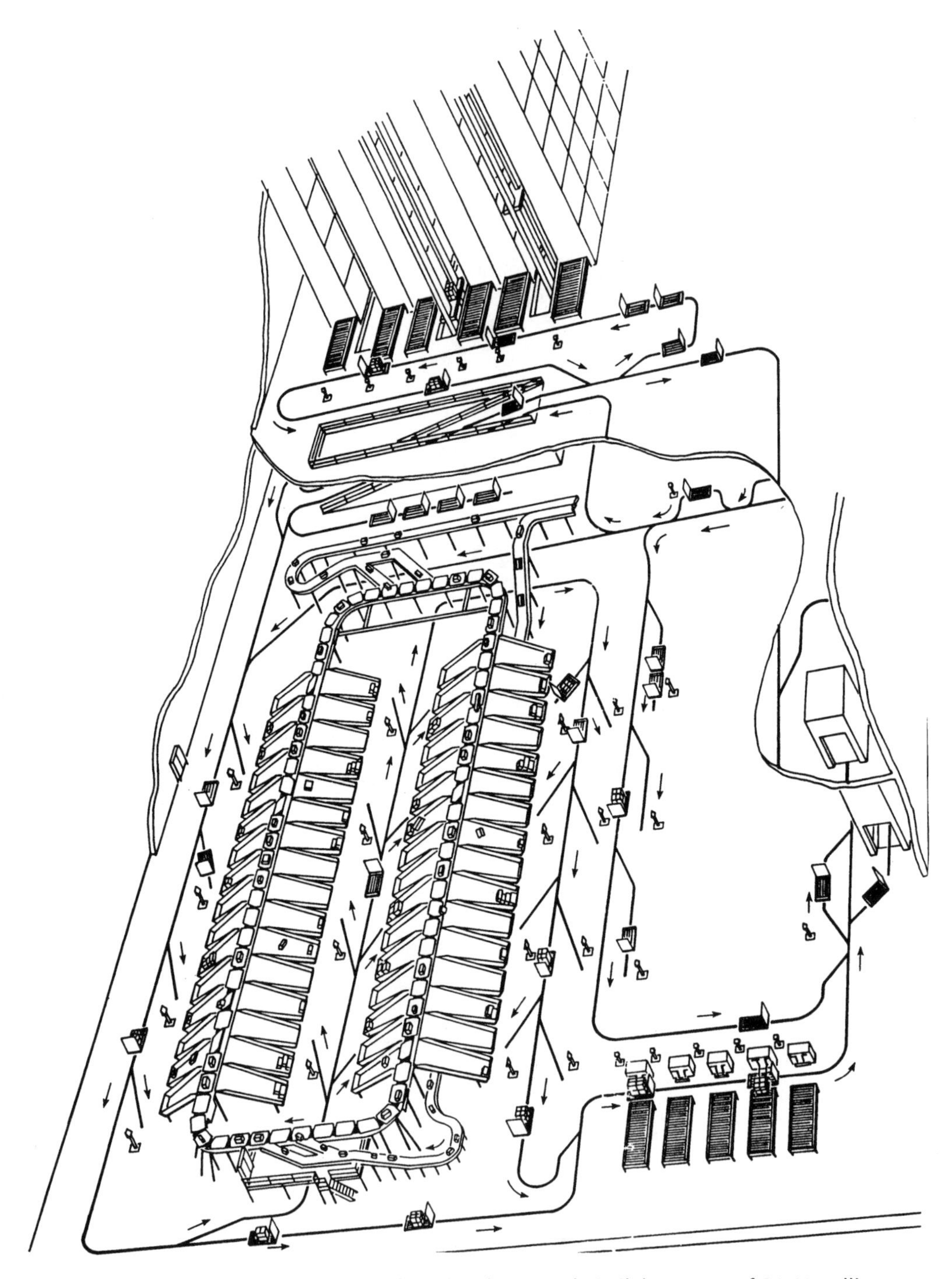

Figure 11–74 Automated storage and retrieval system (ASRS) (courtesy of S.I. Handling Systems, Inc.).

Figure 11–75 ASRS (courtesy of MHIA).

Figure 11–76 ASRS building construction (courtesy of S.I. Handling Systems, Inc.).

arrive in truckloads. Instead of being placed in storage for picking at a later time, the products are moved across the facility to the point of use, or in the case of a distribution center, moved directly into a waiting truck for dispatch to their ultimate destination.

The benefits of cross-docking include

- Reduced warehousing and inventory costs
- Reduced processing costs
- Reduced handling and labor costs
- Reduced storage and warehouse space
- Improved productivity
- More efficient flow of material

Figures 11–77 and 11–78 display the schematics of a computer-integrated cross-docking system.

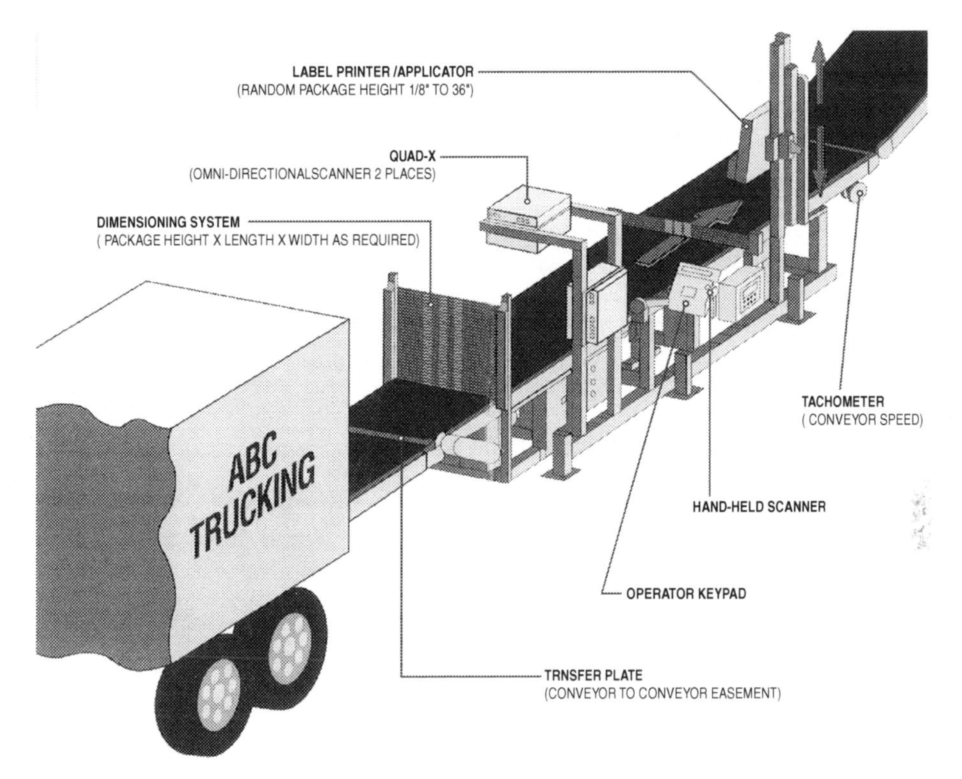

Figure 11–77 Computer-integrated cross-docking system (courtesy of Accu-Sort Systems, Inc.).

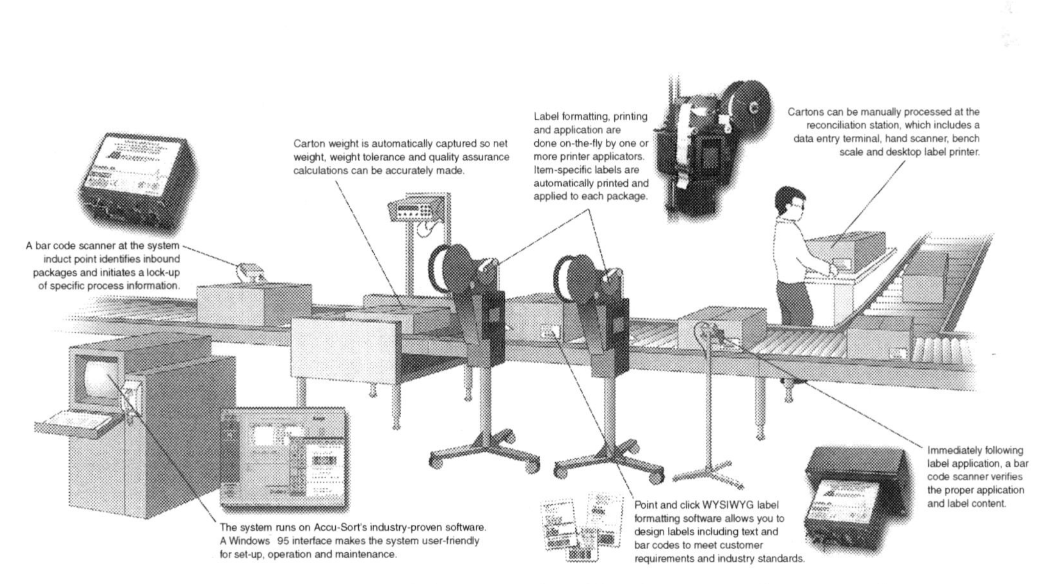

Figure 11–78 Computer-integrated labeling and verification system (courtesy of Accu-Sort Systems, Inc.)

QUESTIONS

1. What are the four basic classifications of material handling equipment?
2. What is a pneumatic conveyor and why would it be classified as fixed-path equipment?
3. A problem dealing with unloading heavy bags of material was discussed in the previous chapter. Did you discover any equipment in this chapter that may resolve this problem? Discuss your solution.
4. What is a magnetic conveyor? What are the advantages and disadvantages?
5. What is a manipulator?
6. Explain cross-docking. Are there any advantages to cross-docking?
7. Do you agree with the statement, "Automated material handling systems are the solution to all material handling problems"? Explain your answer.

CHAPTER 

Office Layout Techniques and Space Requirements

OBJECTIVES:

Upon the completion of this chapter, the reader should:

- Understand the importance of office planning
- Be able to identify various types of office space and advantages and disadvantages of each
- Understand the systematic approach to office planning
- Be able to perform office space calculations

The office layout process is very similar to the manufacturing plant layout procedure. Many of the techniques used to study the flow of material are used to study the flow of paper, information, and people in an office. The activity relationship diagram, worksheet, and dimensionless block diagram studied in Chapter 6 are even more useful in office layouts because there is less variation in the size of the offices than there is in the dimensions of the manufacturing departments. The collection and the analysis of data as discussed in Chapters 2, 3, and 4 will be an important part of the office layout. Instead of studying material flow, we will focus on information and paperwork flow. An office layout designer must learn and understand office systems and procedures in order to establish proper placement of offices. Thus, this chapter will also discuss a systems and procedure analysis technique.

Who works in the office, what tasks are performed in the office, how people are organized into departments, and how these departments relate to each other are all extremely important questions to keep in mind when creating an office layout. An *organizational chart* is an informative tool used to communicate the relationships among the departments and their people.

GOALS OF OFFICE LAYOUT DESIGN

The goals of office layout design will help the designer keep on track and will give the designer a way to evaluate the many alternatives. Some of the most common goals are as follows:

1. Minimize *project cost.* Cost consciousness is important. The layout planner must be responsible for recommending facilities that are cost effective. Buying the cheapest desk may not be cost effective if you need to replace it soon. There is value to good-looking facilities for customer and employee morale and attitude. Cheaper furniture seems to have harder surfaces that add to noise. Being cost conscious means that you want your money's worth and are willing to shop around for the best facilities for the money.
2. The *productivity* of employees is important. You do not want them walking long distances, performing useless work, and using slow equipment, all of which can make life unpleasant. You want to promote an effective use of your employees.
3. Office layouts must be *flexible.* One fact is certain—office layouts will change. Designers must have the ability to expand or shrink overnight. We will discuss specific flexible furniture later in this chapter.
4. *Cleaning* and *maintaining* office space is costly. The type of layout and the equipment you buy will affect this cost.
5. *Noise* must be kept to a minimum. The fabrics on the walls, floors, and ceilings affect the noise level.
6. *Material flow* (paper and supplies) as well as people flow distances must be held to a minimum. The farther they walk or move material, the more the cost is. Good flow analysis will minimize these distances.
7. Create a *pleasing atmosphere* in which to work in order to promote pride and productivity.
8. Minimize *visual distractions.* Panels and furniture can be used to provide at least semiprivate offices.
9. Create a *pleasing reception area.* First impressions or opinions of the company are produced in the visitors' reception area. Does the area appear organized, efficient, and neat, or sloppy and disorganized?
10. *Energy costs* can be affected by the layout and must be minimized wherever possible. Windows, full walls, doors, and the like will all affect energy costs.
11. Each employee requires adequate *work space* and *equipment.* Office layouts must address every office worker's needs.
12. Provide for the *convenience* of employees. Restrooms, lockers (or coat racks), lunchrooms, and lounges must be conveniently located to prevent long trips away from their offices.
13. Provide for the *safety* of employees. Aisle sizes, stairways, machines, and clutter can cause safety problems. The layout plan must consider the safety aspects of the office.

TYPES OF OFFICE SPACE

Office layouts vary in complexity from supervisors' stand-up desks located in the middle of a production department to an office complex housing hundreds of office employees. Office space costs more per square foot than does manufacturing or distribution space, so space use is very important. The median cost per square foot of office space ranges from $75 for low-rise structures (1- to 4-story), $78 for mid-rise (5- to 10-story), to $100 for high-rise (11- to 20-story) office buildings. Of course, other factors such as the size of the city and the location will greatly influence the cost. The advantage of being "uptown," "downtown," or in highly populated areas is being close to many other businesses, transportation, communications, and services. The disadvantages are congestion and costs. Many corporate offices are located in major business centers for convenience to other businesses, but their manufacturing plants and supporting offices are located in rural areas where space costs (and living costs) are usually less. Our discussions are focused on manufacturing plant office layouts instead of corporate offices, but the techniques and processes are the same.

Supervisors' Offices

Manufacturing plant supervisors' offices are good starting points for office discussions because they are small and a "feel" for space can be developed early. A 10 × 10-foot portable office located in the middle of a production department is shown in Figure 12–1. These portable offices can be moved as one large unit. Air conditioning units are attached to one wall because a closed area that is so small can soon become very uncomfortable. Shipping, receiving, and maintenance as well as production supervisors could use this type of office construction.

Supervisors should be located where they are immediately accessible to their employees. Having a line-of-sight view can improve communication. Supervisors also need to meet with employees in confidence at times and this type of office provides the necessary privacy. Discipline should always be carried out in private. If private offices are not available, conference rooms must be provided where private meetings can be held.

Some supervisors use stand-up desks located in the middle of their production area. Figure 12–2 shows such a facility. A stool may also be provided, but the stool should be high enough to allow the supervisor to work either standing up or sitting down.

Open Office Space

Open office space (see Figure 12–3), also called *bull pen,* is large rooms that house many people. Open offices are very popular for the following reasons:

1. *Communications* are easier. To talk with someone requires only the action of lifting the head and speaking. To know if someone across the room is available, just look.

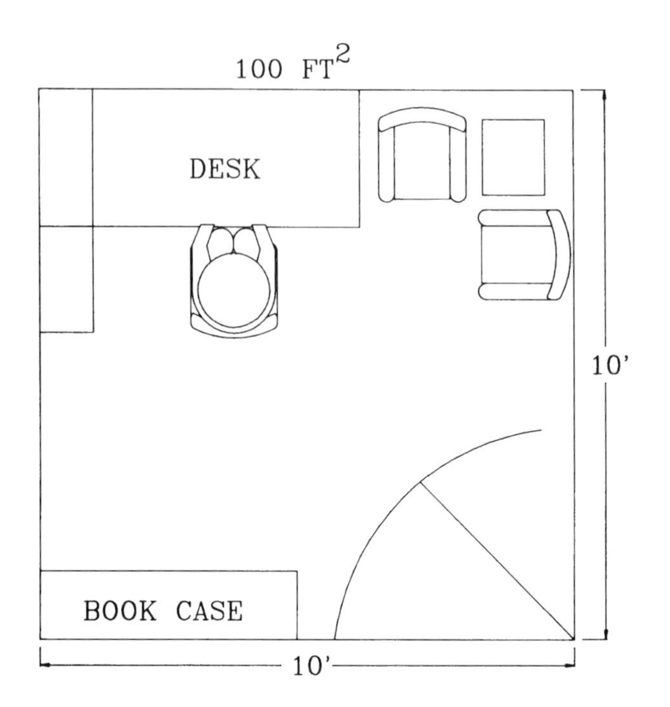

Figure 12–1 Supervisor's office (courtesy of Globalindustrial.com).

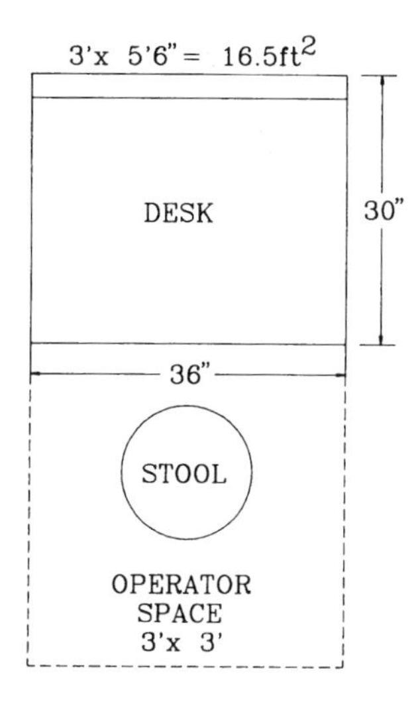

Figure 12–2 Stand-up desk.

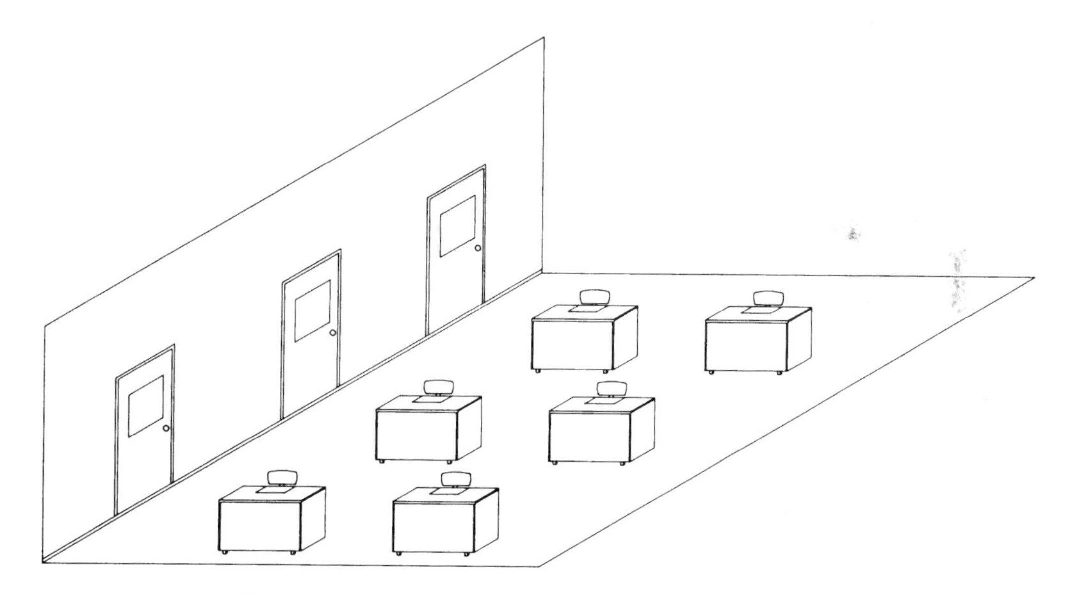

Figure 12–3 Open office (bull pen).

2. Common *equipment* is accessible to more people.
3. Less *space* is required, as compared to private offices.
4. *Heating, cooling,* and *ventilation* costs and problems are minimized because one big room is easier to work with than the same area divided into private offices. Walls are the enemy of good circulation. Open office construction eliminates walls.

5. *Supervision* of the people in an open office is easier. Doors and walls make supervision more difficult.
6. *Layout changes* are quicker and less costly in open offices. Moving desks around a large room is easier than negotiating aisles and doors.
7. *Files* and *literature* are accessible to all requiring fewer files and copies of magazines and journals.
8. *Cleaning, vacuuming,* and *sweeping* work is reduced.

The disadvantages of the open office concept include the following:

1. The *lack of privacy* is probably the biggest problem with open offices. Fellow employees can, very innocently, interrupt the thinking or concentration on a difficult task that then requires starting over. If people are too close and accessible to each other, nonbusiness discussions can eat up great amounts of time. This process is called *coffee klatching.* It can decrease productivity and quality and must be discouraged.
2. *Noise* is another problem with open offices. Equipment that produces most of the noise can be isolated for noise control, but open offices are noisier than private offices.
3. Open office space does not have the *status* that a private office carries. The recruitment of a good potential employee may be missed because of the office space quality.
4. *Confidentiality* of some work may require private space.

The choice of open office space or private office space depends on balancing the advantages and disadvantages for each position. Each company can have both open and private offices, but who gets a private office is an important decision that cannot be made without high-level planning.

Conventional Offices

Conventional offices, also known as *fixed walled offices,* are the opposite of open offices. A conventional office has independent furniture, four walls, and a door. More than one person can be assigned to an office, and at what point it becomes an open office is a little unclear to most layout designers, but if more than one function is performed in this space, it is an open office. A *function* may be accounting, purchasing, personnel, engineering, data processing, sales, or production. Conventional office layouts are older than open offices, but both can be improved upon. A combination of open office concepts and conventional office advantages would allow for the best of both techniques. We will call this the *modern office concept.*

The Modern Office

The modern office design concept (see Figures 12–4 and 12–5) tailors individual work areas to satisfy the needs of the organization. The modern office will provide private office space where needed without negatively affecting the cost of utilities, maintenance, and accessibility. Figure 12–4 shows modern offices and Figure 12–5 a typical layout. Notice the equipment:

1. Panels do not go to either the ceiling or the floor, allowing air to circulate. Panels are padded with soft material to hold down noise.

Figure 12–4 Modern offices (courtesy of American Seating).

Figure 12–4 (continued) Modern offices (courtesy of American Seating).

2. Cabinets are built into the panels to make a better use of the space over desks and tables.
3. Tables are built into the panels to save space and costs.
4. Drawers under the tables allow for storage of supplies just as a desk would.
5. Utility (electrical, computer, and phone) lines can be carried in the panels. This will give the office a cleaner look and also improve safety.

Modern offices can be arranged and rearranged to meet the changing needs of the organization. Organizations develop teams to solve problems. Although these problems and the makeup of teams continue to change, the office needs must be met. The modern office is very flexible. When the company moves, the walls move too.

Modern offices have been described using various terms such as *clustered offices, landscaped offices,* and *free-standing offices.* Whatever the term, the purpose of modern office design techniques is to eliminate the disadvantages of open offices and traditional offices and to promote cost-effectiveness in the long run. Figure 12–6 shows a comparison between conventional and modern office space.

Modern offices should be pleasing to look at, convenient for the users, comfortable, and efficient. The justifications vary from employee relations, to customer opinion, to cost-consciousness.

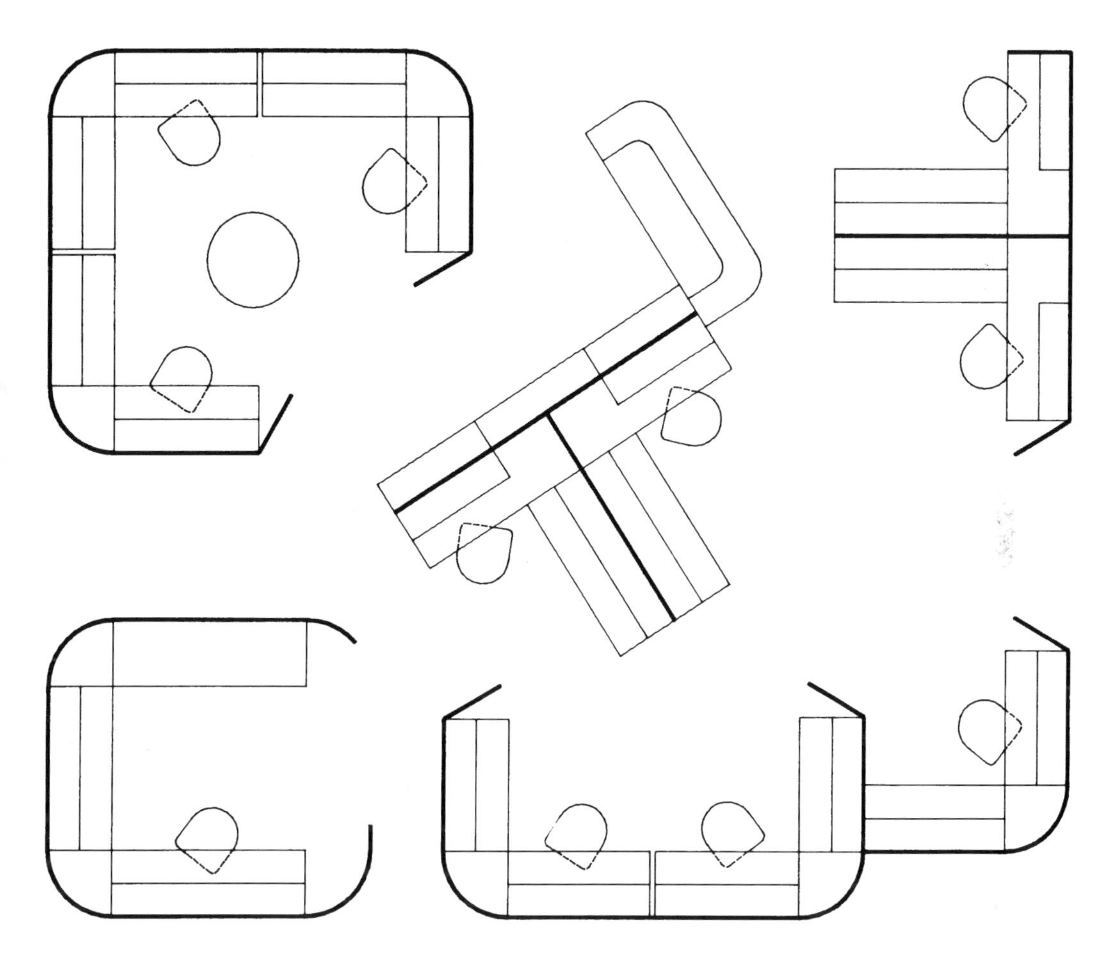

Figure 12–5 Modern office layout.

SPECIAL REQUIREMENTS AND CONSIDERATIONS

Keep these points in mind when designing offices:

1. *Privacy* may be required by some office employees. Personnel problems should be discussed in private. Many financial matters are confidential. Corporate planning can consider many alternatives that will never come to pass, so to prevent harmful rumors, privacy is needed.

2. *Point of use storage* is a layout principle that requires the storage of supplies close to the point of use. Office supplies vary from department to department. Engineering supplies are not the same as accounting supplies; personnel forms are not at all like purchasing forms. Therefore, every office department needs a supply room or controlled area. In small offices, a desk drawer may be used, but

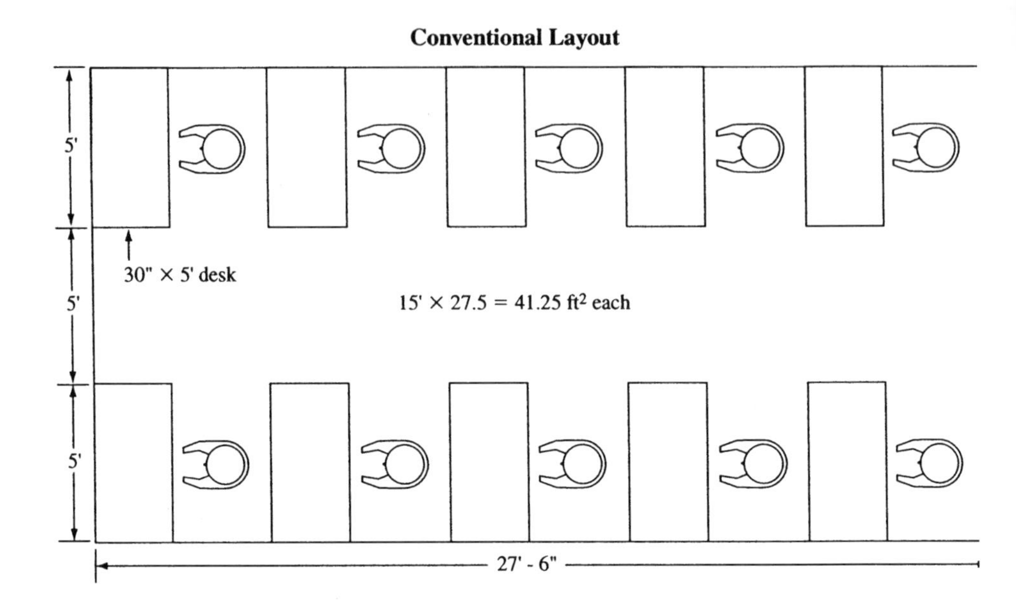

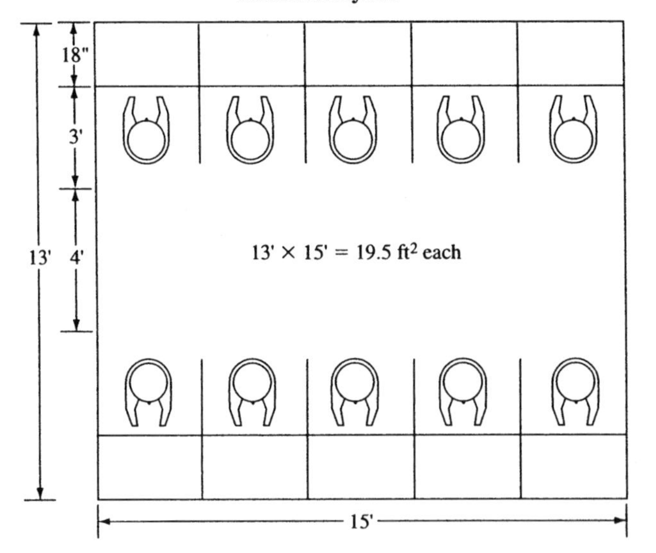

Figure 12–6 Conventional space vs. modern space—Ten data entry clerk workstations.

Note: At a rental rate of $25.00 per square foot per year, the conventional space would cost $5,625.00 per year, a savings of $4,687.50 per year.

in large facilities, large controlled areas are needed to handle these valuable commodities.

3. Offices in manufacturing plants often have a *second floor.* Typically, an office is built within the manufacturing plant. The ceilings of a manufacturing plant are often 20 feet or higher, so using only one floor is a waste of the building cube. Building a second floor is good cube utilization. The functions of the department placed on the second floor should not require outside visitors or much traveling during the day. Personnel, purchasing, and sales have a lot of visitors, so they would be on the ground floor. Engineering, accounting, marketing research, and data or order entry (phone) do not have as many visitors, so they could be on the second floor.

4. *Centralized* or *decentralized?* Where do you place offices? Where they are needed may be the best answer. The question usually is, do you have one big office in the front of the building (centralized) or several smaller offices throughout the plant? The advantages to a centralized office are

a. Single office area construction, including common air conditioning and other utilities and a block wall
b. The convenience of having all office people in one area
c. The convenience to outside visitors without disturbing production
d. Common files and equipment

The disadvantages of a centralized office are that it is not convenient for other operations departments, such as receiving, shipping, maintenance, stores, warehousing, and production, which all have important relationships with the office.

5. Office *flexibility* is an important consideration from the very early stages of planning. When building an office, immediate consideration should be given to expansion. Footings and supports for a second story will be much cheaper if they are installed in the beginning. If you have to break up floors and walls in order to build a second floor, you will never expand upward. Walls must be flexible also—most offices will grow. Partitions and panels are better (more flexible) than walls for this reason. Utility flexibility is also important. Several methods are used to create utility flexibility.

a. *Q floors* are like corrugated metal placed on the ground before the concrete for office floor is poured. The Q floor allows for the running of electricity, computer cable, telephone lines, and the like every fourth foot for the full length of the office (see Figure 12–7). If a desk is moved, the old plug-ins can be plugged up and new plug-ins added.
b. *Drop ceilings* and hollow panel supports keep the utility cords and cables concealed overhead in the ceiling. They can be dropped anywhere.

6. *Conference rooms* can be used to provide privacy when required in open office areas. Privacy may be needed by supervisors holding disciplinary sessions or by salespeople with customers. The important thing is that privacy is available in open office layouts. Boardrooms are special conference rooms where the board of

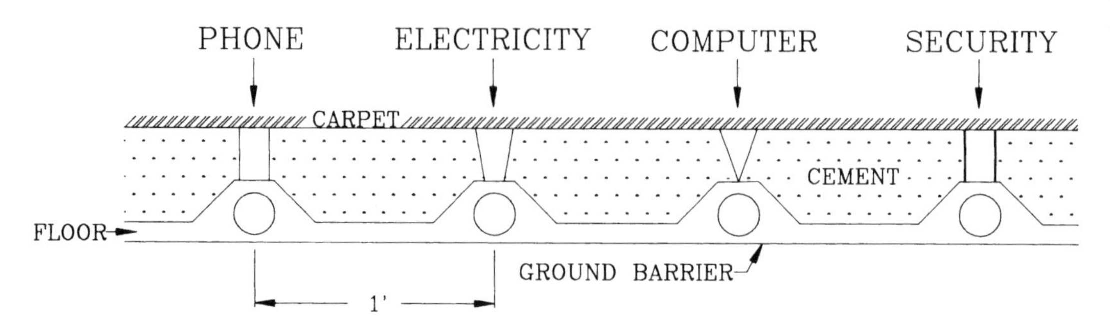

Figure 12–7 Q flooring.

directors of a public corporation meet and, therefore, must be laid out for privacy and noise reduction.

7. *Libraries* are special needs areas where reference books and magazines are kept. This is a cost reduction idea. Instead of purchasing books for individuals to put in their bookcases, books are purchased by the library and kept in a central, convenient area. A *Wall Street Journal* subscription costs about $200 per year. If 10 managers can share one copy, $1,800 per year can be saved. The paper is then available to everyone else as well. Professional journals, handbooks, and many other publications all add to the value of a reference library.

8. A *reception area* is the visitors' center. The front door of your company is where visitors enter. A receptionist will greet the guests and ask how they may be helped. While the receptionist searches out assistance, the guests need a place to wait. This should be a comfortable, attractive area to create a favorable opinion about the company. The best way to make a good impression is not to keep the guest waiting. But unannounced visitors may need to wait, so this area should be equipped with chairs, desk, telephone, magazines, and company information. Product displays are an excellent means by which new or existing products can be introduced to visitors and allow visitors the opportunity to visualize how the products might be used.

9. *Telephone systems* are becoming automatic, but some personal attention is always necessary. If the volume of the incoming calls is not too great, the receptionist can handle the telephone as well as the reception area. Telephone equipment requires space. The central board and exchange may be big enough to have their own room, but these areas can be remote and out of the way.

10. *Copy and fax machines* can be major pieces of equipment. This equipment needs special material, operation instructions, and a clean environment. Like every other piece of equipment, a workstation layout is needed. Smaller copiers may be a part of a small office, but larger reproduction centers may be stand-alone

departments. Storage areas, work-in-process areas, and finished work storage areas may also be required.

11. Incoming and outgoing mail can be big business. A company's mail comes into the *mailroom* and is sorted. Mail is then either delivered or picked up by employees. Outgoing mail will require postage, weighing, and sometimes folding inserts, stuffing envelopes, and sealing. These are called *mass mailings*. Special equipment is available to do this automatically, so the mailroom layout will need equipment layouts as well.

12. Companies create and receive many kinds of documents. Legal requirements force companies to keep many of these documents for years. This creates a need for *file storage areas*. Also, blueprints, processing information, purchase orders, and the like are required by many people. Central files reduce the need for many copies. Computers and microfilm are reducing the space requirements and the configuration of file rooms, but file rooms are still necessary.

13. A *word processing pool* is a group of clerical workers or secretaries in a central area who receive work from many sources. This is an alternative to private workers or secretaries and is, in general, a more efficient use of people.

14. *Aisles* are big users of space. In open offices, the smallest aisles are 3 to 5 feet and the larger aisles are 6 to 8 feet. The traffic during peak periods will determine aisle sizes.

15. More equipment and systems are being controlled by a *computer* every day. Main frames and central processing units are kept in special temperature- and humidity-controlled rooms. Computer security is also important.

16. Other areas and considerations to keep in mind are (a) lighting, (b) vaults, (c) standardization, and (d) expansion.

TECHNIQUES OF OFFICE LAYOUT

The techniques used for creating an office layout are the

1. Organizational chart
2. Flowchart (systems and procedures analysis)
3. Communications force diagram
4. Activity relationship diagram
5. Activity worksheet
6. Dimensionless block diagram
7. Office space determination
8. Detailed master layout

Analyzing organizational needs, paperwork flow, who works with whom, and the relationships among departments lead to a master plan. Each technique will be described in detail in this section. Follow the procedures given and the result will be an efficient and effective layout.

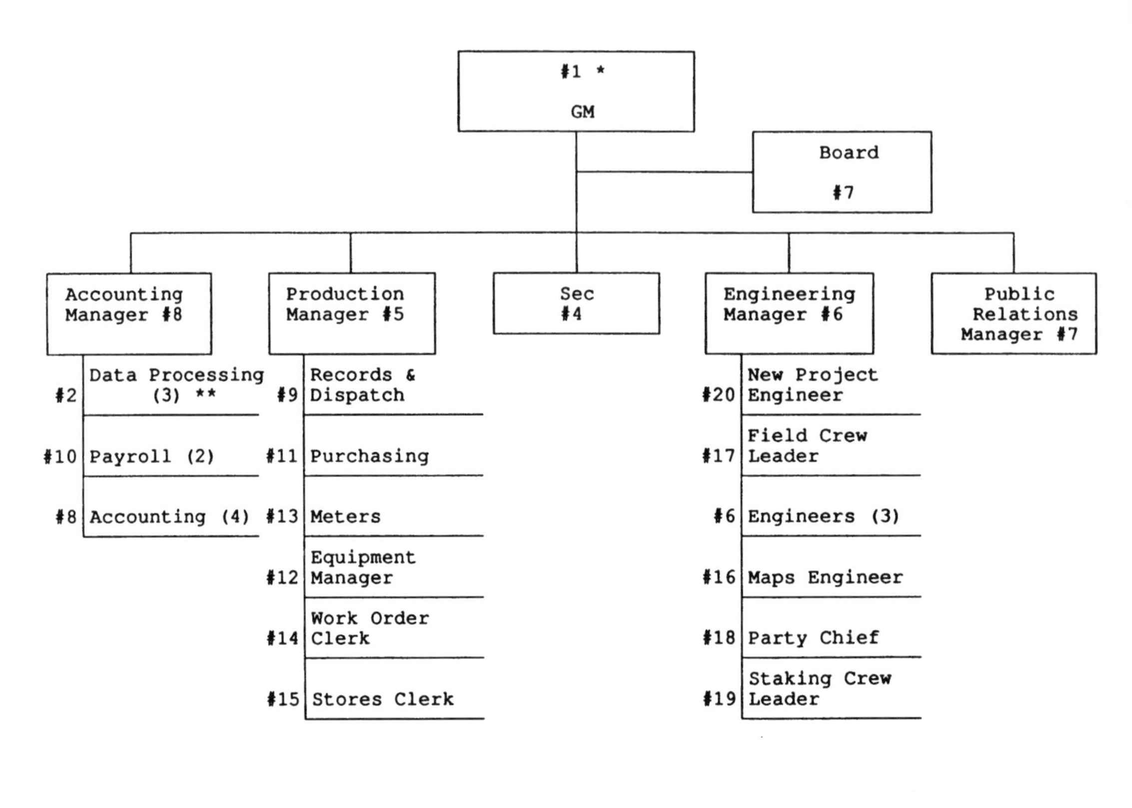

Figure 12–8 Organizational chart—Electrical power co-op.

Organizational Chart

The organizational chart (see Figure 12–8) gives the layout planner an idea of the office size. The organizational chart tells us how many people work in each area and level of the company. Each department must be considered and the space determined. The total number of people is the best indication of office size required. A rough estimate of office space needs can be calculated by multiplying the number of people requiring office space by 200 square feet each. For example, the overall size of an office for 100 people would be 20,000 square feet. This is a good initial planning tool, but it should be used only for determining the total office space, not the departmental space.

Who works in the office, who reports to whom, how many people are in a department, what functions are performed, and similar questions are all covered in the organizational chart. Determining the number of people at each level of the company is another way of calculating space.

Employees	*Square Feet*
General managers and senior executives	200–300
Managers	150–250
Supervisors	100–200
Accountants	75–150
Engineers	100–150
Clerks	75–100

Flowchart

Procedures diagramming is very much like flow process charting, but instead of following the flow of a product, you follow the flow of every copy of a form. Sometimes, you follow many forms (such as the example in Figure 12–10) because one form causes the creation of another form, and so on. To analyze paperwork flow, the *procedures diagram* technique, or flowcharting, was developed. Standard symbols (process chart symbols) have been developed to help explain the standard steps such as those shown in Figure 12–9.

Figure 12–10 illustrates the purchase order (P.O.) procedure. The people or departments are listed down the side. Starting with the requestor asking for something (submitting a purchasing request) and receiving approvals, the purchase order is created and the copies are sent to four other areas (one copy is filed with the requisition in an open file in purchasing). One copy goes to the requestor, one to accounting, one to receiving, and one copy to the vendor supplying the item. Once the order is shipped and received, a copy of the vendor's packing list is matched with the purchase order and a receiving report is made. Five copies of the purchase order, two copies of requisitions, four copies of receiving reports, a packing list, and an invoice all have to end up in the files.

Figure 12–10 shows the movement of purchase order forms around the office. This movement has an effect on the office layout. When all the forms are analyzed, relationships between departments will become clearer and relationship codes can be developed. A from-to chart could be developed resulting in a most efficient layout. The from-to chart has not been included in this section, but it could be the best tool for optimizing paperwork flow.

Communications Force Diagram

The *communications force diagram* is another way to determine office relationships (see Figures 12–11 and 12–12). The procedures diagram (flowcharting) method requires analyzing all paperwork flow and diagramming all the procedures. This can be such a big job that years of analysis may be needed. The results of flowcharting will be extremely valuable, but for office layout needs, the communications force diagram is much faster.

The communications force diagram requires office planners to talk with each person involved in the office and to find out with whom they work the most. Each

Operation	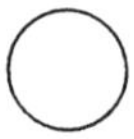 =	Perform some function such as match, review, fill orders, input data, and so on.
Form	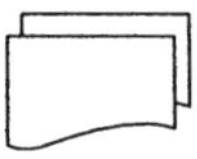 =	Generate a form or a document. A special operation. If more than one copy is used, another page is shown behind the first page for each copy.
File	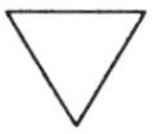 =	File documents. A "T" could be placed inside the triangle to indicate a temporary file or a follow-up file and a "P" could be used for permanent file or completed file.
Transportation	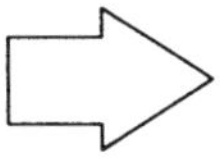 =	Physically moving something such as material (not paperwork).
Decision	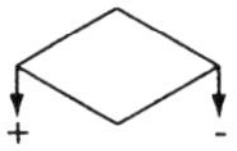 =	Yes/No, Go/No Go at any point where the direction of flow might change.
Approval	=	Used when management approval is required.
Paperwork Flow	 =	Shows flow of information.
Telephone	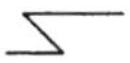 =	Shows flow of information by telephone or computer.
Processing	 =	Used for computer processing.
Delay	 =	Indicates delay in the process, such as waiting for approval.

Figure 12–9 Flowchart symbols.

person you talk with will be the center of the diagram, and each person she works with is placed on the periphery (see Figure 12–11). The number of lines connecting the subject person to the periphery people will indicate the importance of the relationship as follows:

1. If there are *four lines,* it is absolutely necessary that these two people be close together. This code should be reserved for people who communicate several times an hour. This will be coded an A relationship.
2. If there are *three lines,* it is especially important that these two people be close to each other. This code should be reserved for people who need to communicate with each other at least once an hour. This will be coded as an E relationship.

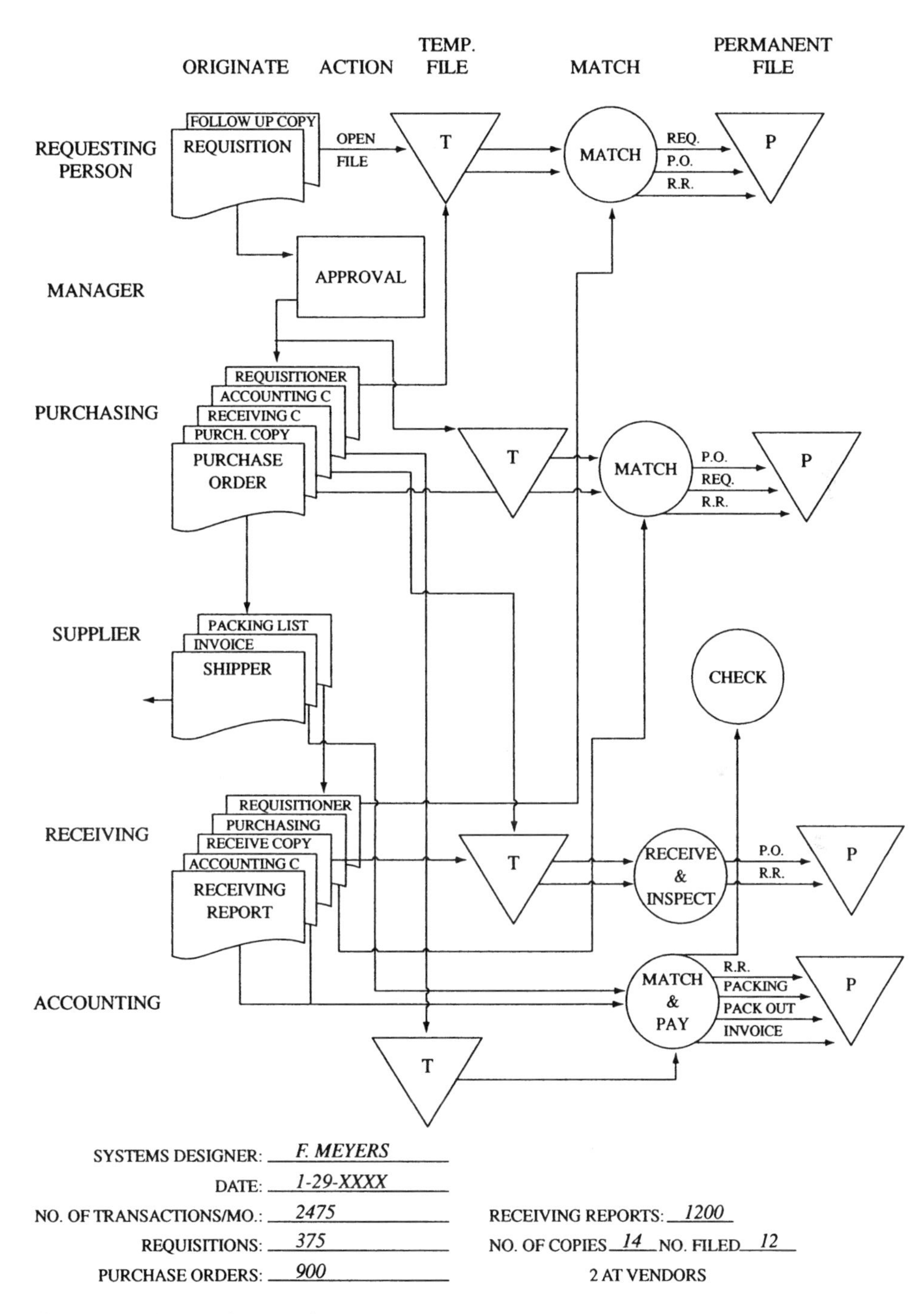

Figure 12–10 Purchase order payment system.

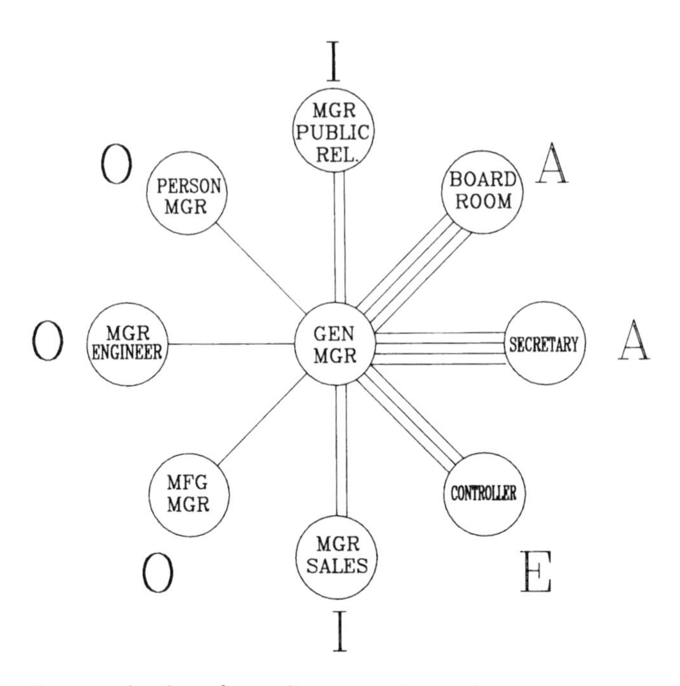

Figure 12–11 Communications force diagram—General manager.

3. If there are *two lines*, this is an important relationship and these two people should be close. This relationship is reserved for people working with each other several times a day. This will be coded an I relationship.
4. If there is *one line*, this is an ordinary relationship and is reserved for people who interface on a daily basis. This will be coded an O relationship.

Figure 12–11 is a communications force diagram for one person. With an office of 22 people, 22 communications force diagrams will be needed. These 22 diagrams must be summarized into one large diagram with 22 circles with all the lines between every circle. Figure 12–12 shows an example of this. Notice the *long* lines. These are departments that you need to put closer together. Those people or departments that have the most contact outside the office are on the perimeter of the diagram; people with a lot of contact within the office are located in the middle of the diagram. You may think of the lines between the individuals or offices as rubber bands or forces that pull these individuals closer to each other. Therefore, the longer the number of the lines, the stronger is the force pulling these offices close together. The relationships established here will be carried forward to the activity relationship diagram.

Activity Relationship Diagram

The activity relationship diagram was discussed in Chapter 6. In brief, it shows the relationship of every department or person with every other department or person. A simple code (*A E I O U* or *X*) is used to tell the importance of the relationship (see

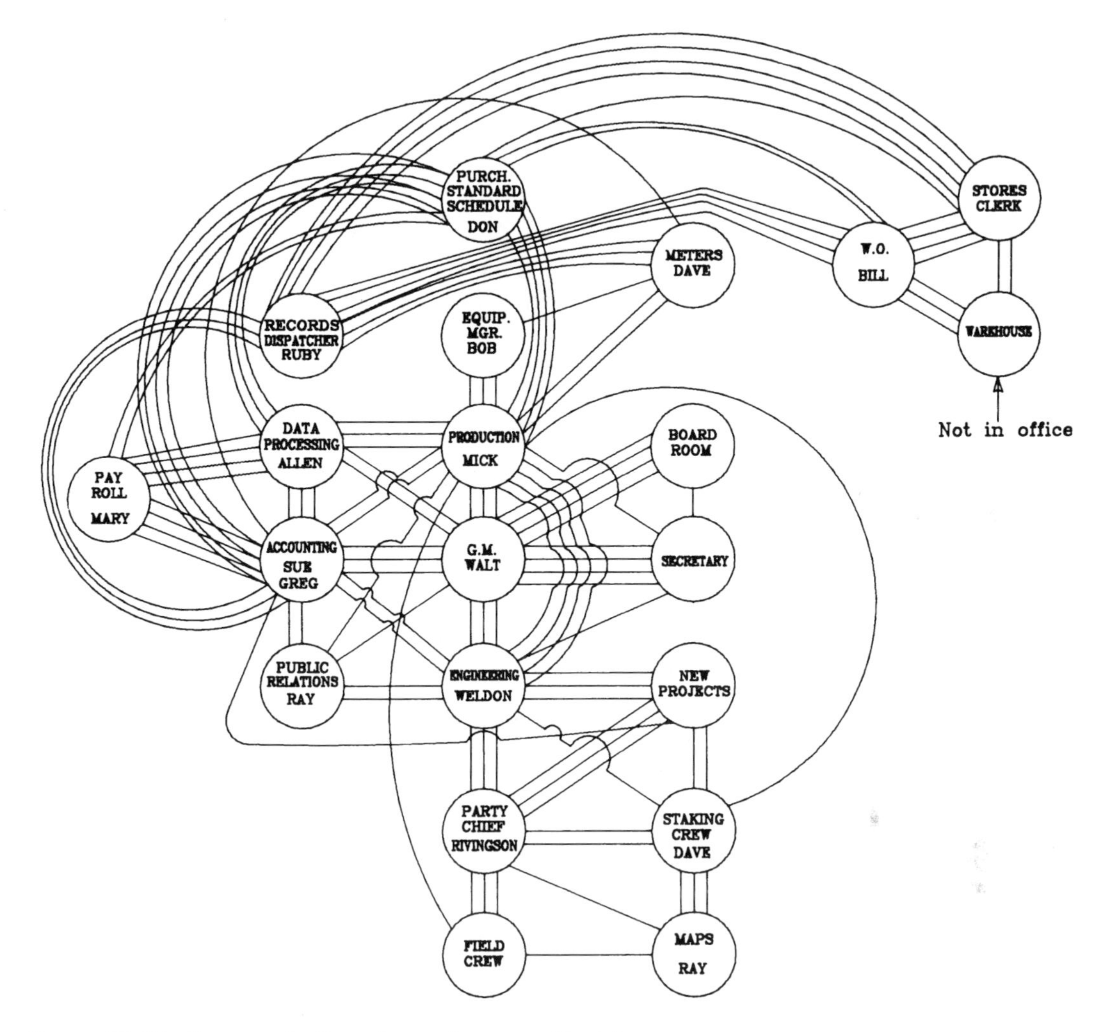

Figure 12–12 Communications force diagram—Electric power co-op.

Figure 12–13). In the office layout, the communications force diagram is used to establish these important codes—or you can talk to each department or person included in the study and have each one record the codes. Figure 12–13 was developed from the communications force diagram (see Figure 12–12).

Activity Worksheet

The activity worksheet was also discussed in Chapter 6. The data in Figure 12–14 was taken from the chart in Figure 12–13 to create 20 individual blocks. This worksheet moves from the activity relationship diagram to the dimensionless block diagram.

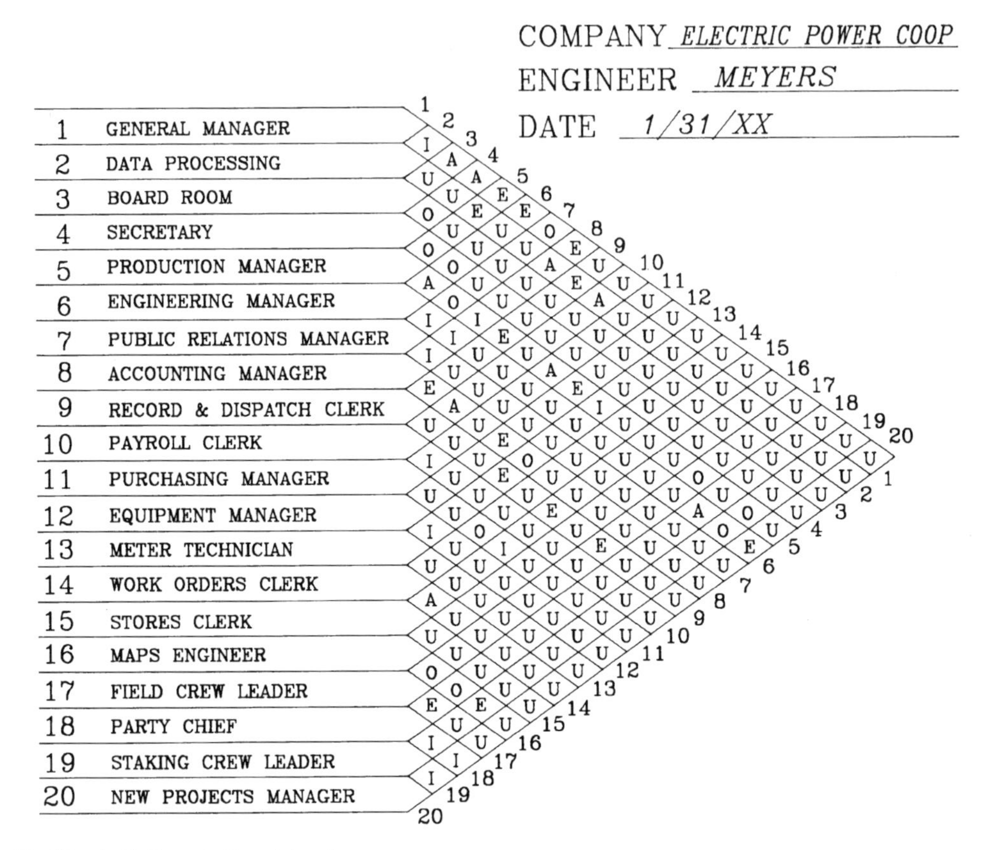

Figure 12–13 Activity relationship diagram.

Dimensionless Block Diagram

To create a dimensionless block diagram, cut out 20 square paper blocks of about 2 × 2 inches. On the worksheet, starting at line 1, place the number of the line and the name of the department in the middle of the block (see Figure 12–15). Then, starting at the top left-hand corner, place the A relationships that this department has with other departments. For example:

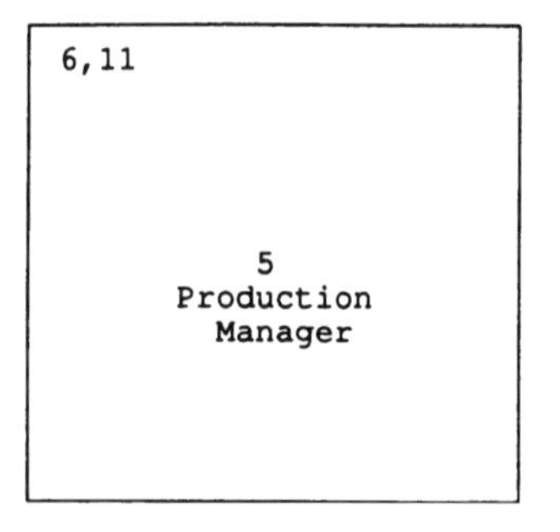

	ACTIVITY	DEGREE OF CLOSENESS					
		A	E	I	O	U	X
1.	GENERAL MANAGER	3,4	5,6,8	2	7		
2.	DATA PROCESSING	8,10	5,9	1	–		
3.	BOARD ROOM	1	–	–	4		
4.	SECRETARY	1	–	–	3,5,6		
5.	PRODUCTION MANAGER	6,11	1,2,9,12	8,13	4,7,17,19		
6.	ENGINEERING MANAGER	5,18	1,20	7,8	4.19		
7.	PUBLIC RELATIONS MANAGER	–	–	6,8	1,5		
8.	ACCOUNTING MANAGER	2,10	1,9,12	5,6,7	13		
9.	RECORDS & DISPATCH	–	2,5,8,13,14,17	–	–		
10.	PAYROLL CLERK	2,8	–	11	–		
11.	PURCHASING CLERK	5	–	10,14	15		
12.	EQUIPMENT MANAGER	–	5,8	13	–		
13.	METER TECHNICIAN	–	9	5,12	8		
14.	WORK ORDER CLERK	15	–	11	–		
15.	STORES CLERK	14	9	–	11		
16.	MAPS ENGINEER	–	19	–	17,18		
17.	FIELD CREW LEADER	–	9,18	–	5,16		
18.	PARTY CHIEF	6	17	19,20	16		
19.	STAKING CREW LEADER	–	16	18,20	5,6		
20.	NEW PROJECTS MANAGER	–	6	18,19	–		

Figure 12–14 Worksheet for activity relationship diagram—Electric power co-op.

Now place the E relationships on the top right corner, the I relationships in the bottom left, and the O relationships on the bottom right. Finish all 20 blocks.

When all 20 blocks are completed, find the one block with the most important (A's and E's) relationships and place it in the middle of your desk.

Now place the most important offices around this center office until the A relationships are satisfied. When you place offices 6 and 11, they will have A relationships to satisfy. Keep working with the A relationships until all offices with an A relationship have a full side contact with each other. Now you can start working with the E relationships, the I relationships, and finally, the O relationships. Although you may try to accommodate the O and I relationships, they are often prevented from being close due to many other important relationships. As you may try many different layouts, be sure you keep track (by making a small plot plan) of your layouts before going on to a change. There are hundreds of possibilities, and the one best answer is the one that satisfies the most relationships.

Once you have the final dimensionless block diagram, you can identify where outside walls, the shops, and warehouses or stores departments (nonoffice) will be, so an orientation can be developed. This will be your plan for where each office goes.

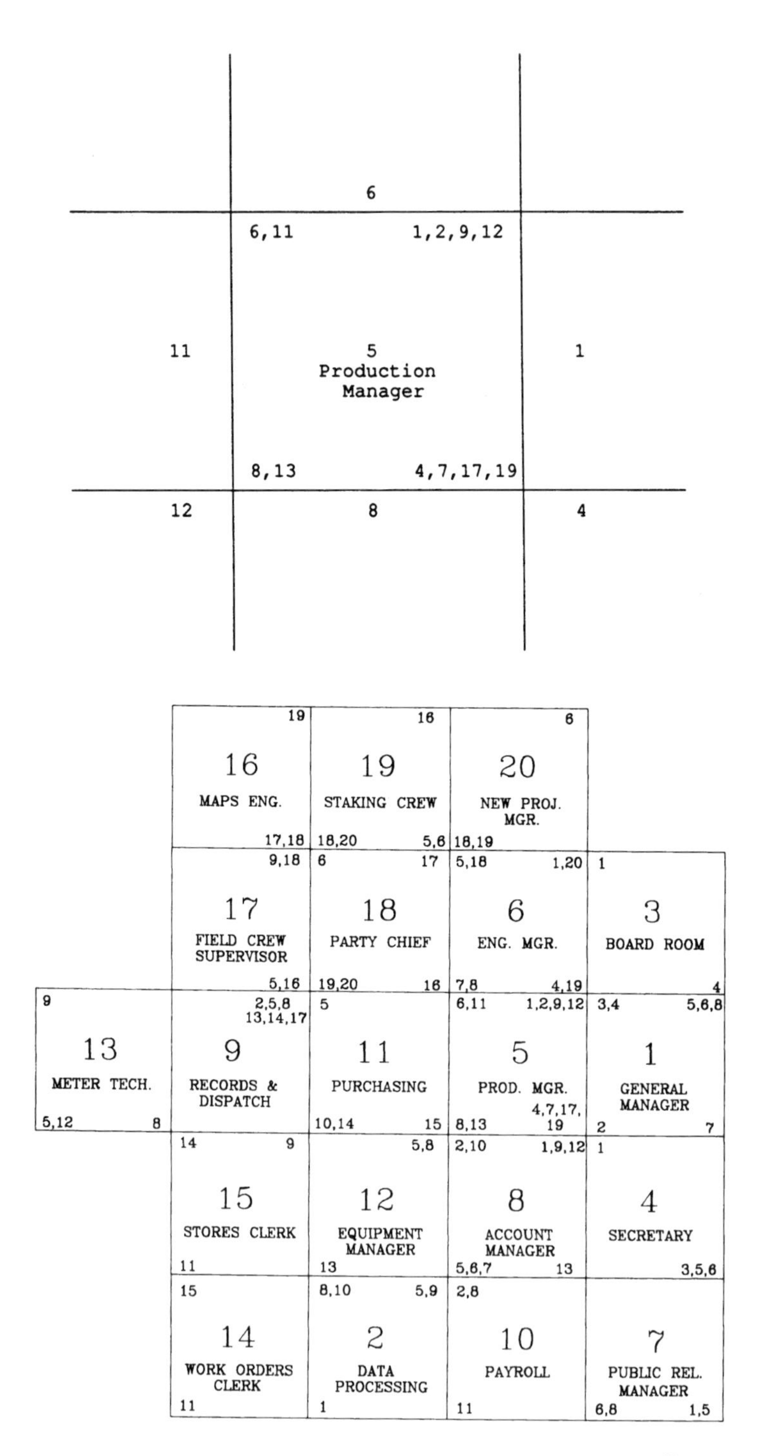

Figure 12–15 Dimensionless block diagram—Electric power co-op office.

Office Space Determination

To calculate the space requirement for the office, use the following techniques:

1. *The 200 square feet per person technique.* The 200 square feet per person technique is used for establishing the overall office space. If you look at Figure 12–8 (the organizational chart for the Electric Power Co-Op), you will count 36 people requiring office space, so 36 × 200 = 7,200 square feet. So, the office needs to be 7,200 square feet.

2. *The level of the organization technique.* Looking at Figure 12–8, you find the following information:

No. of People	*Position*	*Sq Ft Each*	*Total Sq Ft*
1	Senior executive	250	250
4	Manager	200	800
4	Clerical	100	400
9	Accounting	100	900
5	Engineer	125	625
6	Supervisor	150	750
	100 percent extra space allowance		3,725
		Total	7,450 ft^2

Therefore, you need a total of 7,450 square feet.

3. *The workstation technique.* The workstation layout approach is the most detailed and will include restrooms, lockers, cafeterias, reception areas, boardrooms, conference rooms, and anything else that takes up space. This technique adds 25 percent extra space for expansion. A full layout is not required before construction begins. Square footage of an office is simply the length multiplied by the width of the office measured in feet. Thus, a 20 × 20-foot office would be 400 square feet.

Detailed Master Layout

No matter which technique you use, you need to know the length and width of the office for your layout. These measurements may be enough for construction to begin, but more detail will be required. The department layouts are the next level of detail, which will include the internal walls or department boundaries. The final level of detail is where desks, chairs, and all the other equipment are placed. This detailed plan will be needed before space assignments can be achieved. A comprehensive guide and frame of reference for making sound real property decisions, especially in regard to office space planning, is issued by the General Services Administration (GSA). These asset management principles were developed to guide the federal government's real property ownership enterprise. GSA, office of Governmentwide Policy, Office of Real Property, using collaboration, partnership, and customer involvement, developed a set of recommendations for management of the federal real property portfolio. The use of this document for any serious office space planner and developer is strongly recommended. The most current version of this document is posted online by the Office of Governmentwide Policy under the subject heading "Space Use Study" at http://www.gsa.gov.

In creating your detailed master layout, keep in mind the following rules:

- Desks should face the same general direction.
- In open areas, desks should be placed in rows of two.
- For desks in one row, there should be 6 feet from the front of one desk to the front of the desk behind it.
- For desks in rows of two or more and where ingress and egress are confined to one side, 7 feet should be allowed from the front of one desk to the front of the desk behind it.
- If employees are back to back, allow a minimum of 4 feet between chairs.
- Inside aisles within desk areas should be 3 to 5 feet wide.
- Intermediate aisles should be 4 feet wide.
- Main aisles should be at least 5 feet wide.
- Natural lighting should come over the left shoulder or the back of an employee.
- From 50 to 75 square feet are required for a work space consisting of a desk, shelf space, a chair, with a 2-foot space allowance on the length and width.
- Desks should not face high-activity aisles and areas.
- Desks of employees doing confidential work should not be near entrances.
- Desks of employees having much visitor contact should be near entrances with extra space provided.
- The desk of the receptionist should be near the visitor's entrance.
- Supervisors should be positioned adjacent to the clerical workers.
- Supervisors in open areas should be separated from their group by 3.3 feet.
- The flow of work should take the shortest distance.
- People who have frequent face-to-face conferences should be located near each other.
- Employees should be adjacent to those files and references they use frequently.
- Employees should be placed near their supervisors.
- Five-drawer file cabinets should be considered in lieu of four-drawer cabinets.
- Open-shelf filing or lateral file cabinets should be considered in lieu of standard file cabinets.
- Four- or five-drawer file cabinets should be considered as a substitute for 2 two-drawer cabinets.
- The reception area should create a good impression on visitors and an allowance of 10 square feet should be used per visitor if more than one arrives at a given time.
- The layout should have a minimum of offsets and angles.
- Large open areas should be used instead of several small areas.

- Open areas for more than 50 persons should be subdivided by use of file cabinets, shelving, railings, or low “bank-type” partitions.
- Office space should not be used for bulk storage or for storage of inactive files.
- Conference space should be provided in rooms rather than in private offices.
- Conference and training rooms should be pooled.
- The size of a private office will often be determined by existing partitions.
- Private offices should have a minimum of 100 square feet to a maximum of 300 square feet.
- A 300-square-foot private office should be used only if the occupant will confer with groups of eight or more people at least once per day.
- Related groups and departments should be placed near each other.
- Minor activities should be grouped around major ones.
- Work should come to the employees.
- Water fountains should be in plain view.
- Layouts should be arranged to control traffic flow.
- Heavy equipment generally should be placed against walls or columns.
- Noise-producing workstations should be grouped together.
- Access to exits, corridors, stairways, and fire extinguishers should not be obstructed.
- All governmental safety codes should be followed.
- In planning the office, consider the floor load, columns, window spacing, heating, air conditioning and ventilation ducts, electrical outlets, and lighting and sound.
- The scale of the layout should be either 1/4 inch = 1 foot or 1/8 inch = 1 foot.
- Plastic reproducible grid sheets and plastic self-adhesive templates should be considered.

QUESTIONS

1. What are the goals of office layout?
2. What are the four types of office space?
3. What are the advantages of the open office layout concept?
4. What are the disadvantages of the open office layout concept?
5. List 19 special office requirements and considerations.
6. What are the techniques of office layout?
7. How does the organizational chart help in office layout?
8. How much space is required in the office (rough estimate)?
9. What are the standard symbols of the procedures diagram?
10. What is a communications force diagram?
11. What symbols are used in the communications force diagram?
12. What is the basic source of information to create an activity relationship chart?

CHAPTER 

Area Allocation

OBJECTIVES:
Upon the completion of this chapter, the reader should:

- Understand the concept of the building cube
- Be able to utilize various levels of the building cube for the appropriate function
- Be able to determine the total space requirement and building size
- Be able to utilize various facilities planning tools that were introduced throughout the textbook
- Be able to allocate appropriate space and location for each function within the enterprise

Area allocation is a process of simply dividing up the building's space or allocating space among the departments. To allocate space, of course, you need to know how much space is required. Since Chapter 4, we have been developing space requirements for a toolbox plant. Let us continue with that example in order to illustrate area allocation.

SPACE REQUIREMENTS PLANNING

A total plant size and shape is needed very early in the project in order to design the building. Each department's space needs are analyzed and listed on a *total space requirements worksheet.* The manufacturing space (Chapters 4 and 7), production services space (Chapter 8), employee services space (Chapter 9), office space (Chapter 12), and outside area space (Chapters 8 and 9) are all determined separately and then listed on the worksheet. Figure 13–1 shows a recap of the space requirements

	Stations × *W* × *L*	*(Figure No.)*	= *Square Feet*
I. Manufacturing			
A. Fabrication			
Strip shear	2 × 8.5 × 12	(7-5)	204
Chop shear	4 × 5 × 15	(7-7)	300
Punch press	3 × 8 × 11	(7-6)	264
Press break	6 × 8 × 11	(7-8)	528
Roll former	1 × 6 × 18	(7-9)	108
Fabrication Total:			1,404
B. Spot weld	1 × 26 × 30	(4-12)	780
C. Paint	1 × 28 × 100	(7-11)	2,800
D. Assembly & P.O.	1 × 16 × 38	(4-13)	608
Subtotal			5,592
50 percent allowance (mostly aisles)			2,796
Manufacturing Total:			8,388
II. Production Services			
Receiving—steel	13 × 25	(8-2)	325
Receiving—cartons	17 × 19	(8-3)	23
Stores	18 × 25	(8-16)	450
Warehouse	64 × 68	(8-24)	4,352
Shipping	20 × 20	(8-7)	400
Maintenance & tool room	(two people @ 400 ft each)		800
Utilities (estimate only)*	100		
Production Services Area Total: (aisles are included in each layout in this area)			6,750
III. Employee Services			
Employee entrance	10 × 20	(9-3)	200
Locker room	(3.5 ft^2/employee × 50 employees)		175
Toilets	10 × 20	(9-7)	200
Cafeteria	(10 ft^2/employee × 50 employees)		500
Drinking fountain	(6 fountains × 15 ft^2 each)		90
Medical services	(first aid room only 10 × 10 feet)	(9-12)	100
Services Area Required Total:			1,265
IV. Office Area (11 people from organizational chart) (11 people × 200 ft^2 each)			2,200
Total Building Space			18,603
V. Outside Areas			
Receiving, parking, and maneuvering area			
Shipping, parking, and maneuvering area			
Employee parking (50 employees)			
1.5 employees per parking space			
250 ft^2/parking place (9-2)			
$\frac{\text{50 employees}}{\text{1.5 employees/spaces}} = \text{34 spaces}$			
34 spaces × 250 ft^2/space = 8,500 ft^2			

Figure 13–1 Total space requirements worksheet for toolbox plant.

for the toolbox plant. The numbers in parentheses after the length and width dimensions on the total space requirements worksheet are figure numbers or paragraph numbers from which these space requirements come. It is important that space requirements, when presented in the summary form, be documented with appropriate design data or calculations. The design data and calculations are needed to support the stated requirements and will also serve as references should discrepancies occur or clarifications become necessary.

The space requirement for the fabrication area is a total of all machines and workstations. The area for one machine is the maximum length times the maximum width. This makes a rectangle out of each machine and space may be saved by fitting irregularly shaped workstations or machines more creatively. Any space saved in this way can be used in expansion plans for the future. Also, it is nice to have a little extra space because the most common error of plant layout is omission (you forgot something). The size and shape of a department may change to fit into the final building shape. The size should be very close because you minimized space needs while designing that department, but the shape almost always changes a little to fit with other departments in the newly designed plant shape. In Figure 13–1, the office space calculation is based on the average space requirement of 200 square feet per office employee and, therefore, has a built-in allowance for aisle space.

Before converting the space requirements from Figure 13–1 into plant space, you must review the cube utilization. Most of the layout design has concentrated on floor space, but not everything needs to be placed on the floor. Other levels within the plant may be suitable. Consider the following areas.

Under the Floor

Basements are the biggest user of under-the-floor space. Almost anything can be placed in a basement area. Walkways can also be placed underground, especially between buildings. The disadvantages to basement areas are additional construction costs, stairs (safety), elevators (flow restrictions), and maintenance costs. But utilities (electrical, compressed air, and water) can be placed under the floor in small trenches, keeping the overhead areas clear for material handling equipment. This is a cost saver.

Overhead or Clear Space Areas

Clear space is that space from 8 feet above the floor to the ceiling (also called the *truss*). If a building has 22-foot-high ceilings and you use shelves that stack material only 6 feet high, you have used only 27 percent of the available height. A mezzanine could more than double this utilization. A stepladder and 8-foot shelves would increase utilization even more. Racks commonly use the full height of the building. In the paint department, you can stack two dryers on top of each other and move material by overhead trolley conveyors. Overhead conveyor movement of material is a good use of the building cube in manufacturing. You have made good use of the building cube in warehousing, stores, paint, and manufacturing, but what about locker rooms, restrooms, cafeterias, and offices?

If you could place locker rooms over restrooms, you could save floor space. If you could build a two-story office, you could cut the floor space in half. Building on the second floor increases the use of the building cube and decreases the total square footage of land use on the ground floor, therefore reducing the land cost. Furthermore, the construction costs per square foot are lower on the second floor than on the ground floor. Two of the most expensive systems in office building construction are the foundation and the roof. Sharing these two systems in a multiple-story office building significantly reduces the cost per square foot of the office space.

Truss Level

A truss is a rafter. The size of the space in the trusses depends on the width of a bay. The wider the bay (span) is, the thicker the truss is. Trusses vary from 2 to 10 feet. Depending on the size of the truss, many things can be placed in this area. Offices (supported on the floor) are built in the trusses of aircraft plants. Walkways are built in the trusses of steel mills. Many plants run utilities in the trusses. Heaters, blowers, sprinklers, ovens, and the like can be placed in the trusses.

Roof

The roof, although not inside the plant, could be used for recreational purposes, the central air conditioning system, as a silo for material storage, for water towers, cooling towers, quality control testing, parking, and the like. Anything you can get off the floor will reduce building size; always review for cube utilization before determining building size.

BUILDING SIZE DETERMINATION

The toolbox plant building needs to be 18,735 square feet. A standard building is cheaper than custom-designed buildings. No one would build an 18,735-square-foot building because it would be too expensive. Standard buildings come in many size increments such as 100 × 100 feet, 50 × 50 feet, 40 × 40 feet, and even 25 × 50 feet. This refers to column spacing, so a 25 × 50-foot building would come in multiples of 25 feet in width and 50-foot increments in length. A rectangular building results. A 2:1 length-to-width ratio is a very desirable building shape because of material flow and convenient accessibility. Nearly any ratio of length to width is possible (even squares), but you should start with a 2:1 ratio first.

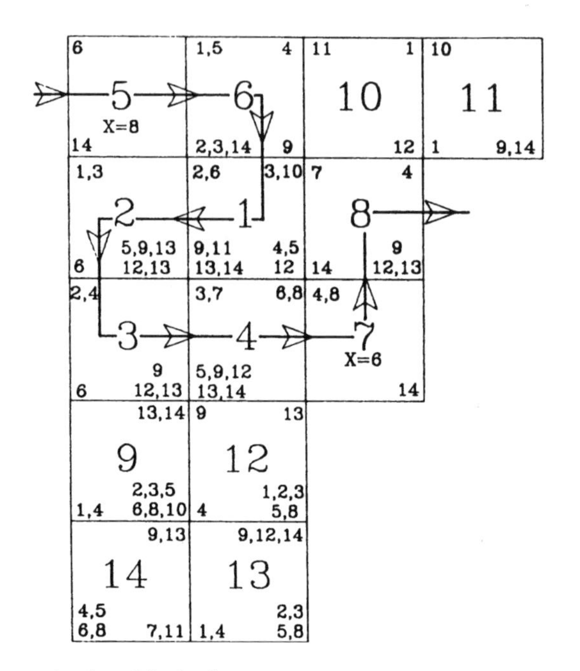

Figure 13–2 Dimensionless block diagram.

To establish a 2:1 ratio, divide the total number of square feet needed by 2 (giving two equal squares). Then take the square root of one half of that figure. The plant needs to be 18,735 square feet. Divide that by 2 to get 9,367.5 square feet. The square root of 9,367.5 is 97 feet. Round up 97 feet to 100 feet (making multiples of 25 and 50 feet). Now you have the size of your building, 100 × 200 feet; this is two 100 × 100 feet areas. A square building would be 137 × 137 feet, rounding up to 150 × 150 feet or 22,500 square feet. The size of 100 × 200 feet is 20,000 square feet, a 2,500-foot savings. Remember, 150 feet across a building can put an employee farther from an emergency exit.

The shape of a building is a unique variable where many answers are correct, but a good starting point is a length-to-width ratio of 2:1.

DIMENSIONLESS BLOCK DIAGRAM

Now that you have determined the starting size and shape of the toolbox plant building (100 × 200 feet), the question is: How are we going to divide this 20,000-square-foot building? The dimensionless block diagram developed in Chapter 6 is the orientation layout plan shown in Figure 13–2. The dimensionless block diagram's relationships must be maintained. A common error is the lack of agreement between the dimensionless block diagram and the final detailed layout.

AREA ALLOCATION PROCEDURE

With the space requirements planning worksheet (see Figure 13–1) and the dimensionless block diagram (see Figure 13–2), the building can now be divided into departments.

1. The first step in area allocation is to establish a 100 × 200-foot grid using something like 1/2-inch graph paper. A scale of 1/2 inch = 20 feet will make each 1/2 × 1/2-inch square equal to 400 square feet. Figure 13–3a shows the first attempt. All that is needed to start are the walls (external only) and the columns (25 × 50 feet).

Department	*Square Feet*	*#400 ft² Blocks*
Fabrication	2,238	6
Spot weld	1,170	3
Paint	4,200	11
Assembly and P.O.	912	3
Receiving	648	2
Stores	450	1
Warehouse	4,352	11
Shipping	400	1
Maintenance and tool room	800	2
Utilities	100	1/4
Employee entrance	200	1/2
Locker room	175	1
Toilets	200	1/2
Cafeteria	500	1½
Drinking fountain	—	—
Medical	100	1/4
Office	2,200	6
Total		50

2. The second step of area allocation is to calculate the number of squares (400 square feet) needed by each department:

You need a total of 50 squares of 400 square feet each or 20,000 square feet. All 50 spaces were assigned by rounding up.

3. The third step is to place these blocks into the area allocation layout (Figure 13–3a) using the dimensionless block diagram as the guide. Figure 13–3b shows an allocation of the (50) 400-square-foot squares. A few open squares are possible because you built 1,300 square feet more than needed, but the example used all 50 blocks because of rounding up. You now know where the departments are going and their shape.

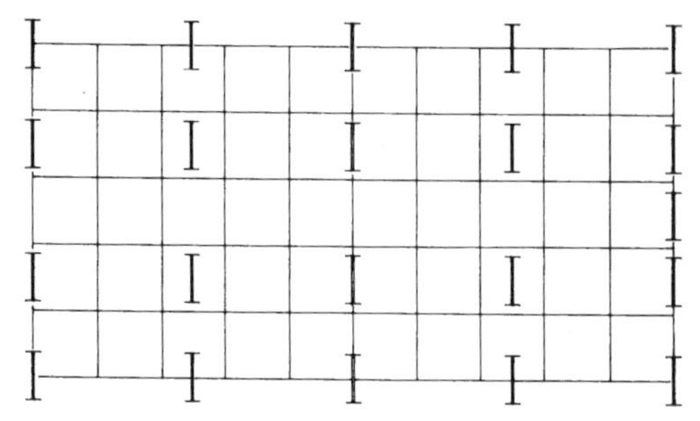

Figure 13–3a 100 × 200-foot grid (2.5 × 5 inch) scale 1/2 Inch = 20 feet.

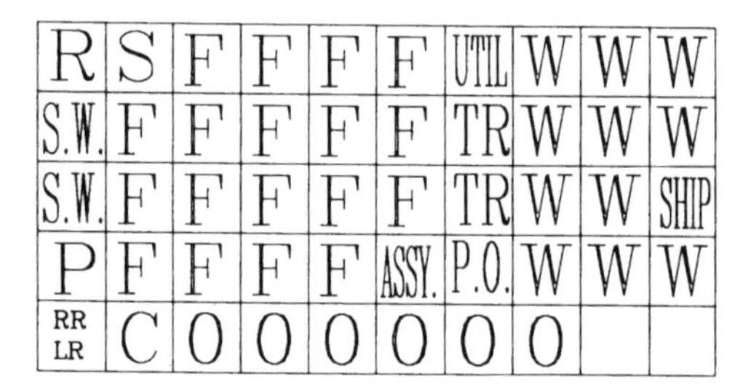

Figure 13–3b Square allocation.

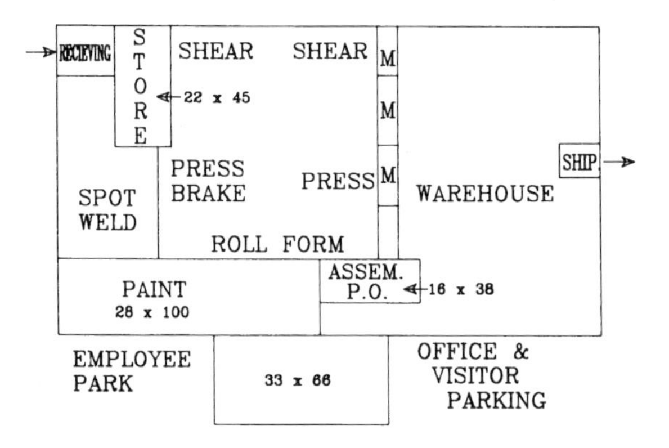

Figure 13–3c Area allocation layout.

4. The fourth step of the area allocation procedure is a layout with the internal wall or (better) the area boundaries. Figure 13–3c illustrates the first full plant layout produced in this book, but there is still much detailed work to do. The detailed layout with the placement of every piece of equipment will be discussed in the next chapter. Once the area allocation procedure produces a final plan, the architect can start on the building design and construction.

OFFICE AREA ALLOCATION

The office area allocation procedure is the same as the plant area allocation procedure. The organizational chart (see Figure 12–10) for the Electric Power Co-Op office and the dimensionless block diagram (see Figure 12–17) will be the basic source of information for a second example.

The organizational chart shows that you need room for 36 people (29 employees and 7 board members). The preliminary estimate of 200 square feet of space per person would require 7,200 square feet of office space (200 × 36 feet). The level of the organization technique shows a need for 7,450 square feet. These two figures are very close, which would make any planner comfortable with the fact that about 7,300 square feet of office space will be adequate.

The office size will be

$$\sqrt{\frac{7{,}200}{2}} = 60 \text{ ft}^2 \text{ or } \sqrt{\frac{7{,}450}{2}} = 61 \text{ ft}^2$$

The building (office) will be 60 × 120 feet (two 60 × 60-foot squares); 60 × 120 feet = 7,200 square feet.

The dimensionless block diagram, as discussed in the office layout chapter, and as shown in Figure 12–18, is the relationship plan. The closeness relationships incorporated into the dimensionless block diagram must be maintained. Those offices and employees referred to in Figures 12–11 and 12–18 are the primary purpose of an office layout. But none are included in the personnel services such as restrooms, cafeterias, supply stores, or files. Nor are office service functions included such as conference rooms or reception areas.

Office space for the seven members of the board of directors was determined using the number of employees method, instead of the level of the organization technique. As you may recall, the number of employees method allows an average of 200 square feet of office space per employee, whereas the level of the organization technique allocates square footage of office space based on the individual's position in the organization chart. When using the level of organization technique, be sure to allow 100 percent allowance for aisle space, and so on.

The area allocation procedure works as follows:

Step 1. Establish a 60 × 120-foot grid using 1/2-inch graph paper. A scale of 1/2 inch = 10 feet will make each 1/2 × 1/2-inch square equal to 100 square feet. Figure 13–4a shows an outline of the office. Allowing 30 × 40-foot column spacing would be a good plan. These columns must be placed on the grid to ensure that aisles or equipment are not placed on them.

Step 2. The second step of area allocation is to calculate the number of squares (100 square feet) needed per office or service function. Figure 13–5 lists offices as developed from Figure 12–10 (the organizational chart). The first number before the name indicates the position number on the organizational chart. These position numbers are also used on the dimensionless block diagram. The square feet required for each function came from Chapter 12. Remember, every area space requirement must depend on something. The number in parentheses behind the area description indicates the number of people in this space, if there is more than one. The total square feet calculated in Figure 13–5 is only 5,790 and the layout calls for 7,200 square feet. The difference is the aisle space. It may be tight (need more space), so aisle space must be used efficiently.

Step 3. The third step starts by placing the dimensionless block diagram (Figure 12–17) and the office area space requirements (Figure 13–5) next to the block diagram (Figure 13–4a). You allocate space by placing the position

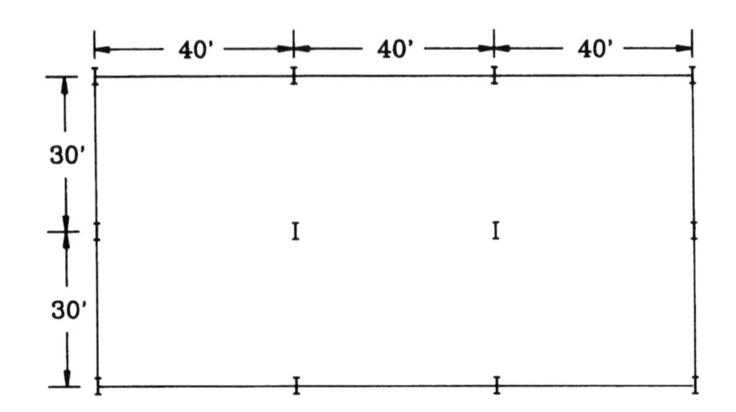

Figure 13–4a Office outside walls—Four columns (30 × 40 feet).

number(s) in the 100-square-feet squares per the dimensionless block diagram. The service areas must be worked in by placing the services conveniently for most of the people. The results of this process will be something like Figure 13–4b. We say "something like" because if four designers did this simultaneously, they would have four different but good answers. The main thing is to be true to the dimensionless block diagram.

The final step in the area allocation procedure is to develop a final area allocation diagram. This step requires placing aisles and specific boundaries. Aisles should be straight and run the full width and length of an office. Establishing aisles is an important first decision in this last step. This is a small office, so figure on 5-foot main aisles and 4-foot cross aisles. Figure 13–4c shows the final layout. Four previous layouts were discarded because improvements kept coming. Do not be afraid of trying many different arrangements. The best arrangement satisfies the most relationships, as shown on the dimensionless block diagram.

16	16/17	18	3	3	3	6	3	3	3	1	1
17	19	18/19	20	3	3	6	3	3	3		4
	26	26		23	25	21	23	11	5	5	7
13	26	26	23	23	25	21	23	12	8		7
13	14		2		24		23		8	10	8
9	15	2	2	10	10	22	22	22	8	10	8

Figure 13–4b Calculate the number of squares.

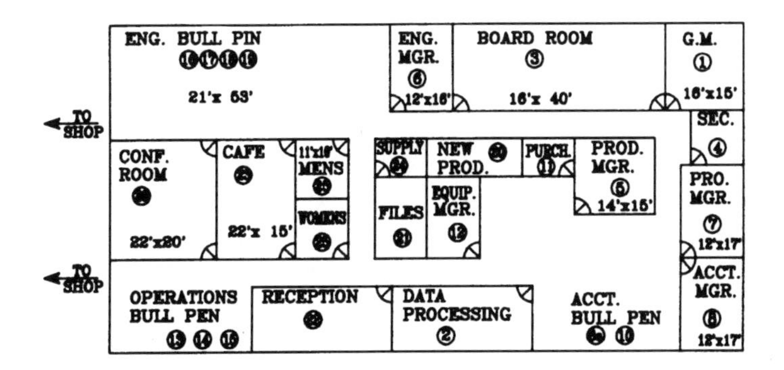

Figure 13–4c Area allocation diagram.

Position Number	*Area Description*	*Square Feet Size*	*No. of 100 Foot Spaces*	*Approximate Size*
1	General manager	250	2.5	15 × 16
2	Data Processing (3)	300	3	19 × 16
3	Boardroom (7)	640*	6.5	40 × 16
4	Secretary	100	1	10 × 10
5	Production manager	200	2	12.5 × 16
6	Engineering manager	200	2	12.5 × 16
6A	Engineers (3)	450	5	28 × 16
7	Public relations manager	200	2	12.5 × 16
8	Accounting manager	200	2	12.5 × 16
8A	Accounting (4)	400	4	25 × 16
9	Records & Dispatch	100	1	10 × 10
10	Payroll (2)	200	2	12.5 × 16
11	Purchasing	125	1	10 × 12.5
12	Equipment manager	150	2	12.5 × 12.5
13	Meter technician	150	1	12.5 × 12.5
14	Work order clerk	100	1	10 × 10
15	Stores clerk	100	1	10 × 10
16	Maps engineer	125	1	10 × 12.5
17	Field crew leader	150	1	12.5 × 12.5
18	Party chief	150	1	12.5 × 12.5
19	Stake crew leader	150	2	12.5 × 12.5
20	New project engineer	150	2	12.5 × 12.5
21	Restrooms (8-D)(2)	200	2	10 × 10 (2)
22	Reception*	300	3	12 × 25
23	Cafeteria	300	3	15 × 20
24	Stores*	100	1	10 × 10
25	Files*	200	2	10 × 20
26	Conference room	400	4	20 × 20
		5,790	61	

*See layout in Figure 13-4c.

Figure 13–5 Office area space requirements—Electric power co-op.

QUESTIONS

1. What is area allocation?
2. What is the total space requirements worksheet?
3. What are the different levels within the plant?
4. How do you convert square footage to building size?
5. What is the area allocation procedure?
6. What is the end result of the area allocation procedure?
7. How can you better use the clear space?
8. Which of these two areas would you place upstairs and why?
   a. Restrooms or locker rooms
   b. Accounting or purchasing
   c. Old files or current files
9. What is a column? Why is it important?
10. What is column spacing?
11. Using the golden rule of architecture, what would be the length and width of the buildings with the following space requirements?
    a. 825,000 square feet
    b. 250,000 square feet
    c. 87,500 square feet
12. Once the length and width of the building have been determined, how do you know where to place the departments?
13. Discuss the advantages or disadvantages of placing offices on the second floor.
14. What are the most expensive systems in building office spaces?

CHAPTER 

Facilities Design—The Layout

OBJECTIVES:

Upon the completion of this chapter, the reader should:

- Be able to develop the layout based on the procedures followed throughout the textbook
- Be able to evaluate the layout based on flow efficiency and space utilization

"Layout" is a simple term that must communicate the complex results of many months of data collection and analysis. The layout is only as good as the data backing it up. The layout is the visual presentation of the data and the subsequent analyses by the facilities planner. The combination of the accuracy and credibility of data and the logical analysis of the information can result in a good layout. Often poor or incomplete data, or poor judgment on the part of the planner, or a combination of both factors can lead to less than desirable outcomes.

The term "layout" will be applied to plot plans and master plans. The layout is the facility planner's biggest selling tool. As the plan is presented to management, it will be regularly referred to in order to show how products flow through the plant. The flow diagram, as discussed in Chapter 5, is a great aid in illustrating material flow in the facility. The flow diagram, however, cannot be produced until a layout has been developed. Of course, it can also be used on the existing layout as the present method is compared with the proposed method as a basis for productivity improvement and cost reduction.

PLOT PLAN

A *plot plan* shows how the building(s), parking lot(s), and driveways fit on the property (see Figure 14–1). The main highways, utilities, drains, and the like are important to the construction project as well. City and county building codes also affect

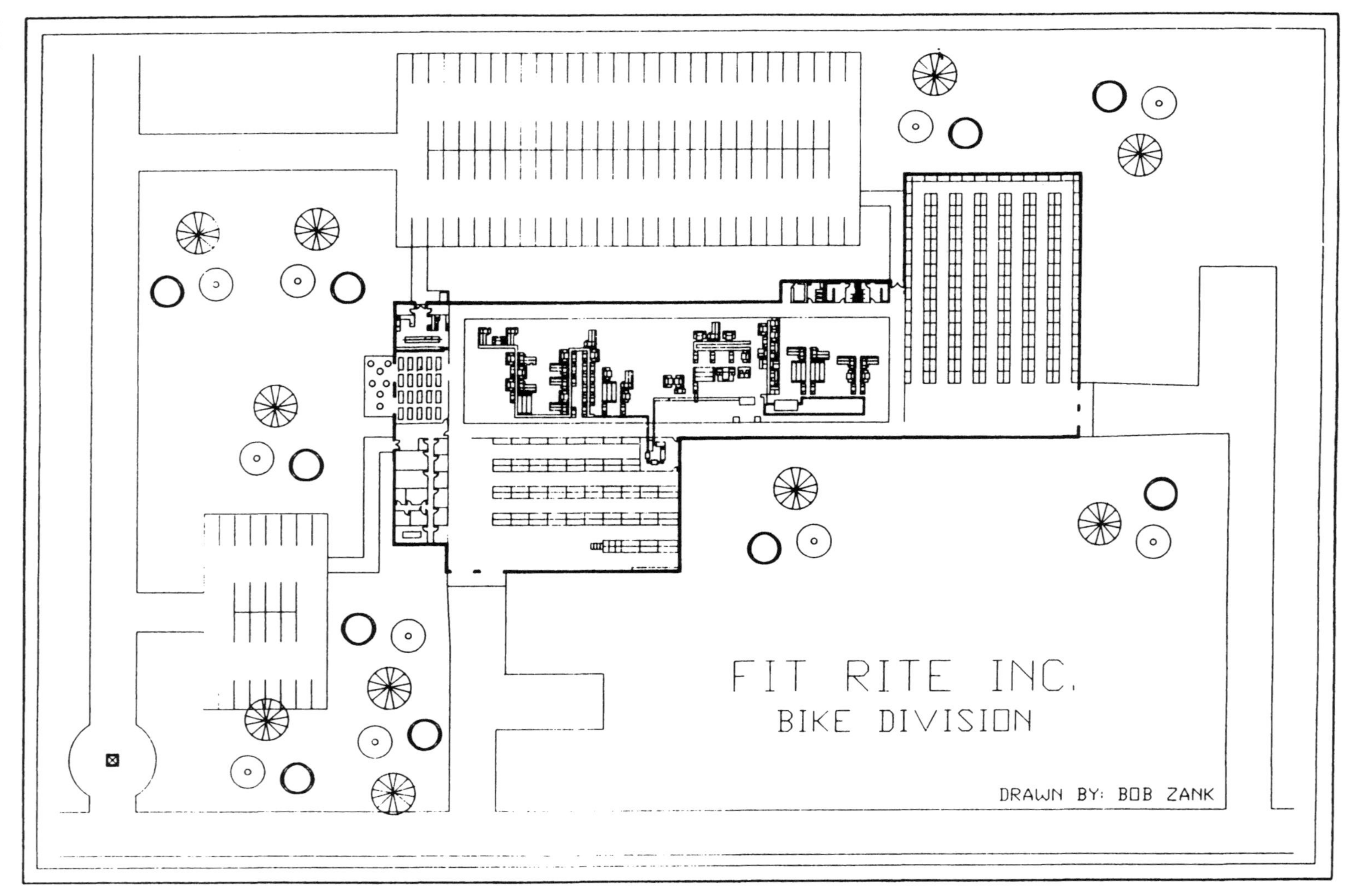

Figure 14–1 Plot plan.

the plot plan. The driveways (entrances) may require frontage roads, whereas the parking lot may require a road to be set back.

Step 1. Start with a layout of the property showing the lot lines.
Step 2. Place in the layout the main roads that border the property or where the access road will enter the property.
Step 3. Show sources of water, power, gas, and phones.
Step 4. Place the building where the front faces the road and the long side faces the road. Expansion plans will go to the rear of the building.
Step 5. Show receiving and shipping (consider where the expansion will go).
Step 6. Connect receiving and shipping to the main road.
Step 7. Show where employee and public entrances will be located.
Step 8. Provide parking for visitors and employees.

Figure 14–2 shows the plot plan for the toolbox plant.

A plot plan must also show expansion possibilities. It is extremely important to consider expansion even before buying property. Property prices vary due to many factors, but one factor that affects the plot plan is frontage cost versus cost of

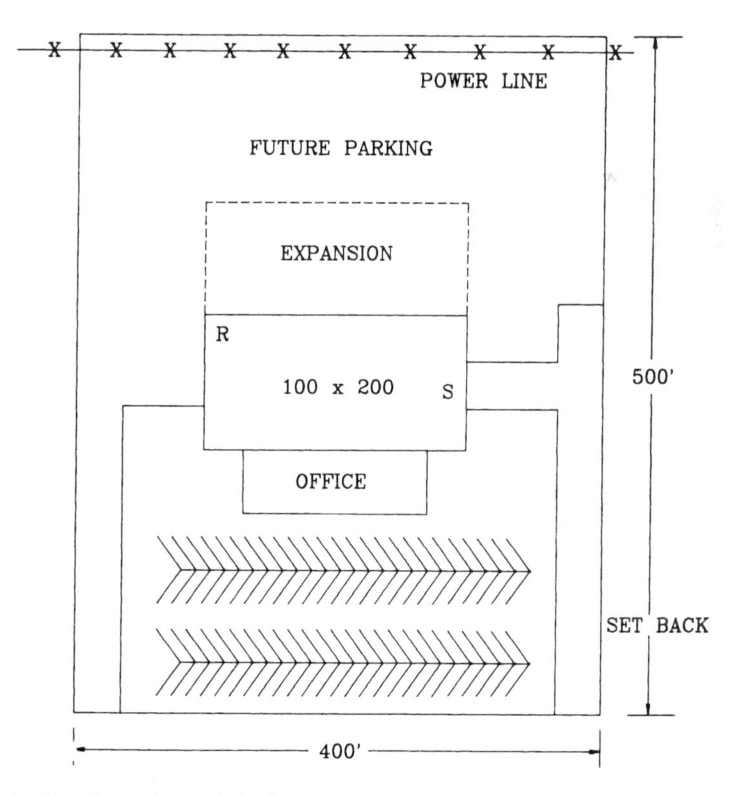

Figure 14–2 Toolbox plant plot plan.

the depth of the lot. The property on a main road is called *frontage*. The cost of a piece of property will vary proportionally with the size of the front footage. The depth of the lot is not as big a factor. Of course, you need adequate size for your plant, but you can purchase additional land behind your lot for a much cheaper price than you can buy land on the road. So, plan for expansion behind the building. Be sure not to place any permanent or costly facilities in the way of expansion. Receiving and shipping docks are two facilities that should not be located in the expansion area.

Plant Layout Methods

There are three methods of depicting the plant layout:

1. Template and tape technique
2. Three-dimensional models technique
3. Computer-aided design (CAD) technique

Template and Tape Method of Facilities Design

This is almost a history lesson. Before the introduction of CAD, the template and tape method was the preferred technique of facilities designers who did numerous plant layouts. There are still planners who use this technique, but CAD has taken over. The *template and tape technique* is a layout made of transparent templates and rolls of various tapes placed on a mylar (plastic) grid base. The mylar has a 1/2-inch grid lightly printed in blue to allow the designer to place walls, aisles, and machines without the use of a straightedge or a ruler. The wall tape is placed first, creating an outline of the building.

Office expansion may be up (adding a second or even a third floor) and the initial construction must allow for the additional floor. Parking can be expanded back onto the property, and an additional employee entrance may be needed. This may require moving the locker rooms, restrooms, and cafeterias as well, but if new facilities are necessary, you must consider all services to maintain good employee traffic.

When buying property, a general rule of thumb is to buy 10 times more property than the building size. A 100 × 200-foot building of 20,000 square feet will require 200,000 square feet of land (about 5 acres). Another economic factor is the cost of frontage (i.e., property adjacent to the road) compared to property off the road. The property gets cheaper as you move back from the road. Frontage property on the road is usually sold by the frontage foot, whereas back property is normally sold by the acre, so buy as much back property as you can justify economically. Due to certain economic considerations such as the cost of land, property taxes, zoning restrictions, and local ordinances, most new plant construction occurs outside the city or township limits.

The plot plan communicates a great deal of information about how the new plant will fit on the lot and what external facilities are required. The architect can now design the driveways, parking lots, and the building. The facilities designer can go back to the internal plant layout problems and create a master plan.

MASTER PLAN

The *master plan* is the finished product of the facilities design project. The term "plant layout" refers to the master plan most of the time. The master plan shows where every machine, workstation, department, desk, and all other important items are located.

Before the proliferation of computers and computer-aided design (commonly referred to as CAD) and drafting, layouts were produced using various templates and other manual aids. Today, however, the task of producing the layouts is accomplished much more efficiently by the aid of various CAD systems.

Whether manual or electronically, standard symbols or templates are used to designate walls, aisles, pneumatic and hydraulic lines, various equipment, and even operators, or any other conceivable entity on the production floor or the entire facility. A sample of these icons or templates is shown in Figures 14-3 and 14-4.

Computer-aided plant layout packages, and most computer-aided drafting or design software systems, contain a wealth of three-dimensional (3-D) as well as flat two dimensional (2-D) templates to aid the facilities planner. Extensive libraries of templates make every conceivable piece of manufacturing equipment, from a basic grinder to CNC lathes and mills to injection molding machines. Material handling systems such as jib and bridge cranes, fork lifts, and a vast array of conveyors are also available to the facilities planner with a simple click of the mouse. Architectural templates, such as building components, receiving shipping and docks, as well as office furniture and equipment, operating personnel, even landscaping templates for the grounds and interior decoration are made available by these state-of-the art facilities planning tools. Should the need or the imagination of the planner exceed the vast extent of these libraries, the menu-driven software tools can be utilized easily and quickly to create the necessary templates and then to store them as part of the system or private library for future use.

Three-Dimensional (3-D) Models

Three-dimensional model layouts have the big advantage of illustrating and highlighting any height problems. New 3-D models are being commercially developed every day. The engineering monthly journals are the best source of where to find these models. Three-dimensional models can be placed on a clear plastic sheet with a grid network of 1 inch = 1-foot square. Three-dimensional models are nice, but the expense, the difficulty of copying, and the problem of storage space make them less desirable. The layout procedure for the 3-D model technique is the same as that of the template and tape technique. The scales for template and tape and the 3-D model technique are the same. One-quarter inch equals one foot is the most popular scale for the plant layout, followed by 1/8 inch = 1 foot. Many commercial

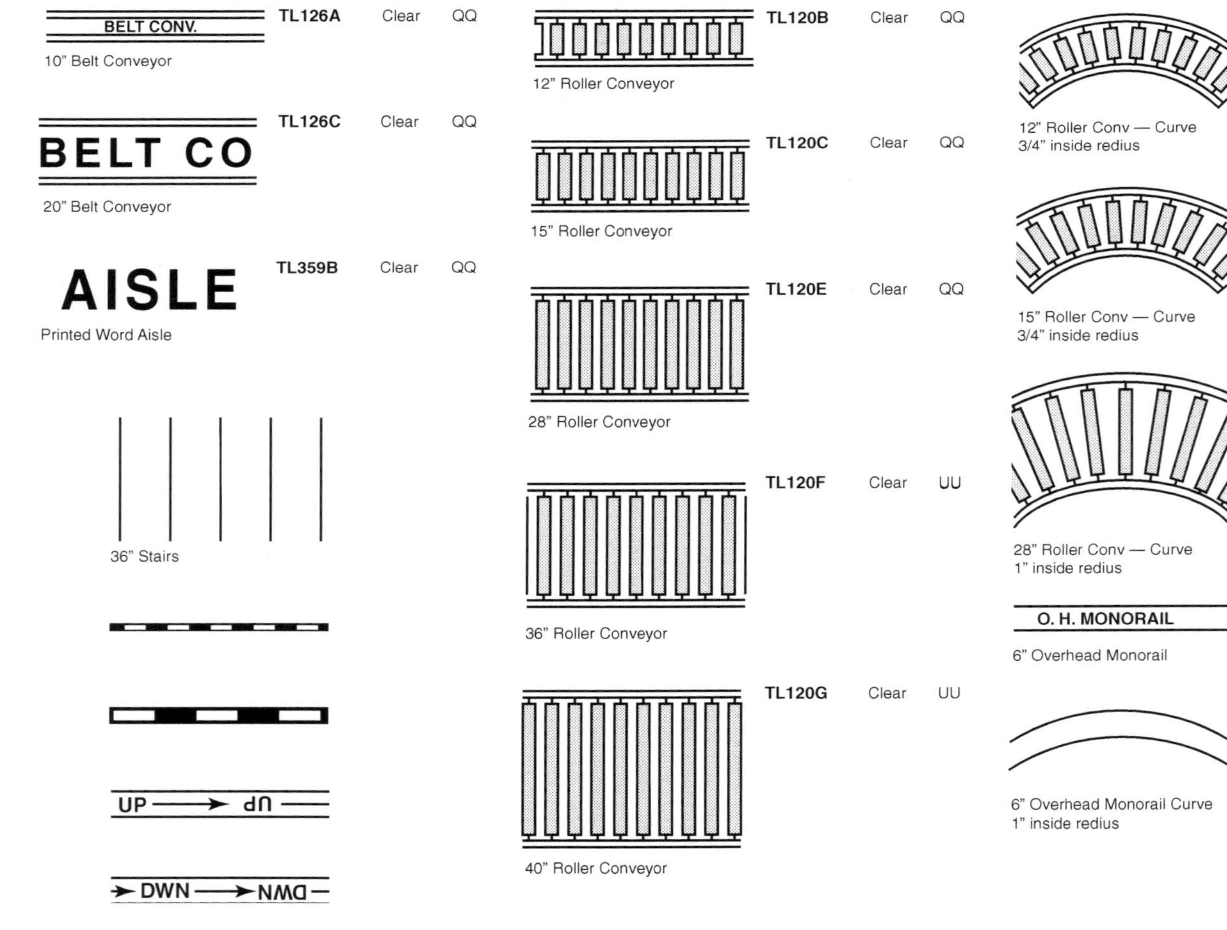

Figure 14–3 Plant and office layout templates.

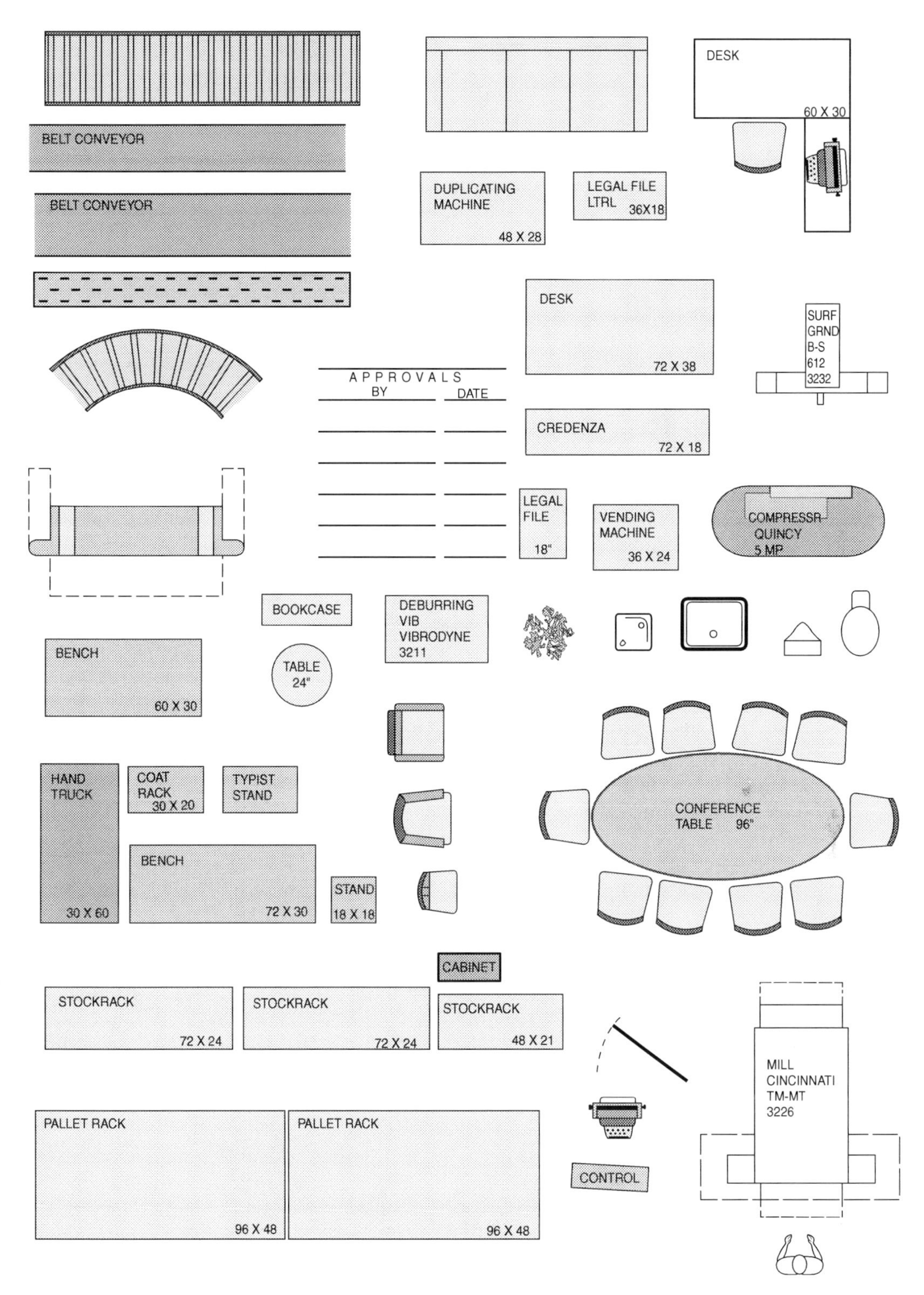

Figure 14–4 Sample templates.

templates and 3-D models are available with these two scales. Any other scale can be used if outside materials are not needed.

A toy company uses a 1/2-inch = 1-foot scale and is able to write much more information on the templates. A 4 × 4-foot pallet is reduced to a 2 × 2-inch label and the part number, name, and quality per pallet are written right on the template.

Computer-Aided Design (CAD) Technique

Computer-aided plant layout design is the state-of-the-art technique. The advantages of all the previous techniques are improved with CAD and the disadvantages have been minimized. This is assuming a trained CAD operator, the equipment, and the program are available to the company. New planners armed with CAD expertise and a knowledge of plant layout will be very valuable to any company.

AutoCAD is one of the software packages used to create the majority of the drawings in this book. There are, however, many other packages and options available to facilities planners.

Although the initial cost of a computer-aided facilities planning software may be considered a disadvantage, less costly CAD packages are quite capable of producing great results of a highly professional caliber. Furthermore, once the initial equipment and software costs are absorbed, the continuous efficiency and cost-effectiveness are most impressive. Changes, corrections, and modifications to layouts can be made very quickly, the quality of drawings are outstanding, especially if plotters are used, and all drawings are stored electronically for future use and can be transferred and shared around the world instantly. As more layouts are completed, the task becomes easier by importing all or parts of a drawing into a new drawing. Three-dimensional layouts and layered (overlay) layouts can assist in visualization and spatial relationships.

Figure 14–5 compares the three layout techniques to assist in the selection of the best technique for you.

Advanced Computer Systems

Facilities design has undergone gradual changes since the 1940s. It has become more efficient, more useful, and (even though more complex) better in every way. From the architectural drawings of the early days, to cutouts of architectural drawings, to templates, to 3-D models, to CAD systems of today, designers look enthusiastically to the future.

Figure 14–6 is the top view of a manufacturing facility drawn with the help of computer-aided drafting software. Once the dimensions of the plant are determined with the aid of the layout tools, as discussed in the previous chapters, the layout is drawn. The location of each activity center is selected in accordance with the activity relationship chart and the block diagram and is marked on the drawing. In order to facilitate visualization, the designer can use a few on-screen, menu-driven commands to manipulate easily and rotate the drawing so that the layout can be viewed from different perspectives and angles.

	Template	*Model*	*CAD*
1. Skill needed	Medium	Low	High
2. Cost of equipment	Medium	Highest	High
3. Time to set up	Low	High	High
4. Time to redraw	Longest	Medium	Fastest
5. Drawing time (once established)	Medium	Medium	Low
6. Changeable	Easy	Easy	Easy
7. Availability of material	Moderate	Poor	Fastest
8. Scales available	Few	Very Few	Any/All
9. Selling tool	Good	Excellent	Excellent
10. Ease of building file	High	Low	Low/Moderate
11. Ease to generate alternatives	Good	Good	Best
12. Storage space required	Moderate	High	Low
13. Ability to copy	Easy	Difficult	Fastest

Figure 14–5 Rating of layout techniques.

A vast library of icons representing various manufacturing and material handling equipment, tools, and operators can be selected with a click of the mouse and placed on the drawing. Should the location or the orientation of the equipment prove unsatisfactory, the placement of the object can be modified with equal ease. Figure 14–7 shows a 3-D presentation of the manufacturing plant. Substituting the traditional flat and 2-D (two-dimensional) plans with this computer-generated 3-D layout should alleviate any difficulty that you may have in visualizing the arrangement of the facility.

The next generation of technology to aid the facilities planner is the virtual reality technology. In addition to its significant contributions to the entertainment industry, this technology has already proven its tremendous power in the training of fighter pilots, physicians, and surgeons, and in underwater exploration and mining, to name a few frontiers. This technology will allow the planner to "walk through" the facilities before the facility ever exists. Whereas computer simulation has already taken a giant leap in answering many what-if questions and scenarios, virtual reality will allow the planner to take the facility for a true "test drive." Imagine inspecting the warehouse or the shipping and the receiving departments to examine the efficiency of their operations, or walking through a new office complex to see if it is aesthetically pleasing and functionally adequate.

Virtual reality has already arrived on the facilities layout scene. Matsushita Works Ltd. of Japan has kitchen showrooms that allow customers to walk through different kitchen designs without leaving their chairs. They wear a special set of goggles hooked up to a computer. A 3-D picture of the newly designed kitchen appears on the inside of the goggles. An electronic glove allows the wearers to steer themselves through the kitchen—it is almost like being in the new kitchen. This technology will revolutionize facilities design. It will not only assist designers but it will also help them to sell their plans. Virtual reality technology is devel-

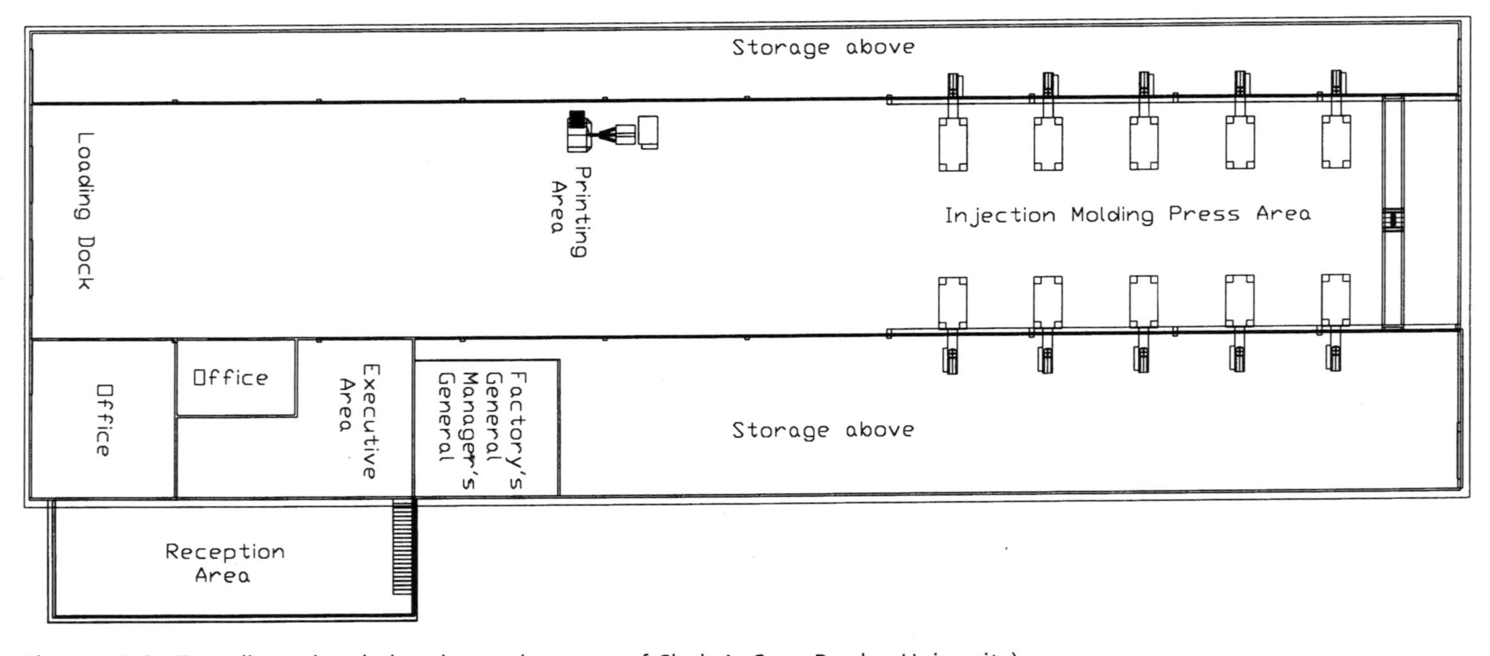

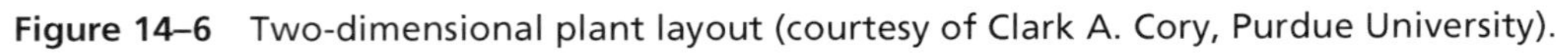

Figure 14–6 Two-dimensional plant layout (courtesy of Clark A. Cory, Purdue University).

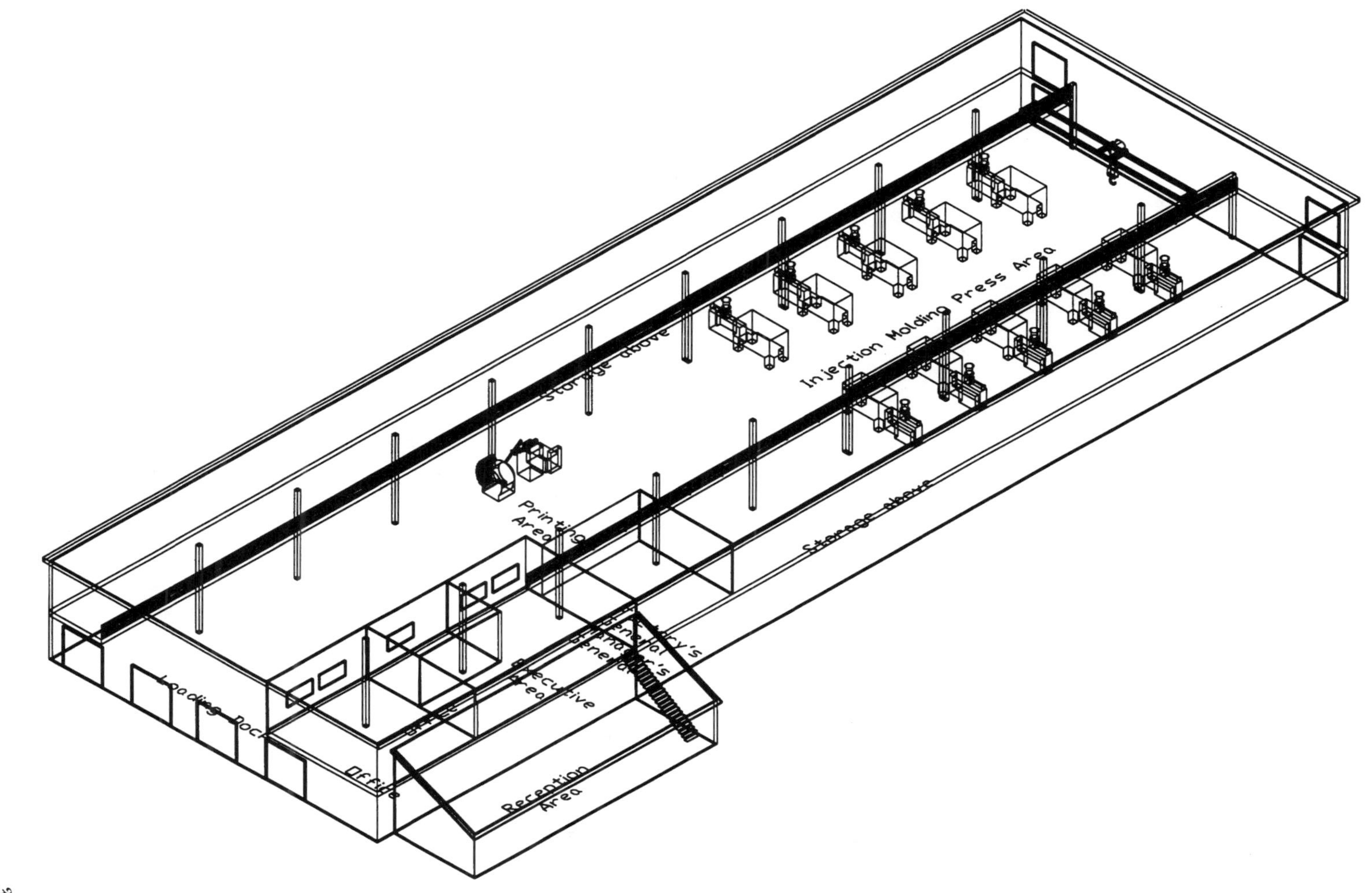

Figure 14–7 Three-dimensional plant layout (courtesy of Clark A. Cory, Purdue University).

oping fast, but cost-efficient plant layout systems will require years of work. The direction of facilities planning technology is clear and the future promises to be very exciting.

Computers and facilities planning software are both becoming more affordable and user-friendly. Electronic data interchanges are facilitated due to the development of universal standards. All these factors lead to a significant reduction in facilities design lead time and expense. Better layouts can be developed faster, more accurately, and more economically, and are easier to "sell" to management.

Figures 14–8 and 14–9 show two state-of-the-art CAD drawings provided by S.I. Handling Systems, Inc.

PLANT LAYOUT PROCEDURE—TOOLBOX PLANT

This is where everything comes together. The area allocation diagram, illustrated in Chapter 13, shows the shape and position of every department and service area. Many of the departments listed below have already been laid out earlier in the text, but they now must be fitted to the area allocation diagram considering material flow and size constraints. Some modifications may be needed to the area allocation diagram or the department layout.

Figure or Page No.	*Department*
4–12	Spot weld
4–13	Assembly and packout
7–5 to 7–10	Workstations (fabrication)
7–11	Paint department
8–2	Steel receiving
8–3	Parts receiving
8–7	Shipping
8–16	Stores layout
8–24	Warehouse
p. 211	Maintenance
9–3	Employee entrance
9–5	Locker room
9–7	Toilets
p. 257	Cafeteria
9–12	Medical service
pp. 339, 370	Office

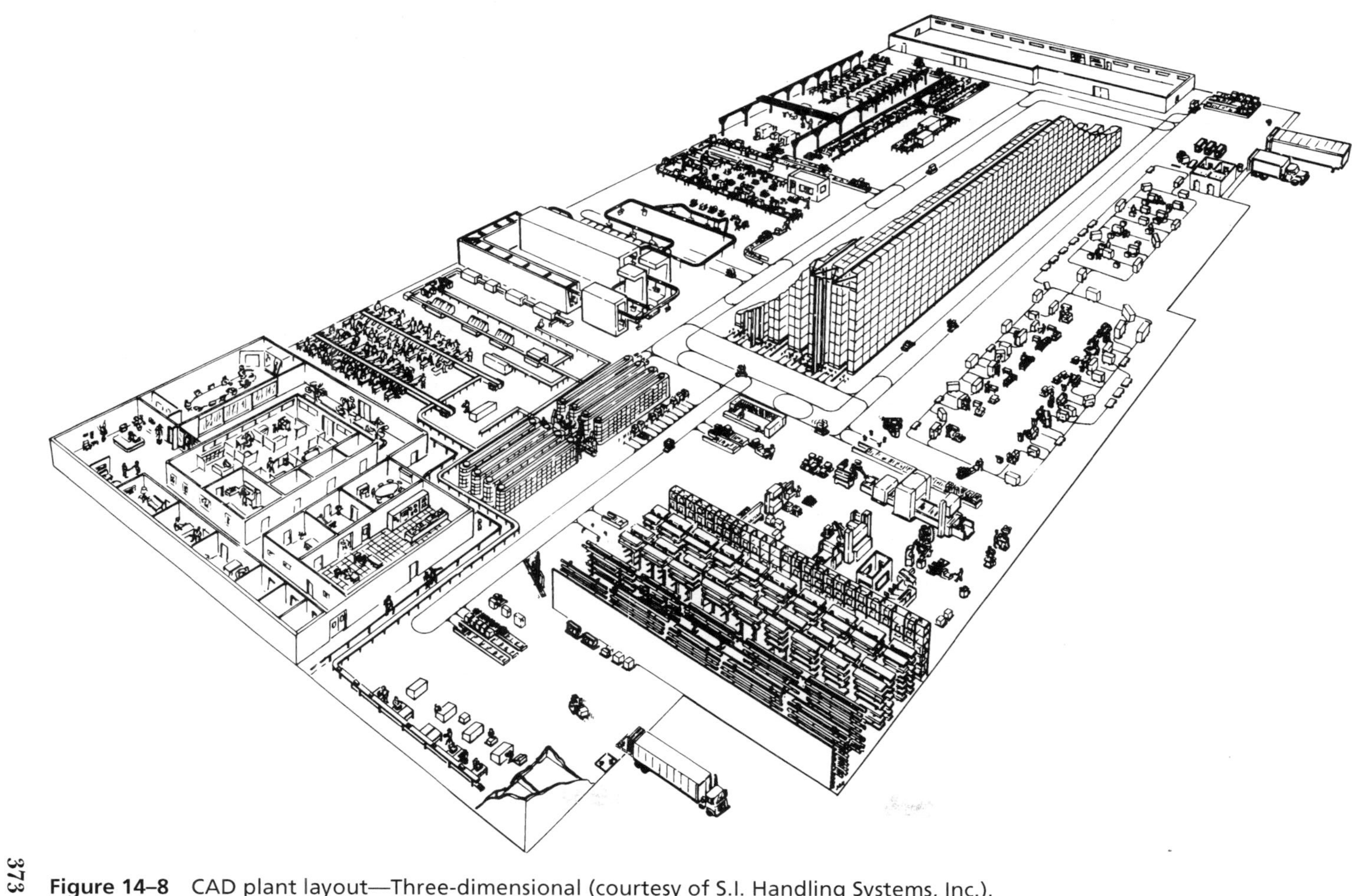

Figure 14–8 CAD plant layout—Three-dimensional (courtesy of S.I. Handling Systems, Inc.).

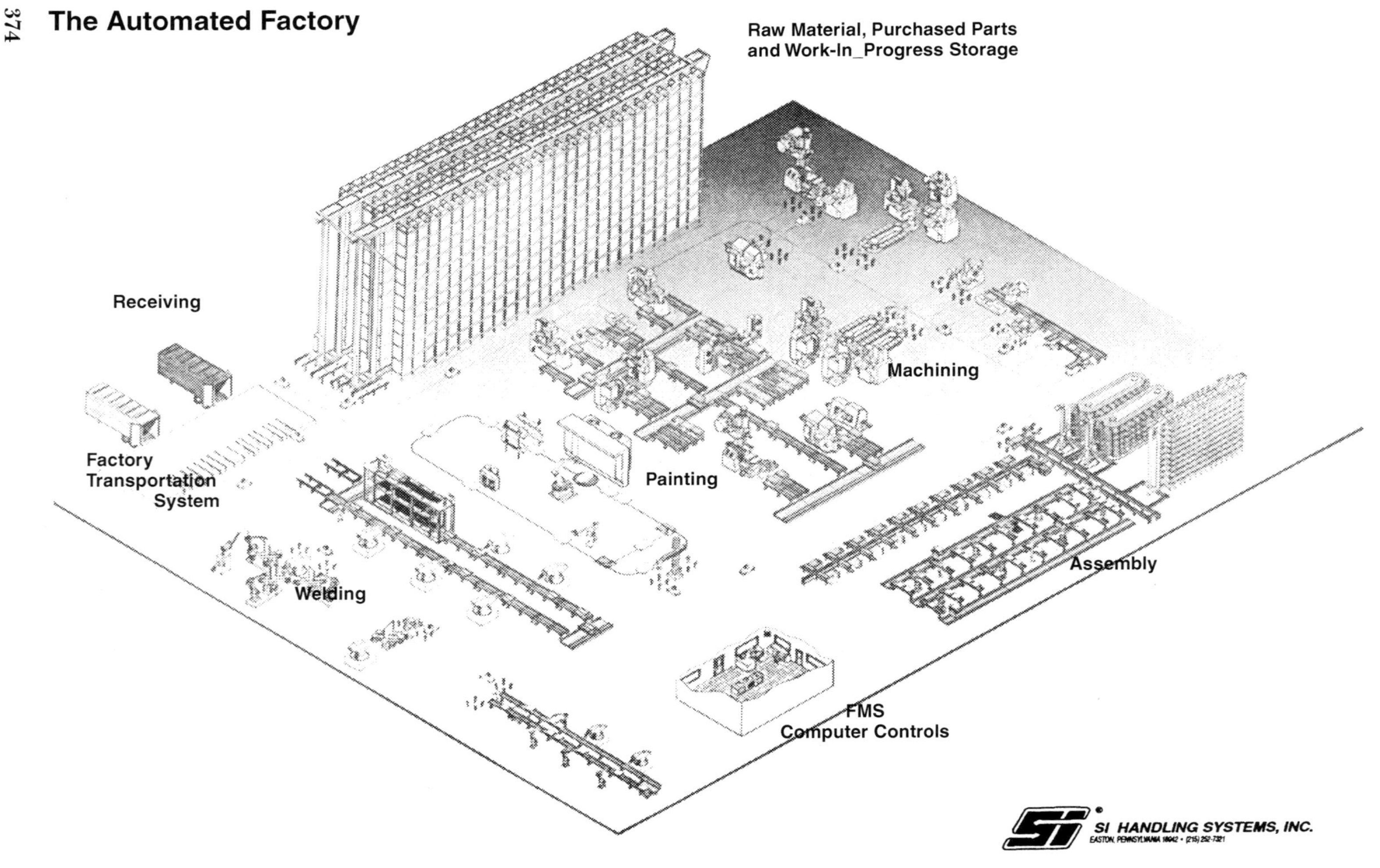

Figure 14–9 CAD plant layout (courtesy of S.I. Handling Systems, Inc.).

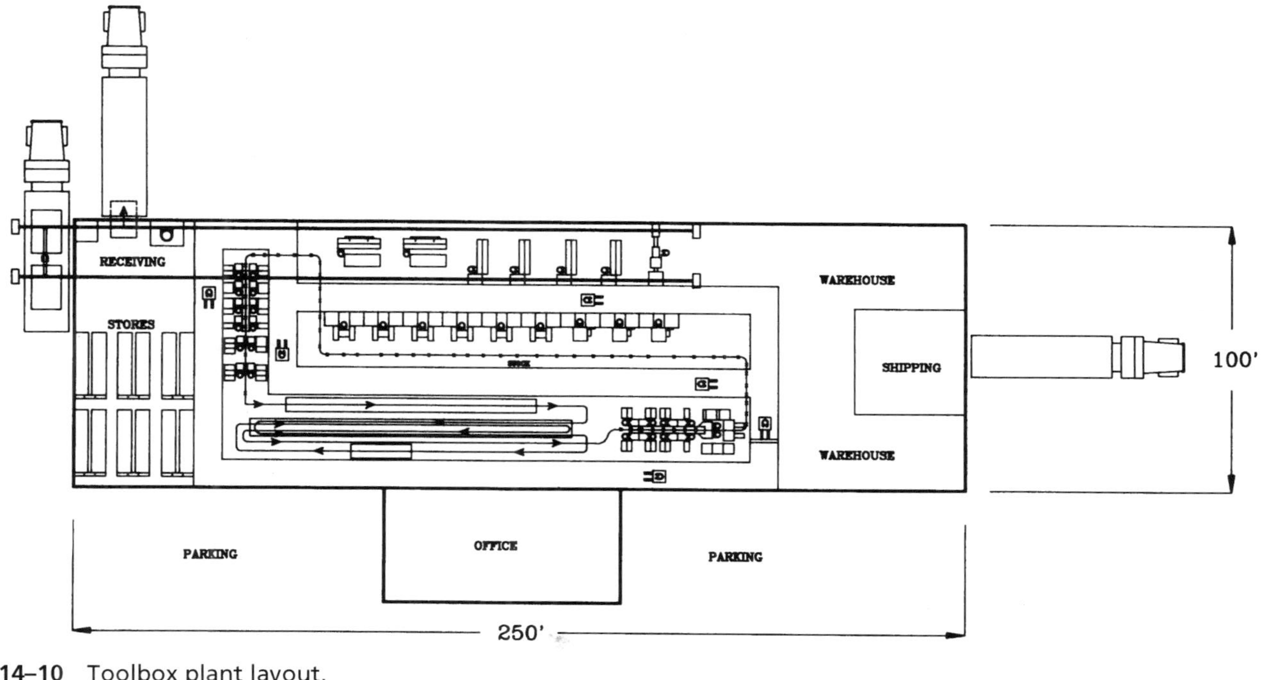

Figure 14–10 Toolbox plant layout.

Using the area allocation diagram as a guide, you must now coordinate these layouts into a final master layout.

The plant layout procedure starts with the exterior walls being located. This is, of course, a constraint. Once the exterior walls have been established, exterior doors, columns, and aisles are located according to the area allocation diagram. Now, one department at a time plus all the equipment and facilities are put in place.

The flow of material must always be considered. We considered material and people flow at every step along the process of laying out the plant, but the flow of material out of one department must line up with the starting point of the next department.

The final entry to the plant layout is material space. Everything must have a place; otherwise it will be in the aisle. Once everything is in place on the layout, the facilities planner should follow the flow of every part from receiving to shipping to ensure that every requirement has been met. This is the flow diagramming technique shown in Chapter 5. Figure 14–10 shows the final layout of the toolbox plant. See how it compares to Figure 5–14, which illustrates the existing layout. Is it better? Evaluation is needed to answer this question.

Office Layout for the Toolbox Plant

Starting with the organizational chart in Figure 2–7, you can determine that the number of employees in the office is 11. Eleven people times 200 square feet per person equals 2,200 square feet. The level of the organization technique would require the following:

Plant manager	200
Secretary	100
Controller	150
Accountant	75
Production manager	150
Manufacturing engineer	100
Supervisors	75
Supervisor	75
Purchasing manager	150
Plant engineer	150
Maintenance supervisor	75
Total:	1,300 square feet
100 percent allowance	1,300 square feet
Total needed:	2,600 square feet

Between 2,200 and 2,600 square feet are needed.

$$\sqrt{\frac{2{,}200}{2}} = 33 \text{ feet} \qquad \sqrt{\frac{2{,}600}{2}} = 36 \text{ feet}$$

Because it is a round number, 35 × 70 feet will be chosen; 35 × 70 feet = 2,450 square feet.

The power plant office layout offers a better example of a detailed office layout. Figure 13-4c depicts the area allocation and Figure 13–5 lists the office area requirements summary. With these two resources, the requirements must be fitted into a 60 × 120-foot space. Figure 14–11 is the resulting layout.

EVALUATION

When deciding upon which method or which alternative is best, measurements of performance must first be developed. In the beginning of the book, facilities planning objectives were established. Did you meet the objectives? Which alternatives met the objectives best? Performance measurement techniques were discussed throughout the book. They are listed here one more time to stress their importance. Figure 14–12 is a collection of facilities design control graphs, which include

1. *Minimize distance traveled.* How many feet does a part travel through the plant? The shorter it travels, the better. Some travel is not as bad as other methods.
   a. How many feet are traveled automatically? This is expressed as a percentage of the total feet traveled and would express the efficiency of the material movement.

   Example: 1,525 of 2,000 is 76 percent. Seventy-six percent may be graphed (and it should be). The 1,525 feet of automatic travel out of 2,000 feet show how well you have done and how much room for improvement is left.

$$\text{Automatic travel ratio} = \frac{\text{automatic feet}}{\text{total feet}} = \frac{1{,}525}{2{,}000} = 76 \text{ percent}$$

   b. Gravity movement is free power. If you want to promote the use of gravity, you will calculate the percentage of footage being traveled by the use of gravity and will graph the progress month after month.

$$\text{Gravity ratio} = \frac{\text{gravity feet}}{\text{total feet}}$$

2. *Maximize space utilization.* This can be measured, charted, and improved. You can increase this utilization in many ways.
   a. Aisle space can be calculated by the square footage of aisle space and divided by the total space available.

$$a = \text{percent of aisle space} = \frac{3{,}150 \text{ square feet of aisle}}{10{,}000 \text{ square feet of plant}} = 31.5 \text{ percent}$$

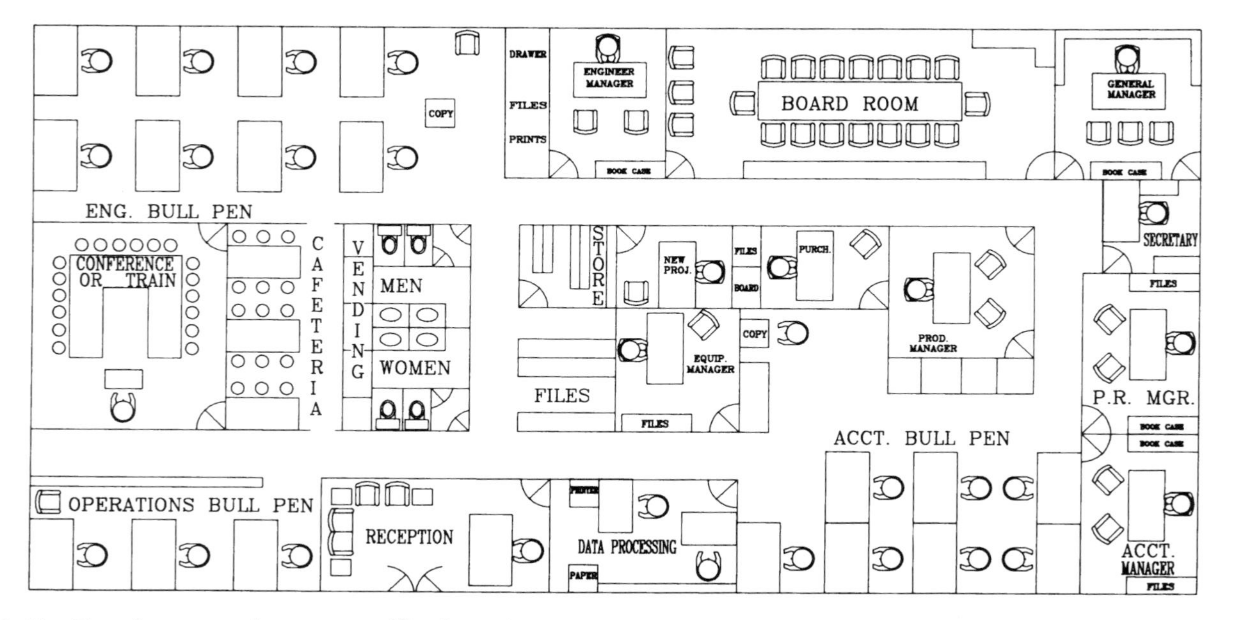

Figure 14–11 Electric power plant co-op office layout.

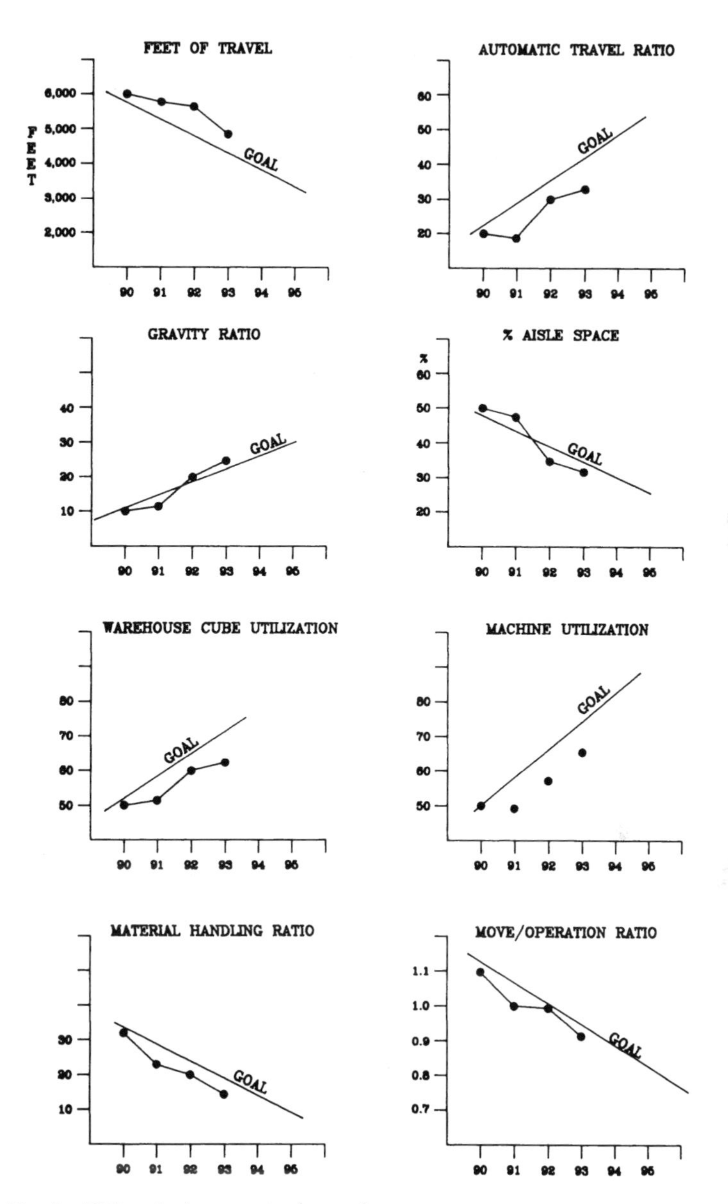

Figure 14–12 Facilities design control graphs.

Plot this percentage on a graph and measure it month after month to show improvement. An improvement would be a lower percentage rate.

b. Stores and warehouse cube utilization is total storage available. Length times width times height of your stores or warehouse equals the total cubic feet of storage available. You may approach 100 percent, but aisle space, space between materials, and not stacking to the full height create use at the 30 to 40 percent level. It should be your goal to improve cube utilization. A measure of this is

$$b = \text{percent of cube utilization} = \frac{\text{storage cubic feet}}{\text{total cubic feet}}$$

c. Machine space utilization is calculated as

$$c = \text{machine space utilization} = \frac{\text{machine space required}}{\text{total plant space}}$$

An increase in this percentage would show a lessening of material in process, aisle space, and services.

3. *Machine utilization ratio.* A machine may have the capacity to cycle 1,000 to 2,000 times per hour, but an operator must unload the aside, get the next part, load, and trip the run buttons. This lowers the standard to 250 to 500 per hour (25 percent utilization). An automatic loading can increase the output by 400 percent.

$$a = \text{percent of automatic loading machine} = \frac{\text{no. of automatic loaded machines}}{\text{total machines}}$$

$$b = \text{percent of machine utilization} = \frac{\text{time standard}}{\text{theoretical maximum}}$$

To promote this goal, b can represent a single machine or the entire department or plant. You want to approach 100 percent.

4. *Control material handling costs.*

a. $$\text{Percent of material handlers} = \frac{\text{no. of material handlers}}{\text{production people}}$$

or

$$\text{Material handling ratio} = \frac{\text{no. of material handling hours}}{\text{total hours worked}}$$

b. Manual move-to-operations ratio equals the number of moves divided by the number of operations. This will promote the combining of operations or the mechanization of moves to eliminate manual moves.

5. *Just-in-time manufacturing ratios measure how long a product is in process* (in the plant). You want to move material through the plant as fast as possible to reduce inventory and inventory carrying costs. If you add up all the time standards in hours per unit, you would have the theoretical shortest time that a product would be in a plant. An appliance manufacturer makes nearly every part and assembles the washing machine in 3 1/2 hours or less, yet it has millions of dollars' worth of inventory in the plant. If the inventory dollar value is divided into the total annual sales, you get the number of turns a year (turnover of inventory). Two turns a year equals 6 months' worth of inventory. Three and one-half hours divided by 2,000 hours (6 months with two shifts per day) is a very small percentage. Through the use of many of the discussed techniques, this can be increased to over 10 percent. The cost savings will be fantastic!

$$\text{In-process time ratio} = \frac{\text{cycle time (total)}}{\text{total time in process}}$$

6. *The from-to chart is an evaluation technique that is quantitative and results in usable measurable efficiency.* The from-to chart is a good example of why measurement and evaluation techniques are so valuable.

7. *The cost evaluation technique is the most complete and most used evaluation technique.* The total cost of the project, the operative costs, the sales price, and the forecasted sales must all be determined with great accuracy and a return on investment (ROI) needs to be calculated. This results in budgets and operations plans that result in the profit goals of the company. The cost evaluation technique is mandatory for new plants, and good management is a "must" for continuing operations.

All the above measurements can be evaluated on a continuing basis and charted. Figure 14–12 shows an example of the use of ratios and key indicators of layout efficiency improvement.

It is important to point out at this time that, as with any evaluation process, it may be wise to take a holistic approach in considering the overall efficiency of the operations instead of considering any single metric in isolation. For example, whereas the overall goal may be to maximize the use of machinery and equipment in order to justify our investment and hence aim to attain 100 percent equipment utilization ratio, such expectations may be reasonable only if there is a demand for the output. For instance, in a just-in-time (JIT) environment, excess output may be more undesirable than equipment idle time, since in a lean and JIT environment, excessive and unwanted inventory constitutes waste.

QUESTIONS

1. The layout is only as good as what?
2. What are the two types of layouts?
3. What flow analysis technique depends on the plant layout?
4. What is a plot plan?
5. Which is more costly, a front foot or an additional foot of depth?
6. To where will you expand the factory? The office?
7. How much property should you buy?
8. What is the master plan?
9. What are the four methods of constructing a master plan?
10. Which is the most expensive? Why?
11. What are the most common scales? List two.
12. Is there a time when the architectural technique would be best?
13. What are 10 measurements of performance used in evaluating layout alternatives?
14. What should the trends be for the following:
    a. Distance traveled
    b. Automatic feet ratio and gravity feet ratio
    c. Aisle space
    d. Cube utilization
    e. Machine space utilization
    f. Percent automatic loading
    g. Machine utilization
    h. Percent material handlers
    i. Material handling ratio
    j. In-process time
15. What are the advantages (or disadvantages) of computer-aided facilities planning?
16. How does standardization in electronic data interchange technology affect facilities planning?
17. What role does virtual reality play in facilities planning?

CHAPTER 

Application of Computer Simulation and Modeling

OBJECTIVES:

Upon the completion of this chapter, the reader should:

- Understand the concept of simulation
- Be able to use some basis simulation software program
- Be able to apply simulation to evaluate layouts and material flow

INTRODUCTION

Recent advances in computer hardware and software development have affected most areas of business and industry, and the arena of facilities planning is no exception. The use of computers in facilities planning is by no means a new or novel idea. A variety of software packages has been around for a number of years. Some older and more traditional programs have given way to state-of-the-art packages and have dropped from the scene. The advent of faster and more powerful computers and the development of user-friendly, menu-driven software packages have made the use of technology much more appealing and, therefore, more commonplace.

Today computer simulation and modeling are becoming an integral part of the planning and decision-making process of the manufacturing and service segment of American industry. As a result of market dynamics and fierce global competition, manufacturing and service enterprises are forced to provide a better quality product or service on a more cost-effective basis while trying to reduce significantly the production or service lead time. The quest for the competitive edge requires continuous improvement and changes in the processes and implementation of new technologies. Unfortunately, even the most carefully planned, highly automated, sophisticated manufacturing systems are not always immune from costly design blunders or unanticipated failures. Among the common examples of these costly mistakes are insufficient space to hold in-process inventory, mismatches in machine capacities, inefficient material flow and unexpected bottlenecks, miscalculations in expected ROI (return on investment) for a particular piece of equipment, congested paths for automatic guided vehicles (AGVs), and the list goes on.

Several generations of computer simulation and modeling have been used to solve complicated mathematical problems or to provide insight into sophisticated statistical distributions. The power of the new-generation software has dramatically increased the application of computer modeling as a problem-solving tool and has created new opportunities for productivity improvements in the area of facilities planning. Simulation packages that are currently available no longer require a strong background in mathematics or computer programming languages in order to perform real-world and interactive simulations. A number of user-friendly advanced simulation packages are available that allow the user to simulate the working of a factory, performance of various material handling equipment, a just-in-time inventory environment, a warehousing and logistics problem, or the behavior of a group technology system. These simulation packages have demonstrated to be a valuable aid in the decision-making processes. They also require a relatively small investment of time on the part of the novice in order to develop a working knowledge of the simulation process.

The use of computer simulation is not limited to the manufacturing environment. The health care industry, as managed health care, public policy issues, and reform initiatives gain momentum, is facing pressure to reduce costs and to provide better service. Many health care facilities are turning to computer simulation as a route toward salvation. Models to study emergency room activities, patient-tracking procedures, outpatient surgery systems, and physician and other resource allocations are primarily, but definitely not exclusively, the focus of concern.

DEFINING COMPUTER SIMULATION

"Simulation" is defined as an experimental technique, usually performed on a computer, to analyze the behavior of any real-world operating system. Simulation involves the modeling of a process or a system where the model produces the response of the actual system to events that occur in the system over a given period of time.

Simulation can be used to predict the behavior of a complex manufacturing or service system by actually tracking the movements and the interaction of the system components. The simulation software generates reports and detailed statistics describing the behavior of the system under study. Based on these reports, the physical layouts, equipment selection, operating procedures, resource allocation and utilization, inventory policies, and other important system characteristics can be evaluated.

Simulation modeling has two important characteristics that set simulation apart from other forms of analysis. Simulation modeling is dynamic, in that the behavior of the model is tracked over simulated time. A simple what-if analysis is static in nature. The state of a static model does not change as a function of time. If you were to simulate the roll of a die, then the output of the model would not be affected by time. However, if you were to simulate the utilization or breakdown of a machine, or the accumulation of work-in-process inventory at a workstation, then these phenomena would not be static in nature. Equipment utilization or breakdown, material handling and transportation systems' behavior, and interaction among various activities in a manufacturing cell, for instance, are dynamic in nature and the output of such models is a function of time.

Second, simulation is a stochastic model rather than a deterministic one. If, for example, the mean time to failure (MTTF) for a piece of equipment is 1,000 hours, it does not mean that the equipment will necessarily fail once every 1,000 hours. Such an expectation would create a deterministic model. In the real world, however, the breakdown follows a particular statistical distribution, that is, exponential, weibull, and so on. A random simulation model allows for these real-life breakdowns or other random occurrences.

ADVANTAGES AND DISADVANTAGES OF SIMULATION

The most obvious and clear advantage of simulation can be construed from the meaning of the word itself: to make a pretense of, or to mimic an entity. Simulation allows us to observe the behavior of a system, a group of individuals, the layout of a facility, or a cluster of equipment without actually disturbing the real individuals, the layout, or the equipment. With the aid of the simulation software, a representation or a model is developed. This model can then be easily manipulated and studied for various effects. A number of scenarios can be played or "what-if" questions asked. What if a new or a different material handling system is added, what if an additional lathe is employed, what if the current arrangement of the equipment is reconfigured, or what if the current number of operators is changed? Consider the tremendous expenditures of time and money and interruption to the production activities if we were to actually implement these changes to the shop floor to study their effect. Further, consider whether these changes actually result in a greater deterioration of productivity. These and many other questions can be easily asked and the effects of these changes on the system can be observed and measured without actually changing the real work environment, hence avoiding unnecessary expenses and disruption to the production schedule. Through simulation, once a feasible solution is found, it can be implemented with a well-understood and expected outcome.

Some advantages that have given simulation widespread acceptability are that simulation is relatively flexible and straightforward. It can be used to analyze large and complex models that may not easily lend themselves to mathematical models.

Furthermore, simulation allows the study of the interactive effects of many components in a dynamic and stochastic environment, with the distinct advantage of providing the investigator with a clear visual effect. For example, the effects of adding an additional operator in a manufacturing cell, or the advantage (or the disadvantage) of an additional piece of equipment on the machining center and its overall effect on the plant output, can be studied visually in real time. Besides its technical advantages, simulation's basic concepts are easily comprehended. Thus, a simulation model is often easier to justify to management and customers than most analytical and mathematical models.

The main disadvantage of simulation is that development of some very complex models may be quite expensive and time consuming. Indeed, a corporate planning model, or large manufacturing plant with all its components, activities, and services, may actually take years to develop. An analyst may, therefore, settle for a quick and dirty estimate, which may not reflect all the essential facts. Another disadvantage is that some simulations do not generate optimal solutions to problems and generate results based only on the model presented for analysis. It is then incumbent upon the planner to study under simulation various scenarios in order to find the best alternative. Allowance for the randomness of the process coupled with the trial-and-error approach can produce different results in repeated runs that may lead to difficulty in interpretation of the output. However, an astute planner can take advantage of the randomness of the output to emphasize the randomness of most real-life occurrences and to put in place remedies for these uncertainties that are sure to occur.

SIMULATION IN FACILITIES PLANNING

Facilities planners can use simulation to study various aspects of facilities design, capacity planning, inventory policies, office and parking lot layouts, quality and reliability systems, warehousing and logistics planning, and maintenance scheduling, to name a few possibilities. They can evaluate alternatives in material handling systems such as fork trucks, AGVs, automated storage and retrieval systems (ASRS), and transport and accumulation conveyors. By using simulation, the planner can compare different alternatives and study various scenarios to determine, for example, whether in a given situation a conveyor would be more effective than a robot or an AGV.

Currently, a number of user-friendly advanced simulation and layout planning packages are available to facilities planners at affordable costs. These software packages offer tremendous potential in aiding in the process of planning and optimizing the entire facility, a complete production system, or only a small department, or as a tool in balancing a simple assembly line. The limitations go only as far as the planner's imagination.

Simulation can be used to plan a flexible manufacturing system (FMS) environment. The purpose of an FMS is to produce a wide variety of parts where the production schedule can change quite often. An FMS consists of complex software and an integrated network of material handling systems. The system assigns different parts to different machines and allocates different resources to obtain maximum efficiency. Facility planners' understanding of the system can greatly improve by observing, through simulation, what kind of products are selected and how the resources are allocated. Furthermore, one is made aware of the problems that can arise and the corrective actions to be taken when the schedule or the quantity of parts is changed.

The use of computer simulation and modeling can also facilitate an understanding for non-normal probability distributions such as the exponential, the poisson, or the binomial. Contrary to popular belief or wishes, not all phenomena in a manufacturing facility, or in industry in general, have a normal probability distribution. Because most simulation packages are capable of analyzing the preliminary data to determine the most appropriate probability distribution for a given situation, a more accurate scenario can be developed for such stochastic processes. Machine utilization ratios, space requirements, inventory policies, material handling systems, and manufacturing cell capacities can be evaluated in virtual reality before costly implementation and blunders.

HOW SIMULATION WORKS

The purpose of simulation is to help the decision maker solve a particular problem. A basic outline for building a simulation model may be proposed as follows. This process of model building can be modified and restated to meet the planner's need. The approach can be used to approach systematically a facility planning problem and to work toward a logical solution.

1. *Problem definition.* Clearly define the problem and state the goals of the study so that the purpose is known; that is, why am I studying this problem, what do I hope to find out, and to what questions do I wish to find answers?

2. *System definition.* Define the boundaries and the restrictions of the system in terms of resource availability. You need to remember that every real-life system faces time, space, and financial constraints among others.

3. *Conceptual model.* Develop a graphical model to define system components, variables, and their interactions that constitute the system. Here the planner has an opportunity to use logic to construct the behavior of the system under study and to determine how these components will perform in concert or disarray.

4. *Preliminary design.* Decide on and select those factors that you think are critical in the performance of the system, and select the levels at which these factors are to be investigated; that is, what data need to be gathered from the model, in what form, and to what extent? Simulation studies can generate a vast "sea" of data in which the planner can drown without seeing the critical information. Do not obscure critical data with trivia.

5. *Input data preparation.* Remember the cliché, "garbage in, garbage out." Ascertain the integrity of input data. Identify and collect the data required by the model and understand that the output of the system is only as reliable as the input data.

6. *Model translation.* At this point, the planner will develop a working knowledge of the simulation package by formulating the model in the appropriate simulation language.

7. *Verification and validation.* The facilities planner must confirm that the model indeed represents the system it intended to represent and operates as expected, and the output is representative of the real system.

8. *Experimentation.* Now you can truly learn the power of experimentation and investigation. The planner can manipulate the system in a real-time environment and learn how various changes can affect the output of the process. Adding or deleting resources, or using a different type of resource, will affect the outcome of the process. These modifications and their long-term impact can be studied.

9. *Analysis and interpretation.* The planner can draw inferences from the data generated by the simulation. Once again, you can appreciate the conditions under which the input data were collected and realize to what extent the validity of the output is dependent on the validity of the input data.

10. *Implementation and documentation.* The results can now be recorded, documented, and implemented along with its uses and limitations.

Simulation modeling practices can be performed for a variety of reasons:

1. *Evaluation.* Determine and measure how well a proposed system design performs in an absolute sense when compared against set criteria. Does the system meet these criteria; that is, does it meet the production requirement, can it perform within the budget, and so on?

2. *Comparison.* Compare alternative designs to carry out a specific function. Planners can select from various alternatives by critically comparing them for cost, performance, and other factors.

3. *Prediction.* It allows the planner to investigate the performance of a proposed system under specific conditions over a period of time. Under the stipulated conditions, the performance of a system can be simulated over a period of hours, days, or even years, in a matter of minutes or hours.

4. *Sensitivity analysis.* Although there may be many variables operating in a system, only a few may critically affect the performance of the process. Sensitivity analysis helps determine which of the many factors and variables have the greatest effect on the overall operations of the system.

5. *Optimization.* Once the critical factors have been isolated, you can attempt to optimize the plan by establishing what factors or which combination of factors produces the best overall system response.

6. *Bottleneck analysis.* The facilities planner can discover the nature and location of bottlenecks affecting the flow of the system.

AN OVERVIEW OF LAYOUT AND SIMULATION SOFTWARE

Facilities planners today can greatly benefit from two distinct categories of software programs. The first classification consists of software packages that can aid in the planning and design of the layout. The second category includes simulation and performance analysis systems. Software systems such as STORM, FactoryCAD, FactoryPLAN, and SPIRAL can be placed in the former group of planning tools, whereas PROMODEL, FactoryFLOW, FACTOR/AIM, and ARENA may be included in the latter category.

In the following sections of this chapter, we will briefly examine some computer-aided facilities planning tools.

COMPUTER-AIDED LAYOUT DESIGN

Among the user-friendly, state-of-the-art layout tools that are available to the planner today is FactoryCAD. FactoryCAD is a powerful drawing tool that can be used for industrial and manufacturing layout. By customizing AutoCAD, FactoryCAD makes it easy to create, refine, improve, and edit either new or existing drawings. The package contains a tutorial, which introduces you to FactoryCAD using hands-on exercises. FactoryCAD allows the fixed-size items to appear at their actual size. These items may be user-generated or obtained from existing libraries and added to libraries for future use. Other FactoryCAD blocks may represent variable-size objects such as doors, windows, pallets, desks, and so on.

Toolbars are used for common commands that can produce detailed menus. Among many such toolbars, of special interest to the facilities planner may be the industrial and conveyor toolbars. The commands on the FactoryCAD industrial toolbar includes the bridge crane, jib crane, detail rack, pit, platform, booth, mezzanine, and guard rail. The conveyor toolbar features a variety of conveyors such as the automotive floor conveyor, parametric take-up, and traction wheel turn. These conveyors can be placed on the drawing by simply specifying the path and then selecting the desired size and type. FactoryCAD will do the rest. In addition to a vast library of object blocks, a variety of architectural blocks are also made available to the facilities planner.

The planner may also find the "animate" command of particular interest. Used with various materials handling equipment, the animate command tracks the path of the equipment to assure that sufficient clearances have been allowed.

FactoryCAD allows the drawing of objects in either two-dimensional (2-D) or three-dimensional (3-D) formats. By using the "2-D to 3-D convert" command, users can also convert 2-D objects drawn by FactoryCAD to 3-D objects by simply selecting the object. Model rotation can be accomplished with the same ease. Figures 15–1 and 15–2 illustrate 3-D presentations created using FactoryCAD. Figure 15–1 shows a conveyor system and Figure 15–2 depicts an assembly line area. Both drawings were created with the help of on-screen menus and existing libraries of icons and blocks.

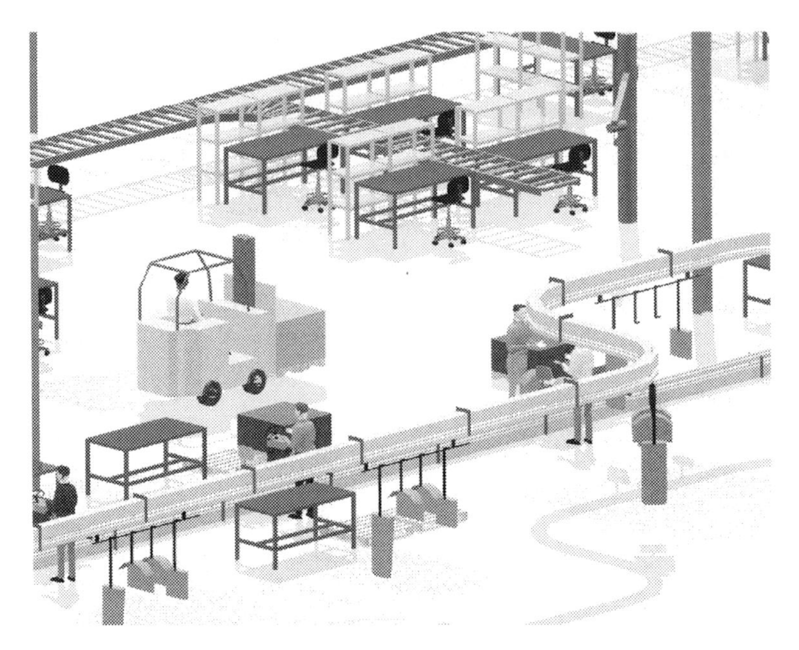

Figure 15–1 Three-dimensional conveyor system drawn with the aid of FactoryCAD (courtesy of Engineering Annomation, Inc.).

Figure 15–2 Three-dimensional illustration representing an assembly operation created using FactoryCAD (courtesy of Engineering Annomation, Inc.).

Layers and layer manipulations such as copy, move, import, export, freeze, and so on are among the features.

FactoryPLAN is another computer-aided facilities planning tool. FactoryPLAN is a tool for designing and analyzing layouts based on the desirability of closeness of different departments, work areas, offices, storage areas, or manufacturing cells. Through a series of on-screen, menu-driven options, the designers assign codes based on desired proximity, intensity of material flow, or an aggregate of the two values.

Whereas FactoryCAD can aid in the drawing of the plant layout, FactoryPLAN is a planning tool that can be used to analyze and optimize the layout. The most important aspect of this software is to assist with the analysis of the relationships among the different work areas in the plant. The program can be simply used to take the drudgery out of the manual method of constructing an activity relationship chart, or it can be used, in an interactive environment, to dynamically add, define, and modify work areas. Activity relationship codes such as A, E, and X and their respective weights are assigned. Furthermore, reason codes, such as shared equipment, personnel movements, or noise and dirt, and so on, can be placed on the graph or the drawing itself. With the relationship codes in place, FactoryPLAN calculates a quantitative measure or a score for the layout. FactoryPLAN can also draw flow lines on the drawing using various line sizes and colors to illustrate the heavy traffic patterns. The work areas can then be moved by simply clicking and dragging the object or the activity center and placing it at a different location to show how this may affect the facility in search for a better layout. The new drawing is scored. The score can be compared with that of the previous layout for a quantitative comparison.

Through the systematic analysis of activity relationships, the FactoryPLAN can be used to create designs for a new building or to analyze and redesign an existing layout. Integrating FactoryCAD and FactoryPLAN would allow the planner to move easily and quickly between drafting, planning, and evaluating various alternatives.

FactoryOPT, working with FactoryPLAN, determines the optimum locations of the activity centers, hence the optimum plant layout. The program creates an adjacency graph based on the data proximity and flow relationship data entered by the designer. Along with space information, FactoryOPT automatically creates a block diagram. The designer, however, has extensive control over the block diagram generated by FactoryOPT. The algorithms used by FactoryOPT can be manipulated by setting various variables. Through the various settings of these variables, 324 combinations are possible. Surely, facilities planners should be able to find the right combination for their desired layout algorithm.

Computer-Assisted Layout Performance Analysis

In the following paragraphs, two software packages that can aid the designer in the evaluation and analysis of various alternative layouts are briefly examined.

FactoryFLOW is probably the first layout analysis tool to integrate actual facilities drawings and material flow paths with production and material handling data.

As a result of this integration, FactoryFLOW provides the planner with the ability to view and manipulate spatial problems in a spatial medium. The software incorporates large amounts of data, including product and part files, production volumes, part routings, path distances, material handling data, and fixed and variable costs. Critical paths, potential bottlenecks, and production flow efficiency can, therefore, be readily and realistically determined. The system also provides an array of detailed text reports for inspection, including the cost of individual and combined moves.

The planner can, in real time, easily make changes to the model, routings, production volumes, material handling equipment, and other system variables in order to examine various alternatives. The analysis can help the designer to eliminate or reduce non-value-added steps, to reduce travel distances, to increase product throughput, to reduce work-in-process inventory, and to determine the material handling requirements.

FactoryFLOW creates graphs using "intelligent" flow lines that are ideal for flow problem analyses and can help illustrate total move distances, intensities, and costs—a very convincing justification to the management for a layout change and improvement. FactoryFLOW automatically generates numerical comparisons between flow paths and alternate layouts of machines and other work areas. FactoryFLOW places flow directionality and path legends on the drawing for ease of visualization. These intelligent lines can then be queried. The system calculates Euclidean and actual path and distance calculations. Detailed reports show individual and total move distances, costs, and the number of moves and move times. Other reports include distance-intensity charts and flow reports.

Figure 15–3 shows a manufacturing layout and the flow lines that are generated by FactoryFLOW based on the relationship codes supplied by the planner. Pay close attention to the length and the thickness of these flow lines, which illustrate the intensity of material flow between activity centers. Based on this type of flow analysis, activity centers can be rearranged to improve and optimize the layout. Figure 15–4 shows a significant improvement as illustrated by fewer and thinner flow lines. It is interesting to note that, for the same product demand, the total part travel distance was reduced by 65 percent.

FactoryFLOW can also perform aisle congestion reports. The report can categorize aisles based on usage and congestion. Each category, for example the top 25 percent or the next 25 percent, shows the number of trips per year and dollar costs of the trips per year. After the initial analysis, alternatives can be produced based on the results calculated by FactoryFLOW.

ProModel is a user-friendly simulation and analysis tool that is available to the facilities planner. The software can aid the planner to analyze an existing facility or to develop a new plant.

The simulation package, through an elaborate library of icons and on-screen menus, allows the planner to define an entire manufacturing facility, a distribution center, or a simple production cell. The planner defines the critical operating parameters or variables within the facility such as machines and buffers, parts and raw materials, routings, and part and material arrivals. Various icons that clearly

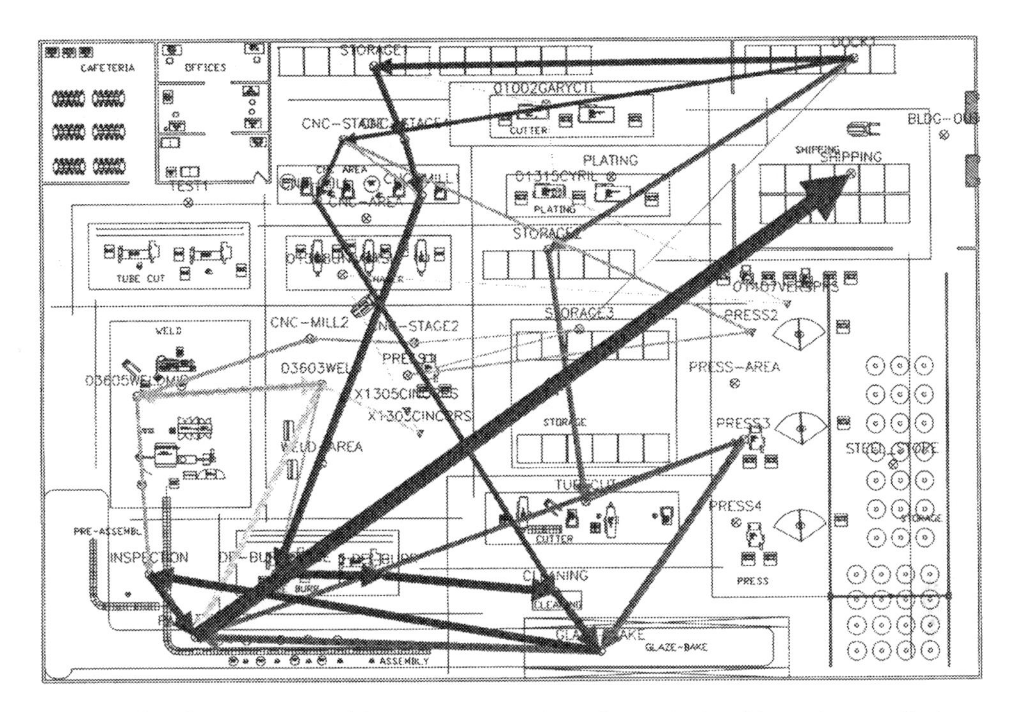

Figure 15–3 Flow lines generated by FactoryFLOW based on relationship codes supplied by planner (courtesy of Engineering Annomation, Inc.).

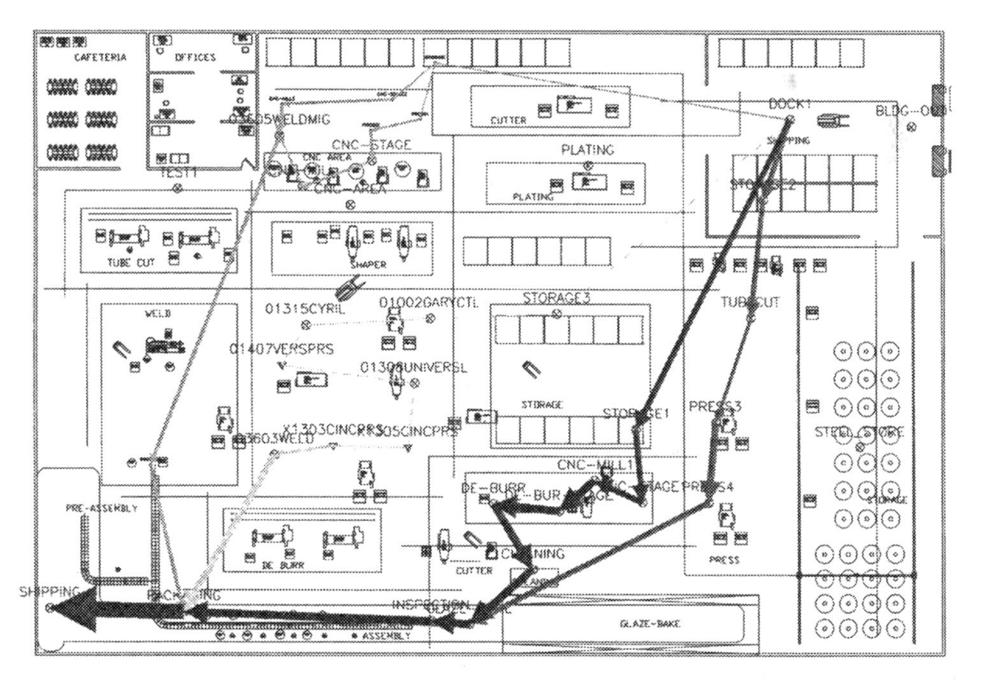

Figure 15–4 Flow lines after the analysis by FactoryFLOW (courtesy of Engineering Annomation, Inc.).

define the equipment, materials, and different parts and components represent these entities.

Figure 15–5 displays a representative sample of icons that are available to the modeler. Using realistic facsimiles of equipment, material handling systems, and parts, the facility planner can define the physical layout and arrangement of the plant. Furthermore, with the use of the autobuild feature, the planner is guided to define the quantity, the routing, and finally, the destination of each part.

Once everything is in place, the planner runs the simulation step. Although simulation can be run without animation, the animation adds a special dimension to the entire process of simulation. One can observe the entire, or a selected part of, the facility in motion on the computer screen. In addition to evaluating the layout based on a number of factors such as space and resource utilization, cost analysis, material flow, and overall plant throughput, a variety of what-if scenarios can be played in order to arrive at the ultimate, or at least near-ultimate, solution. Should we invest in a new machine? What is the effect of batch size reduction on the overall system? Would a change in routings affect the throughput? How? How would a change in the material handling system affect the process? These and many other scenarios can be easily played out and the long-term results can be obtained in a matter of minutes. Extensive statistics showing the results of the simulation runs are produced by the system.

Figure 15–6 displays a simple work cell featuring NC machining operations, a degreaser, and an inspection center, and Figure 15–7 shows a simplified kanban system.

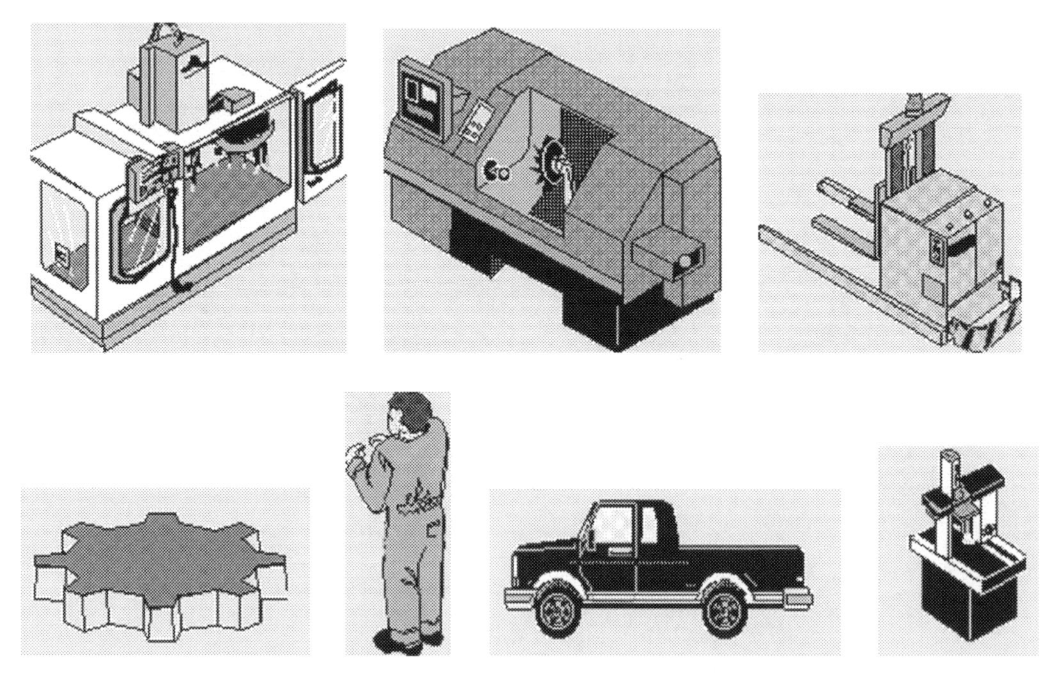

Figure 15–5 Sample of icons available to modeler (courtesy of ProModel Corp.).

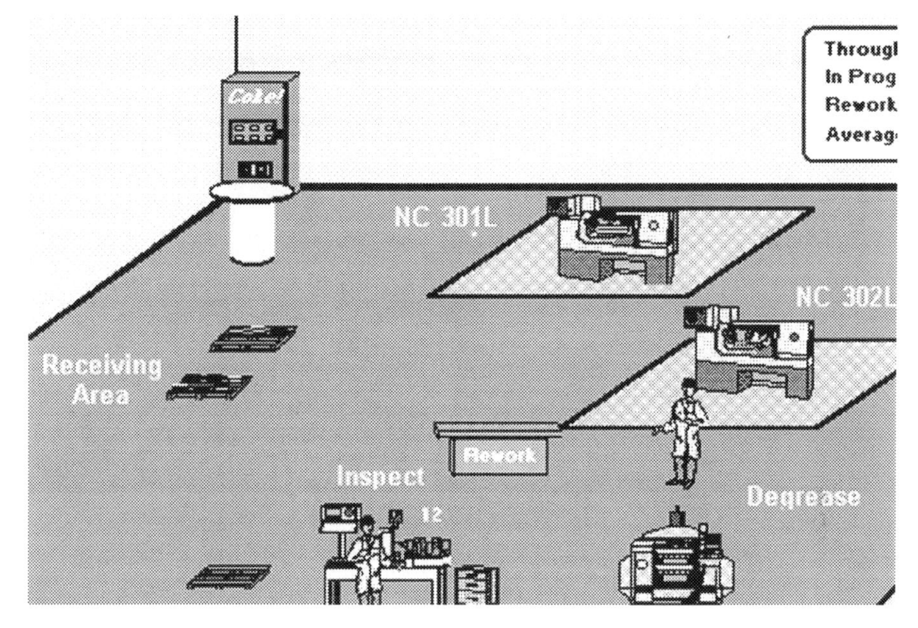

Figure 15–6 CNC machining center (courtesy of ProModel Corp.).

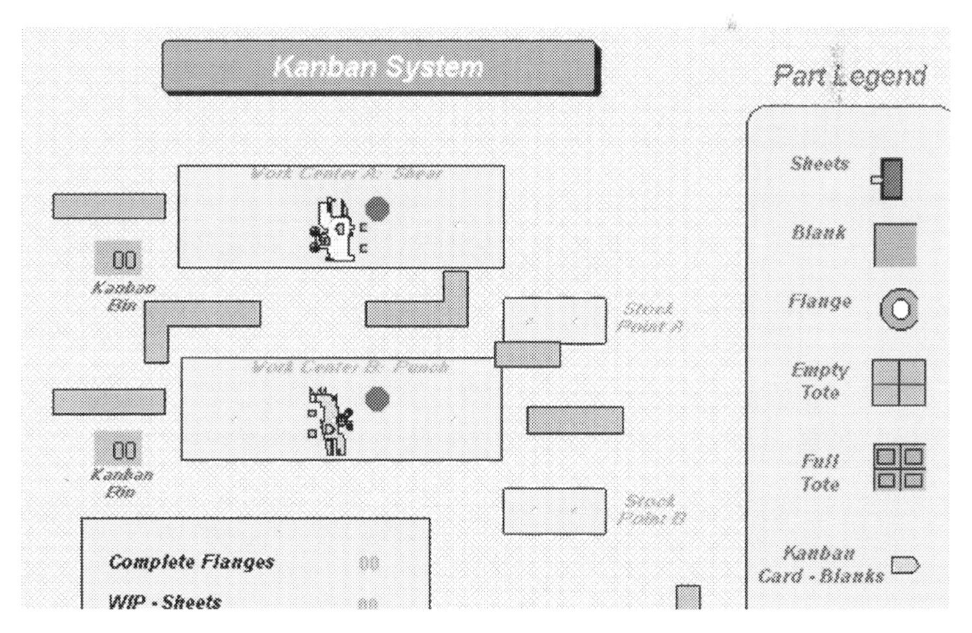

Figure 15–7 Kanban inventory management system (courtesy of ProModel Corp.).

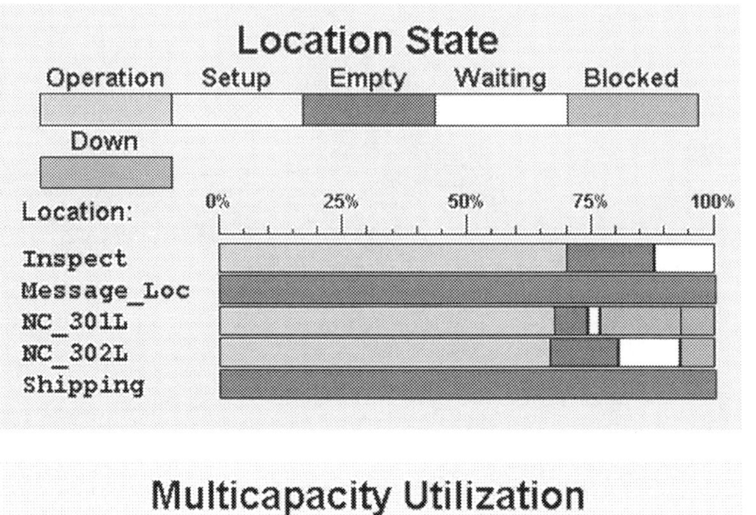

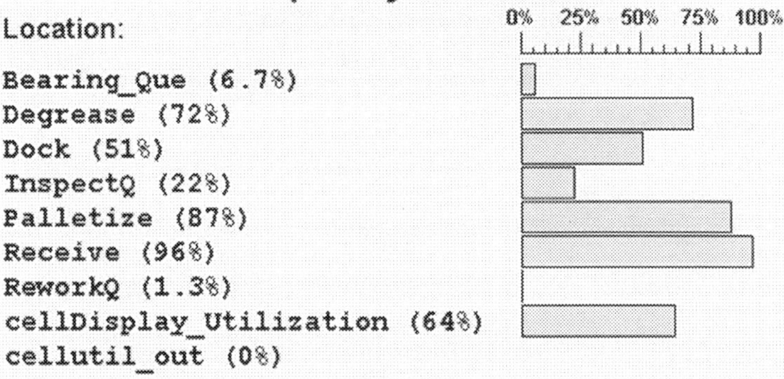

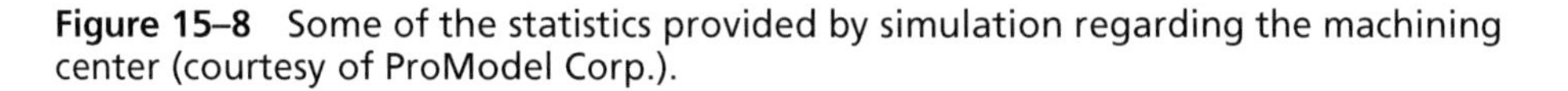
Figure 15–8 Some of the statistics provided by simulation regarding the machining center (courtesy of ProModel Corp.).

The data that are collected and analyzed by the program allow the user to make an informed decision regarding layout improvements and modification. Figure 15–8 provides just a brief glimpse of some of the statistics that are collected by the system.

If the planners propose to improve the efficiency of the layout by the addition of personnel or a certain piece of equipment, then they can make these modifications to the model. By playing various what-if scenarios and running the simulation, one can determine whether the proposed changes indeed have a positive effect on the manufacturing facility before implementing such changes on the factory floor.

LAYOUT-IQ: COMPUTER-BASED WORKSPACE PLANNING

This new edition of the textbook provides the reader with a link to Layout-iQ. Layout-iQ is one of the newest and most advanced computer-based workspace planning software, which can be used as an instructional tool in the classroom or by the professional layout designer. The software diagrams flow through the workspace using "Straight Line" and "Actual Path" models, and the system calculates travel distances for both models. Layout-iQ does not require AutoCAD in the background, so therefore, the user does not have to be well versed in AutoCAD or other CAD software in order to use Layout-iQ. The most efficient file format for Layout-iQ is the windows Meta file (wmf) format. However, since the software supports common graphical file formats, the user can import CAD drawings from most CAD systems.

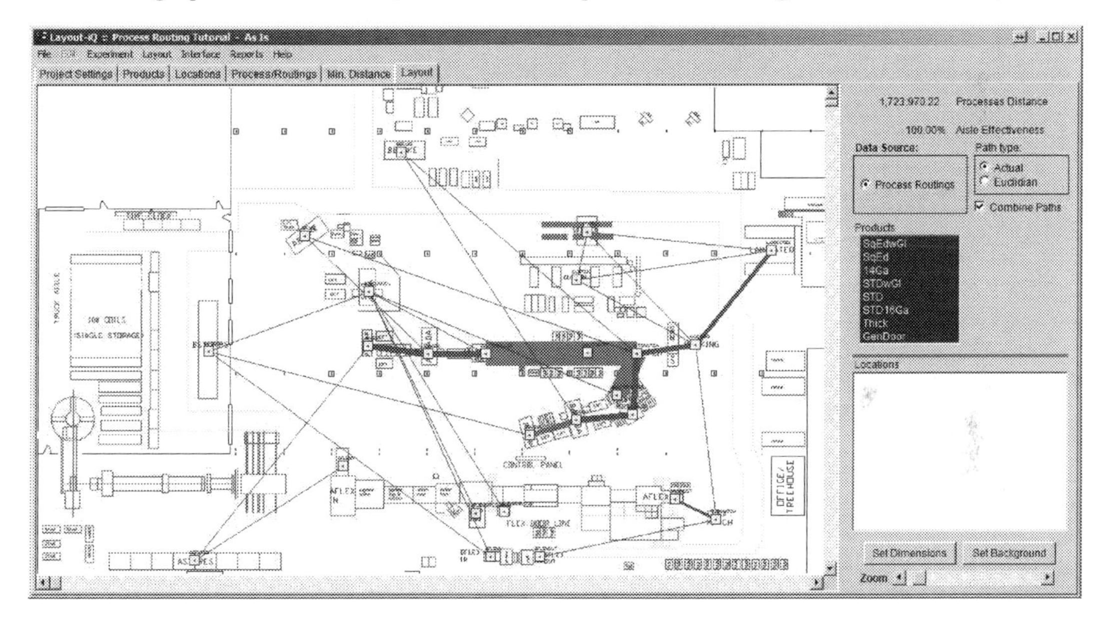

Figure 15–9 Layout-iQ Diagramming Engine.

The main design objective of Layout-iQ is to minimize travel distance and material handling cost in the design of workspaces. Layout-iQ calculates travel distances using both Euclidian (as the crow flies) and actual path distances. The software also calculates the efficiency of the aisle plan. The aisle plan efficiency is used to calculate the potential for improvement in a layout. It measures the efficiency of the proposed plan for material flow and determines the potential for improvements in the layout. Aisle plan efficiency is calculated by dividing the Euclidian travel distance by the Actual travel distance.

An additional objective, which is equally as important, is to provide a method to comply with design rules and regulations of the U.S. Government and to implement best practices. Best practices are often developed by industry associations, or are the result of research from academic institutions or government agencies, or they can be developed and maintained by companies such as architectural and design firms. Layout-iQ includes a built-in methodology to ensure that the final layout complies with the known rules and best practices of industry. Layout-iQ includes three unique modeling methods.

Process-Routing

The first method is called "Process-Routing" and it is designed to capture data directly from a manufacturing material requirement planning (MRP) system. The data template is similar to a routing sheet in an MRP system. This method of modeling is the most flexible in Layout-iQ because it allows the user to filter the flows based on product name and it provides the most information to the diagramming engine. Users can export the data directly from their existing computer systems into this file format. Due to the added flexibility of the "Process-Routing" module, users from those industries that do not utilize routings, such as healthcare, prefer this modeling method.

Product Name	Part Name	Flow Factor	From Location	To Location	Device	Unit Load
14 Gauge Door	A Panel	1	ASTORES	ASHEAR	FORK	50
14 Gauge Door	A Panel	1	ASHEAR	AAMADA	CRANE	50
14 Gauge Door	A Panel	1	AAMADA	AAUTOROLLER	CRANE	50
14 Gauge Door	A Panel	1	AAUTOROLLER	AUTOMATCH	CONV	1
14 Gauge Door	A Panel	1	AUTOMATCH	MARKING	FORK	10
14 Gauge Door	A Panel	0.07	MARKING	CUSTOM	FORK	10
14 Gauge Door	A Panel	0.87	MARKING	LAMINATEQ	FORK	10
14 Gauge Door	A Panel	0.06	MARKING	STIFFEN	FORK	10
14 Gauge Door	A Panel	0.01	CUSTOM	STIFFEN	FORK	10
14 Gauge Door	A Panel	0.06	CUSTOM	LAMINATEQ	FORK	10
14 Gauge Door	A Panel	0.06	STIFFEN	LAMINATEQ	FORK	10
14 Gauge Door	B Panel	1	BSTORES	BSHEAR	FORK	50
14 Gauge Door	B Panel	1	BSHEAR	BAMADA	CRANE	50
14 Gauge Door	B Panel	1	BAMADA	BBRAKE	FORK	50
14 Gauge Door	B Panel	1	BBRAKE	AUTOMATCH	CONV	1
14 Gauge Door	B Panel	1	AUTOMATCH	MARKING	FORK	10
14 Gauge Door	B Panel	0.07	MARKING	CUSTOM	FORK	10
14 Gauge Door	B Panel	0.87	MARKING	LAMINATEQ	FORK	10
14 Gauge Door	B Panel	0.05	MARKING	STIFFEN	FORK	10
14 Gauge Door	B Panel	0.01	CUSTOM	STIFFEN	FORK	10
14 Gauge Door	B Panel	0.06	CUSTOM	LAMINATEQ	FORK	10
14 Gauge Door	B Panel	0.06	STIFFEN	LAMINATEQ	FORK	10

Figure 15–10 Process-Routing spreadsheet ready to import into Layout-iQ,

From-To Trips

The second modeling method is called "From-To Trips" and this is the simplest method for defining flow in a workspace. The "From-To Trips" Table, as shown in Figure 15-11, is a matrix created in a spreadsheet in which all locations in the layout are listed in the row labels and in reverse order in the column labels. The number of trips between two locations is entered in the intersecting field. This Table supports variable bi-directional flow since each location-to-location relationship also has a special field in the Table to maintain data on return trips. Because of the simplicity of this method, it is often used for a quick and rough analysis. When populating the data Table for a rough analysis or examining a what-if scenario, since not all actual data are available, reasonable assumption can be used.

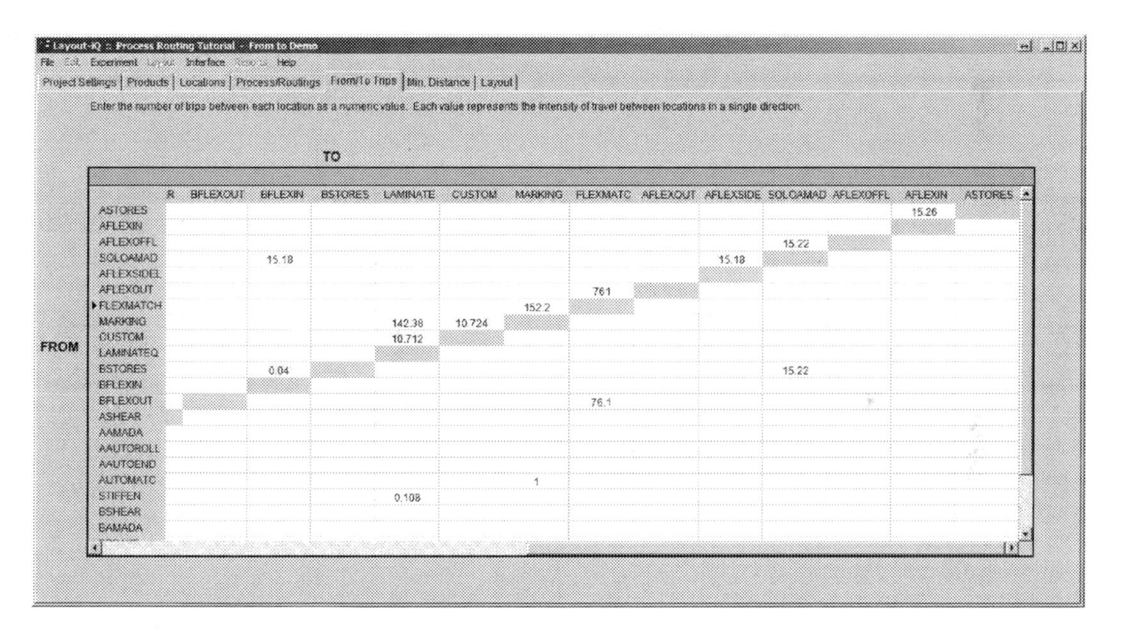

Figure 15–11 From-To Trips Screen

Subjective Analysis

The third modeling method is called a "Subjective Analysis" because it relies on the opinions of the workers who use the workspace. This methodology is often referred to as Systematic Layout Planning or "SLP". Users rank the importance of location relationships using an A, E, I, O, U, X ranking. Layout-iQ allows an unlimited number of people to provide opinions for the analysis. Once the opinions are entered in the system, the diagramming engine uses the collective data and calculates a subjective score for the layout. Users can diagram each opinion individually or select all or any combination of opinions in the analysis.

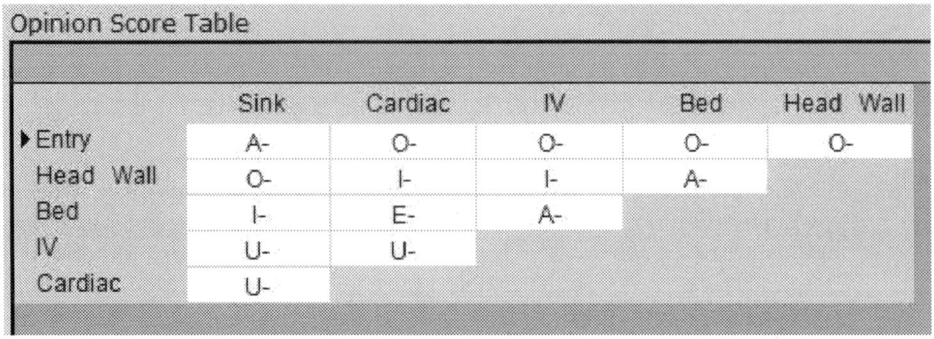

Opinion Score Table

	Sink	Cardiac	IV	Bed	Head Wall
Entry	A-	O-	O-	O-	O-
Head Wall	O-	I-	I-	A-	
Bed	I-	E-	A-		
IV	U-	U-			
Cardiac	U-				

Figure 15–12 Subjective Analysis Screen

Model Building Wizard

Layout-iQ includes a dual interface for model building. The default interface is a wizard that guides the user through the model building process. Each module requires information that needs to be entered into the system in a specific order because of the interdependencies of these data elements. The wizard guides the modeler through various data entry screens automatically, ensuring that all the data requirements are completed successfully and no data elements are missing. Once the data entry stage is completed, the system defaults to a tab interface where any section of data can be accessed directly without having to go back through the wizard. The organization of the tabs is in the same order that the wizard flows through the data entry templates, so the interface reinforces the methodology.

Most diagramming software do not allow dynamic changes to the layout and "force" material flow via the shortest path, even though this may not always be the most practical approach for the layout. These limitations dramatically simplify the diagramming engine, but they also significantly limit the accuracy and usefulness of the tool. In this respect, the Layout-iQ modeling engine is exponentially more complex than other modeling software, allowing the user more flexibility and control over the flow lines.

Developing alternative scenarios and testing alternative layouts is an important part of any layout and design project. Layout-iQ was designed to optimize the development and reporting of experiments and scenarios. Layout-iQ stores and maintains all the data for each experiment or what-is scenarios, which can be opened, analyzed, and referred to at any time during the project.

Layout-iQ is useful for the following tasks.

1. Layout-iQ can be used to find the best arrangement of equipment and locations in a given workspace.

2. Layout-iQ can be used to select the most efficient material handling devices for specific moves.

3. Layout-iQ can be used to make an informed and educated decision as to whether arrange equipment in a mass production orientation, in cells, or in a job shop orientation.

The software also includes reports that compare the results of various experiments and what-if scenarios side-by-side and recommends the best alternative. The efficiency of the aisle plan is a secondary metric that informs the designer on the potentials for improvement of a layout.

Tutorials and Modeling Exercises

There are three tutorials for use in the classroom that will allow students to learn the principles of layout planning. Each tutorial is designed to teach the student the principles of the three modeling methods (Process/Routing, From-To Trips, and Subjective Analysis). The tutorials also allow the students to develop new layouts and make improvements using the experimentation process built into the software. The tutorials are located in the help menu in the software and the software can be downloaded from the internet using the following link.

http://www.rapidmodeling.com/FileDN/Layout-iQInstaller.exe

CASE STUDIES

Three case studies from distinctly different areas of manufacturing and health care industries are presented below to illustrate the applicability of computer simulation and modeling.

Simulation in Manufacturing

A tire and rubber manufacturing company used computer simulation to assist in the implementation of a scheduling package in a high-volume facility. The purpose of the project was to develop an analysis tool with which the production planning team could pilot and evaluate a production schedule. Other issues such as storage capacity and utilization, tooling constraints, and the need for additional equipment were also under study. The model simulated various stages of tire building and curing operations and the relevant work-in-process storage needs. The model was also capable of varying the production schedule and the product mix, as well as the critical production parameters that would allow for the reduction of critical changeover costs, especially those costs due to labor and scrap.

The model allowed the manufacturer to compare alternative scheduling scenarios and to provide for testing and debugging of the schedules prior to implementation.

Simulation in Health Care

Health care systems likewise can benefit from computer simulation and modeling. A simulation study to assess and improve the operations of the emergency department at Florida health care facilities is a case in point. The facility, which handles nearly 60,000 patients per year, is composed of 33 rooms and is divided into three units. Each unit is staffed separately and has different hours of operation. The purpose of the simulation was to examine the sequencing of triage and registration activities, to examine the effect of bedside registration on nurse and physician utilization, and to provide for a more timely decision support system.

The model examined several different scenarios focusing on sequencing and location of triage and registration functions; using X-ray equipment, operation hours, and standing physician orders; and improving laboratory turnaround times. The model provided some primary results. First, both triage and registrations were shown to be activities on the critical path. That is, the amount of time required for these activities significantly affected the overall turnaround time. In addition, the model showed that the location of these activities did not affect the overall performance of the system.

The model also showed that additional X-ray facilities were not required for nonurgent patients regardless of the general belief that these patients were constantly bumped back for X-rays. The third point demonstrated by the model was that the reduction of operating hours in two of the units did not affect the third unit. The study showed that the hours could be cut back and adequate capacity for proper patient flow could still be maintained.

The findings of the study aided in implementing decisions that improved the overall utilization of the resources, reduced the overall length of stay for the patients, and positively affected patients' perception of the facilities.

Simulation in Waste Handling

After Congress changed the regulatory requirements and the U.S. Department of Energy (DOE) decided to eliminate the startup period for the Waste Isolation Pilot Plant, the need for an analytical tool capable of simulating material handling activities under varying conditions became apparent. A simulation model was developed in order to study and provide for safe and permanent disposal of defense-generated waste materials.

An initial 5-year startup and test period in preparation for full-scale operation was planned that would have provided an opportunity to evaluate and make necessary design modifications. The startup period, however, was eliminated and a fully operational start date was established.

A simulation model was successfully designed and used to determine the optimum configuration and utilization of the existing facility; to identify necessary equipment and process modification; and to determine the required resources to meet an initial reduced waste receipt rate.

QUESTIONS

1. Define *simulation.*
2. How does a mathematical model differ from a computer simulation model?
3. What is meant by a "stochastic" model and how does it differ from a "deterministic" model?
4. What is the difference between a dynamic and a static model?
5. Discuss some advantages and some disadvantages of computer simulation.
6. How do you envision using computer simulation in planning a manufacturing facility?
7. Explain how, and why, what-if scenarios are played when designing a facility.
8. What are the two distinct categories of computer-aided facility layout software packages?
9. When using simulation, why are problem and system definitions important?
10. Why is the integrity of input data significant?

C H A P T E R

Selling the Layout

OBJECTIVES:

Upon the completion of this chapter, the reader should:

- Be able to develop an appropriate presentation highlighting the strengths of the layout
- Understand the weaknesses of the layout
- Be able to make adjustments to remedy the weaknesses

The easy part is over. Now is the time to seek the approval of your months of hard work. This entire book has been devoted to collecting and analyzing the data to produce the best layout possible. If management can follow your reasoning, it will come to the same decision you did. The planner's job in "selling the layout" is to lead management through the reasoning process. The written project report should do exactly that: Lead the reader to your conclusion. The biggest mistake made by facilities planners is assuming management knows more than it does about the project. Assume it knows nothing (just as you did when you started this project) and show management the systematic approach you took.

THE PROJECT REPORT

The project report outline was introduced in Chapter 1 in the 24-step plant layout procedure. Now that you have laid out the toolbox plant, the following could be the specific outline for the project report:

1. The goal is to lay out a manufacturing plant and to support services in order to produce 2,000 toolboxes per 8-hour shift and to achieve the subgoals of
   a. minimizing unit cost
   b. optimizing quality

c. promoting the effective use of resources such as people, equipment, space, and energy
d. providing for the employees' convenience, safety, and comfort
e. controlling project cost
f. achieving the expected production start date
g. minimizing work-in-process inventory

2. Set a volume and plant rate (*R* value or takt time):
   a. 2,000 units per day
   b. 10 percent personal fatigue and delay allowance
   c. 80 percent historical performance
   d. an *R* value of .173 or 5.8 sets of parts per minute from every operation in the plant
3. Drawings of the product should include
   a. blueprints (Figure 2–1)
   b. an assembly drawing (Figure 2–2)
   c. an exploded drawing (Figure 2–3)
   d. a parts list (Figure 2–4)
4. Set a management policy. It should include
   a. an inventory policy—maintain a 30-day supply
   b. an investment policy—50 percent ROI
   c. a startup schedule—date
   d. a make or buy decision (Figure 2–5)
   e. an organizational chart (Figure 2–7)
5. The process design should include
   a. a route sheet for each "make" part (Figures 4–2 and 4–3), including time standards
   b. the number of machines required (Figure 4–4)
   c. the assembly chart (Figure 4–8)
   d. assembly time standards (Figure 4–9)
   e. conveyor speeds (paint 17.34 feet per minute, assembly 11.56 feet per minute) (page 109)
   f. assembly line balance (Figure 4–11)
   g. the subassembly line layout (Figure 4–12)
   h. the assembly and P.O. layout (Figure 4–13)
   i. the process chart (Figure 5–11)
   j. the flow diagram (Figure 5–14)
   k. the operations chart (Figure 5–17)
   l. the flow process chart (Figure 5–18)
6. The activity relationship should include
   a. the activity relationship diagram (Figure 6–1)
   b. the worksheet (Figure 6–2)
   c. the dimensionless block diagram (Figure 6–4)
   d. the flow analysis (Figure 6–4)
7. The workstation design should include
   a. machine layouts (Figure 7–5 to Figure 7–9)
   b. area determination (Figure 7–10)

   c. paint layout (Figure 7–11)
   d. aisles
8. Auxiliary services should include
   a. receiving (Figures 8–2 and 8–3)
   b. shipping (Figure 8–7)
   c. stores (Figure 8–16)
   d. warehouse (Figure 8–24)
   e. maintenance (see Chapter 8)
9. Employee services should include
   a. parking lots (see Chapter 9)
   b. employee entrances (Figure 9–3)
   c. locker rooms (see Chapter 9)
   d. toilets (see Chapter 9)
   e. the cafeteria (see Chapter 9)
   f. medical services (Figure 9–12)
10. The office should include
   a. an organizational chart (Figure 2–7)
11. Area allocation should include
   a. the total space requirements worksheet (Figure 13–1)
   b. the building size (Chapter 13)
   c. the dimensionless block diagram (Figure 13–2)
   d. the area allocation diagram (Figure 13–3)
12. Material handling systems and requirements
   a. state types and the number of material handling units
   b. calculate all conveyor speeds (feet per minute)
13. The layout should include
   a. the plot plan (Figure 14–2)
   b. the master plan (Figure 14–11)

THE PRESENTATION

The presentation of the project occurs at a management meeting where the project engineer (or engineers) presents the project plan. The presentation should be visual. Otherwise, the managers could read the report and there would be no need for a meeting. The two most visual items are the product model and the layout.

Using the product model, the presenter can cover

1. The goal and subgoals
2. The volume and plant rate
3. The product
4. The make or buy decisions
5. The process design

With the layout, the presenter can cover

1. The process design (further description on flow of each part)
2. Assembly and packout
3. The operations chart or flow process chart
4. The activity relationships and dimensionless block diagrams
5. The auxiliary services
6. The employee services
7. The office
8. The area allocation diagram

The plot plan will show how the plant is positioned on the lot. The presentation should include a cost budget; however, budgeting and cost allocation are beyond the scope of this book.

ADJUSTMENTS

The facilities planner should present the layout to every person who will listen. Your friends will criticize your project to help prevent costly errors, your enemies will tell you "great job, go get them" (meaning take it to management and make a fool of yourself). With every presentation, you will adjust the layout, making it better and better.

APPROVAL

Once you have a completed project (the schedule probably dictated the date), a formal presentation (or presentations) is required. The first presentation would be to your supervisor and the production manager. Their great experience will almost always point out problems with your plan. Depending on the magnitude of the problems, they may "sign off on" (approve) the project subject to the suggested changes.

More major changes may require a second presentation. Most companies will require many levels of approval depending on the amount of money being requested.

One of the authors presented a layout to the general manager of a plant and he approved. However, he did not have the authority to approve the $75,000 in expenses. After a trip to Los Angeles and then New York to present the proposal to top management, the project was finally approved.

The approval process is important, and those top managers did not get there without a lot of experience. Their input is valuable and will only serve to make a better project. When the project is successful, you will get credit because you made it happen. Whenever a top manager makes a suggestion that you incorporate, you make that individual a part of the project, and you have recruited another person who has an interest in the success of your project. Involve everyone to ensure everyone's cooperation. What management is approving most of all is an expenditure budget (limit). Successful project engineers and managers will make every effort not to exceed the budget. Project managers who come in underbudget are promotable.

THE REST OF THE PROJECT

Although this book is coming to an end, it would be a mistake to overlook a few important topics that have not yet been covered.

Sourcing

Sourcing is the process of finding suppliers that can provide the equipment, materials, and supplies needed for a project. These suppliers can be very helpful to the project engineer. Not only can they provide information on exact models, speeds, feeds, cycle times, and cost but they will also help with special design requirements and calculations and even do some layout work. It is quite normal to work with several suppliers of each piece of equipment, but these suppliers expect to get some of the jobs—not every one, every time. If suppliers feel that you are using them, they will be unwilling to help in the future. The result of sourcing is a list of equipment and supplies needed to create the layout you designed, and a specific source and price. The total dollars are a major part of your project budget. The day the budget is approved, you can spend 70 to 80 percent of the dollars because you have chosen the supplier and had a purchase order waiting for approval.

The purchasing department normally does all the company's buying, but sometimes (especially in building a new plant) the purchasing function is delegated to a project manager. This project manager is totally responsible for getting the job done and within budget. Either way, the purchasing department should be involved because of its special skills and knowledge. If the project manager works through purchasing, the purchasing person will want to know your desires and needs, and will appreciate the help.

Installation

Once the new plant is built or the existing plant is readied, the equipment starts arriving. This equipment must be placed and connected to power, water, or air. The delivery time varies from purchase to purchase, and some special pieces of equipment can take months to arrive. Once the equipment enters the plant, *installation* can also take months. A chrome plating machine or powder paint system are good examples of this process. The installation costs money, so it must be part of the budget. The installation takes time and must be a part of the schedule. Installation ends with the project engineer (or an engineer from the supplier) trying out the machine.

Engineering Pilot

An *engineering pilot* is a tryout of all the tools, equipment, and raw materials to see if the plant can make the product. At least one of every workstation must be available. The first small order of parts or raw materials should be available, and a few production people are asked to run every operation. There are always problems when starting up anything new, and the engineering pilot finds the problems with machines, tools, and materials so that they can be corrected. The results of an engineering pilot may be a few new products, but mostly, a list of problems that must be fixed before production begins.

The product engineers (designers of parts), the purchasing management (providers of raw material and finished parts), the quality control engineers (to anticipate quality problems), the tooling engineers (designers of tooling), the industrial engineers (designers of workstations and standards), and the facilities design project manager (the boss) all want to be part of the engineering pilot. After the pilot, a meeting is conducted where all the problems are reviewed, discussed, and assigned. This needs to be a very closely knit group of people.

Production Start

Within 2 weeks to a month of an engineering pilot, *production* will start. This is the most exciting and challenging day of a facilities planner's life. Everything up until this point has been fun. Seeing the plan come together is great, but when the production people show up en masse wanting to go to work, you, the supervisor and lead person, must train everyone. Everything is supposed to work as planned, but it never does, so you need to direct maintenance work, get parts reworked so they will fit, adjust machines, retrain people, and most important, make a list of what needs to be fixed before tomorrow morning. When the people go home at the end of the shift, your day is about half over. You must get everything fixed by tomorrow. This is a hectic time and most project engineers feel that they are the most productive during production's start.

Production efficiency for a second-year product averages 85 percent in a plant with a performance control system. First-year products average about 70 percent for the entire year, which means that in the beginning of the production year, performance may be as low as 50 percent or less. This low performance is normal and must be anticipated in order to maintain the delivery schedule. It also increases costs and must be part of the startup budget. Use the first-year efficiency of 70 percent when calculating the R value (plant rate).

Debugging and Follow-Up

Debugging is commonly used to describe the process of making the plan work—getting the bugs out of every operation in order to perform properly. Depending on the complexity of the product and the processes, debugging can last from 2 months to a year. After the debugging period is the *follow-up* period. The dividing line between debugging and follow-up is invisible, and there is no ending to the follow-up. Once you stop following up, improvement stops and the productivity and quality will start on a down slide.

CONCLUSION

The plant layout procedure described in the first section of this chapter is a good outline for most plant layout projects. Not every step is used in every project, but skipping a step must be done after thoughtful consideration. The toolbox plant did not need a from-to chart because every part flowed through the same sequence of machines. The results were obviously 100 percent, so why bother? This is an example of thoughtful consideration to eliminate a step.

Plant layout projects are mostly fun. The greatest influence you will ever have on a plant's effectiveness and efficiency (doing right things right) is laying out a new plant. A relayout is second. Industrial managers do not hand out large projects to project engineers unless they have proved their ability. Project engineers must prove themselves on small projects before they earn the right to work on large projects. Accept every project that is offered to you with enthusiasm and do the best job possible. You will earn the big jobs sooner than you imagine!

Answers

Chapter 1

1. Plant layout is the organization of the company's physical facilities to promote the efficient utilization of equipment; material, people, and energy.
2. Facility design includes plant location, building design, plant layout, and material handling.
3. Material handling is defined simply as moving materials.
4. The cost reduction formula is in fact a word, not a mathematical, formula. It consists of six questions regarding everything that can happen to a part as it moves through the facility. The questions are: why, who, where, what, when, and how. The purpose is to determine whether any given step can be eliminated, combined with another operation, moved to a different point in the sequence of the operation, or simplified. This procedure requires that you study the product in order to identify every step in the process and that you are able to justify the necessity of each step.
5. **a.** 50%
   b. 40–80%
6. **a.** Minimize unit cost.
   b. Optimize quality.
   c. Promote the efficient utilization of
      1) people
      2) equipment
      3) space
      4) energy
   d. Provide for employees'
      1) convenience
      2) safety
      3) comfort
   e. Control project costs.
   f. Achieve production start date.
7. A mission statement is a simple statement of quantity, quality product, and cost goals used to keep our minds on track.
8. Items 10 and 11 (product 1670).
9. This approach is a systematic approach that results (like magic) in a great plant layout.
10. The 24 steps (pages 11–13).
11. New plant, new product, design change, cost reduction, and retrofit.
12. It is best to treat both retrofit and new facility design similarly until the final layout and to make as few compromises as possible.
13. See pages 4 and 18.
14. See pages 4, 17, and 18.
15. Simulation is a technique by which a real-life situation can be mimicked. In the area of facilities planning, it can be used to play a variety of what-if games or scenarios. For example, how adding or deleting a piece of machinery or personnel may affect the overall outcome of the line or the facility.
16. A major aspect of ISO 9000 is to complete documentation and data collection. Various facilities planning tools will satisfy this requirement.
17. Random processes are those occurrences that take place without prior warning or planning such as machine breakdown. Simulation can be used both to understand and prepare better for these events.
18. The most common of these devices are the wands and scanners used at the checkout counters in the grocery stores. In the plant, they can be incorporated in various activities such as material handling to track inventories, WIP, equipment status, and so on.
19. Such changes are necessitated as the result of product changeovers; production volume increase or decreases; and adding, changing,

or deleting various operations and processes to the shop floor activities.

Chapter 2

2. Selling price, sales volume, seasonality, replacement parts.
3. The plant rate in decimal minutes (how fast workers must produce every part).
4. Working minutes, efficiency history, downtime, and the number of units to produce.
5. It determines the speed of the entire plant.
6. Blueprints, parts list or bill of materials, model shop samples.
7. Investment policy, inventory policy, startup schedule, make or buy decision, organizational chart, feasibility studies.
8. Figure 2–6 shows a list of parts that the company will make and a list of parts that it will buy.
9. Purchasing, because they will buy the part if it is cheaper on the outside.
10. The six causes are listed on page 39.
11. The indented bill of material shows various levels of assembly and subassemblies and the required parts that form various components. Also, see page 30.
12. In addition to the data provided by the flat bill of material, the indented bill of material shows the hierarchy of parts and components.
13. It can be used to construct the assembly chart and it can aid in visualizing the overall relationship between parts and assemblies.
14. It deals with concurrent planning of all aspects of product development, design, and manufacturing. The concept can be used in facilities planning to develop the proper relationship among various departments.

Chapter 3

1. To determine the number of machines and operators, direct labor cost, assembly line balancing, and scheduling; evaluate individual performance and incentive wages; and develop human resource budget. See also pages 52 and 53.
2. The time required to perform a task by a qualified, well-trained operator, working at a normal pace, doing a specific task.
3. Time standards are communicated by decimal minutes, pieces per hour, and hours per piece.
4. Productivity is defined as output divided by input. Labor productivity = earned hours divided by actual hours.
5. Predetermined time standard systems, stopwatch method, work sampling, standard data, expert opinion, or historical data.
6. Predetermined time standards system.
7. Stopwatch or time study.
8. Work sampling or expert opinion.
9. Standard data.
10. Predetermined time standards system.
11. 60%, 85%, 120%.
12. Takt time is the amount of time available to produce one unit in order to meet the product schedule.
13. **a.** Takt time *R* value or = 480 minutes per shift − 48 minutes downtime = 432 minutes
    432 @ 75% = 324 minutes divided by 3,000 units = .108 minute per unit
    Time standard = .284 minute per unit divided by .108 minute per unit = 2.63 or 3 machines.
    b. Time standard of .284 minute divided by 60 minutes per hour = .00473 hour per unit, divided by 75% = .00631 hour per unit, times $15.00 per hour = $.095 per unit.
    c. 480 minutes per shift divided by .108, times 75% = 3.495 parts.
14. 200, .005; 30; .033; 133.33; .0075; 1,200; .00083.
15. Measured observed or time is the result of the time study, the amount of time a particular operator has taken to perform a task. Normal time is the observed time adjusted by the pace rate or the operator's rating. Standard time is the normal time after allowances have been added.
16. Allowances are given for nonproductive factors such as fatigue, operator personal needs, and unavoidable delays to make the time standards practical.
17. Standard time = normal time + allowances.
18. 10 hours = 600 minutes − 30 minutes downtime = 570 minutes. At 85% you will get only 484.5 minutes from an average person, divided by 2,500 units = .194 minute per unit.

Line Balance Improvement

Operation No.	Time Standard	No. of Stations	Average Time	% Load	Hours per Unit	Units per Hour
SSSA1	.306	2.00	.153	99	.00517	194
SSA1	.291	2.00	.146	94	.00517	194
SSA2	.260	2.00	.130	84	.00517	194
SA1	.356	3.00	.119	77	.00775	129
A1	.310	2.00	.155	100	.00517	194
A2	.555	4.00	.139	90	.01033	97
A3	.250	2.00	.125	81	.00517	194
SA2	.415	3.00	.138	89	.00775	129
SA3	.250	1.61	.250	Sub	.00417	240
P.O.	.501	4.00	.125	81	.01033	97
		25.61			.06618	

19. **a.** .542
b. .637
c. 94
d. 10.63
e. $.127
f. $.134

20. 5,260.

Chapter 4

1. Determining how you are going to make each part, with what equipment, what time standard, tools, sequence of assembly, and so on.
2. Fabrication and assembly/packout.
3. A sequence of operations to make a part.
4. Part number, part name, quantity to produce, operation numbers, operation description, machine numbers, machine games, tooling needed, and time standard.
5. How many units per day are needed, what machine runs what parts, and what is the time standard for each operation.
6. Decimal minute.
7. The assembly chart shows the sequence of operations in putting the product together.
8. Number of units needed per minute times the distance between the leading edge of one unit and the next unit.
9. Hook spacing and parts per hook.
10. Equalize work, identify bottlenecks, establish line speed, determine number of workstations, determine product cost, establish percent load of each person, assist in layout, and reduce production costs.
11. See table below.
    a. .06618
    b. 3,104
    c. 25.44
    d. A1
    e. Yes, because the total hours per unit is less.
    f. $23,030 per year.
12. Mass production and job shop.
13. See table below.

Operation No.	Time Standard	Actual No. of Stations	Average Stations per Cycle	% Load	Hours per Unit	Pieces per Hour
1	.390	2	.195	78	.00834	120
2	.235	1	.235	94	.00417	240
3	.700	3	.233	93	.01251	80
4	1.000	4	.250	100	.01668	60
5	.240	1	.240	96	.00417	240
6	.490	2	.245	98	.00834	120

14. 94% (assuming 2, 1, 3, 4, 1, and 2 stations for operations 1–6, respectively). Standard time = 3.055 minutes. 13 operators times .25 (the 100% station) = 3.25 minutes. 3.055 divided by 3.25 minutes = .92 times 100 = 94%. Six percent inefficiency is caused by making these 13 people work together on a line instead of letting each work at his or her own pace (standard time). This is the cost of line balance, which must be offset by savings in material handling, lower work-in-process inventory, and reduction in product damage resulting from excessive handling and storage.

15. Adding stations to bottleneck operations often results in reduction in idle time, hence a reduction in cost per unit.

If the 100% (bottleneck) station has considerably more work than the next closely loaded station, adding another person here may reduce the average station time. With 10 people on the line, this would require a 10% difference. With 100 people, only 1% difference between the highest average station time and the second highest average time is necessary to pay for an additional person.

16. In process-oriented layout, similar equipment are grouped together (job shop); in product-oriented layout, machines are arranged to accommodate specific operation sequence per the routing sheets. Also, see page 124.

17. Group technology takes advantage of similarity in part geometry and processes without regard to the part's final destiny. Also, see page 124.

18. A group of machines may form a "cell" to perform a series of operations more efficiently. This may work best with the group technology concept. Also, see page 124.

19. One possible solution is as follows:

1	.455	2	.228	56	.01357	74
2	.813	2	.407	100	.01357	74
3	.233	1	.233	57	.00678	147
4	.081	1	.081	20	.00678	147
5	.945	3	.315	77	.02035	49
Totals	2.527	9			.06100	

Total standard minutes = 2.527. Nine times the 100% station's average time = 3.663. Efficiency = 2.527 divided by 3.663 = 69%. This is considered a very poor line balance. How would you improve this line? What about combining stations 3 and 4? How about adding a person to station 2, the bottleneck station?

1	.455	2	.228	72	.01015	95
2	.813	3	.271	86	.01575	63
3 & 4	.314	1	.314	100	.00508	190
5	.945	3	.315	100	.01575	63
Totals	2.527	9			.04723	

Efficiency = 2.527 divided by 9 times .315 (the 100% station time) = 89%. The initial balance used a total of .061 hour per unit. This line requires only .04723 hour per unit, a savings of .01377 hour per unit. At $15.00 per hour, you are saving approximately .21% per unit, or at the rate of 190 units per hour, $314 per 8-hour shift.

Another alternative would result in 76.5% efficiency (assuming 1, 2, 1, 1, and 2 stations for operations 1–5, respectively.)

Chapter 5

1. The path a part takes through a plant.
2. Minimize distance traveled, backtracking, cross traffic, and cost.
3. Fabrication and total plant.
4. String diagram, multicolumn process chart, from-to chart, and process chart.
5. 65%. See Figure A–1.
6. Using a standard form, show all the operations from the route sheet (Figure 4–1), adding transportation. Taking distances from flow diagram Figure 5–13, if place inspections, delays, and storages where they belong.
7. Flow diagram, operations chart, and flow process chart.
8. See Figure A–2.
9. Operations chart and process chart.
10. Arguments can be given for worker efficiency, reducing walking distances (and the resulting distractions), safety, and so on.
11. FactoryFLOW is a software program that aids with various aspects of flow analysis.

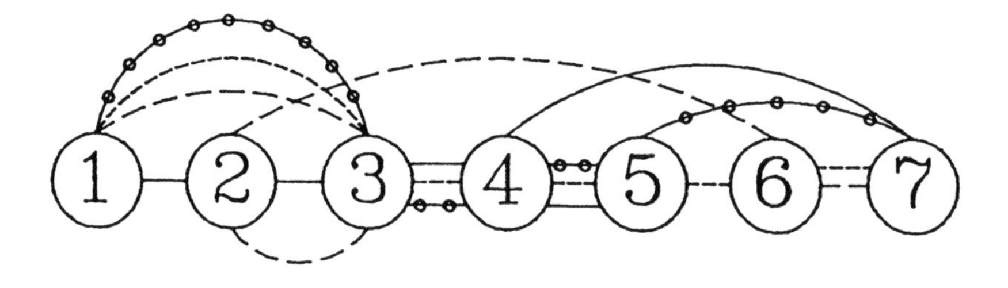

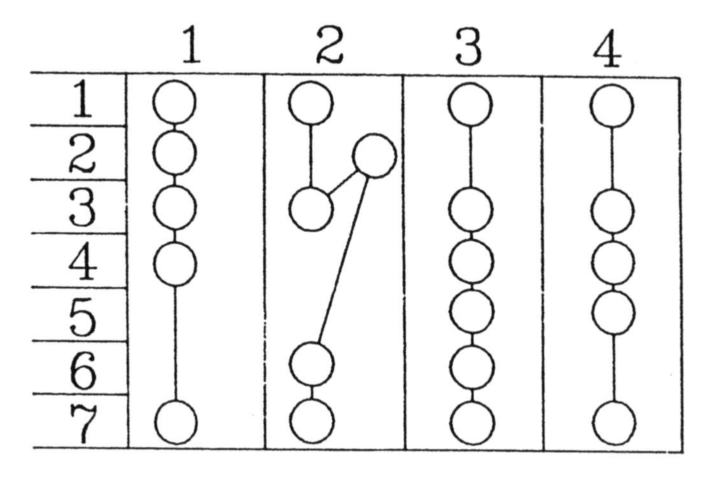

	1	2	3	4	5	6	7	N.P.	P.P.
1		① 1	(18) 2+3+4					10	19
2			① 1			⑧ 2		3	9
3		④ 2		8 1+3+4				10	12
4					⑦ 3+4		③ 1	6	10
5						③ 3	⑧ 4	7	11
6							⑤ 2+3	5	5
7								43	66

Figure A–1

12. Faster and more efficient. Analysis can be performed in real time without actually rearranging the facility. Major problems can arise from inaccurate or incomplete input into the systems.

Chapter 6

1. A = absolutely necessary that these two departments be close to each other
E = especially important
I = important
O = ordinary importance
U = unimportant
X = undesirable

2. These codes are set as a result of discussions with managers and other personnel.

3. **a.** A reminder as to why a particular relationship code was assigned.
b., c. See top of page 183.

4. A worksheet helps transfer information from the activity relationship diagram to the dimensionless block diagram without committing errors.

5. Templates in this case aid with the construction of the dimensionless block diagram. They are -inch squares with the closeness codes in the proper place.

6. Compare your solution with the class. Look for the best answer: the one with the least number of checkmarks.

7. 105.

8. 5, 10, 16.

9. Figure 6–7 shows the traffic patterns with the associated intensity and distances. Figure 6–4 shows the interdepartmental relationships.

Chapter 7

1. Anywhere, because you will always make improvements so the first sketch will be wrong.

2. The cheapest way to get into production because any additional expense must be cost-justified.

3. See Chapter 7, page 205.

4. Guidelines for efficient and effective workstation designs.

5. Doing right things.

6. Doing things right.

7. Aisles, work in process, and small miscellaneous extra room.

8. Students draw workstations for instructor's review.

9. Attention to motion economy, worker and equipment efficiency, human factor considerations all account for a well-designed job.

10. Ergonomic considerations will result in designing the workstation to "fit" the human body rather than attempting to fit the human body to the workstation.

11. Physical dimensions and measurements of the human body. Designing work and workstations while keeping these physical attributes in mind.

Chapter 8

1. Receiving, stores, warehouse, shipping, maintenance, tool room, and utilities.

2. Similar people, equipment, and space requirements.

3. Common equipment personnel, spaces are used, and reduced facility costs.

4. Space congestion and material flow.

5. No! What is most efficient.

6. Morning delivery and afternoon pickup is standard.

7. Less than truckload quantity (common carrier business) uses break bulk stations.

8. Only one (or a few) trucks would show up in the morning instead of many trucks showing up around the clock.

9. See Chapter 8, page 225.

10. A sequential numbering stamp and system to record the order of receiving.

11. The actual day number of the year, January 1 being number 1 and July 1 being number 183.

12. An over, shortage, or damage report made out by the receiving clerk sent to purchasing for resolution.

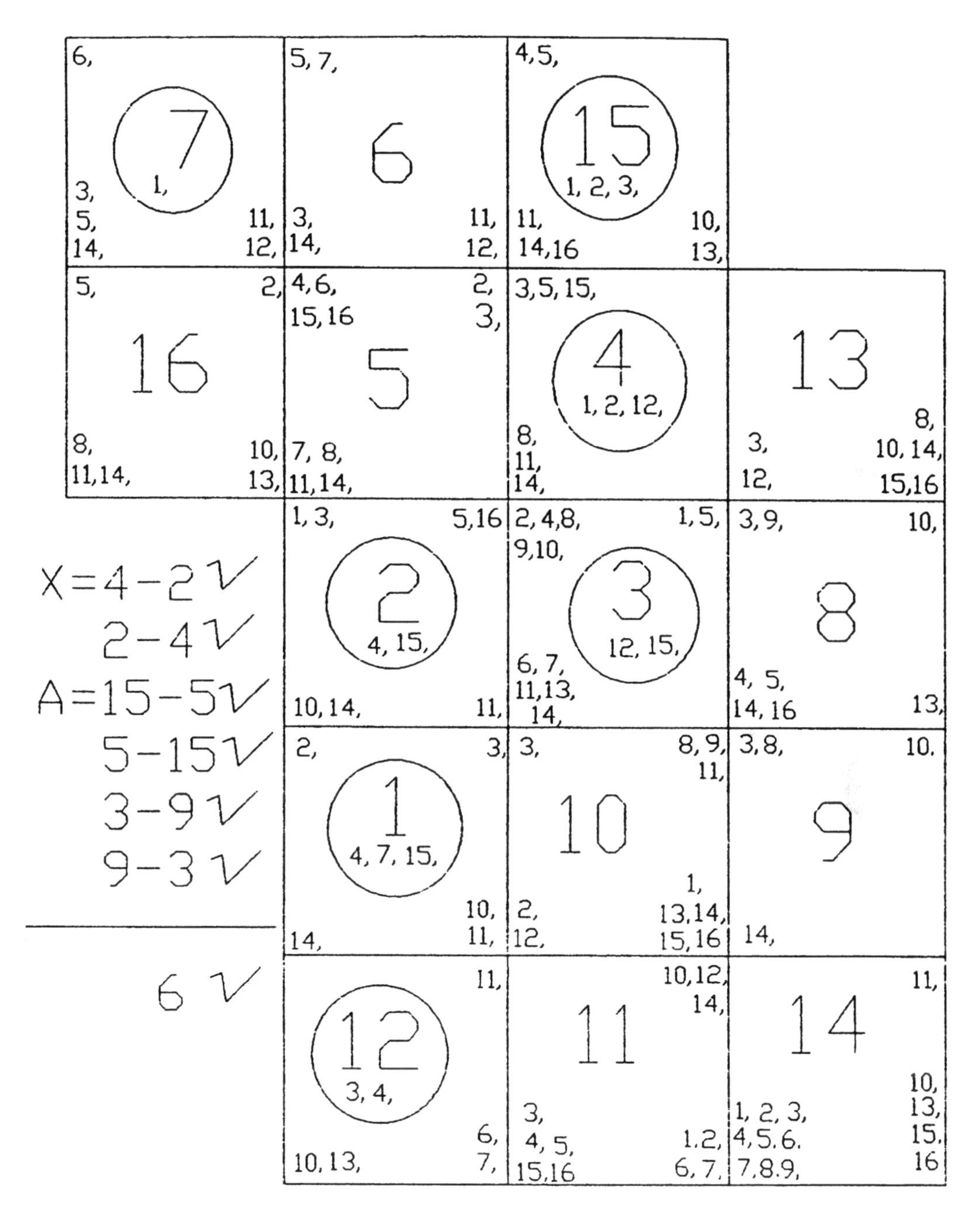

6,
7
1,
3,
5,
14,
11,
12,
5, 7,
6
3,
14,
11,
12,
4,5,
15
1, 2, 3,
11,
14,16
10,
13,
5,
2,
16
8,
11,14,
10,
13,
4,6,
15,16
2,
3,
5
7, 8,
11,14,
3,5,15,
4
1, 2, 12,
8,
11,
14,
13
8,
10, 14,
15,16
3,
12,
X=4-2 √
2-4 √
A=15-5 √
5-15 √
3-9 √
9-3 √
6 √
1, 3,
5,16
2
4, 15,
10, 14,
11,
2, 4,8,
9,10,
1, 5,
3
12, 15,
6, 7,
11,13,
14,
3, 9,
10,
8
4, 5,
14, 16
13,
2,
3,
1
4, 7, 15,
10,
11,
14,
3,
8, 9,
11,
10
1,
13,14,
15,16
2,
12,
3,8,
10.
9
14,
11,
12
3, 4,
6,
7,
10, 13,
10,12,
14,
11
3,
4, 5,
15,16
1.2,
6, 7.
11,
14
10,
13,
15,
16
1, 2, 3,
4,5.6.
7,8.9,

Figure A–2

13. A notice to the rest of the company that a product has been received.
14. Depends on arrival rate (trucks per hour) at peak and service rate (unloading time).
15. How many trucks must be serviced per hour.
16. Parking space, maneuvering space, and roadway.
17. Comparison of the pounds of finished product produced in one day to the size of a semitrailer that holds 40,000 pounds. If you produced 100,000 pounds of product per day, 2 1/2 trailers could bring the raw material into the plant.
18. See Chapter 8, page 229.
19. Trucking companies charge by the pound. Also helps determine the ratio of tare weight to the weight of goods shipped.
20. The trucker's authorization to remove the product from the plant and a part of the trucking company's billing process.
21. A place to hold raw material and supplies.
22. Raw material, finished parts, office supply, maintenance supply, janitorial supply, and so on.
23. Size of parts, number to be stored.
24. A inventory items are the 20% of the parts that account for 80% of the value of material. B items are 40% of the parts that account for 15% of the value, and C items are 40% of the parts that account for 5% of the value. If you can reduce the inventory level of A items, you can reduce the space requirements and inventory carrying costs.
25. About 25% of the value of the inventory each year and includes interest, taxes, insurance, space, utilities, damage, and obsolescence.
26. Just-in-time is an inventory policy that stresses having only enough inventory to run a few hours.
27. Maximize the utilization of the cubic space, provide for immediate access to everything, and provide for the safekeeping of the inventory.
28. Review Figure 8–8.
29. The inventory curve (Figure 8–8) shows that on the average only ½ the inventory is on hand (it is full on the first day of receipt and empty on the last day), so when material comes in, it is placed anywhere an empty spot exists.
30. Every location in the storeroom is identified with a location number. When something is placed in this location, the location number is recorded in the locator system.
31. One foot of aisle access, on both sides of the aisle. You have two aisle feet of access with every foot of length of the aisle.
32. See layout in Figure A–3.
33. A location for storage of a finished product.
34. Fixed locations and a small amount of everything.
35. Safekeeping of finished goods and maintaining some stock of every product the company sells.
36. A function of the warehouse that collects a customer's ordered goods.
37. By identifying the most popular items sold by the company and locating those items more conveniently.
38. The layout of cartons on a pallet to ensure safe loads and maximum cube utilization.
39. A form of balcony built over an area to use overhead space.
40. From 2 to 4% of the plant's personnel.
41. Item tracking and inventory management and control.
42. Aids with categorizing inventory items from the most frequently used or accessed to the least frequently needed items.
43. Items are not assigned a specific location but are placed where space is available. It reduces wasted and idle space.
44. In a random location storeroom, a locator file keeps track of quantity and location of various inventory items.
45. Fluctuation in order lead time and usage rate.
46. **a.** For an average inventory level, 55, 12, 141, 56 pallets.
    b. 22 racks, 308 aisle feet.

Chapter 9

1. See Chapter 9, page 264.
2. Depends on the number of employees and the employees to parking space ratio.

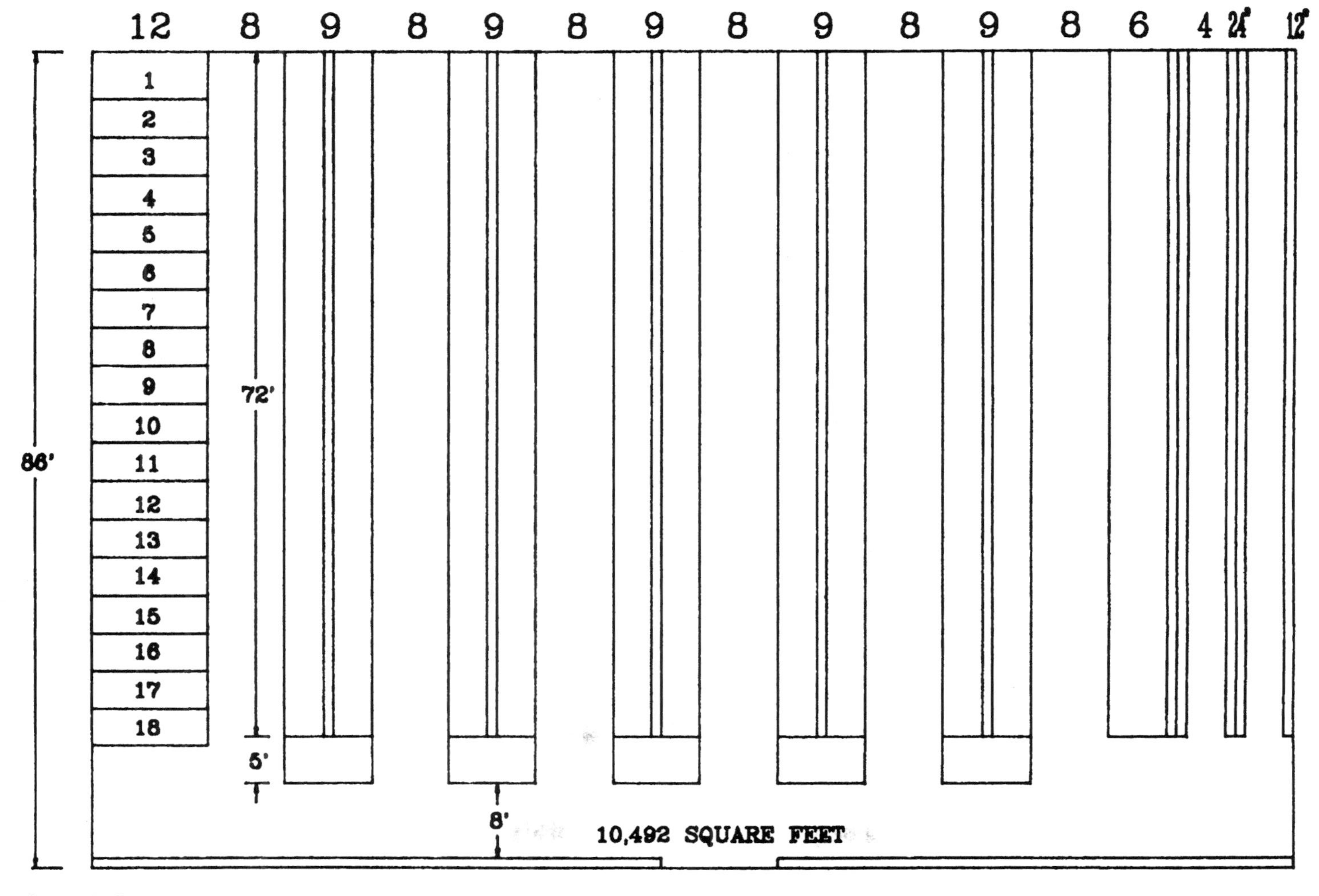
12
8
9
8
9
8
9
8
9
8
9
8
6
4
24"
12"
1
2
3
4
5
6
7
8
9
10
11
12
13
14
15
16
17
18
86'
72'
5'
8'
10,492 SQUARE FEET

Figure A–3

3. 250 square feet.
4. Security, time cards, bulletin boards.
5. Parking, locker room, restroom, and cafeteria.
6. Gives employees space for street clothes, work clothes, personal things such as lunches, coats, and so on.
7. Four square feet per employee.
8. No farther than 500 feet from any employee. No fewer than two (one men's, one women's).
9. The building code will tell you.
10. Normally 15 square feet is required for toilets, washbasins, and doorways and 50% extra for the aisle. A good rule of thumb is 60 square feet per toilet, and this would include everything.
11. Cafeteria, vending machine, mobile vendors, dining rooms, off-site.
12. Activity relationship diagram plus an outside wall close to restrooms and locker rooms.
13. 10 square feet per employee.
14. Located within 200 feet of every person.
15. 15 square feet each—including drinking space.
16. As little as possible, but 25% would be outstanding.
17. 500 employees equal one nurse.
18. From 36 square feet to 300 square feet per nurse per shift. Three nurses could occupy one 300-square-foot area if one nurse were assigned to each shift.
19. Americans with Disabilities Act (ADA) of 1989 requires that employers provide special and properly designed parking spaces and create a barrier-free environment in all aspects of the facility. The ADA requirements are a matter of the law.
20. Child care, outlet store, gym, and so on.

Chapter 10

1. Material handling is the function of moving the right material to the right place at the right time in the right amount in the sequence and in the right position to minimize costs.
2. Part numbering, location, inventory control, standardization lot size, order quantity, safety stock, labeling, and automatic identification techniques.
3. The one that produces the lowest unit cost.
4. See "Goals of Material Handling," page 290.
5. The College Industry Committee on Material Handling Education, sponsored by the Material Handling Institute, Inc.
6. A summary of generations of experience in material handling engineering. A guideline for the application of sound judgment.
7. See Figure 10–3.
8. See "The Material Handling Problem Solving Procedure," page 301.
9. Some of the activities that can possibly be incorporated with material handling may include sorting, counting, inspection, and inventory management to name a few.
10. Via the use of automatic and built-in scanners and other automatic data collection devices.
11. Automatic routing of parts and inventory management. Can you discuss others?
12. Safety and ergonomic and human factor considerations.
13. The cost of ownership (purchase, maintenance, etc.) and the cost of operation (training, energy usage, etc.).
14. Avoidance of work-related injury and cumulative trauma disorders.
15. By providing efficient and timely material flow and eliminating the need for storage between operations.
16. Movement, quantity, time, space, and control.
17. The adverse long-term effect of an activity (lifting, bending of the wrist, etc.) on the operator is referred to as CTD.
18. Through proper design so that stress and strain on the body are reduced.

Chapter 11

1. **a.** Point-to-point or fixed path
   b. Fixed-area, variable path
   c. Variable-area, variable path
   d. Auxiliary equipment
2. Such as canisters used at the bank drive-in windows. It can only serve the path where it is installed.

3. See Figures 11–10a and 11–10b as one possible solution.
4. The bed is magnetized and can be used for ferrous materials.
5. Material handling device that helps with lifting, turning, rotating, and positioning items.
6. It attempts to eliminate storage steps between receiving and shipping items by moving the product from receiving through the facility to its ultimate destination uninterrupted.
7. No. Automation is not necessarily the solution to every material handling problem. The primary goal should always be to eliminate the move! Simple mechanical devices may just do the trick without resorting to automation.

Chapter 12

1. See Chapter 12, pages 399–400.
2. Supervisor's, open space, conventional, and modern.
3. Easy communication, common equipment, less space needed, easier to heat and cool, easier to supervise, easier to change, common files and literature, easier to clean and maintain.
4. Lack of privacy, noise, status, confidentiality.
5. Privacy, point of use storage, second floor, centralized or decentralized, flexibility, conference room, library, reception area, telephone system, copy machine, mail, file storage, word processing, aisles, computers, lighting, vaults, expansion.
6. Organizational chart, procedures diagram, Communications force diagram, activity relationship chart worksheet, dimensionless block diagram, office space determination, and master layout.
7. It would tell you the number of people and their level in the organization.
8. 200 square feet per employee.
9. See Figure 12–11.
10. It tells who talks and works with whom.
11. Circles and lines. 4 lines = absolutely necessary that these two departments (or people) be close together; 3 lines = especially important; 2 lines = important; 1 line = ordinary importance.
12. The communications force diagram.

Chapter 13

1. Dividing the building's space among the departments.
2. See Figure 13–1. A collection of space requirements for every department and service area to develop the total plant space needs.
3. Under floor, overhead, floor, in the trusses, and on the roof.
4. By using the golden rule of architecture. A building is most efficient if it is twice as long as wide. This is a 2:1 ratio, rounded off to the nearest column space. This dimension will be the width, and twice this dimension will be the length.
5. See Chapter 13, pages 430–432.
6. An area allocation diagram, see Figure 13–3c.
7. Mezzanines, racks, shelves, overhead conveyors, and so on.
8. **a.** Locker rooms because they are used less often.
   b. Accounting because it has fewer guests.
   c. Old files because they are used less often.
9. Columns are posts that hold up the roof.
10. The distance between columns.
11. **a.** feet
    b. feet*
    c. feet

*May be one bay longer or wider.

12. The dimensionless block diagram.
13. Advantages are it frees up prime space that can be used for manufacturing or other heavy activities; it separates office areas from noise, dirt, and manufacturing-related hazards. Possible disadvantages may be increased difficulty in communication and supervision.
14. Foundation and roof.

Chapter 14

1. As the data backing it up.
2. Plot plan and master plan.
3. The flow diagram is drawn on a plant layout.
4. Shows how the building, parking, and driveways fit on the property.

5. A front foot.
6. To the *back* of the plant and *up* in the office.
7. Ten times the building space needed.
8. The finished product of a plant layout project. Shows where every machine, workstation, department, desk, and so on are located.
9. Architectural drawing, templet and tape, three-dimensional, and CAD.
10. Architectural because of redraw time.
11. 1/4 inch = 1 foot; 2nd 1/8 inch = 1 foot.
12. Yes, when it is a one-time project.
13. Distance traveled, automatic travel ratio, gravity feet ratio, stores and warehouse cube utilization, aisle space ratio, machine space ratio utilization, automatic machine loading ratio, machine utilization, material handling costs ratio, in-process time.
14. **a.** down
    b. up
    c. down
    d. up
    e. up
    f. up
    g. up
    h. down
    i. down
    j. down
15. Advantages are speed, accuracy, and efficiency. Data can be entered once and shared with different people and for different purposes. Disadvantages may include reliance on inaccurate input or oversimplification of the input variables.
16. Standardization would enable sharing of various electronic files, drawings, and so on.
17. Enables the planner and the users of the facilities to "walk through" the plant and to evaluate its layout during the planning and design stages.

Chapter 15

1. Simulation is an experimental technique that attempts to mimic a real-life situation in order to evaluate various scenarios and to answer what-if questions.
2. Mathematical models are more precise and more clearly defined and formulated. When such exact definitions are not possible or feasible, computer modeling is more helpful.
3. Stochastic events are random in nature, whereas the outcome of a deterministic model is not left to chance.
4. Dynamic models are tracked over a period of time and their behavior is influenced by the passage of time. Static models are unaffected by time.
5. Fast and efficient and allows for examining various scenarios. Complex models may not easily lend themselves to simulation and may result in simplistic solutions.
6. Capacity planning, inventory policies, line balancing, warehousing and logistics, office layout, and so on.
7. Examine various alternatives, that is, adding (or deleting) additional stations, equipment, and so on.
8. **a.** Aid in planning and design
   b. Perform simulation and analysis.
9. Problem definition defines the *goals and the purpose;* system definition defines the *boundaries and limitations* of the system.
10. As the old cliché states, "garbage-in, garbage-out!"

APPENDIX A

S.S. Turbo Manufacturing Group

PROJECT LEADERS:
Manny Cuevas
Michael Thoma
Bryan Orozco
Jarrett Hullinger
Ben Unger

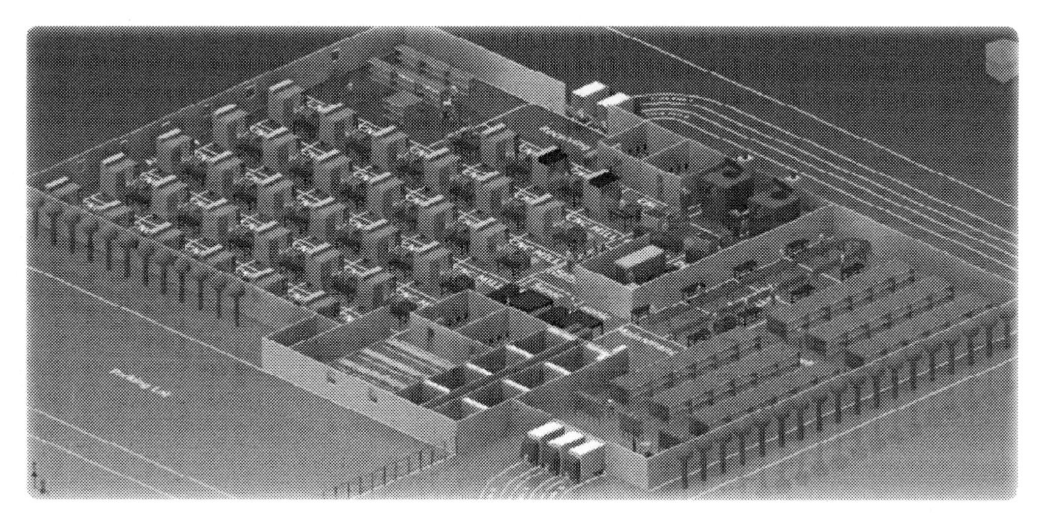

TABLE OF CONTENTS

S.S.T. MANUFACTURING

Mission Statement

Our organization seeks to provide the consumer market with top of the line automotive turbochargers at a reasonable price. While providing our customers with top of the line service, we strive to lead the automotive industry with higher standards of quality and performance. We strive to be the turbocharger of choice.

Project Description

Utilizing the systematic and lean approach to manufacturing facilities planning, our goal is to design an efficient and productive manufacturing facility that is capable of fabricating, assembling, and delivering high quality turbochargers for the automotive industry while embracing lean principles. The final deliverable is a realistic proposal that can be utilized by anyone in the turbocharger market seeking to begin turbocharger production.

Company Structure

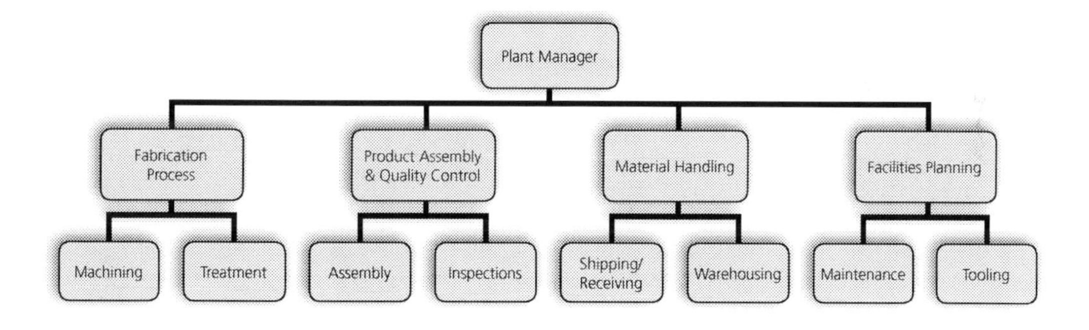

PRODUCT DESCRIPTION

- Horsepower Rating: 325-550 HP
- Actual Weight: 25 lbs
- Spool: 3000-6300 RPM
- Turbo Housing: 0.63 AR. Designed for max air efficiency

A turbocharger is a tool used to increase the horsepower (HP) of an automobile by means of highly efficient forced airflow. The overall power of the automobile is increased due to the pressure differential created within the turbocharger. A turbocharger has a compressor powered by a turbine, which is driven by the engine's own exhaust. This allows the turbocharger to achieve a higher degree of efficiency.

Cutting edge technology has allowed us to manufacture top of the line components that make up our powerful turbo. Our design of the model S.S.1 allows for optimum airflow to the turbo to maximize efficiency across the operating range. We have ported an oversized intake into the turbo housing to greatly reduce the effects of surge, and to provide additional airflow to reach maximum HP at high boost pressure levels.

COMPANY NEWS
New plant proposed in the city of Lafayette, IN
Expected deliverables April 7th, 20xx
Product pricing available on April 1st, 20xx

REPORT PROPOSAL
Manny Cuevas
Michael Thoma
Bryan Orozco
Jarrett Hullinger
Ben Unger

ABOUT US
Unveiling new Turbo SS model in Lafayette, IN plant.
New online store available November 20xx

FACILITY LOCATION

The S.S. Turbo manufacturing facility is located in Lafayette, Indiana, centrally positioned between our target customers, suppliers, and trained workforce. Our raw materials, such as steel and cast iron, comes from U.S. Steel Corporation (D) located in Gary, Indiana. Target markets General Motors (B) and Subaru Automotive (C) are located in Fort Wayne, Indiana and Lafayette, Indiana, respectively. Lastly, our buyout parts are procured from an Indianapolis, Indiana, auto part supplier.

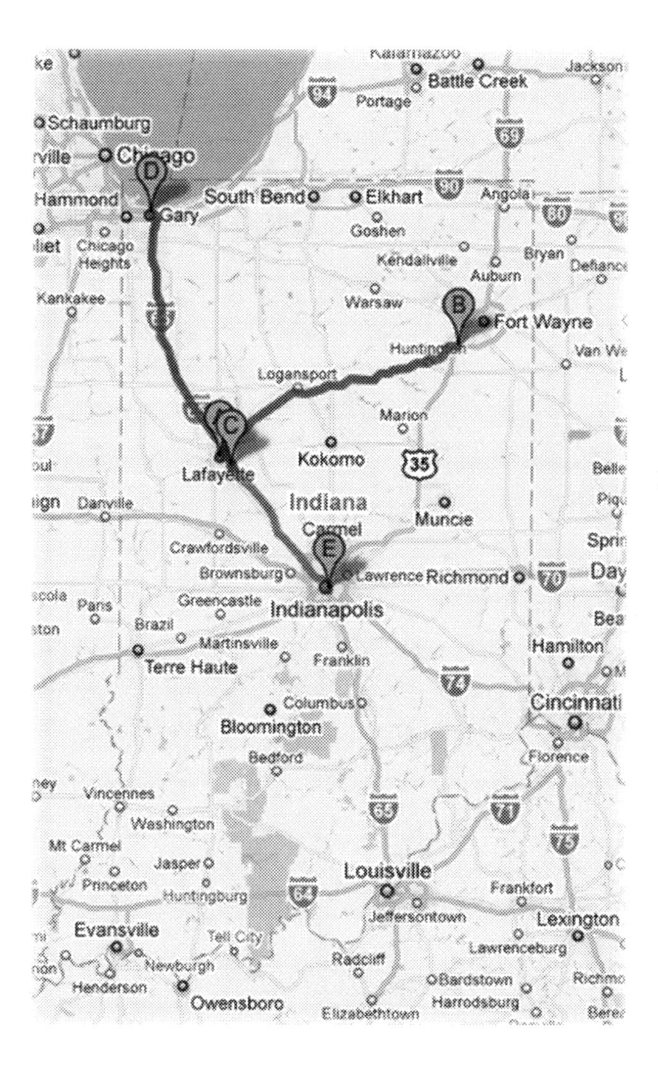

SAMPLE PRODUCT DRAWINGS

Part & Assembly Drawings:

This section contains drawings of several of the parts that are produced in the S.S.T. manufacturing facility. For each part, a brief description and a picture of the part is followed by a detailed engineering drawing. All drawings were hand modeled by the S.S.T. Group at Purdue University using the SolidWorks modeling program.

COMPRESSOR COVER

Part Description:

The part above and on the following page is the compressor cover. It is used to cover the compressor housing and compressor wheel within the turbocharger.

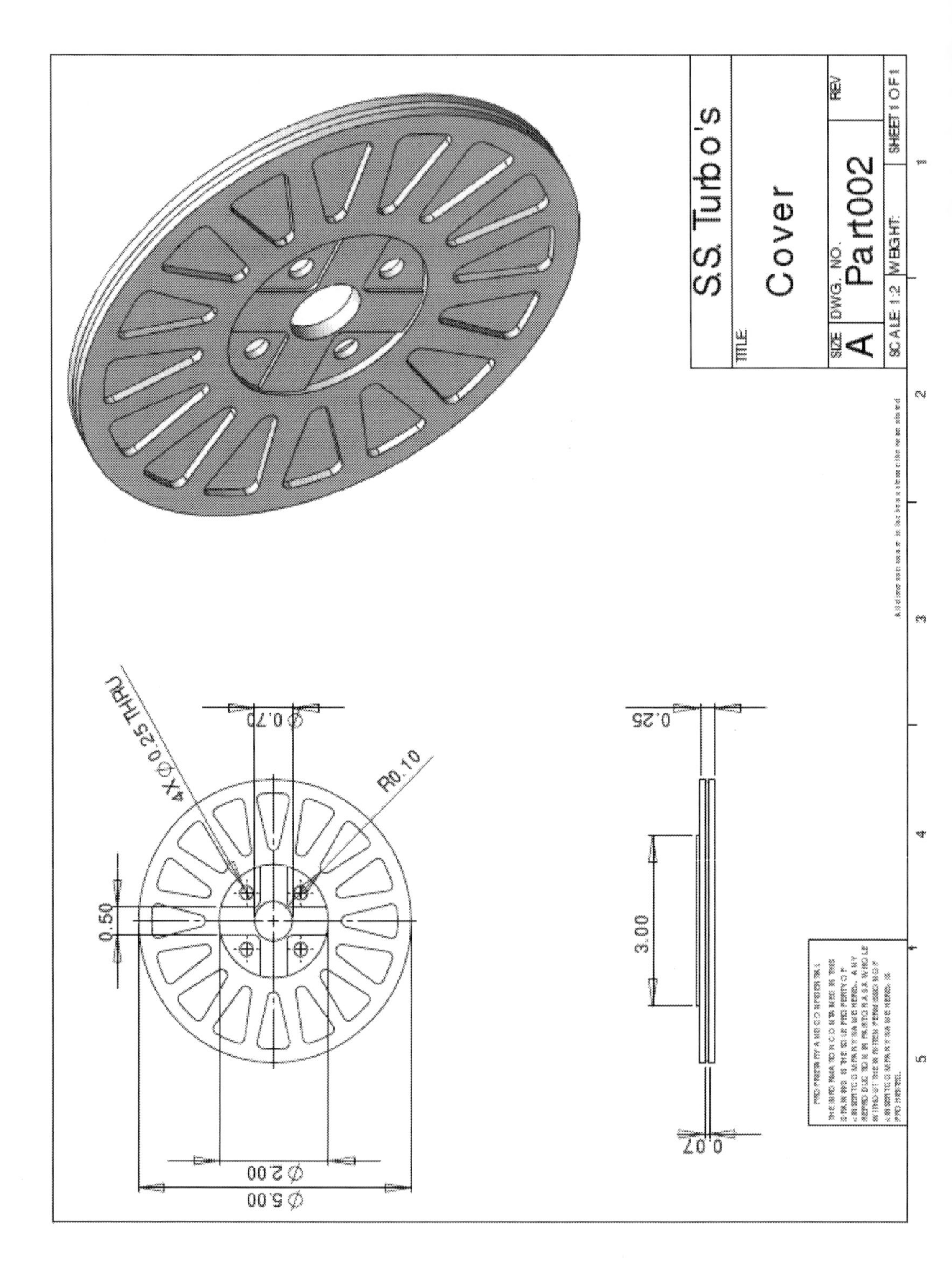

S.S. Turbo's
TITLE:
Cover
SIZE
A
DWG. NO.
Part002
REV
SCALE 1:2
WEIGHT:
SHEET 1 OF 1
4X ⌀ 0.25 THRU
R0.10
⌀ 0.70
0.50
⌀ 2.00
⌀ 5.00
0.25
3.00
0.07

TURBO HOUSING

Part Description:

The turbo housing above and on the following page is used as a cover for the internal components that make up the turbocharger. The compressor wheel fits inside this housing and attaches to the bearing housing.

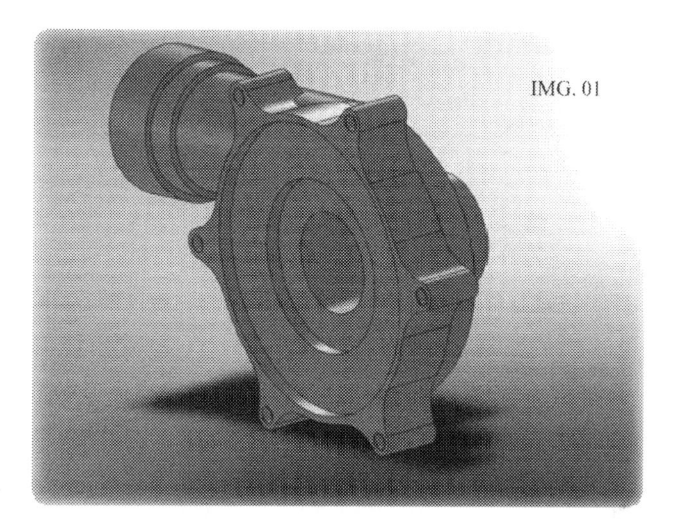
IMG. 01

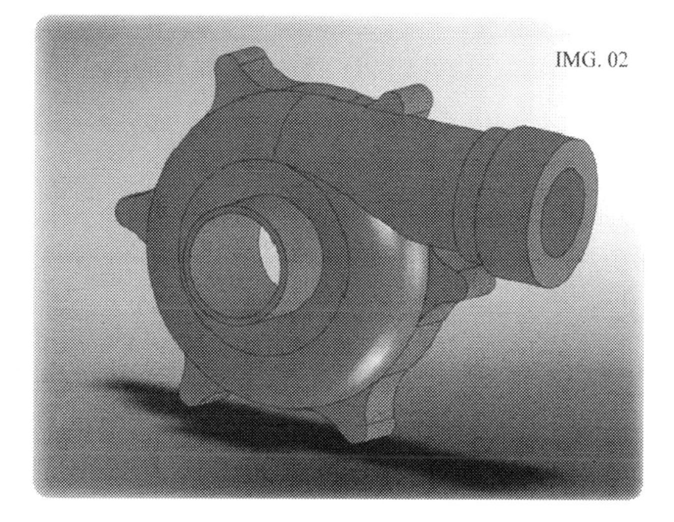
IMG. 02

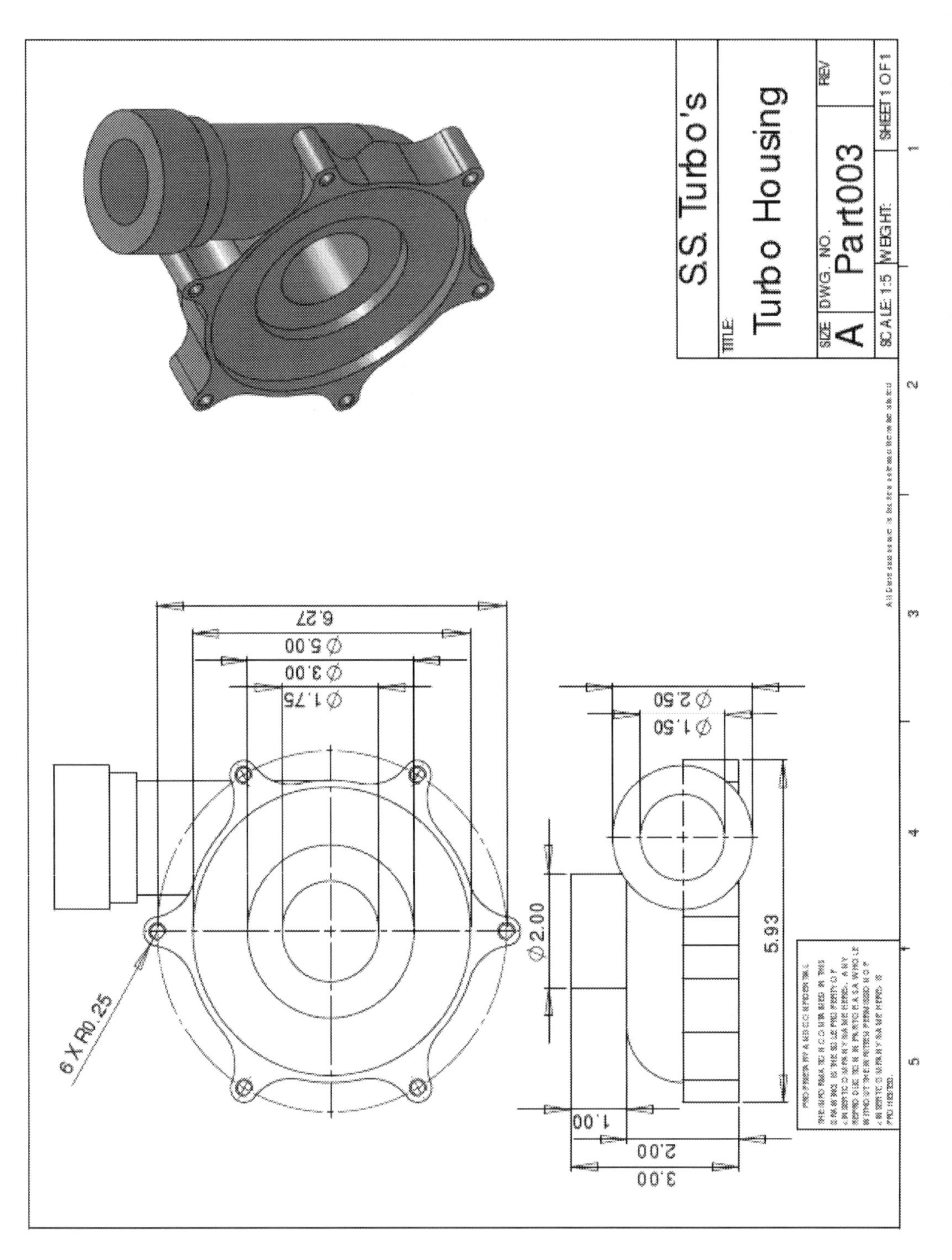
S.S. Turbo's
TITLE:
Turbo Housing
SIZE
A
DWG. NO.
Part003
REV
SCALE 1:5
WEIGHT:
SHEET 1 OF 1
6.27
Ø 5.00
Ø 3.00
Ø 1.75
Ø 2.50
Ø 1.50
6 X R0.25
Ø 2.00
5.93
1.00
2.00
3.00

BEARING HOUSING

Part Description:

The bearing housing above, and on the following page, houses internal bearings and is also the point at which oil enters the turbo charger to lubricate the components through the oil feed line. The oil leaves the bearing housing and is cycled through the entire engine before re-entering the turbo.

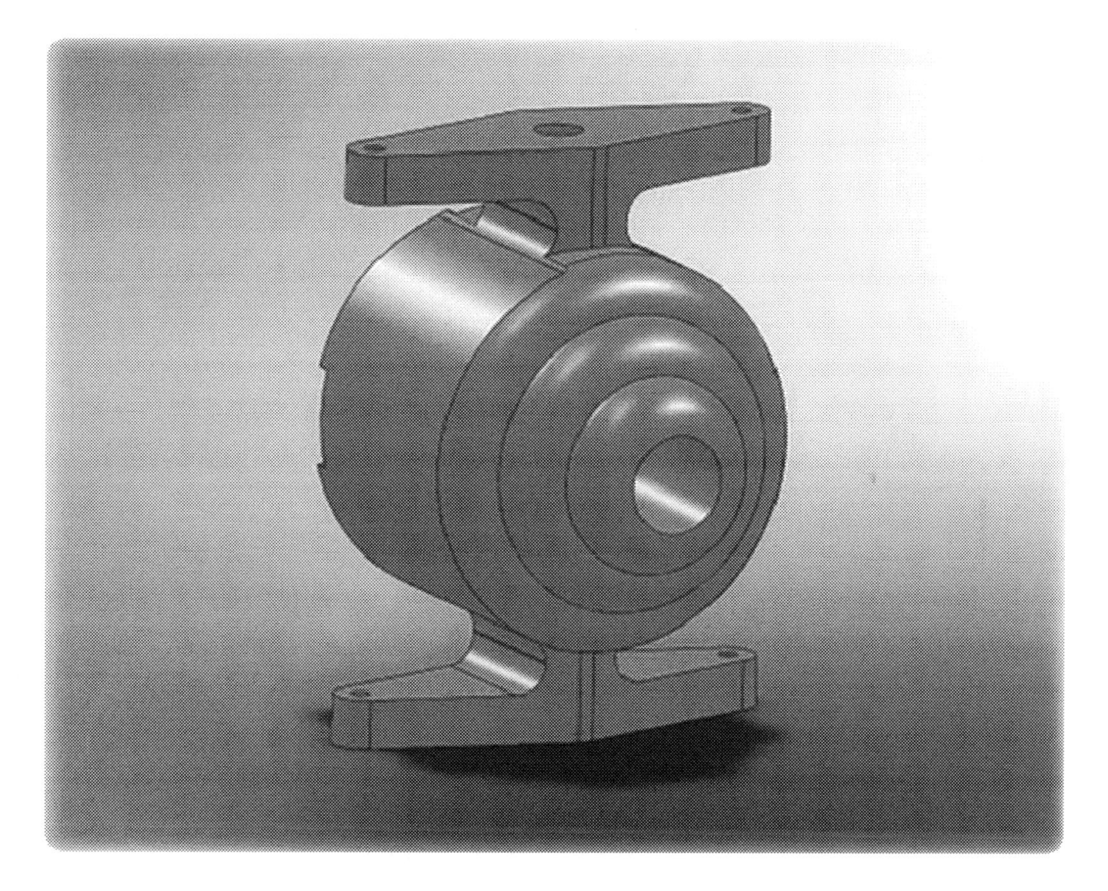

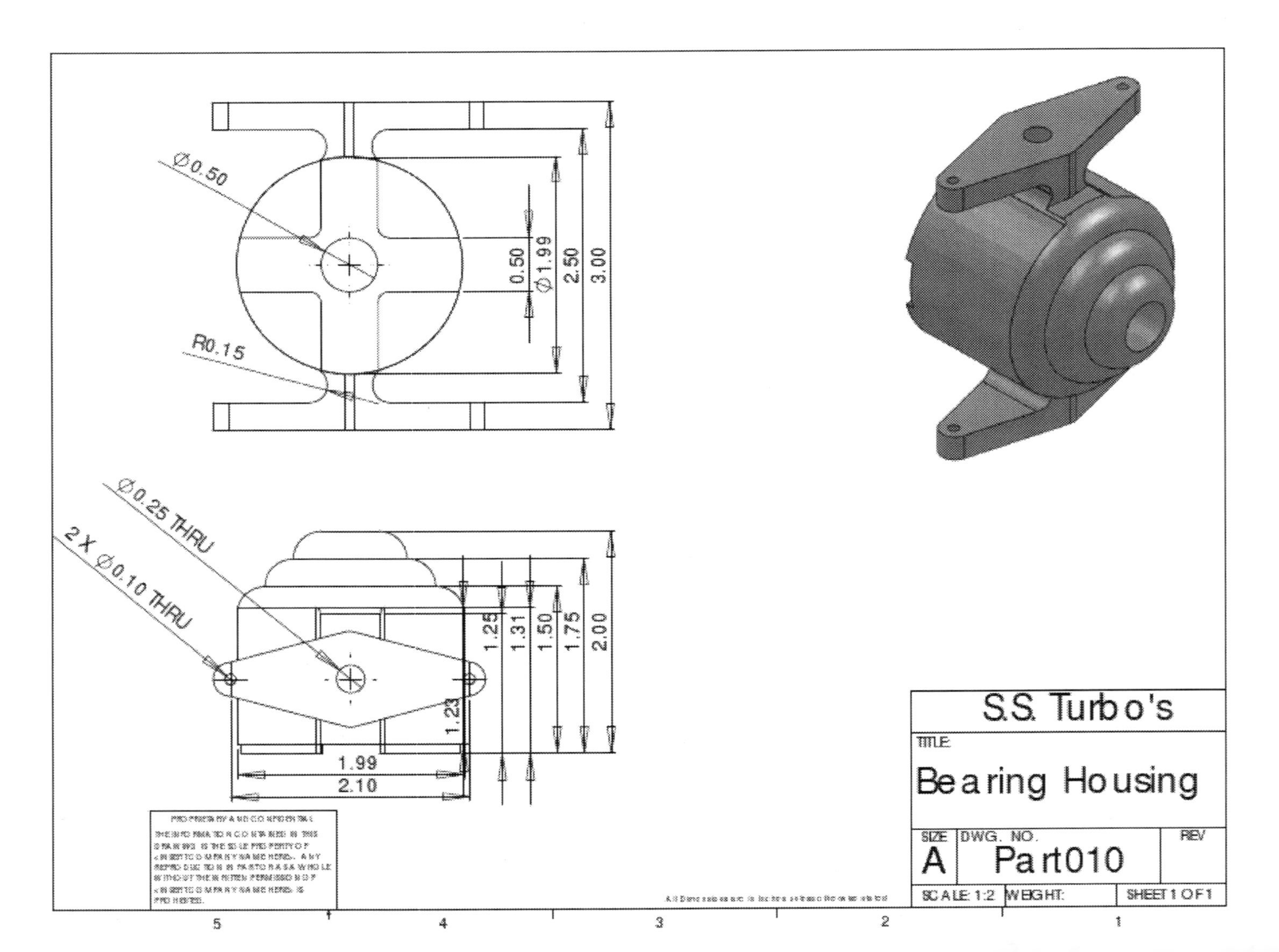
Ø0.50
R0.15
0.50
Ø1.99
2.50
3.00
Ø0.25 THRU
2 X Ø0.10 THRU
1.25
1.31
1.50
1.75
2.00
1.23
1.99
2.10
S.S. Turbo's
TITLE
Bearing Housing
SIZE
A
DWG. NO.
Part010
REV
SCALE 1:2
WEIGHT:
SHEET 1 OF 1
5
4
3
2
1

ROTOR BLADE

Part Description:

The rotor blade above, and on the following page, fits inside the rotor housing. It is turned by exhaust gases which drive the rotor assembly, providing power to the turbocharger.

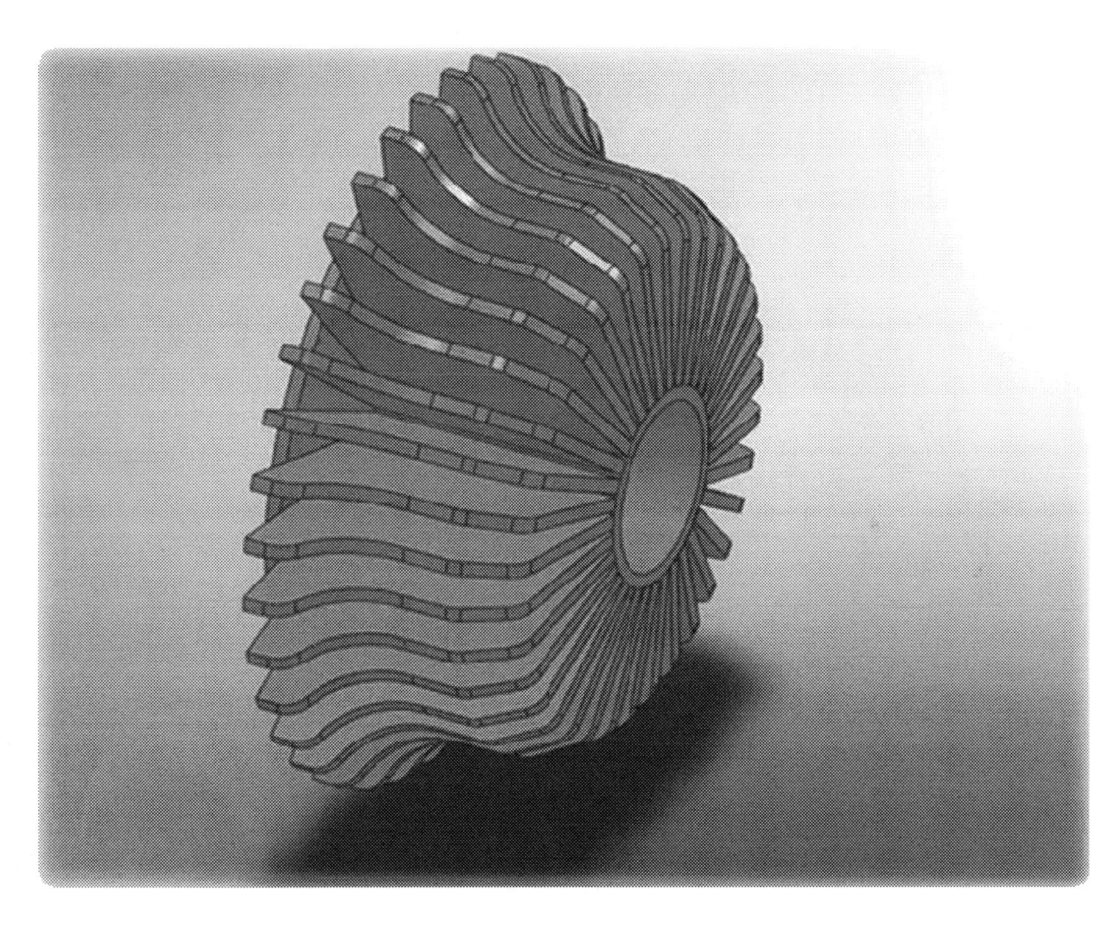

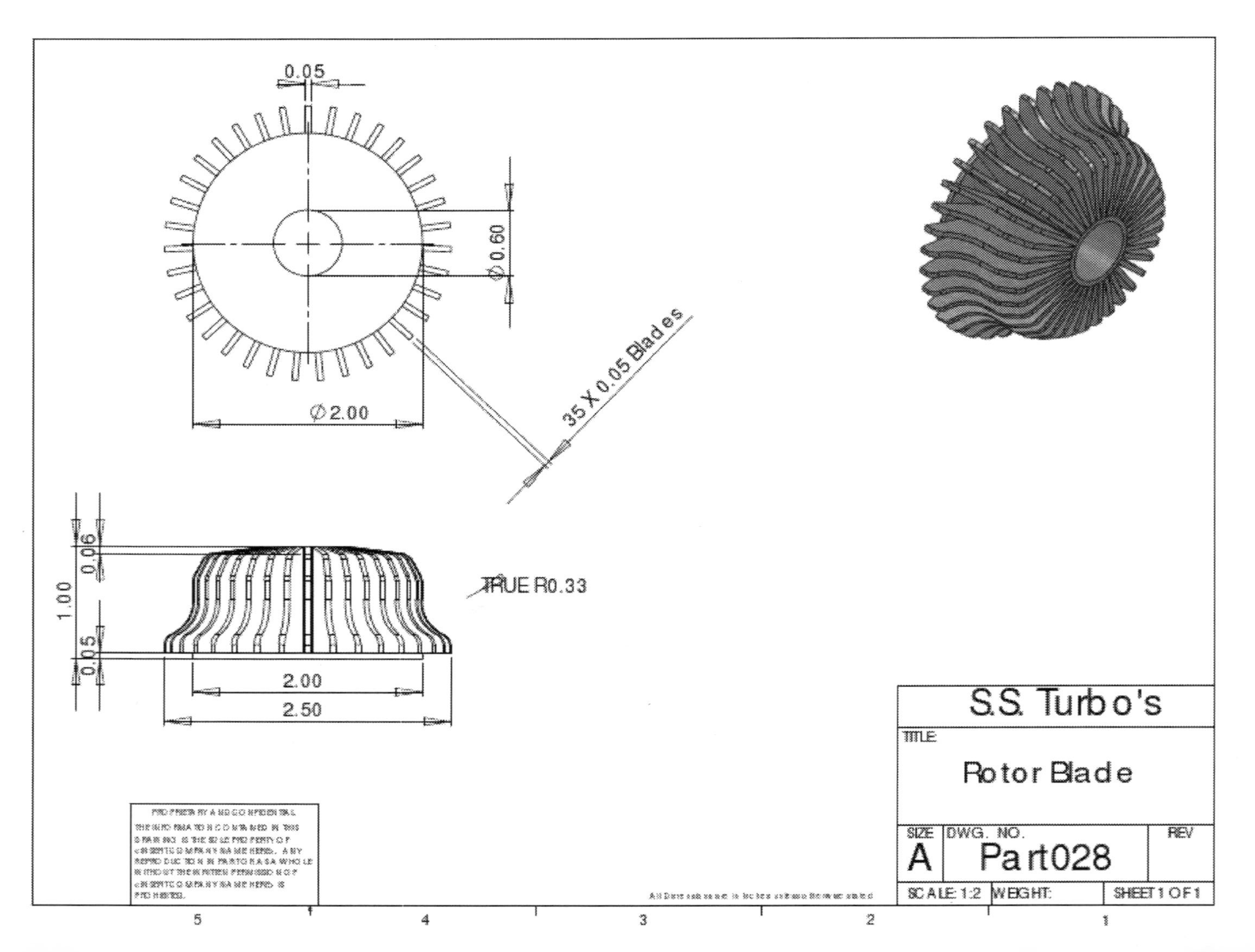

0.05
Ø0.60
Ø2.00
35 X 0.05 Blades
TRUE R0.33
0.06
1.00
0.05
2.00
2.50
S.S. Turbo's
TITLE
Rotor Blade
SIZE
A
DWG. NO.
Part028
REV
SCALE 1:2
WEIGHT:
SHEET 1 OF 1
5
4
3
2
1

COMPRESSOR WHEEL

Part Description:

The compressor wheel above and on the following page is used to compress the air entering the compressor housing, which in turn creates power supplied to the engine of the automobile.

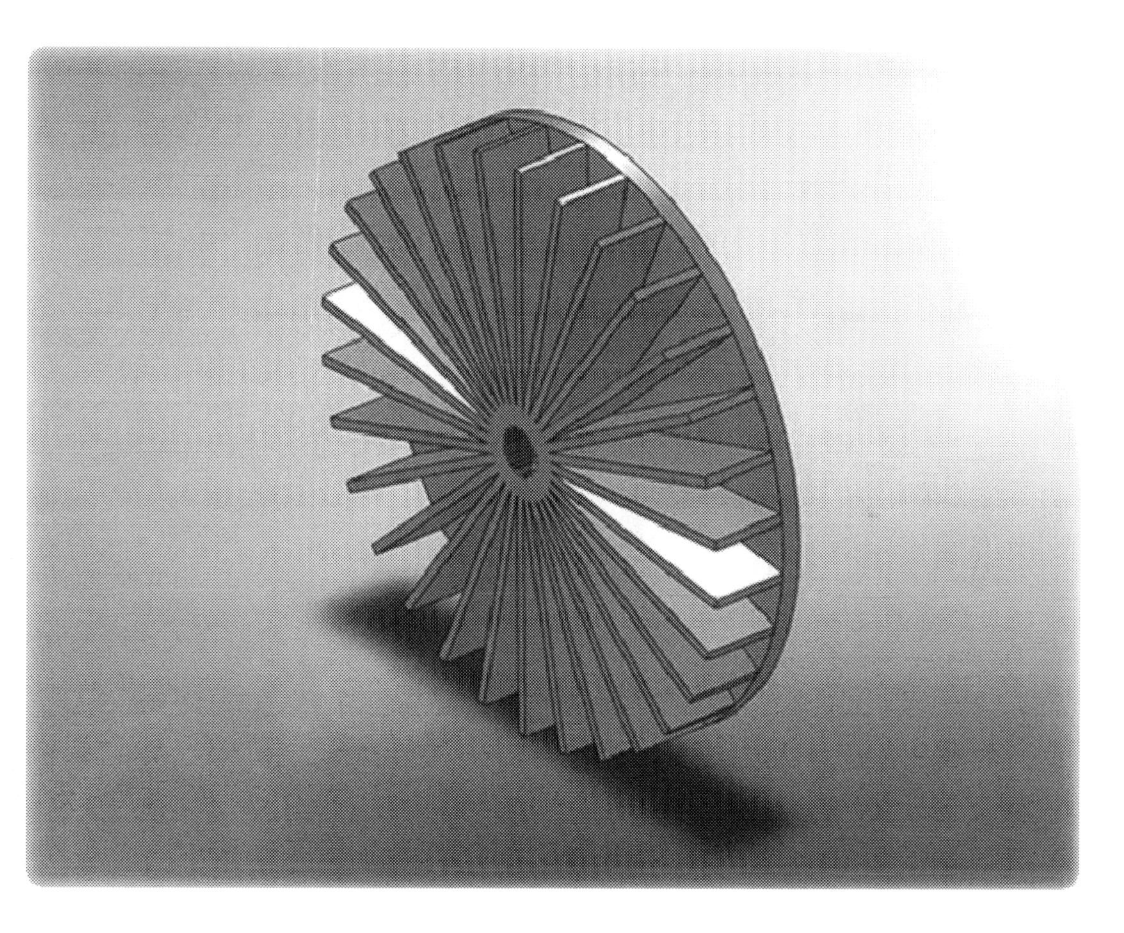

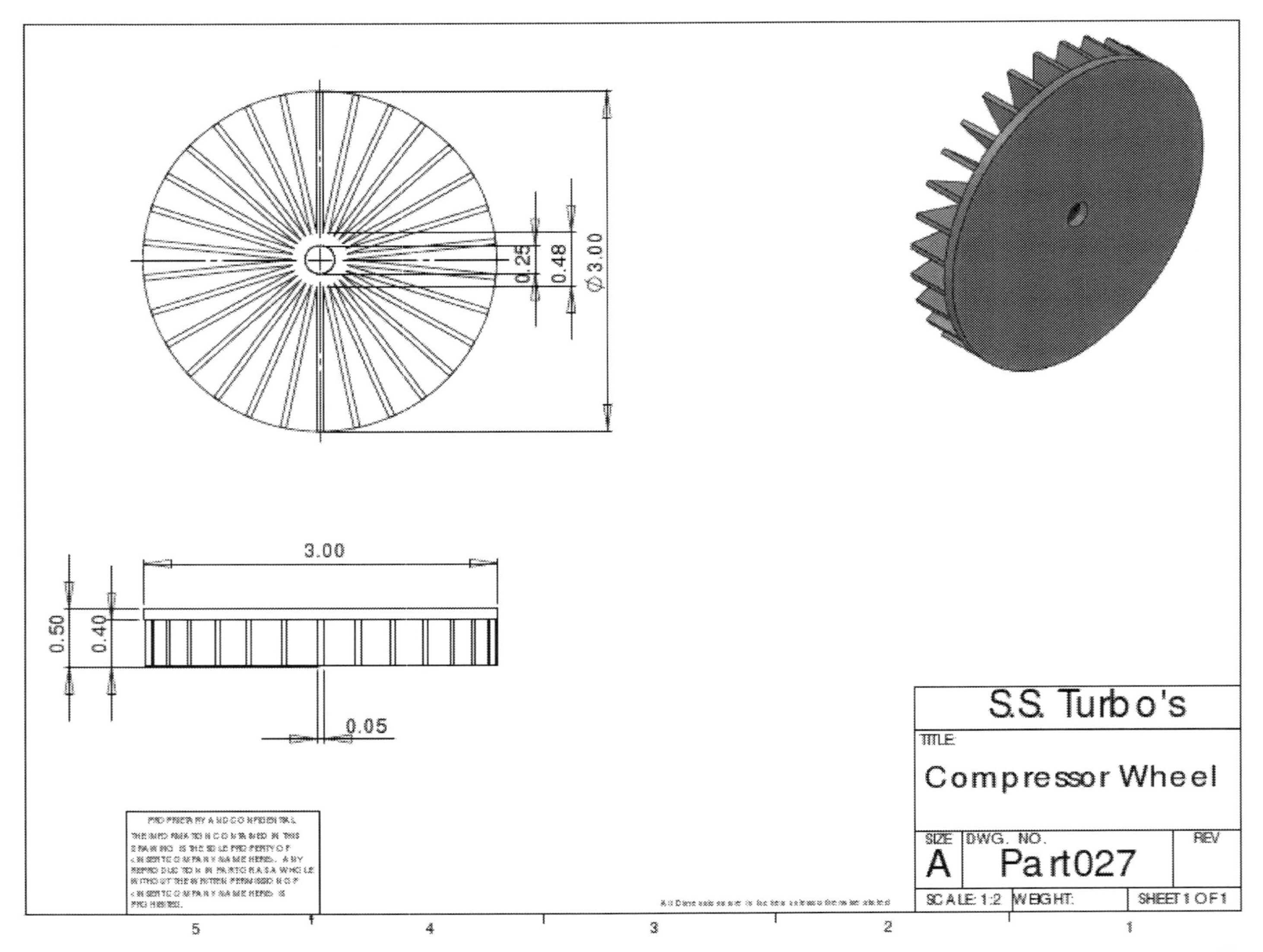
S.S. Turbo's
TITLE:
Compressor Wheel
SIZE
A
DWG. NO.
Part027
REV
SCALE 1:2
WEIGHT:
SHEET 1 OF 1
Ø3.00
0.48
0.25
3.00
0.05
0.40
0.50
5
4
3
2
1

ASSEMBLY DRAWINGS

Part Description:

The assembly drawing is the complete assembly fabricated and assembled in the S.S.T. manufacturing facility. These drawings and models do not include any nuts, bolts, or other sub-assembly purchased components.

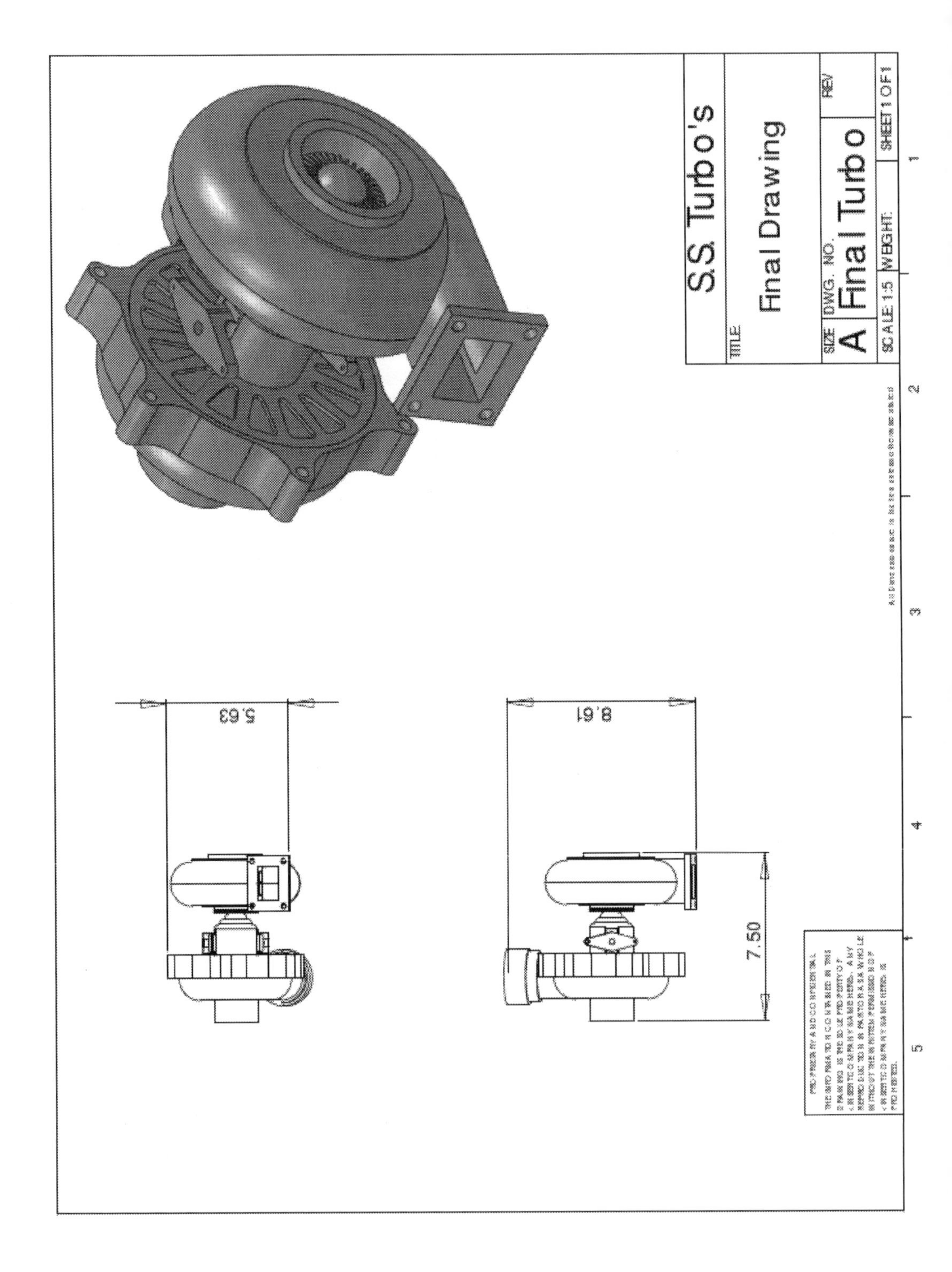

S.S. Turbo's
TITLE
Final Drawing
SIZE
DWG. NO.
A
Final Turbo
REV
SCALE 1:5
WEIGHT:
SHEET 1 OF 1
5.63
8.61
7.50
1
2
3
4
5

EXPLODED ASSEMBLY

Assembly Description:

This view shows all the necessary components that go into making a finished turbocharger. This turbocharger will be ready for installation in a car or truck.

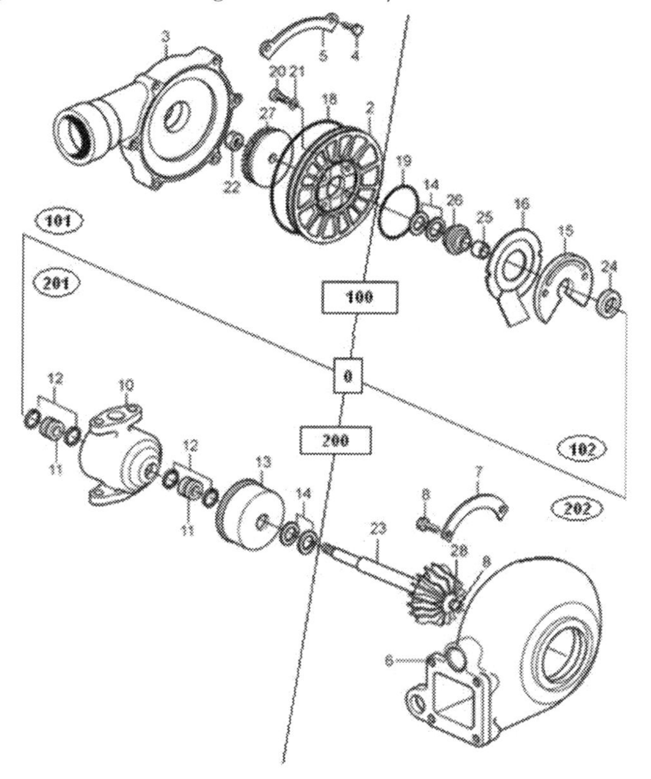

FLAT BILL OF MATERIALS

Flat Bill of Materials				
ITEM	PART #	QTY	DESCRIPTION	MAKE/BUY
1	002	1	Compressor Cover	M
2	003	1	Turbo Housing	M
3	004	8	Turbo Screw	B
4	005	4	Turbo Segment	B
5	006	1	Rotor Housing	M
6	007	3	Rotor Segment	B
7	008	7	Rotor Screw	B
8	010	1	Bearing Housing	M
9	011	2	Bushing	B
10	012	4	Lock Ring	B
11	013	1	Heat Shield	M
12	014	4	Piston Ring	B
13	015	1	Thrust Bearing	B
14	016	1	Oil Deflector	B
15	018	1	Large O-Ring	B
16	019	1	Small O-Ring	B
17	020	4	Housing Screw	B
18	021	4	Washer	B
19	022	1	Nut	B
20	023	1	Rotor Shaft	M
21	024	1	Thrust Ring	B
22	025	1	Sleeve	B
23	026	1	Seal	B
24	027	1	Compressor Wheel	M
25	028	1	Rotor Blade	M

INDENTED BILL OF MATERIALS

Indented Bill of Material				
LEVEL	PART #	QTY.	DESCRIPTION	MAKE/BUY
0	000	1	Turbo	M
. 1	100	1	Turbo Assy	M
. . 2	101	1	Housing Sub Assy	M
. . . 3	022	1	Nut	B
. . . 3	027	1	Compressor Wheel	M
. . . 3	020	4	Housing Screw	B
. . . 3	021	4	Washer	B
. . . 3	002	1	Compressor Cover	M
. . . 3	018	1	Large O-Ring	B
. . . 3	005	4	Turbo Segment	B
. . . 3	004	8	Turbo Screw	B
. 1	003	1	Turbo Housing	M
. . 2	102	1	Oil Sub Assy	M
. . . 3	019	1	Small O-Ring	B
. . . 3	014	2	Piston Ring	B
. . . 3	026	1	Seal	B
. . . 3	025	1	Sleeve	B
. . . 3	016	1	Oil Deflector	B
. . . 3	015	1	Thrust Bearing	B
. . . 3	024	1	Thrust Ring	B
. 1	200	1	Rotor Unit Assy	M
. . 2	201	1	Bearing Sub Assy	M
. . . 3	012	4	Lock Ring	B
. . . 3	011	2	Bushing	B
. . . 3	010	1	Bearing Housing	M
. . . 3	013	1	Heat Shield	M
. . . 3	014	2	Piston Ring	B
. . 2	202	1	Turbine Sub Assy	M
. . . 3	023	1	Rotor Shaft	M
. . . 3	028	1	Rotor Blade	M
. . . 3	008	1	Rotor Screw	B
. . . 3	007	1	Rotor Segment	B
. 1	006	1	Rotor Housing	M

TAKT TIME CALCULATION

Takt Time Calculator			
	SHIFT	**DAILY TOTALS**	**UNITS**
3 Shifts	480	1440	Min.
Lunch	30	90	Min
2 @ 15 min. Breaks	30	90	Min.
Efficiency 85%	357	1071	Min.
Demand/Day	333.33	800	Turbos
S.S.T. TAKT TIME		**1.33875**	**MIN./TURBO**
		Takt Time (including scrap)	**1.285 Min./Turbo**

Takt Time Description:

TAKT time represents the pace (tempo) with which the facility must operate in order to meet the production schedule. The Takt time calculation above indicates that S.S.T. Manufacturing needs to produce one turbocharger every 1.285 minutes.

Scrap affects Takt time. The following page demonstrates how scrap rate was included in Takt time calculation.

OPERATION SCRAP RATES	
Melting	—
Cutting	0.05%
Turning	0.05%
Casting	2.00%
Milling	0.50%
Drilling	0.25%
Tapping	1.00%
Debur	0.05%
Rinsing	—
Heat Treatment	0.15%

I = 833.390

Adjusted Takt = 1.285

SCRAP RATE CALCULATION

Calculation Description:

Each part has an estimated scrap rate for every machining operation. This scrap rate is used to adjust the Takt time to more realistically reflect production expectations.

PART #	PART NAME	QTY	OPERATIONS NEEDED		CALC.SCRAP	PARTS/DAY
002	Compressor Cover	800	Cutting	0.05%	6.84	806.84
			Milling	0.50%		
			Drilling	0.25%		
			Debur	0.05%		
			Rinsing	—		
003	Turbo Housing	800	Melting	—	31.21	831.21
			Casting	2.00%		
			Milling	0.50%		
			Drilling	0.25%		
			Tapping	1.00%		
			Debur	0.05%		
			Rinsing	—		
006	Rotor Housing	800	Melting	—	22.90	822.90
			Casting	2.00%		
			Milling	0.50%		
			Drilling	0.25%		
			Debur	0.05%		
			Rinsing	—		
010	Bearing Housing	800	Melting	—	31.21	831.21
			Casting	2.00%		
			Milling	0.50%		
			Drilling	0.25%		
			Tapping	1.00%		
			Debur	0.05%		
			Rinsing	—		
013	Heat Shield	800	Melting	—	22.90	822.90
			Casting	2.00%		
			Milling	0.50%		
			Drilling	0.25%		
			Debur	0.05%		
			Rinsing	—		
028	Rotor Blade	800	Cutting	0.05%	3.61	803.61
			Drilling	0.25%		
			Rinsing	—		
			Heat Treatment	0.15%		
023	Rotor Shaft	800	Cutting	0.05%	1.20	801.20
			Turning	0.05%		
			Debur	0.05%		
			Rinsing	—		
027	Compressor Wheel	800	Cutting	0.05%	3.61	803.61
			Drilling	0.25%		
			Rinsing	—		
			Heat Treatment	0.15%		

SAMPLE ROUTING SHEETS

The following pages in this report include the routing sheets for several of the fabricated parts at the S.S.T. manufacturing facility. Routing sheets are used to identify individual fabricated components and document the sequential manufacturing processes for each part. Only parts fabricated at the S.S.T. manufacturing facility (excluding purchased components) are listed in the following pages.

Part #002 & #003

PROCESS ROUTING SHEET										
Part No.	Part Name	Drawing No.							Date Approved	3/4/2011
#002	Compressor Cover	#002								
Seq #	Operation Description	Equipment Type	Mach. #	Takt Time (min/pc.)	C.T. (min)	Avg. Cycle Time	Fractional Equipment	Actual Equipment	Pc./Hr	Hr/1000 Pc.
10	Cut Stock to Length	Band Saw	BS001	1.285	0.333	0.333	0.259	1.000	180.180	5.550
20	Milling Notches	CNC	CNC001	1.285	4.150	1.038	3.230	4.000	57.831	17.292
30	Drilling 4# Holes	CNC	CNC001	1.285	0.483	0.483	0.376	1.000	124.224	8.050
40	Drilling Shaft Hole	CNC	CNC001	1.285	0.416	0.416	0.321	1.000	144.231	6.933
50	Cutting Back Side	CNC	CNC001	1.285	1.200	1.200	0.934	1.000	50.000	20.000
55	Debur	Belt Sander	B001	1.285	0.416	0.416	0.324	1.000	144.231	6.933
60	Wash/Rinsing	Sonic Clean	SC001	1.285	0.300	0.300	0.233	1.000	200.000	5.000

PROCESS ROUTING SHEET										
Part No.	Part Name	Drawing No.							Date Approved	3/4/2011
#003	Turbo Housing	#003								
Seq #	Operation Description	Equipment Type	Mach. #	Takt Time (min/pc.)	C.T. (min)	Avg. Cycle Time	Fractional Equipment	Actual Equip-ment	Pc./Hr	Hr/1000 Pc.
10	Melt Cast Iron	Furnace	F001	1.285	0.583	0.583	0.454	1.000	102.916	9.717
20	Cast Housing	Diecasting Machine	M001	1.285	0.583	0.583	0.454	1.000	102.916	9.717
30	Milling	CNC	CNC002	1.285	4.166	1.042	3.242	4.000	57.609	17.358
40	Drilling 6# Holes	CNC	CNC002	1.285	0.566	0.566	0.440	1.000	106.007	9.433
50	Tapping 6# Holes	CNC	CNC002	1.285	0.666	0.666	0.518	1.000	90.090	11.100
60	Debur	Belt Sander	B001	1.285	0.416	0.416	0.324	1.000	144.231	6.933
70	Wash/Rinsing	Sonic Clean	SC001	1.285	0.300	0.300	0.233	1.000	200.000	5.000

Part #023 & #027

PROCESS ROUTING SHEET										
Part No.	Part Name	Drawing No.							Date Approved	3/4/2011
#023	Rotor Shaft	#023								
Seq #	Operation Description	Equipment Type	Mach. #	Takt Time (min/pc.)	C.T. (min)	Avg. Cycle Time	Fractional Equipment	Actual Equipment	Pc./Hr	Hr/1000 Pc.
10	Cut Stock to Length	Band Saw	BS001	1.285	0.416	0.416	0.324	1.000	144.231	6.933
20	Turning	CNC Lathe	CNL001	1.285	1.900	0.950	1.479	2.000	63.158	15.833
25	Debur	Belt Sander	B001	1.285	0.500	0.500	0.389	1.000	120.000	8.333
30	Wash/Rinsing	Sonic Clean	SC001	1.285	0.300	0.300	0.233	1.000	200.000	5.000

PROCESS ROUTING SHEET										
Part No.	Part Name	Drawing No.							Date Approved	3/4/2011
#027	Compressor Wheel	#027								
Seq #	Operation Description	Equipment Type	Mach. #	Takt Time (min/pc.)	C.T. (min)	Avg. Cycle Time	Fractional Equipment	Actual Equip-ment	Pc./Hr	Hr/1000 Pc.
10	Cut Stock to Length	Band Saw	BS002	1.285	0.333	0.333	0.259	1.000	180.180	5.550
20	Cutting	CNC	CNC007	1.285	7.550	1.258	5.875	6.000	47.682	20.972
30	Drilling Shaft Hole	CNC	CNC007	1.285	0.250	0.250	0.195	1.000	240.000	4.167
40	Wash/Rinsing	Sonic Clean	SC001	1.285	0.300	0.300	0.233	1.000	200.000	5.000
50	Heat Treatment	Oven	OV001	1.285	0.666	0.666	0.518	1.000	90.090	11.100

SAMPLE PROCESS LINE BALANCING

In the following pages, the S.S.T manufacturing group experimented with various combinations of process routing in an effort to make our fabrication processes more efficient. The compressor cover and turbo housing had two processes on the same machine, which were consolidated in order to save production time. On the remaining parts, we made the decision to batch various operations to reduce bottlenecks.

While experimenting with these process line balancing operations, we concluded that it would be much more cost effective to batch all the operations. Fabricating on a production line with CNC mills and CNC lathes did not prove to be a cost effective method to manufacture turbo chargers.

Part #002 & #003

PROCESS ROUTING SHEET												
Part No.	Part Name	Drawing No.									Date Approved	3/4/2011
#002	Compressor Cover	#002										
Seq #	Operation Description	Equipment Type	Mach. #	Takt Time (min/pc.)	C.T. (min)	Avg. Cycle Time	Fractional Equipment	Actual Equipment	Line Time	Percent Load	Pc./Hr	Hr/1000 Pc.
10	Cut Stock to Length	Band Saw	BS001	1.285	0.333	0.333	0.259	1.000	1.200	27.75%	180.180	5.550
20	Milling Notches	CNC	CNC001	1.285	4.150	1.038	3.230	4.000	4.800	86.46%	57.831	17.292
30	Drilling 4# Holes	CNC	CNC001	1.285	0.899	0.899	0.700	1.000	1.200	74.92%	124.224	8.050
50	Cutting Back Side	CNC	CNC001	1.285	1.200	1.200	0.934	1.000	1.200	100.00%	50.000	20.000
55	Debur	Belt Sander	B001	1.285	0.416	0.416	0.324	1.000	1.200	34.67%	144.231	6.933
60	Wash/Rinsing	Sonic Clean	SC001	1.285	0.300	0.300	0.233	1.000	1.200	25.00%	200.000	5.000
Comments	Highlighted row consolidated OP 30 & 40				Sum =	4.186		Sum =	10.800		EFF=	38.75%
											% Change =	3.88%

PROCESS ROUTING SHEET												
Part No.	Part Name	Drawing No.									Date Approved	3/4/2011
#003	Turbo Housing	#003										
Seq #	Operation Description	Equipment Type	Mach. #	Takt Time (min/pc.)	C.T. (min)	Avg. Cycle Time	Fractional Equipment	Actual Equipment	Line Time	Percent Load	Pc./Hr	Hr/1000 Pc.
10	Melt Cast Iron	Furnace	F001	1.285	0.583	0.583	0.454	1.000	1.232	47.32%	102.916	9.717
20	Cast Housing	Diecasting Machine	M001	1.285	0.583	0.583	0.454	1.000	1.232	47.32%	102.916	9.717
30	Milling	CNC	CNC002	1.285	4.166	1.042	3.242	4.000	4.928	84.54%	57.609	17.358
40	Drilling 6# Holes	CNC	CNC002	1.285	1.232	1.232	0.959	1.000	1.232	100.00%	48.701	20.533
60	Debur	Belt Sander	B001	1.285	0.416	0.416	0.324	1.000	1.232	33.77%	144.231	6.933
70	Wash/Rinsing	Sonic Clean	SC001	1.285	0.300	0.300	0.233	1.000	1.232	24.35%	200.000	5.000
Comments	Highlighted row consolidated OP 30 & 40				Sum =	4.156		Sum =	11.088		EFF=	37.48%
											% Change =	-2.42%

MACHINE/EQUIPMENT REQUIREMENTS

The equipment requirements for our facility were generated by taking the fractional equipment from the routing sheets needed on each part at each operation and adding them together. This gives the total number of each type of machine needed for our facility.

Since S.S.T. is a 3-shift, 24-hour operation, rounding each fractional piece of equipment up was essential in order to meet TAKT time. The only machine not rounded up was the heat treat oven. Buying another oven for 3.6 percent of its needed use is not justifiable. Heat treating ovens are expensive and they take up a large amount of floor space.

Equipment Requirements

FABRICATION EQUIPMENT SPREAD SHEET										
Equipment	Compressor Cover	Turbo Housing	Rotor Housing	Bearing Housing	Heat Shield	Rotor blade	Rotor Shaft	Compressor Wheel	TOTAL	DECIDED NUMBER
Band Saw	0.259					0.259	0.324	0.259	1.101	2
CNC Milling Machine	4.864	4.2	3.826	3.533	1.722	6.070		6.070	30.285	31
CNC Lathe							1.479		1.479	2
Belt Sander	0.342	0.324	0.324	0.416	0.324		0.389		2.119	3
Furnace		0.454	0.454	0.324	0.259				1.491	2
Diecast Machine		0.454	0.454	0.324	0.259				1.491	2
Sonic Clean	0.233	0.233	0.233	0.233	0.233	0.233	0.233	0.233	1.864	2
Heat Treat Oven						0.518		0.518	1.036	1

ASSEMBLY CHART

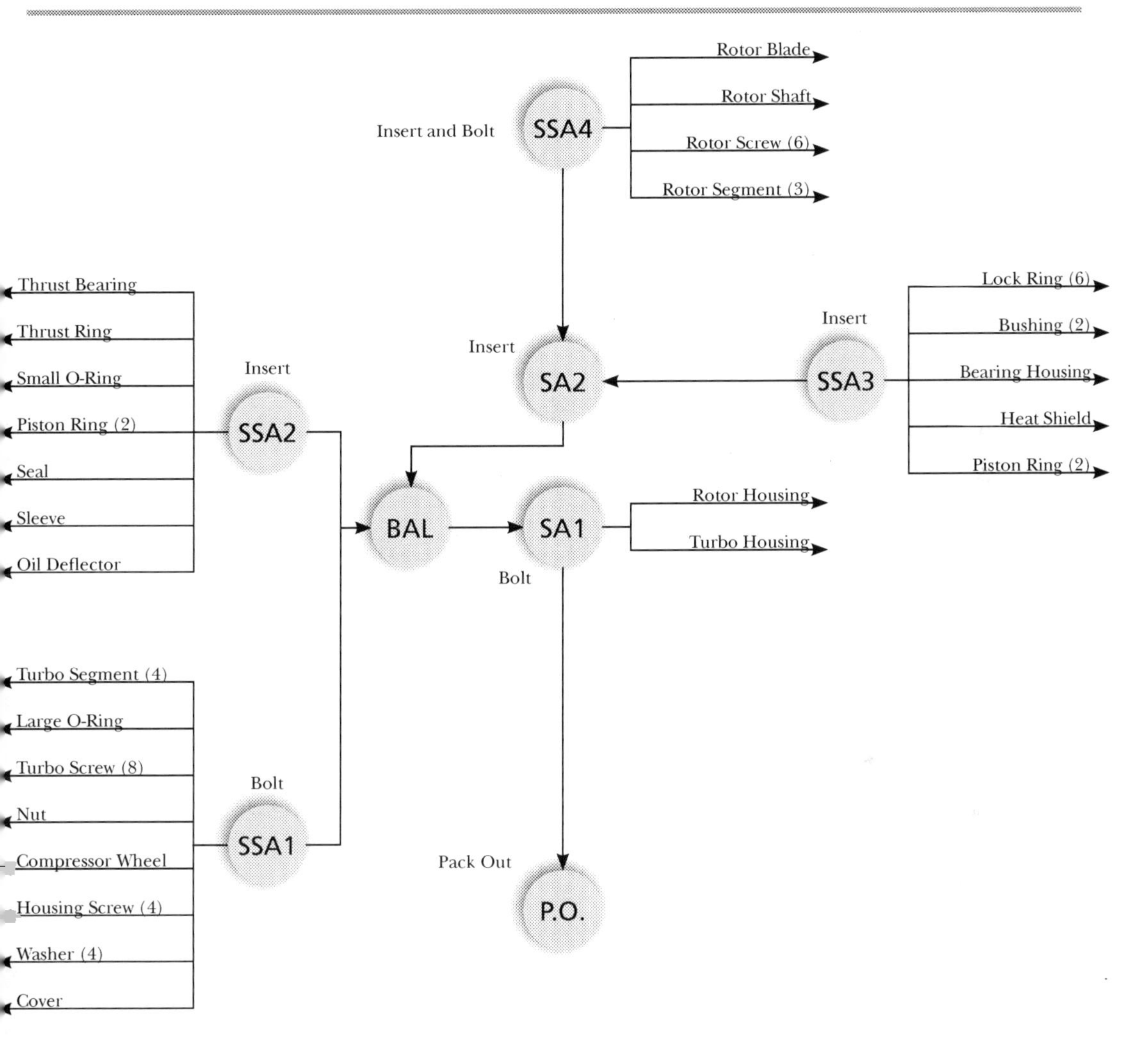

Operation No.	Operation Description	Minute
SSA4	Assemble and bolt shaft and blade together	0.8
SSA3	Assemble by inserting parts together	0.916
SA2	Slide both sub-sub assemblies together on the shaft	0.666
SSA2	Insert parts together	1.25
SSA1	Bolt pieces together	1.5
Bal	Balance the turbocharger	1.083
SA1	Assemble both sub-sub assemblies together and the bolt on both housings	1
P.O.	Pack fully assembled turbo in oil saturated plastic packaging. Form box and insert turbo into box.	

SAMPLE ASSEMBLY LINE BALANCING

The purpose of assembly line balancing is to ensure that the workload is distributed evenly among all the workstations on the assembly line. The sample assembly line balancing chart on the following page details how the S.S.T. group balanced its assembly lines.

Assembly Line Balance

Product No. 111	Product Description: Low end power turbocharger									Takt Time = 1071 Units per day
Date: March 10, 2011										
By I.E.: Jarrett Hullinger	Number Units per day: 834									R = 1.285 Min

No.	Operation Description	Takt Time	Cycle Time	No. Stations	Rounded Up	Avg Cycle Time	Line Time	% Load	Hrs/1000 Line Balance	Pcs/Hr Line Balance
SSA4	Assemble and bolt shaft and blade together.	1.285	0.8	0.623	1	0.8	1.25	64.00	13.333	75.000
SSA3	Assemble by inserting parts together.	1.285	0.916	0.713	1	0.916	1.25	73.28	15.267	65.502
SA2	Slide both sub-sub assemblies together on the shaft.	1.285	0.666	0.518	1	0.666	1.25	53.28	11.100	90.090
SSA2	Insert parts together.	1.285	1.25	0.973	1	1.25	1.25	100.00	20.833	48.000
SSA1	Bolt pieces together.	1.285	1.5	1.167	2	0.75	2.50	60.00	12.500	80.000
Bal	Balance the turbo.	1.285	1.083	0.843	1	1.083	1.25	86.64	18.050	55.402
SA1	Assemble both sub-sub assemblies together and then bolt on both housings.	1.285	1	0.778	1	1	1.25	80.00	16.667	60.000
P.O.	Pack fully assembled turbo in oil saturated plastic packaging. Form box and insert turbo into box.	1.285	1.25	.973	1	1.25	1.25	100.00	20.833	48.000
		Sum:	8.465			Sum:	11.25		EFF:	75.24%

Bal 1

No.	Operation Description	Takt Time	Cycle Time	No. Stations	Rounded Up	Avg Cycle Time	Line Time	% Load	Hrs/1000 Line Balance	Pcs/Hr Line Balance
SSA4	Assemble and bolt shaft and blade together.	1.285	0.8	0.623	1	0.8	1.5	53.33	13.333	75.000
SSA3	Assemble by inserting parts together.	1.285	0.916	0.713	1	0.916	1.5	61.07	15.267	65.502
SA2	Slide both sub-sub assemblies together on the shaft.	1.285	0.666	0.518	1	0.666	1.5	44.40	11.100	90.090
SSA2	Insert parts together.	1.285	1.25	0.973	1	1.25	1.5	83.33	20.833	48.000
SSA1	Bolt pieces together.	1.285	1.5	1.167	1	1.5	1.5	100.00	25.000	40.000
Bal	Balance the turbo.	1.285	1.083	0.843	1	1.083	1.25	86.64	18.050	55.402
SA1	Assemble both sub-sub assemblies together and then bolt on both housings.	1.285	1	0.778	1	1	1.5	66.67	16.667	60.000
P.O.	Pack fully assembled turbo in oil saturated plastic packaging. Form box and insert turbo into box.	1.285	1.25	.973	1	1.25	1.5	83.33	20.833	48.000
		Sum:	8.465			Sum:	11.75		EFF:	72.04%

SAMPLE PROCESS CHART

A process chart is developed for every fabricated part to document every process and step that the part encounters throughout the facility from the receiving department to the shipping department. The charts on the following pages describe what happens as the part travels through the facility. The symbols in the following key correspond to the process through which specific turbo parts go.

PROCESS SYMBOL KEY	
○	Operation, work on the part
⇨	Transportation, moving the part
□	Inspection, quality control, work on the part
D	Delay, very temporary storage, usually at a work station
▽	Storage, storerooms, warehouse, work in process

Process Chart

	Summary	
Part Name: Cover	**Process**	**Number**
Process Description: Fabrication of Turbo Compressor Cover	**Operation**	7
	Inspection	2
Department: Fabrication	**Transportation**	7
	Delay	3
Plant: West Lafayette, IN	**Storage**	2
Recorded By: Jarrett Hullinger **Date:** 03/11/11	**Total Steps**	21

Step	Process Symbol (Darken the appropriate Symbol)	Description or Current Method	Value Added ? (Y/N)	Distance (ft.)	Method	Qty. Moved	Freq. per hr.
1	○ ⇨ ■ D ▽	Inspection of Incoming 5" X 70" Steel Roundstock	N				
2	○ ➡ □ D ▽	Transported to Warehouse	N		ForkLift	15	1/5 Days
3	○ ⇨ □ D ▼	Stored in Warhouse	N				
4	○ ➡ □ D ▽	Barstock Transported to Band Saw	N		ForkLift	1	1/8 hr
5	● ⇨ □ D ▽	Cut Stock to Length on Band saw	Y				
6	○ ⇨ □ ◗ ▽	Delay Until 47 Covers are Cut	N				
7	○ ➡ □ D ▽	Compressor Covers cut are Transported to CNC001	N		Push Cart	47	1
8	● ⇨ □ D ▽	Compressor Covers Milling Notches	Y				
9	● ⇨ □ D ▽	Compressor Covers Drilling 4X Holes	Y				
10	● ⇨ □ D ▽	Compressor Covers Drilling Shaft Hole	Y				
11	● ⇨ □ D ▽	Compressor Covers Cutting Back Side	Y				
12	○ ➡ □ D ▽	Transpored to Belt Sander	N		Conveyer	1	47
13	● ⇨ □ D ▽	Compressor Covers Deburred	Y				
14	○ ⇨ □ ◗ ▽	Delay Until 47 Covers are Deburred	N				
15	○ ➡ □ D ▽	Transported to Wash/Rinse	N		Push Cart	47	1
16	● ⇨ □ D ▽	Sonic Clean/Chemical Treat Cover	Y				
17	○ ⇨ □ ◗ ▽	Delay Until 47 Covers are Treated	N				
18	○ ⇨ ■ D ▽	Inspection of Covers	N				
19	○ ➡ □ D ▽	Transported to Warehouse	N		Push Cart	47	1
20	○ ⇨ □ D ▼	Stored in Warehouse	N				
21	○ ➡ □ D ▽	Transported to Assembly Department	N		Push Cart	47	1

Process Chart

		Summary	
Part Name: Rotor Housing		**Process**	**Number**
Process Description: Fabrication of Rotor Housing		**Operation**	7
		Inspection	2
Department: Fabrication		**Transportation**	8
		Delay	4
Plant: West Lafayette, IN		**Storage**	2
Recorded By: Jarrett Hullinger	**Date:** 03/11/11	**Total Steps**	23

Step	Process Symbol (Darken the appropriate Symbol)	Description or Current Method	Value Added ? (Y/N)	Distance (ft.)	Method	Qty. Moved	Freq. per hr.
1	○ ⇨ ■ D ▽	Inspection of Incoming Cast Iron	N				
2	○ ➡ □ D ▽	Transport Cast Iron to Warehouse	N		Forklift	8	1/5 days
3	○ ⇨ □ D ▼	Stored in Warehouse	N				
4	○ ➡ □ D ▽	Cast Iron Transported to Furnace	N		Forklift	1	1 per day
5	● ⇨ □ D ▽	Melt Cast Iron	Y				
6	○ ⇨ □ ◗ ▽	Delay Until all Cast Iron is Melted	N				
7	○ ➡ □ D ▽	Transport to Die Casting Machine	N		Piping	1043 in^3	1
8	● ⇨ □ D ▽	Cast Housing	Y				
9	○ ⇨ □ ◗ ▽	Delay Until 47 Units Casted	N				
10	○ ➡ □ D ▽	Transport Casted Part to CNC	N		Push Cart	47	1
11	● ⇨ □ D ▽	Mill Housing	Y				
12	● ⇨ □ D ▽	Drill Housing # 4Holes	Y				
13	● ⇨ □ D ▽	Drill Housing #2 Holes	Y				
14	○ ⇨ □ ◗ ▽	Delay Until 47 Units are Machined	N				
15	○ ➡ □ D ▽	Transport 47 Units to Belt Sander	N		Push Cart	47	1
16	● ⇨ □ D ▽	Debur Rotor Housing	Y				
17	○ ⇨ □ ◗ ▽	Delay Until 47 Units are Deburred	N				
18	○ ➡ □ D ▽	Transport 47 Units to Sonic Clean	N		Push Cart	47	1
19	● ⇨ □ D ▽	Sonic Clean Rotor Housing	Y				
20	○ ⇨ ■ D ▽	Inspection of Rotor Housings	N				
21	○ ➡ □ D ▽	Transported to Warehouse	N		Push Cart	47	1
22	○ ⇨ □ D ▼	Stored in Warehouse	N				
23	○ ➡ □ D ▽	Transported to Assembly Department	N		Push Cart	47	1

Process Chart

	Summary	
Part Name: Bearing Housing	**Process**	**Number**
Process Description: Fabrication of Bearing Housing	**Operation**	9
	Inspection	2
Department: Fabrication	**Transportation**	8
	Delay	5
Plant: West Lafayette, IN	**Storage**	2
Recorded By: Jarrett Hullinger **Date:** 03/11/11	**Total Steps**	26

Step	Process Symbol (Darken the appropriate Symbol)	Description or Current Method	Value Added ? (Y/N)	Distance (ft.)	Method	Qty. Moved	Freq. per hr.
1	○ ⇨ ■ D ▽	Inspection of Incoming Cast Iron	N				
2	○ ➡ □ D ▽	Transport Cast Iron to Warehouse	N		Forklift	2	1/5 days
3	○ ⇨ □ D ▼	Stored in Warehouse	N				
4	○ ➡ □ D ▽	Cast Iron Transported to Furnace	N		Forklift	1	1/2.5 days
5	● ⇨ □ D ▽	Melt Cast Iron	Y				
6	○ ⇨ □ ◗ ▽	Delay Until All Cast Iron is Melted	N				
7	○ ➡ □ D ▽	Transport to Die Casting Machine	N		Piping	250	1
8	● ⇨ □ D ▽	Cast Housing	Y				
9	○ ⇨ □ ◗ ▽	Delay Until 47 Units Casted	N				
10	○ ➡ □ D ▽	Transport Casted Part to CNC	N		Push Cart	47	1
11	● ⇨ □ D ▽	Mill Housing	Y				
12	● ⇨ □ D ▽	Drilll Housing # 4Holes	Y				
13	● ⇨ □ D ▽	Drill Housing #2 Oil Hole	Y				
14	● ⇨ □ D ▽	Drill 1# Shaft Hole	Y				
15	● ⇨ □ D ▽	Tapp Bearing Housing	Y				
16	○ ⇨ □ ◗ ▽	Delay Until 47 Units are Machined	N				
17	○ ➡ □ D ▽	Transport 47 Units to Belt Sander	N		Push	47	1
18	● ⇨ □ D ▽	Deburr Bearing Housing	Y				
19	○ ⇨ □ ◗ ▽	Delay Until 47 Units are Deburred	N				
20	○ ➡ □ D ▽	Transport 47 Units to Sonic Clean	N		Push Cart	47	1
21	● ⇨ □ D ▽	Sonic Clean Bearing Housing	Y				
22	○ ⇨ □ ◗ ▽	Delay Until 47 Units are Sonic Cleaned	N				
23	○ ⇨ ■ D ▽	Inspection of Bearing Housings	N				
24	○ ➡ □ D ▽	Transported to Warehouse	N		Push Cart	47	1
25	○ ⇨ □ D ▼	Stored in Warehouse	N				
26	○ ➡ □ D ▽	Transported to Assembly Department	N		Push Cart	47	1

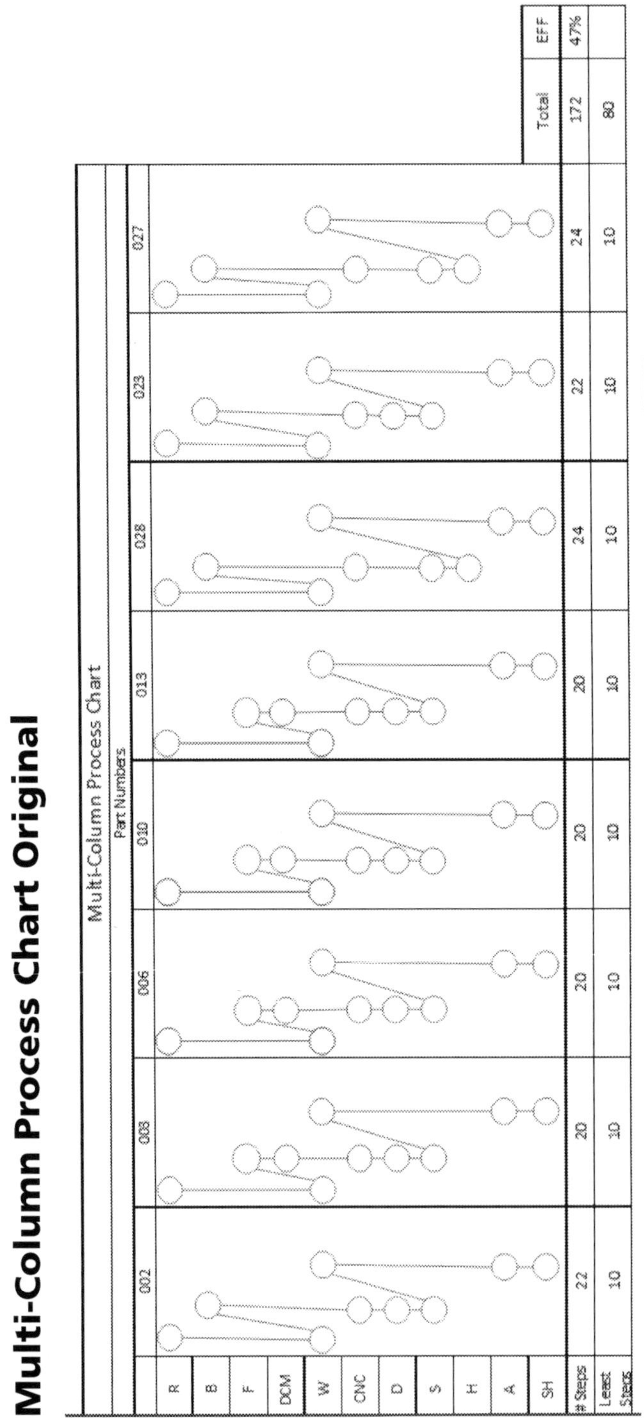

MULTI-COLUMN PROCESS KEY	
R	Receiving
W1	Warehouse 1
B	Band Saw
F	Furnace
DCM	Die Cast Machine
CNC	CNC Machine
D	Belt Sander
S	Sonic Clean
H	Heat Treatment
W2	Warehouse 2
A	Assembly
SH	Shipping

The multi-column chart to the left shows the flow of products through the manufacturing facility. This version of the chart was the first flow possibly that was identified. This flow is only 47% efficient due to extreme backtracking.

Multi-Column Process Chart Optimized

Multi-Column Process Chart										
	Part Numbers									
	002	003	006	010	013	028	023	027		
R	○	○	○	○	○	○	○	○		
W1	○	○	○	○	○	○	○	○		
B	○					○	○	○		
F		○	○	○	○					
DCM		○	○	○	○					
CNC	○	○	○	○	○	○	○	○		
D	○	○	○	○	○		○			
S	○	○	○	○	○	○	○	○		
H						○		○		
W2	○	○	○	○	○	○	○	○		
A	○	○	○	○	○	○	○	○		
SH	○	○	○	○	○	○	○	○	Total	EFF
# Steps	11	11	11	11	11	11	11	11	88	100%
Least Steps	11	11	11	11	11	11	11	11	88	

MULTI-COLUMN PROCESS KEY	
R	Receiving
W1	Warehouse 1
B	Band Saw
F	Furnace
DCM	Die Cast Machine
CNC	CNC Machine
D	Belt Sander
S	Sonic Clean
H	Heat Treatment
W2	Warehouse 2
A	Assembly
SH	Shipping

As shown in this second iteration of our multi-column chart, by adding a second warehouse to our facility we are able to eliminate all the backtracking and raise our flow efficiency from 47% to 100%.

FROM-TO CHART

The from-to chart aids with the analysis of the material flow through the facility based on the weighted factor for each part. The weighted factor for each part is determined based on the quantity and weight, among other considerations.

Based on the S.S.T. Manufacturing's from-to chart, the overall efficiency of the layout was estimated at 80.26%. This efficiency was achieved by adding a second warehouse to store some of the fabricated parts, thus eliminating backtracking to warehouse 1. As a result, S.S.T. has been able to achieve no backtracking and very little out-of-sequence movement.

FROM TO CHART KEY	
R	Receiving
W1	Warehouse 1
B	Band Saw
F	Furnace
DCM	Die Cast Machine
CNC	CNC Machine
D	Belt Sander
S	Sonic Clean
H	Heat Treatment
W2	Warehouse 2
A	Assembly
SH	Shipping

From-To Chart

TO

FROM	R	W1	B	F	DCM	CNC	D	S	H	W2	A	SH	Total	P.P
R		12.9 +3.1+1.7+2.6 +1.0+1.3											49.1	49.1
W1			3.5+2.6+1.0 +1.3 **8.4**	23.0+12.9+3.1 +1.7 **81.4**									49.1	89.8
B						3.5+2.6+1.0 +1.3 **25.2**							8.4	25.2
F					23.0+12.9+3.1 +1.7 **40.7**								40.7	40.7
DCM						23.0+12.9+3.1 +1.7 **40.7**							40.7	40.7
CNC							12.9 +3.1+1.7+1.0 **45.2**	2.6+1.3 **7.8**					49.1	53
D								12.9 +3.1+1.7+1.0 **45.2**					45.2	45.2
S									2.6+1.3 **3.9**	12.9 +3.1+1.7+1.0 **90.4**			49.1	94.3
H										2.6+1.3 **3.9**			3.9	3.9
W2											12.9+3.1+1.7+ 2.6+1.0+1.3 **49.1**		49.1	49.1
A												12.9 +3.1+1.7+2.6 +1.0+1.3	49.1	49.1
SH														
T P.P.		49.1 49.1	8.4 8.4	40.7 81.4	40.7 40.7	49.1 65.9	45.2 45.2	49.1 53.0	3.9 3.9	49.1 94.3	49.1 49.1	49.1 49.1	433.5	540.1
						Stainless Steel Density	.326 lb/in^3	Cast Iron Density	.265 lb/in^3				EFF	80.26%

Turbo Weight = 22.47

Part No.	QTY/Day	Weight (lbs.)	Total Wieght	Rela. Import.
002	834	1.61	1344.2	3.5
003	834	10.52	8769.86	23.0
006	834	5.88	4904.54	12.9
010	834	1.41	1175.56	3.1
013	834	0.80	663.03	1.7
028	834	1.20	996.67	2.6
023	834	0.46	381.00	1.0
027	834	0.61	508.94	1.3

ACTIVITY RELATIONSHIP DIAGRAM

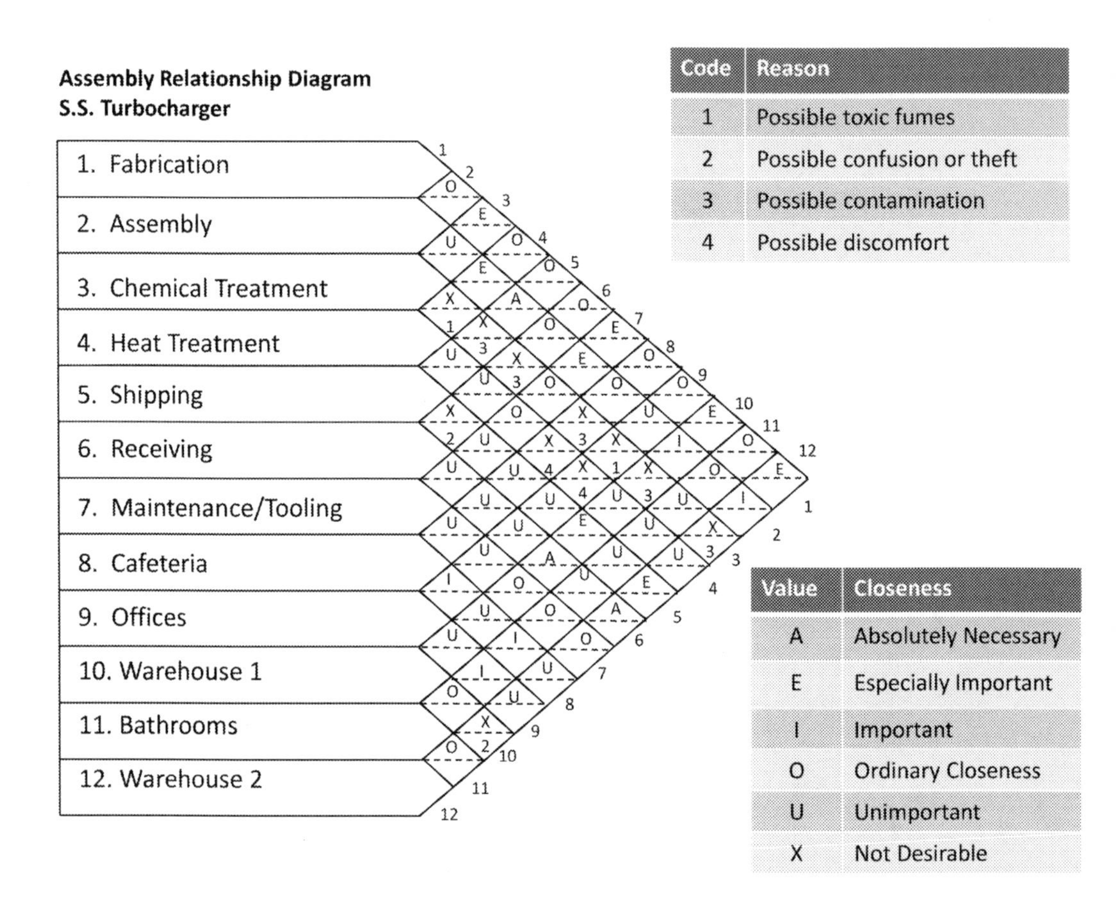

Code	Reason
1	Possible toxic fumes
2	Possible confusion or theft
3	Possible contamination
4	Possible discomfort

Value	Closeness
A	Absolutely Necessary
E	Especially Important
I	Important
O	Ordinary Closeness
U	Unimportant
X	Not Desirable

Activity Relationship Worksheet

ACTIVITIES	A	E	I	O	U	X
1. Fabrication		3, 7, 10, 12		2, 4, 5, 6, 8, 9, 11		
2. Assembly	5	1, 4, 7	10, 12	6, 8, 11	3, 9	
3. Chemical Treatment		1		7	2, 11	4, 5, 6, 8, 9, 10, 12
4. Heat Treatment		2		1, 7	4, 5, 10, 11, 12	3, 8, 9
5. Shipping	2	10, 12		1	4, 7, 8, 9, 11	3, 6
6. Receiving	10, 12			1, 2	4, 7, 8, 9, 11	3, 5
7. Maintenance/ Tooling		1, 2		3, 4, 10, 11, 12	5, 6, 8, 9	
8. Cafeteria			9, 11	1, 2	5, 6, 7, 10, 12	3, 4
9. Offices				1, 8, 11	2, 5, 6, 7, 10, 12	3, 4
10. Warehouse 1	6	1, 5	2	7, 11	4, 8, 9	3, 12
11. Bathrooms			8, 9	1, 2, 7, 10, 12	3, 4, 5, 6	
12. Warehouse 2	6	1, 5	2	7, 11	4, 8, 9	3, 10

This is a summary of the activity relationship diagram.

DIMENSIONLESS BLOCK DIAGRAM

This is the outcome of the activity relationship chart which gives the first visual representation of what the S.S.T. facility layout would look like, without dimensions. It is constructed based on the considerations for the degree of closeness or separation for the critical departments.

After the required space for each department is calculated the dimensional characteristics of this diagram will change, but the overall relative locations of the departments will remain similar to this layout.

10 / 6 Receiving X = 5,3 / 3,1	2 / 4 Heat Treatment X = 3,8,9 / 1,7	5 / 4,7 / 2 Assembly X = N/A / 10 / 1,6,8,11	2 / 10 / 5 Shipping X = 3,6 / 1
6 / 5,1 / 10 Warehouse 1 X = 3 / 2 / 7	3,7,10 / 1 Fabrication X = N/A / 2,4,5,6,8,9,11	6 / 5,1 / 12 Warehouse 2 X = 10,3 / 2 / 11,7	10 / 9 Office X = 4,3 / 3,1
	1 / Chemical Treatment X = 4,5,6,8,9,10 / 7	2,1 / Maintenance/Tooling X = N/A / 4,3,10,11	9,11 / 8 Cafeteria X = 4,3 / 2,1
		11 Bathroom X = None / 9,8 / 10,7,2,1	

SAMPLE MATERIAL HANDLING SYSTEM

The following page presents a sample piece of material handling equipment needed in the facility. Detailed equipment specifications and their functions should be presented for *every* piece of material handling equipment.

Following the specification page for equipment, a thorough and a detailed "Planning Principle Analysis" is presented. Material handling operations and equipment for the facility are scrutinized using the material handling principles.

Material Handling System

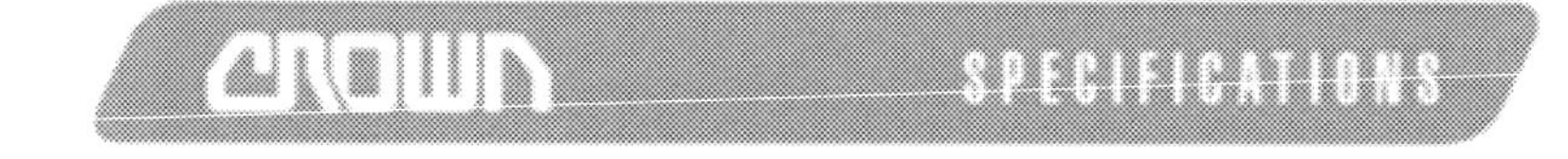

General Information	1	Manufacturer	Crown Equipment Corporation		
	2	Model			C5 1050-40
	3	Load Capacity		lb	4000
	4	Load Center	Fork Face to Load CG	in	24
	5	Power (Fuel) Type	Gas, LPG, Diesel		LPG
	6	Operator Type			Sit-down Rider Counterbalance
	7	Tire Type			Resilient
	8	Wheels (x = driven)	Number Front/Rear		2x / 2
	9	Steering Type			Hydrostatic
	10	Mast	Lifting Height (MFH)	in	188
			Free Lift Height	in	32.2

Source: http://www.crown.com/usa/products/pdfs/specs/ic_pneumatic_c5_spec.pdf

Material Handling System Planning Principle Analysis

<table>
<tr><th colspan="2">PART NAME: COVER</th></tr>
<tr><td>
Transported to Warehouse
Why = Steel needed to create cover
What = 5" × 70" stainless steel bar stock
Where = From Receiving to Warehouse 1
Who = Forklift operator
When = One time per 24-hour work day
How = Forklift

Bar Stock Transported to Band Saw
Why = Bar stock needs to be cut to size
What = Hot rolled stainless steel bar stock
Where = From Warehouse to Band Saw
Who = Forklift operator
When = One time per 8-hour shift
How = Forklift

Compressor covers transported to CNC001
Why = Covers need to be milled and drilled
What = Cut compressor covers
Where = From Band Saw to CNC
Who = Band Saw operator
When = One time per hour
How = Push cart

Transported to Belt Sander by conveyor
Why = Covers need to be sanded
What = Machined compressor covers
Where = From CNC to Belt Sander
Who = CNC operator
When = 47 times per hour
How = Conveyor
</td><td>
Transported to Wash/Rinse
Why = Covers need to be chem. treated
What = Sanded compressor covers
Where = From Belt Sander to Chemical Treatment
Who = Belt Sander operator
When = One time per hour
How = Push cart

Transported to Warehouse
Why = Covers should wait to be assembled
What = Treated compressor covers
Where = From Chem. Treat to Warehouse 2
Who = Chemical Treatment operator
When = One time per hour
How = Push cart

Transported to Assembly Department
Why = Covers required for turbo assembly
What = Finished compressor covers
Where = From Warehouse 2 to Assembly
Who = Assembly Line operator
When = One time per hour
How = Push cart
</td></tr>
</table>

SAMPLE SPACE CALCULATIONS

The following pages in the report entail space calculations for individual work cells, storage, warehousing, receiving and shipping areas, bathrooms, fabrication, assembly, maintenance, aisles, office, cafeteria, and the total S.S.T. manufacturing facility. Dimensions for individual machines and equipment were attained from original manufacturer specifications.

Machinery & Equipment Layout Data

Machinery & Equipment Layout Data

Name/Type Band Saw

File

Plant: Chyun Yow Machinery Industry Co			Steam	N/A	Drains Yes
Prepared by: Michael Thoma			Gas	N/A	Level/Lag N/A
Water Yes			Pit	N/A	
Compressed Air No			Sp'l Electric	N/A	
Foundation: Yes			Max Height	9.5'	
Exhaust N/A			Weight	200kgs	
Electrical	H.P	Volts	Cycle	Phase	Amps
Drive Motor	1.5	400	-	-	-
Auxiliary Motor	N/A	N/A	N/A	N/A	N/A
Auxiliary Motor	N/A	N/A	N/A	N/A	N/A

Manufacture: Chyun Yow Machinery	
Speed/Capacity	45-90 mph
Left-Right	2 ft
Front-Back	3 ft
Net Floor Area	6 ft²
Scrap Bin	15 ft²
Work Bench Area	21 ft²
Area Aisle	90 ft²
Work Area/Service	231.36 ft²
Gross Area	396 ft²

Size/Model	George 275A
Sgnf. Ident	BS001-002

Co. Mach/Eqpt. Ident. Numbers Covered By this Sheet

Pallets	26.64 ft²

Aisle

In

Out

Saw

Scrap

Saw

Workbench

Elevation Sketch or Photos

Source: http://www.bandsawmachine.com.tw

Plan View

one square = one foot

Date: March, 25 2011

Machinery & Equipment Layout Data

Machinery & Equipment Layout Data

Name/Type Furnace — File

Plant: Thermtronix Plant			Steam	-	Drains N/A
Prepared by: Michael Thoma			Gas	Yes	Level/Lag N/A
Water No			Pit	N/A	
Compressed Air No			Sp'l Electric N/A		
Foundation: N/A			Max Height 10'		
Exhaust Yes			Weight	5,600lbs	
Electrical	H.P	Volts	Cycle	Phase	Amps
Drive Motor	240	-	-	-	480
Auxiliary Motor	N/A	N/A	N/A	N/A	N/A
Auxiliary Motor	N/A	N/A	N/A	N/A	N/A

Manufacture: Thermtronix	
Speed/Capacity	300lbs/hr
Left-Right	9.95 ft
Front-Back	14.25 ft
Net Floor Area	142 ft²
Scrap Bin	N/A
Work Bench Area	21 ft²
Area Aisle	95 ft²
Work Area/Service	179 ft²
Gross Area	437 ft²

Size/Model	GC2000T
Sgnif. Ident	F001

Co. Mach/Eqpt. Ident. Numbers Covered By this Sheet

Aisle Space

Furnace

Work Bench

Elevation Sketch or Photos
Source: www.cortechmachines.com/hef.html

Plan View one square = one foot
Date: March, 25 2011

Warehouse 1 & 2 Receiving & Shipping Space Calculation Sheet

WAREHOUSE #1 SPACE CALCULATION SHEET

Name	Length (Inches)	X	Width (Inches)	X	Height (Inches)	=	Cubic Inches	Qty (5 days) 2/3	Cubic ft. Needed	Roundstock Shelf (10' X 3' X 8')	Pallet (4' X 3.33' X 3')
5" × 70" Steel Roundstock	70	×	5	×	5	=	1750	10	10.13	2	
3" × 47" Steel Roundstock	47	×	3	×	3	=	423	10	2.45	1	
0.7" × 83" Cast Iron Roundstock	165	×	0.7	×	0.7	=	80.85	40	1.87	2	
3" × 24" Cast Iron Roundstock	24	×	3	×	3	=	216	10	1.25	1	
1' × 1' × 13" Cast Iron Ingot	12	×	12	×	13	=	1872	18	19.50		9
									Total:	6	9

WAREHOUSE #2 SPACE CALCULATION SHEET

Part No.	Part Name	Length (Inches)	X	Width (Inches)	X	Height (Inches)	=	Cubic Inches	Qty (5 days) 2/3	Cubic Ft. Needed	Shelf (2' X 2' X 3')	Pallet (4' X 3.33' X 3')
002	Cover (Boxed 80)	10.00	×	10.00	×	10.00	=	1000.00	34.75	20.11	3	
003	Turbo Housing (Not stackable)	8.00	×	6.00	×	3.00	=	144.00	2780.00	231.67		78
004	Turbo Screw 1/4 20 (Boxed 2000)	10.00	×	10.00	×	6.00	=	600.00	11.12	3.86	1	
005	Turbo Segment (Boxed 75)	6.00	×	5.00	×	6.25	=	187.50	148.27	16.09	2	
006	Rotor Housing (Not stackable)	8.00	×	6.00	×	2.00	=	96.00	2780.00	154.44		78
007	Rotor Segment (Boxed 75)	6.00	×	7.00	×	6.25	=	262.50	111.20	16.89	2	
008	Rotor Screw 1/4 20 (Boxed 2000)	10.00	×	10.00	×	6.00	=	600.00	9.73	3.38	1	
010	Bearing Housing	4.00	×	5.00	×	3.00	=	60.00	2780.00	96.53		3
011	Bushing (Boxed 200)	9.00	×	9.00	×	9.00	=	729.00	27.80	11.73	2	
012	Lock Ring (Boxed 200)	9.00	×	9.00	×	8.00	=	648.00	55.60	20.85	4	
013	Heat Shield	3.00	×	3.00	×	20..	=	18.00	2780.00	28.90		1
014	Piston Ring (Boxed 200)	9.00	×	9.00	×	80..	=	648.00	55.60	20.85	4	
015	Thrust Bearing (Boxed 64)	7.00	×	3.25	×	8.00	=	182.00	43.44	4.58	1	
016	Oil Deflector (Boxed 72)	7.50	×	3.50	×	9.00	=	236.25	38.61	5.28	1	
018	Large O-Ring (Boxed 500)	14.00	×	10.00	×	10.00	=	1400.00	5.56	4.50	1	
019	Small O-Ring (Boxed 700)	14.00	×	10.00	×	10.00	=	1400.00	3.97	3.22	1	
020	Housing Screw (Boxed 2000)	10.00	×	10.00	×	6.00	=	600.00	5.56	1.93	1	One Shelf
021	Washer (Boxed 2000)	9.00	×	6.25	×	5.00	=	281.25	5.56	0.90	C	
022	Nut (Boxed 500)	7.00	×	8.00	×	4.00	=	224.00	5.56	0.72	C	
023	Rotor Shaft (Bins/Shelf 100 per bin)	10.00	×	7.25	×	10.50	=	761.25	27.80	12.25	2	
024	Thrust Ring (Boxed 500)	7.25	×	5.00	×	8.00	=	290.00	5.56	0.93	1	One Shelf
025	Sleeve (Boxed 600)	7.50	×	6.00	×	8.00	=	360.00	4.63	0.97	C	
026	Seal (Boxed 500)	8.00	×	6.00	×	7.00	=	336.00	5.56	1.08	1	
027	Compressor Wheel (Not stackable)	3.00	×	3.00	×	0.50	=	4.50	2780.00	7.14		14
028	Rotor Blade (Not stackable)	4.00	×	4.00	×	2.00	=	32.00	2780.00	51.48		24
000	Turbo (Not stackable)	8.61	×	7.50	×	5.63	=	363.56	2780.00	584.89		112
									Total:	1305.32	28	310

Aisle Space Calculation Sheets

S.S. TURBO AISLE SPACE CALCULATIONS (WAREHOUSE 1)			
Shelf Type	**Length (ft.)**	**Number Shelves Needed**	**Aisle Space Needed (ft.)**
Roundstock shelf	10.00	6	60.00
Pallet racks	8.50	1	8.50
		Total Aisle ft.	**65.80**

S.S. TURBO AISLE SPACE CALCULATIONS (WAREHOUSE 2)				
Shelf Type	**Length (ft.)**	**Number Needed**	**Shelving Units**	**Aisle Space Needed (ft.)**
Pallet rack	8.50	36	36	306.00
Shelf units	2.00	28	14	28.00
			Total Aisle ft.	**334.00**

STANDARD MATERIAL STORAGE DIMENSIONS					
Name	**Length (ft.)**	**X**	**Width (ft.)**	**X**	**Height (ft.)**
Roundstock shelf (W1)	10.00	X	3.00	X	8.00
Shelf (W2)	2.00	X	2.00	X	3.00
Pallet rack (W1 & W2)	8.50	X	8.00	X	14.00

Machinery & Equipment Layout Data

S.S. TURBO PRODUCTION SPACE REQUIREMENT SHEET												
No.	Activity Dept., Area, or Item	OPER No.	Machine or Equipment	Space Requirements								
				Machine ETC. L x W = A	Auxiliary Equip L x W = A	Operator Space L x 3 = A	Material Space L x W = A	Subtotal	Sub. x 1.5 = Allow	No. MAC	Total sq. ft. / Oper.	Total per Area
1	Area	1	Furnace	142	21	9	0	172	258	2	9	534
2	Area	1	Diecastor	117.3	21	9	41.64	283.41	283.41	2	9	584.82
3	Dept.	2	Band Saw	12	21	18	41.64	138.96	138.96	1	9	147.96
4	Area	1	CNC Lathe	56.3	21	9	41.64	191.91	191.91	2	9	401.82
5	Area	1	CNC Milling Machine	55.2	21	9	41.64	190.26	190.26	31	9	6177.06
6	Dept.	3	Belt Sander	10.56	21	27	41.64	150.3	150.3	1	9	159.3
7	Area	1	Sonic Clean	60	21	9	100	285	285	2	9	588
8	Area	1	Heat Treat Oven	52	21	9	113.32	292.98	292.98	1	9	301.98
9	Area	1	Balance	15	29	9	39	138	138	1	9	147

Auxiliary Equipment is workbench space

1 operator has 3’ x 3’. We are not expanding.

Total Facility Space Calculations

	Ft^2/Employee	Employees	Stations	X	Width (ft.)	X	Length (ft.)	=	Total Ft^2
Manufacturing									
Fabrication									
Band Saw		2	1		18		22		396
Belt Sander		3	1		18		23		414
CNC Lathe		2	2		21		22		924
CNC Mill		31	31		21		22		14,322
Die Caster		2	2		21		27		1,134
Furnace		2	2		19		23		874
Heat Treat		1	1		23		29		667
Sonic Clean		2	2		20		29		1,160
Assembly									
Line		8	7		20		13		1,820
Balancer		1	1		20		13		260
Manufacturing Total (Aisles included)									19,891
Production Services									
Receiving									
Steel Bar, Cast Iron, Buyout Stock		1	1		38		21		798
Warehouse 1									
Raw Material		2	1		38		38		1,444
Warehouse 2									
Fabricated and Buyout Parts		2	1		56		138		7,728
Shipping		1	1		38		21		798
Maintenance & Tooling		2	1						800
Production Service Area Total (Aisles Included)									11,568
Employee Services									
Employee Entrance			1		10		20		200
Bathrooms	10								441
Cafeteria			71						710
Drinking Fountain			9		(15 ft^2 per fountain)				135
Medical services		1	1		10		10		100
Services Area Total									1,586
Office Area	200	8							1,600
Total Building Space									**34,645**
Outside areas									
Receiving					9		65		0
Parking			1						0
Maneuvering			2						0
Shipping					9		65		0
Parking			1						0
Maneuvering			1						0
Employee Parking			47.33333		15.8		15.8		11,816
Total Outside Space									**11,816**
Total Employees per Shift		71							

Bathroom Calculations

	X MEN			X WOMEN		
	Units	Ft^2/Unit	Total Ft^2	Units	Ft^2/Unit	Total Ft^2
Toilets	3	15	45	4	15	60
Washbasins	4	15	60	4	15	60
Urinals	1	9	9			
Reclining Area				2	15	30
Door	1	15	15	1	15	15
Total			**129**			**165**
X 150%			**193.5**			**247.5**

LAYOUT IQ FACILITY ANALYSIS

The following pages exhibit the facility drawings. These drawings were produced utilizing *Layout IQ*. The S.S.T. Manufacturing group optimized the layout using the three modules, 1) Routing sheets, 2) From-To charts, and 3) Subjective analysis of the LayoutIQ software.

Author's Note: In the interest of brevity, a significant number of drawings have been eliminated from the original project.

Layout IQ

Part Name	Flow Factor	From	To	Unit Load
Compressor Wheel	1	Bandsaw	CNC	1
Compressor Wheel	1	CNC	Belt Sander	1
Compressor Wheel	1	Belt Sander	Sonic Clean	1
Turbo Housing	1	Furnace	Die-Castor	1
Turbo Housing	1	Die-Castor	CNC	1
Turbo Housing	1	CNC	Belt Sander	1
Turbo Housing	1	Belt Sander	Sonic Clean	1
Rotor Housing	1	Furnace	Die-Castor	1
Rotor Housing	1	Die-Castor	CNC	1
Rotor Housing	1	CNC	Belt Sander	1
Rotor Housing	1	Belt Sander	Sonic Clean	1
Bearing Housing	1	Furnace	Die-Castor	1
Bearing Housing	1	Die-Castor	CNC	1
Bearing Housing	1	CNC	Belt Sander	1
Bearing Housing	1	Belt Sander	Sonic Clean	1
Heat Shield	1	Furnace	Die-Castor	1
Heat Shield	1	Die-Castor	CNC	1
Heat Shield	1	CNC	Belt Sander	1
Heat Shield	1	Belt Sander	Sonic Clean	1
Rotor Blade	1	Bandsaw	CNC	1
Rotor Blade	1	CNC	Sonic Clean	1
Rotor Blade	1	Sonic Clean	Heat Treament	1
Rotor Shaft	1	Bandsaw	CNC	1
Rotor Shaft	1	CNC	Belt Sander	1
Rotor Shaft	1	Belt Sander	Sonic Clean	1
Compressor Wheel	1	Bandsaw	CNC	1
Compressor Wheel	1	CNC	Sonic Clean	1
Compressor Wheel	1	Sonic Clean	Heat Treament	1

Layout IQ

	Shipping	Assembly	Warehouse	Heat Treatm	Sonic Clean	Belt Sander	CNC	Die-Castor	Furnace	Bandsaw	Warehouse	Receiving
▸Receiving												
Warehouse 1									2.9	12		
Bandsaw							96					
Furnace								96				
Die-Castor							96					
CNC					48	144						
Belt Sander					144							
Sonic Clean			144	48								
Heat Treatme			48									
Warehouse 2		192										
Assembly	192											
Shipping												

Layout IQ

	Shipping	Assembly	Warehouse	Heat Treatm	Sonic Clean	Belt Sander	CNC	Die-Castor	Furnace	Bandsaw	Warehouse
Receiving	1										
▸ Warehouse 1	|		1	1	1						
Bandsaw											
Furnace	1	1	1								
Die-Castor											
CNC											
Belt Sander											
Sonic Clean	1	1	1								
Heat Treatme	1	1	1								
Warehouse 2											
Assembly											

Layout IQ

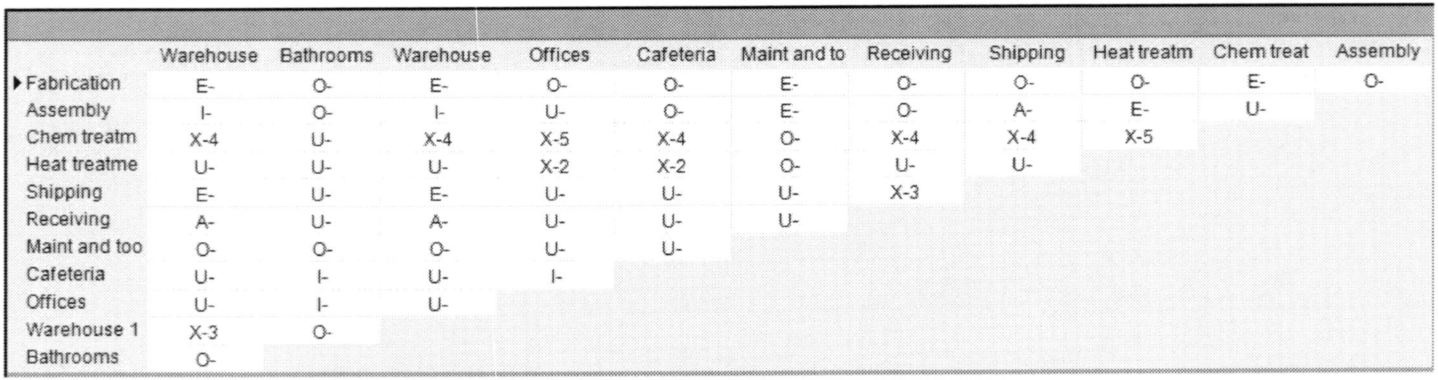

	Warehouse	Bathrooms	Warehouse	Offices	Cafeteria	Maint and to	Receiving	Shipping	Heat treatm	Chem treat	Assembly
▸ Fabrication	E-	O-	E-	O-	O-	E-	O-	O-	O-	E-	O-
Assembly	I-	O-	I-	U-	O-	E-	O-	A-	E-	U-	
Chem treatm	X-4	U-	X-4	X-5	X-4	O-	X-4	X-4	X-5		
Heat treatme	U-	U-	U-	X-2	X-2	O-	U-	U-			
Shipping	E-	U-	E-	U-	U-	U-	X-3				
Receiving	A-	U-	A-	U-	U-	U-					
Maint and too	O-	O-	O-	U-	U-						
Cafeteria	U-	I-	U-	I-							
Offices	U-	I-	U-								
Warehouse 1	X-3	O-									
Bathrooms	O-										

Layout IQ

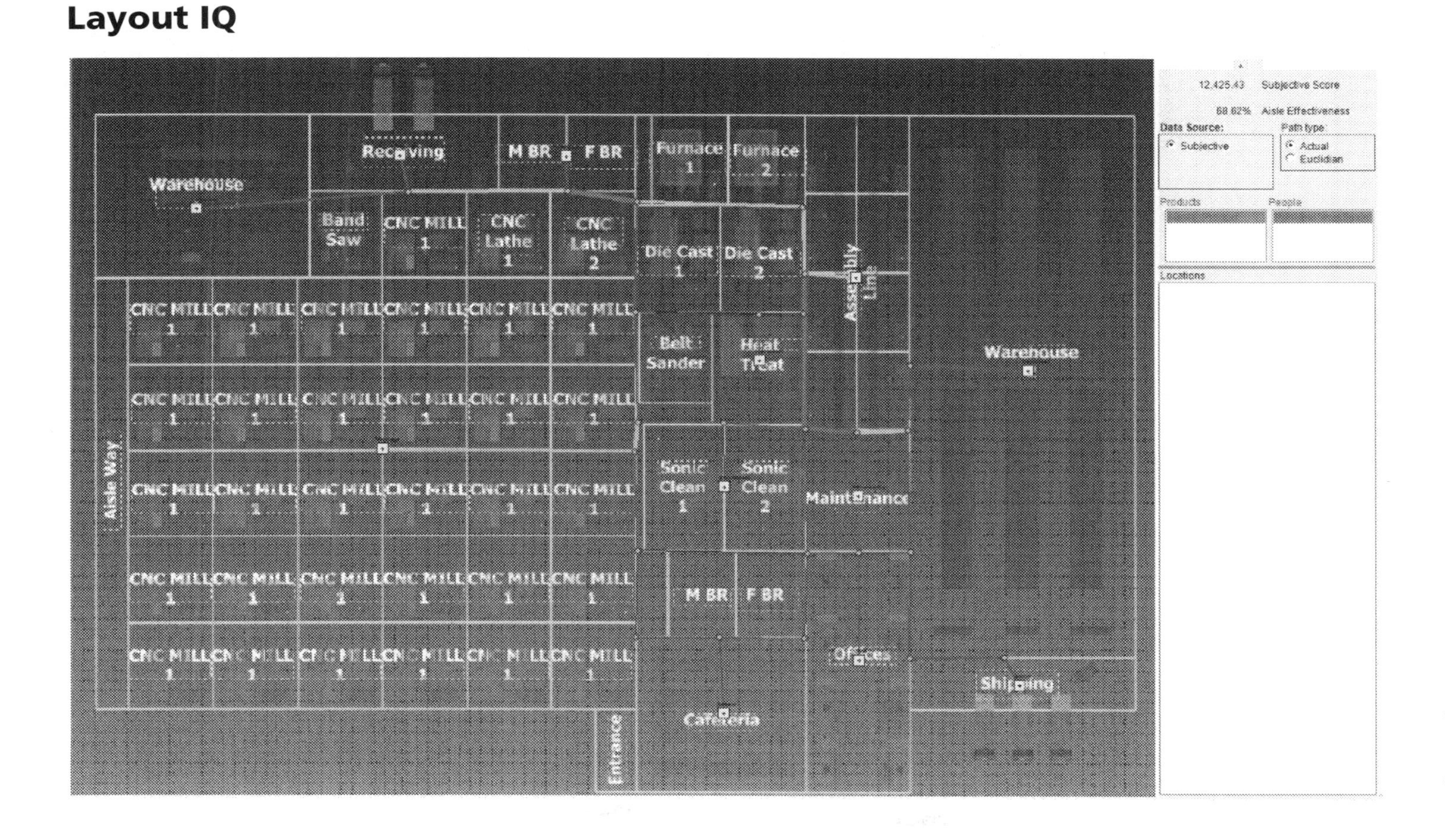

12,425.43 Subjective Score
88.82% Aisle Effectiveness
Data Source:
Subjective
Path type:
Actual
Euclidean
Products
People
Locations
Warehouse
Receiving
M BR
F BR
Furnace 1
Furnace 2
Band Saw
CNC MILL 1
CNC Lathe 1
CNC Lathe 2
Die Cast 1
Die Cast 2
Assembly Line
Belt Sander
Heat Treat
Warehouse
Aisle Way
Sonic Clean 1
Sonic Clean 2
Maintenance
M BR
F BR
Offices
Shipping
Entrance
Cafeteria

LAYOUT IQ OPTIMIZATION EXPLANATION

The S.S.T. facility layout design is optimized using routing sheets, from-to-charts, and the *Layout IQ* software. While building the routing sheets, S.S.T. determined how much equipment was necessary in the manufacturing facility to maintain a Takt time of 1.285 minutes per turbocharger. Turbochargers require very precise tolerances, therefore the manufacturing process requires many CNC machines. While trying to balance the routing sheets, S.S.T. found that assembly line manufacturing would not be possible if high efficiencies were to be maintained. Batch operation was determined to be the best method of fabricating and manufacturing turbo components. This changed the facility layout from conveyer-assembly based manufacturing to work cell batch operation.

The multi-column charts revealed that the process sequence was not at a desirable efficiency level. This was mainly due to extreme backtracking to warehouse 1. S.S.T. decided to remedy this by adding a second warehouse between fabrication and assembly operations to store parts that were fabricated, as well as parts that were needed for assembly. The second warehouse enabled zero backtracking to warehouse 1. Warehouse 2 is strategically located to allow access from assembly, fabrication, and shipping. This improved the multi-column process chart efficiency from 47% to 100%. Zero backtracking and relatively linear sequencing allowed for a high efficiency rating on the from-to-charts. The efficiency rating of the from-to-charts is calculated at 80.26% which is near the target planned efficiency rating.

Layout IQ is a powerful facility design tool that has allowed S.S.T. to simulate efficiencies of a drawn dimensioned layout. The first attempted optimization was the *subjective analysis* which was based on the activity relationship chart as conceived by the project planners. Based on the simulation results, S.S.T. found that the layout design had potential to be optimized. This optimization involved the rearrangements of the furnaces, die-castors, restrooms, cafeteria, and the sonic cleans. The reconfiguration of these departments increased the efficiency on the subjective analysis by 5%.

S.S.T. MANUFACTURING FACILITY MODELS

All fabrication and assembly equipment models, drawings, and assemblies were hand modeled by the S.S.T. Group at Purdue University using the Autodesk Inventor 2011 Factory Utilities modeling program. All equipment was modeled to exact specifications from the equipment manufacturer.

S.S.T. Facility Isometric Views

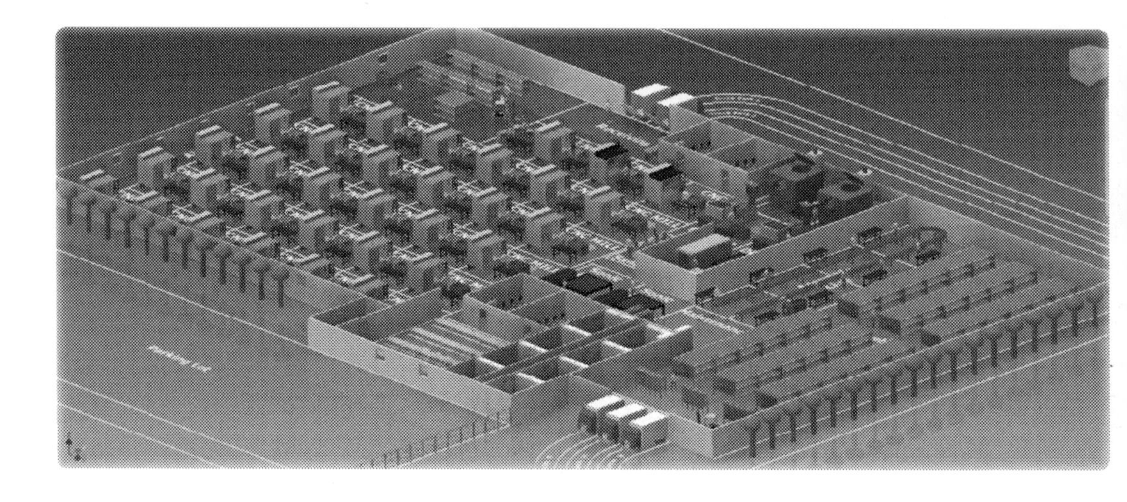

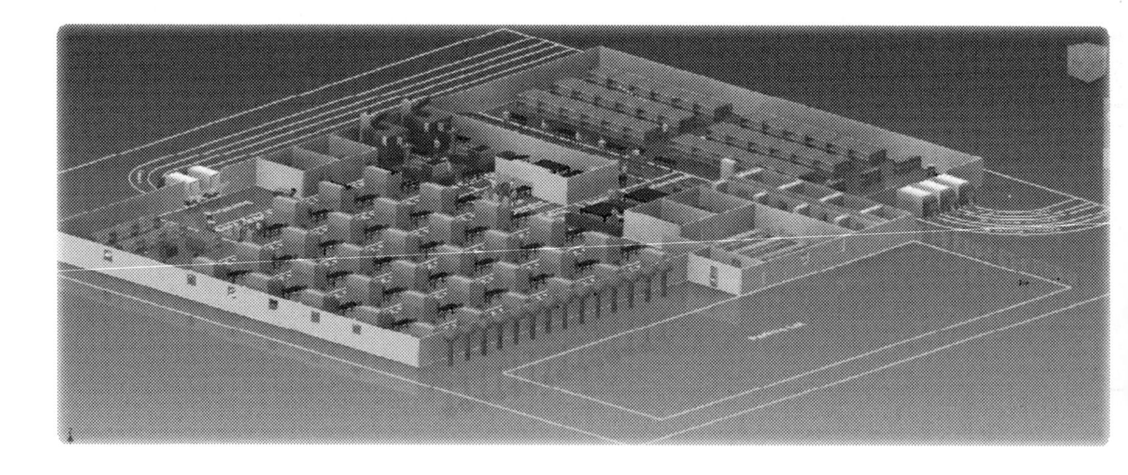

S.S.T. Facility Top View

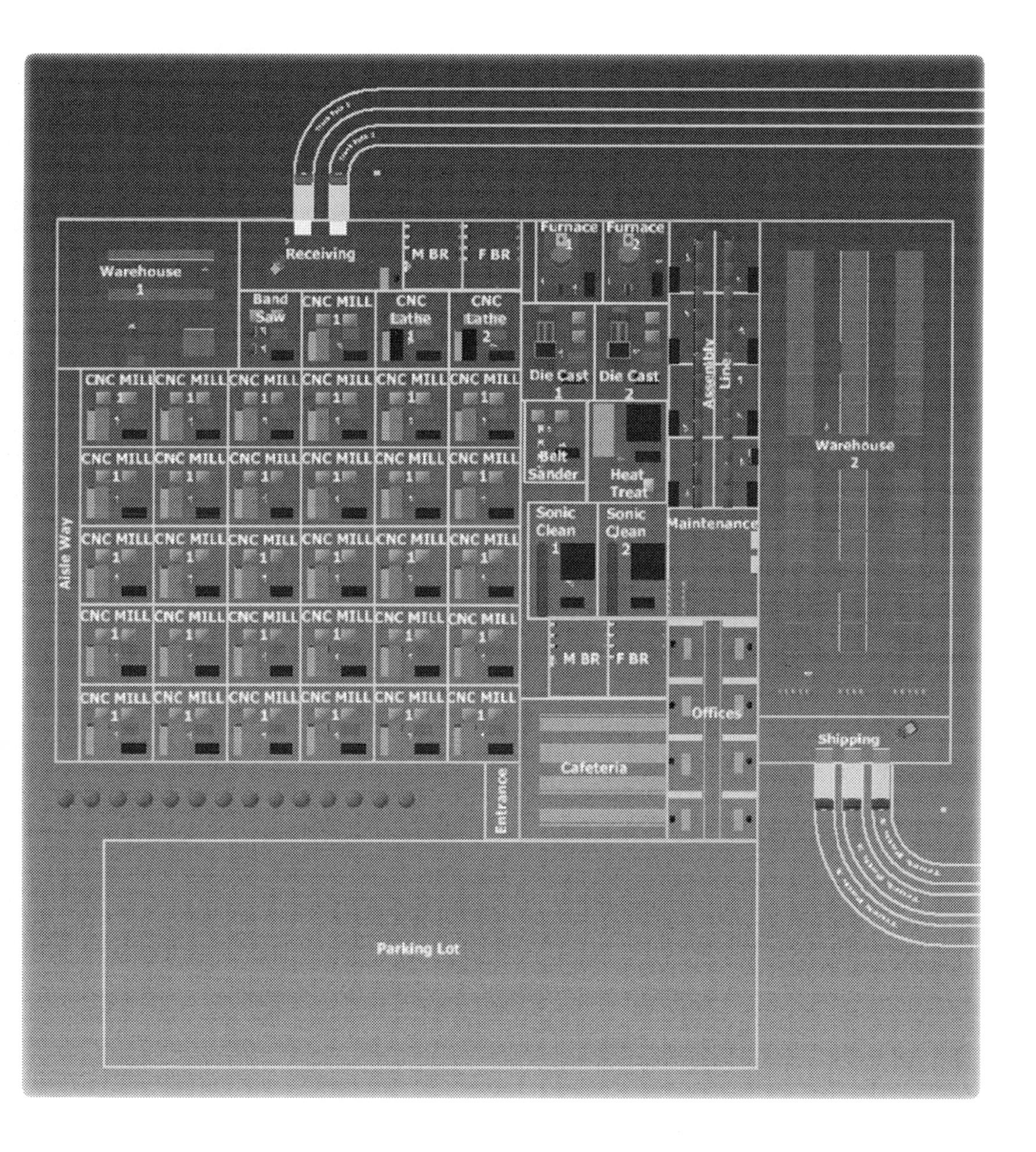

Warehouse 2: Finished Parts & Turbo Buyout Part Storage

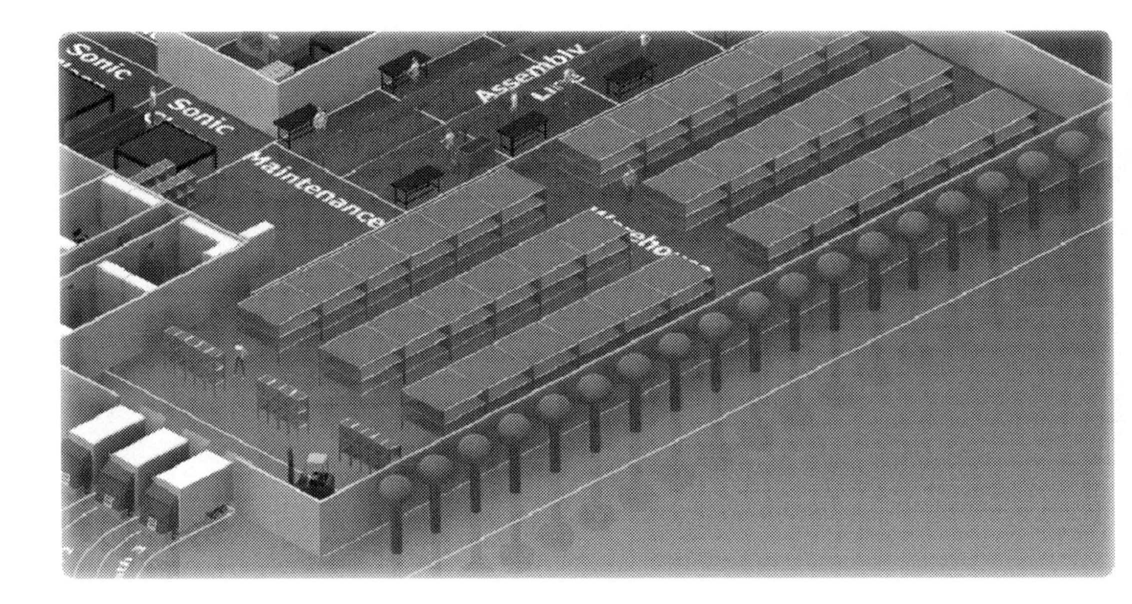

Assembly Department Isometric

Fabrication Department Isometric Views

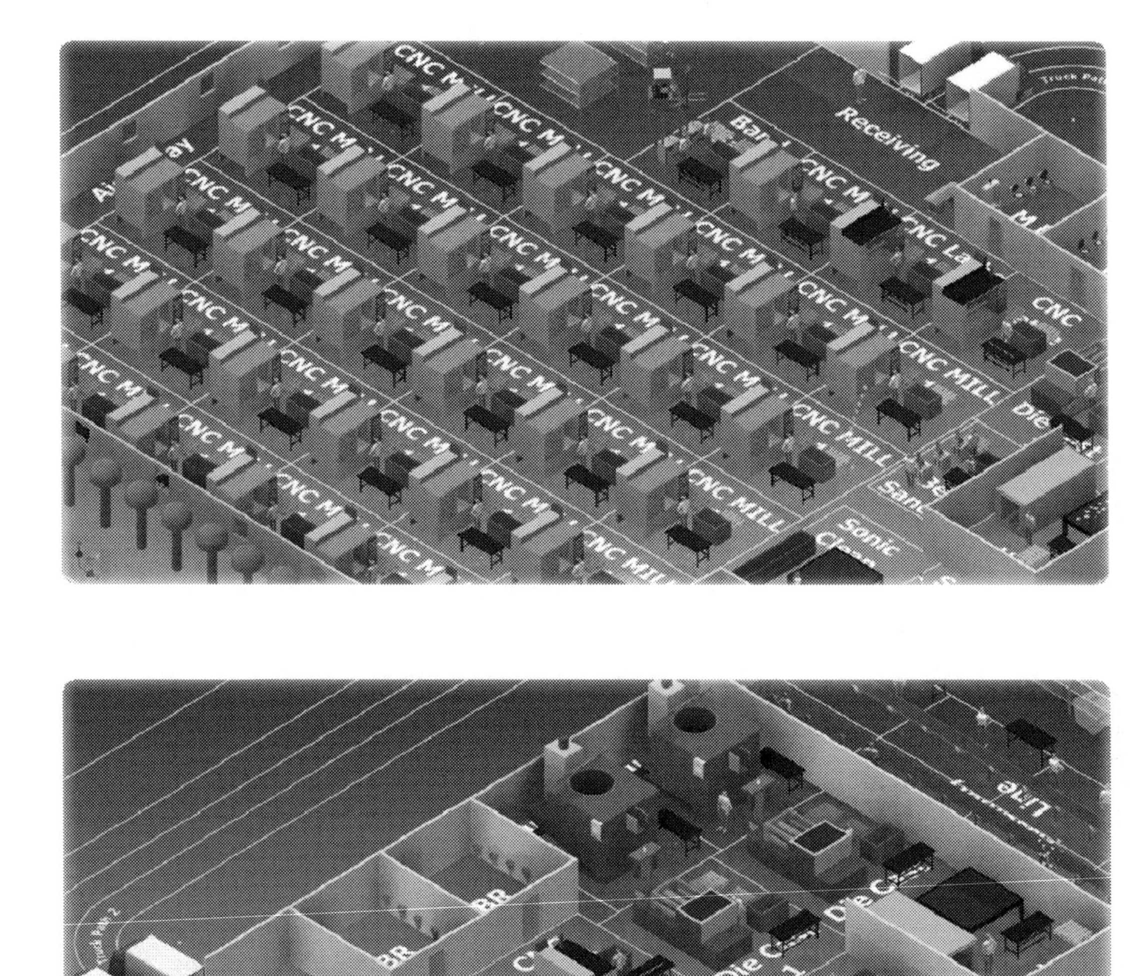

COST ANALYSIS

Total Capital Expense

S.S. TURBO MANUFACTURING CAPITAL EXPENDITURES			
Manufacturing Equipment	**Cost ($ per Unit)**	**Number Eq.**	**Total Cost**
Band Saw	$ 2,800.00	2	$ 5,600.00
CNC Milling Machine	$ 350,000.00	31	$ 10,850,000.00
CNC Lathe	$ 300,000.00	2	$ 600,000.00
Belt Sander	$ 750.00	3	$ 2,250.00
Furnace	$ 130,000.00	2	$ 260,000.00
Die Caster	$ 300,000.00	2	$ 600,000.00
Sonic Clean	$ 75,000.00	2	$ 150,000.00
Heat Treat Oven	$ 115,000.00	1	$ 115,000.00
Balancer	$ 70,000.00	1	$ 70,000.00
	Total Mfg. Equipment Cost		**$ 12,652,850.00**
Material Handling Equipment	**Cost ($ per Unit)**	**Number Eq.**	**Total Cost**
Forklifts	$ 33,000.00	4	$ 132,000.00
Ergonomic Pushcarts	$ 500.00	35	$ 17,500.00
10' Conveyor Belt	$ 1,675.00	8	$ 13,400.00
Metal Transfer Piping	$ 800.00	1	$ 800.00
Shelving/Racks	$ 350.00	57	$ 19,950.00
	Total Mat. Hand. Eq. Cost		**$ 183,650.00**
Facility	**Cost ($ per Unit)**	**Multiplier**	**Total Cost (12 mos.)**
Steel Building (leased @ $9/sqft/mo.)	$ 9.00	46,461	$ 5,017,788.00
Fixtures	$ 200.00	200	$ 40,000.00
Tooling	$ 130.00	6000	$ 780,000.00
	Total Yearly Facility Cost		**$ 5,837,788.00**
	TOTAL MFG. FACILITY COST		**$ 18,674,288.00**

Total Variable Expense

WEEKLY BUYOUT PARTS EXPENDITURE			
Buyout Parts	Cost ($ per Unit)	Unit/Week	Total Cost
Turbo Screw	$ 0.10	32,000	$ 3,200.00
Turbo Segment	$ 0.45	16,000	$ 7,200.00
Rotor Segment	$ 0.50	4,000	$ 2,000.00
Rotor Screw	$ 0.10	28,000	$ 2,800.00
Bushing	$ 0.35	8,000	$ 2,800.00
Lock Ring	$ 0.04	16,000	$ 640.00
Piston Ring	$ 0.78	16,000	$ 12,480.00
Thrust Bearing	$ 1.45	4,000	$ 5,800.00
Oil Deflector	$ 1.35	4,000	$ 5,400.00
Large O-ring	$ 4.50	4,000	$ 18,000.00
Small O-ring	$ 2.75	4,000	$ 11,000.00
Housing Screw	$ 0.12	16,000	$ 1,920.00
Washer	$ 0.04	16,000	$ 640.00
Nut	$ 0.07	4,000	$ 280.00
Thrust Ring	$ 0.20	4,000	$ 800.00
Sleeve	$ 1.10	4,000	$ 4,400.00
Seal	$ 0.75	4,000	$ 3,000.00
		Total Weekly Buyout Parts	**$ 82,360.00**

WEEKLY RAW MATERIALS EXPENDITURE			
Raw Material Type	Number	Cost/Unit	Total Cost
5" × 70" steel round stock	14	$ 2,365.00	$ 33,110.00
3" × 47" steel round stock	14	$ 672.00	$ 9,408.00
0.7" × 83" cast iron round stock	54	$ 80.75	$ 4,360.50
3" × 24" cast iron round stock	14	$ 102.00	$ 1,428.00
1' × 1' × 13' cast iron ingot	24	$ 100.00	$ 2,400.00
		Total Weekly Raw Materials Cost	**$ 50,706.50**

WEEKLY LABOR COST EXPENDITURE			
Labor Classification	Number	Rate/Week	Total Cost
Managers	8	$ 1,500.00	$ 12,000.00
Machinist	46	$ 840.00	$ 38,640.00
Assembly	8	$ 720.00	$ 5,760.00
Warehouse and Inventory Control	6	$ 720.00	$ 4,320.00
Maintenance	2	$ 840.00	$ 1,680.00
Medical	1	$ 880.00	$ 880.00
Total Employees and Weekly Labor Costs	**71**		**$ 63,280.00**

VARIABLE COST PER TURBO					
Unit	Labor/ Week	R. Mat./ Week	Buyout/ Week	Utilities/ Week	Units/ Week
1 Turbocharger	$ 63,280.00	$ 50,706.50	$ 82,360.00	$ 10,000.00	4000
	Total Variable Cost Per Turbo				**$ 51.59**

Break-Even Analysis

STRAIGHT LINE DEPRECIATION OF ASSETS			
Total Equipment Cost	Residual Value	Depreciation Term (Years)	Annual Depreciation Expense
$ 12,836,500.00	$ 640,000.00	7	$ 1,742,357.14

Sales/Week	Variable/Week	Contribution Margin (% Sales)
$ 1,200,000.00	$ 206,346.50	83%

S.S. TURBO BREAK EVEN CALCULATION			
Total Fixed Cost Per Year	Unit Contribution Number (Sell Price - Variable Cost)	Volume Until Break Even	Time Until Break Even (Weeks)
$ 7,580,145.14	248.41	30,514.24	7.63

CITATIONS

AB Volvo. (2012). Marine diesel engines. Retrieved from http://www.volvopentastore.com/Turbocharger-Components-861260/dm/cart_id.566768382--session_id.085457323–store_id.366–view_id.785228

MatWeb, LLC. (n.d.). Overview of materials for alloy cast iron. Retrieved from http://www.matweb.com/search/DataSheet.aspx?MatGUID=4092fbc5374e4874ab58c8db3fead276

MatWeb, LLC. (n.d.). Overview of materials for cast stainless steel. Retrieved from: http://www.matweb.com/search/DataSheet.aspx?MatGUID=dc78ba5a80e347d182805eeb0a50e28c&ckck=1

Turbocharger. (n.d.). In Wikipedia. Retrieved March 4, 2011 from http://en.wikipedia.org/wiki/Turbocharger

Turbonetics, Inc. (n.d.). GT-K turbochargers. Retrieved from http://www.turboneticsinc.com /gtk

Turbotechnicsltd (Poster). (2008, December 9). High-speed turbo balancing machine [Video]. Retrieved from http://www.youtube.com/watch?v=TMA5RFpGRbg.

ACKNOWLEDGMENTS

PROJECT LEADERS:
Manny Cuevas
Michael Thoma
Bryan Orozco
Jarrett Hullinger
Ben Unger

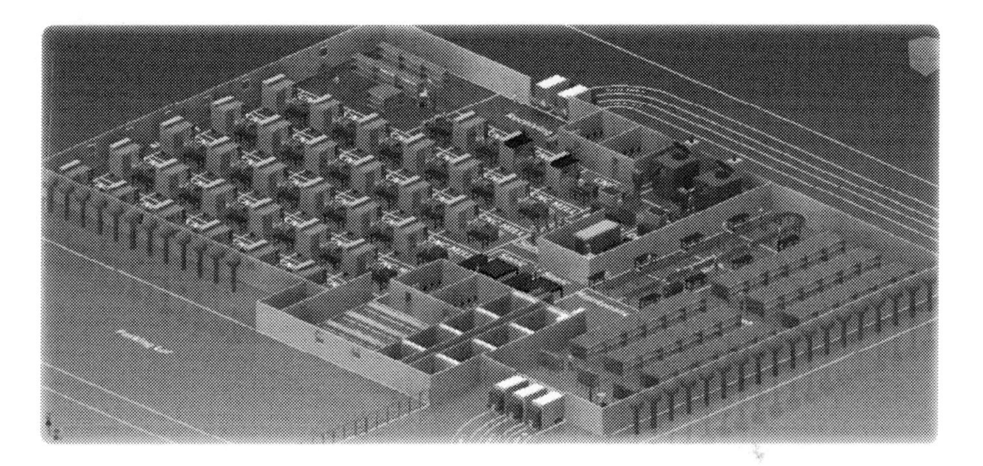

It has been a long, strenuous, and eye-opening semester, but in the end it was all worth it. All the members of the S.S.T. manufacturing team now feel fully prepared to enter the workforce as professionals in their chosen field. Through many hours of dedication and commitment from each and every team member, we have developed a final product that we feel fully encompasses all that we have attained through our studies. We are proud of our product, and we highly believe in everything that it represents.

Sincerely,

S.S.T. Manufacturing Group

Author's Note: These students' full project was much longer, and only a sample of it has been presented here. A full project includes drawings of every single required part, as well as data for all processes and necessary equipment.

Index